CRC Series
in
Analytical Toxicology

Editor-in-Chief

Irving Sunshine
Chief Toxicologist, Cuyahoga County Coroner's Office
Professor of Toxicology, School of Medicine
Case Western Reserve University
Cleveland, Ohio

Handbook of Analytical Toxicology
Irving Sunshine, Editor

Handbook of Mass Spectra of Drugs
Irving Sunshine, Editor
Michael Caplis, Assistant Editor

Handbook of Spectrophotometric Data of Drugs
Irving Sunshine, Editor

CRC Handbook of Mass Spectra of Drugs

Editor

Irving Sunshine

Chief Toxicologist, Cuyahoga County Coroner's Office
Professor of Toxicology, School of Medicine
Case Western Reserve University
Cleveland, Ohio

Assistant Editor

Michael Caplis

Director, Northwest Indiana Criminal Toxicology Laboratory
Assistant Professor, Indiana School of Medicine
Gary, Indiana

CRC Series in Analytical Toxicology

Editor-in-Chief

Irving Sunshine

CRC Press, Inc.
Boca Raton, Florida

Library of Congress Cataloging in Publication Data

Main entry under title:

Handbook of mass spectra of drugs.

(CRC series in analytical toxicology)
Bibliography: p.
Includes index.
1. Drugs—Analysis. 2. Mass spectrometry.
I. Sunshine, Irving. II. Series.
RS189.H27 615'.19015 80-22188
ISBN 0-8493-3572-8

This book represents information obtained from authentic and highly regarded sources. Reprinted material is quoted with permission, and sources are indicated. A wide variety of references are listed. Every reasonable effort has been made to give reliable data and information, but the author and the publisher cannot assume responsibility for the validity of all materials or for the consequences of their use.

Direct all inquiries to CRC Press, Inc., 2000 N.W. 24th Street, Boca Raton, Florida 33431.

International Standard Book Number 0-8493-3572-8

Library of Congress Card Number 80-22188
Printed in the United States

PREFACE

"The old order changeth, yielding place to the new". This truism is attested to by the flood of scientific data on drugs which has been published in the last few years. This plethora should be at the fingertips of scientists so that they can find and use it easily. Thus, two Volumes have been compiled to complement the chromatographic data accumulated in the CRC Handbook Series in Clinical Laboratory Science, Section B: Toxicology. One of these Volumes concerns itself with spectrophotometry and the other with mass spectrometry. In each of these Volumes, a presentation of the particular aspect precedes the tabulation of the assembled data. These data are permuted in several ways so that the analyst may find the particular datum he needs in its sequential arrangement. Using this format may be redundant, but this was done with the user's best interest in mind. His ability to search for the information he requires must be facilitated so that these volumes truly become "desk side" references.

Compiling and collating the various tabulations is a tedious, painstaking, laborious process which can never be complete. There are few comprehensive sources from which one can abstract the desired information. The number of products with which one is concerned in Analytical Toxicology keeps growing, thanks to the ingenuity of medicinal and pharmaceutical chemists. This growth precludes the inclusion of all substances in the tables. Also, data on many older preparations are not included simply because it is very difficult to get all the desired information from original sources. While many laboratories have been most generous in this cooperative effort, a significant number could not find the necessary time and personnel to provide requested facts.

As the compilations such as those included in these volumes demonstrate their value, successive efforts will enlarge and improve them. In the coming age of computer technology, the black box may replace these books. Until that happens, I trust the user will find these volumes helpful. Input is also helpful, so an open invitation is extended to each user of these volumes to submit corrections, complaints, and additional data.

Obviously, all this could not be achieved by one person working alone. To the many scientists who contributed their little bits to these volumes go my and your profuse thanks. Without their help these volumes would never evolve.

Irving Sunshine
Cleveland, 1980

THE EDITOR

Dr. Irving Sunshine is Chief Toxicologist at the Cuyahoga County (Cleveland), Ohio Coroner's Office, Professor of Toxicology in the Department of Pathology and Professor of Clinical Pharmacology in the Department of Medicine at the School of Medicine, Case Western Reserve University; Chief Toxicologist for the University Hospitals in Cleveland, Ohio; Director of the Cleveland Poison Information Center; and Editor-In-Chief for Biosciences for CRC Press, Inc. He is a Diplomate of both the American Board of Clinical Chemistry and The American Board of Forensic Toxicology and is on the Board of Directors of both these organizations.

Born in New York City, he obtained all his formal education in various Colleges of New York University, earning the B.Sc., M.A., and Ph.D. degrees. While earning his Ph.D., he taught chemistry in various colleges in the New York area and during the war, he worked during the "grave yard" shift on a pilot plant for the separation of uranium isotopes as a part of "The Manhattan Project". His development in toxicology was encouraged by two memorable mentors, Dr. Alexander O. Gettler and Dr. Bernard Brodie.

Prior to moving "west" to Cleveland, Ohio, where he has been since 1951, he served as the Toxicologist for the City of Kingston (N.Y.) Laboratory and for Ulster County. Since coming to Cleveland he has developed many interests which resulted in the publication of over 100 papers and several monographs. He is also a member of the Boards of Editors of many of the major toxicology journals.

His educational activities extend beyond the local college campuses. In the course of years he has organized and participated in numerous toxicology workshops which were held in many centers throughout the United States. As a member of the Education Committee of the American Association for Clinical Chemistry, he has been responsible for the National Tour Speaker Program, the Local Section Guest Lecturer Program, and the Visiting Lecturer Program. In recognition of his achievements in clinical chemistry, Dr. Sunshine was presented with the Association's "Ames Award" in 1973. Further recognition was accorded Dr. Sunshine by the Italian Society of Forensic Toxicologists which voted to make him an Honorary Member of that group. In 1978, The International Exchange of Scholars awarded him a Fulbright Visiting Professorship to the Free University of Brussels.

He is also a "has been." He has been President of The American Association of Poison Control Centers, Chairman of the National Council for Poison Control Week, Chairman of the Toxicology Section of The Academy of Forensic Sciences, and Chairman of the Cleveland Section of The American Association for Clinical Chemists, as well as a former member of the Association's Board of Directors.

ASSISTANT EDITOR

Dr. Michael Caplis is Director of Biochemistry and Toxicology at St. Mary's Medical Center, Gary, Indiana, Consultant Toxicologist, Northwest Indiana, Region I, and Adjunct Professor, Indiana School of Medicine — Indiana University Northwest. He is a Diplomate on the American Board of Forensic Toxicology and a Fellow of the American Academy of Forensic Science, American Academy of Clinical Toxicology, National Academy of Clinical Biochemistry.

Born in Ypsilanti, Michigan, he obtained a formal education in chemistry, biology, and biochemistry. He earned a Bachelor of Science degree in chemistry at Eastern Michigan University and a Masters of Science and Ph.D. degrees in biochemistry at Purdue University. While attending college, he taught math, chemistry, and physics at the high school level and worked as an analytical biochemist for the Indiana State Chemist, Purdue University. He has taught biochemistry, toxicology, and forensic toxicology at the university level.

Since coming to Gary in 1969, he has developed a specialized clinical biochemistry laboratory at St. Mary's Medical Center and, through cooperative funding at the federal, state, and local level, a regional Toxicology Center for Northwest Indiana. His interests have resulted in numerous publications in both the clinical and toxicology fields.

He is a member of the American Academy of Clinical Toxicology, American Academy of Forensic Science, National Academy of Clinical Biochemists, American Association for the Advancement of Science, American Association of Clinical Chemists, American Chemical Society, Association of Official Analytical Chemists, Midwest Association of Forensic Scientists, and American Association of Crime Laboratory Directors.

TABLE OF CONTENTS

MASS SPECTROMETRY

Rodger L. Foltz

Within the past few years mass spectrometry has become an important analytical tool in many of the larger toxicology laboratories. Its popularity is destined to increase rapidly as toxicologists become more familiar with its capabilities, since mass spectrometry combines sensitivity and specificity to a degree unmatched by other analytical techniques. These features are particularly valuable in toxicological analyses where the toxicant is usually a trace component in a complex biological matrix.

Mass spectrometry is currently limited by the requirement that the compound be capable of volatilization at temperatures below its decomposition point. Several new techniques of ionization show promise of eliminating this restriction, but their widespread application is at least several years in the future. In practice, the volatility requirement is only a minor limitation to toxicologists since most organic toxicants do have adequate volatility, or they can be converted to volatile compounds by derivatization or other chemical manipulations. A more vexing limitation to wider use of mass spectrometry is the high cost of purchasing and maintaining a mass spectrometer system. Unfortunately, the cost problem appears to be worsening. New mass spectrometry capabilities and techniques are being developed at an awesome rate, and often the new developments require additional instrumentation, such as new ionizers, inlets, and computer hardware. Although a basic mass spectrometer system can be purchased for under $30,000, the lure of expanded capabilities and throughput achievable by inclusion of a computer and other options is often irresistible. As a result, laboratories wanting to stay competitive will find themselves spending upwards from $100,000 for a new mass spectrometer system. In order to justify costs of this magnitude, it is important that the mass spectrometer be operated efficiently and with as high a throughput of samples as possible. This can only be accomplished if the persons involved in operating the facility are knowledgeable and dedicated spectroscopists, willing and able to participate in the preparation of samples, the operation and maintenance of the instruments, and interpretation of the data. Furthermore, it is important to recognize the types of analyses for which mass spectrometry is best suited. Mass spectrometry is appropriately used when no other analytical methods possessing adequate sensitivity and specificity are available. In this regard, mass spectrometry, particularly in combination with gas chromatography (GC-MS) and isotope-labeled internal standards, can form the basis of a definitive quantitative assay which can be used to validate other assays. GC-MS suffers some disadvantages: sample throughput is relatively slow and the system is difficult to fully automate. Consequently, one should always consider whether a particular assay can be done adequately by a cheaper and faster method.

Of the many books on mass spectrometry, those authored by Beynon et al.,[1] Biemann,[2] McLafferty,[3] and Budzikiewicz et al.[4] and edited by Waller[5] have proven most useful. Mass spectrometry research results are published in a wide variety of journals. Fortunately, excellent reviews of the field appear at regular intervals.[6,7] Currently there are three English-language journals devoted exclusively to publishing research involving mass spectrometry: (1) *Biomedical Mass Spectrometry,*[8] (2) *Organic Mass Spectrometry,*[9] and (3) the *Journal of Mass Spectrometry and Ion Physics.*[10] Of these, the first is most likely to be used extensively in a toxicology laboratory, as it is highly applications oriented, while papers appearing in the second and third tend to be concerned with fundamental processes occurring in mass spectrometry. The *Mass Spectrometry Bulletin*[11] is the most current and comprehensive guide to the mass spectrometry literature. This monthly publication lists the titles, key subject terms, and references for articles containing mass spectrometry data appearing in over 250 journals. Each issue contains indexes based on subject, author, compound classification, and elements. Also, an Elemental Composition Index is published annually.

Basic Components of a Mass Spectrometer

The basic processes in any mass spectrometer include introduction and volatilization of the sample, ionization of the sample molecules, separation of the resulting ions according to their masses, and measurement of the ion current at each mass. Numerous books[1-5,12-15] and review articles[16-20] contain detailed descriptions of the different methods and types of instrumentation that have been used to accomplish each of these processes. This discussion will be limited to those instrumental methods which are particularly useful to toxicology laboratories.

Sample Ionizer

A mass spectrum is most often represented by a bar graph in which the height of each bar or line represents the relative intensity of ion current at a particular mass (Figure 1). Actually the units of the abscissa are mass to charge rations (M/e); however, the charge is normally one and, therefore, the ratio is often loosely referred to as mass. The appearance of a mass spectrum is determined primarily by the structure of the sample molecules and the ionization process employed. Electron impact (EI) is the most widely used method of ionization for organic molecules. In this method, sample molecules in the gas phase are

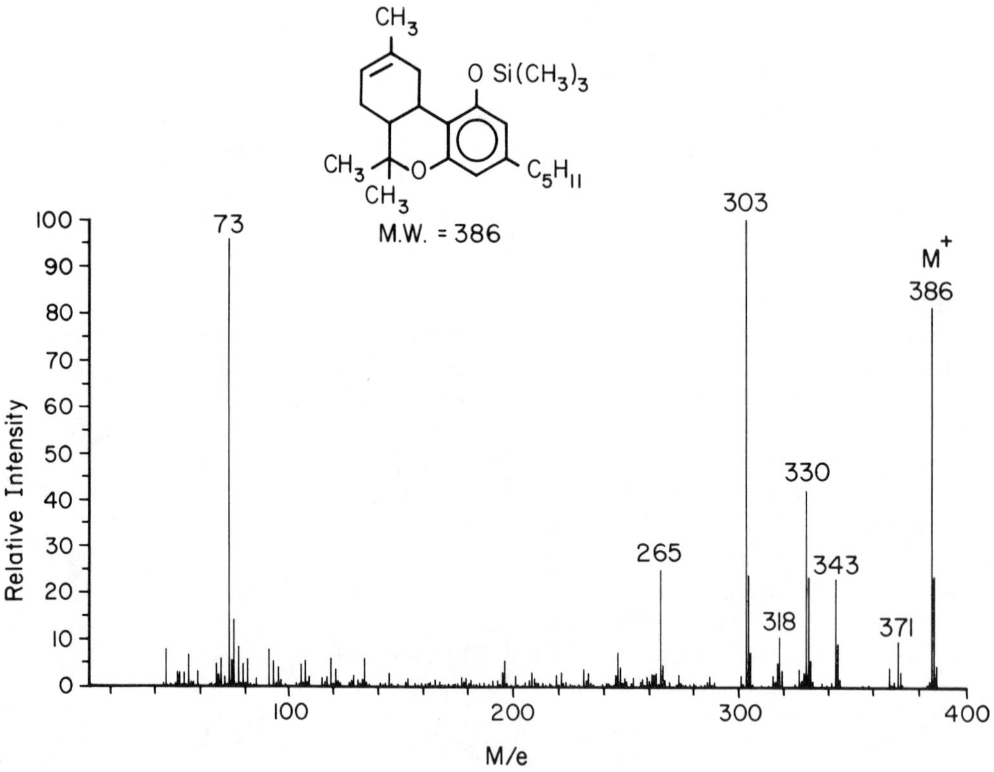

CH3
O Si(CH3)3
CH3
O
C5H11
CH3
M.W. = 386
Relative Intensity
100
90
80
70
60
50
40
30
20
10
0
73
265
303
318
330
343
371
M+
386
100
200
300
400
M/e

FIGURE 1. A representative mass spectrum in which the line heights represent the relative ion current intensities at each M/e value (mass to charge ratio). The compound represented corresponds to the trimethylsilyl derivative of Δ^8-tetracannabinol.

passed through a beam of electrons. A small fraction of the molecules is ionized by electron bombardment. Typically, electron beams with energies of about 70 eV are used. Since the ionization potentials of organic compounds lie between 8 and 12 eV, the excess energy usually causes extensive fragmentation to lower mass ions. The resulting pattern of ion currents vs. M/e is reproducible and characteristic for each organic molecule. Even though both positive and negative ions are formed, the former predominate and constitute the normal EI mass spectrum. EI mass spectra are often complex and potentially contain considerable structural information. However, our ability to extract this information is severely restricted by our limited understanding of the fundamental processes that govern how a molecular ion fragments. Interpretation of EI mass spectra is still based primarily on a collection of empirical observations relating structural features to specific fragmentation processes.[1-4]

Chemical ionization (CI) is an alternative method of generating gas-phase organic ions which is rapidly gaining in popularity.[21-23] In this technique, a reagent gas is introduced into the ion source to give a pressure of about 1 torr. Some of the reagent gas molecules are ionized by electron impact, and they subsequently cause ionization of the sample molecules by means of ion-molecule reactions. These reactions include proton transfer, hydride abstraction, ion attachment, and charge transfer, all of which are relatively low energy processes. As a consequence, CI mass spectra typically show intense peaks in the molecular ion region and relatively little fragmentation. The ability to clearly indicate the molecular weight of a compound is the most notable feature of CI mass spectra. However, the technique offers other useful features. Different reagent gases can be used, each generating a different spectrum.[24,25] Methane was the first reagent gas to be used and still is the most popular.[26] Methane CI mass spectra often show a moderate amount of fragmentation, but the fragment ions can be easily identified and often provide useful structural information. Isobutane[27] and ammonia[28] are "milder" reagent gases which typically generate CI mass spectra containing little or no fragmentation. When aprotic gases such as argon[29] and helium[30,31] are used as reagent gas, the sample molecules are ionized primarily by charge exchange and the resulting spectra are very similar to conventional EI mass spectra. Nitric oxide[32,33] is a reagent gas which has also been shown to be useful for certain classes of organic compounds.

Until recently, only the positive ions generated by chemical ionization were recorded. However, it has been shown that negative ions can also be generated in high abundance and that the resulting negative ion CI mass spectra contain additional, useful, structural information.[34,35] What is more, a commercial GC-MS system has been modified to permit simultaneous acquisition of both negative and positive CI mass spectra.[36]

A CI ion source is sufficiently similar in design to an EI ion source that it is possible to build a single instrument capable of performing well in either mode of ionization. In view of the complementary nature of CI and EI ionization, it is likely that in the near future all new organic mass spectrometers will have both capabilities.

Other ionization methods such as field ionization,[37] field desorption,[38] atmospheric pressure ionization,[39] and californium-252 plasma desorption ionization[40] have exciting potential for special applications but do not currently have general applicability in toxicology laboratories.

Mass Analyzer

The vast majority of mass spectrometers presently in use achieve mass analysis by either magnetic deflection or a quadrupole mass filter. Each type of analyzer has its advantages. Magnetic instruments have a higher mass range and generally are capable of greater resolution. Metastable ion peaks can be observed on magnetic instruments but are not detected with quadrupole mass spectrometers. Metastable ions are those which undergo fragmentation between the ion source and the ion detector.[41] The ability to detect the products of metastable ion decompositions is very useful in the study of fragmentation mechanisms. Quadrupole mass spectrometers tend to be lower priced, are capable of very rapid scans, and are better suited to operation under computer control.

Most of the major differences in performance capabilities between magnetic and quadrupole instruments have been largely overcome by design improvements. The early quadrupole analyzers were limited to ion masses below about 500, whereas the newer instruments are capable of detecting ions up to M/e 1000 to 1200. The less stringent vacuum requirements of the quadrupole analyzer and the absence of high accelerating voltages made it easier to adapt quadrupole instruments to chemical ionization. However, manufacturers of magnetic instruments have now re-engineered their products so that they can also offer CI capability. Magnetic instruments tend to have a better inherent sensitivity (particularly at high mass), but computer control of quadrupole analyzers permits optimization of the scan parameters to the point where comparable sensitivities can be achieved.

Ion Current Detector

The electron multiplier is almost universally used as the primary ion current detector in organic mass spectrometers. This device converts the impinging ions to electrons and amplifies the electrical current by as much as 10^7. The output of the electron multiplier is further amplified and passed to a high-speed recorder or a digital data system. The overall gain of this system can be sufficient to observe single ions reaching the detector. Other types of detectors, such as photographic plates,[42] are useful for special applications.

Sample Inlets

Sample inlets provide a means of volatizing the sample and introducing it into the ion source of the mass spectrometer. The type of inlet that should be used depends on the volatility and stability of the sample, as well as the amount of material available and its state of purity. Every mass spectrometer used for analysis of organic materials should have at least three separate inlets: (1) a direct insertion probe, (2) a controlled leak inlet, and (3) a gas chromatographic inlet.

Direct Insertion Probe

Solids and high-boiling liquids can be introduced into the mass spectrometer by means of a "direct insertion probe." The sample is placed in a small glass capillary which is seated in a cavity at the end of a heatable probe. The probe is then introduced via a vacuum lock into the ion source, where it is heated to a temperature sufficient to give a vapor pressure of about 10^{-6} torr. The entire operation is simple and rapid (<5 min); therefore, it is usually the inlet used if the sample is relatively pure. It is also the preferred inlet if the sample material is thermally unstable or has insufficient volatility to be introduced via the gas chromatographic inlet. It is an efficient method of sample introduction with respect to sample utilization. Mass spectra can be obtained on quantities as small as 0.1 μg. However, when working with such small sample quantities, contaminants can be a problem. If the sample and the contaminants have different vapor pressures, some fractionation can be achieved by slow, controlled heating of the probe. Nevertheless, it is highly desirable to minimize contaminants by keeping the sample probe and glass capillaries scrupulously clean. Since the latter are difficult to clean, many laboratories simply use readily available melting point capillary tubes which can be easily cut to the desired length and discarded after use.

An alternative to placing the sample inside the glass capillary is to evaporate a solution of the sample on the outside of the capillary tube or a glass rod of similar dimensions. This technique has several advantages. First, the deposited film tends to evaporate more uniformly when heated than do crystals placed inside the capillary. Second, it is less likely that too much sample will be used, since the amount is limited by the quantity of residue which will adhere to the outside of the capillary. Finally, it has been reported[43] that certain compounds which are difficult to volatilize may give usable spectra if they are deposited on the outside of the capillary and introduced directly into the ion chamber of a chemical ionization ion source.

The direct probe technique is often used to obtain mass spectra of compounds isolated by paper or thin-layer chromatography (TLC). Some success has been achieved by scraping the portion of the TLC adsorbent containing the material of interest directly into the glass capillary tube.[44] However, the presence of the solid adsorbent tends to lower the volatility of the sample and can catalyze decomposition when the probe is heated in the mass spectrometer. Consequently, it is usually preferable to elute the sample from the TLC adsorbent, concentrate the eluent, and then deposit it on or in the glass capillary sample holder.

Mass spectra obtained on samples isolated by TLC inevitably show the presence of contaminants, often in such high concentration that the ions due to the sample are masked by the more abundant contaminant ions. This is particularly likely when the material of interest is located on the TLC plate by a selective method of visualization, such as UV absorption, fluorescence, or radioactivity. Consequently, before submitting a TLC-isolated sample for mass spectral analysis, it is often helpful to subject a duplicate plate to a general visualization process, such as exposure to iodine vapor or acid-charring, in order to determine if the TLC spot of interest is free of other organic materials.

Controlled Leak Inlet

Gases and volatile liquids can be introduced into the mass spectrometer by means of a reservoir connected to the ion source via a controlled leak. This type of inlet is normally used for introducing a reference material for mass calibration and tune-up of the instrument. It can also be used for introducing CI reagent gases or when a relatively steady sample flow rate into the ion source is required. The amount of sample needed depends on the size of the reservoir and the conductance of the leak. However, in general, this type of inlet is not used if less than about 1 mg of sample is available.

Gas Chromatographic Inlet

The development of techniques for coupling the gas chromatograph to the mass spectrometer has done more to expand the usage of mass spectrometry than any other single development. Whether one views the gas chromatograph as an inlet for the mass spectrometer or the mass spectrometer as a detector for the gas chromatograph depends on one's personal perspective and bias. The important fact is that the combination of the two instruments constitutes an analytical system of unprecedented capabilities. In most respects, the gas chromatograph and the mass spectrometer complement each other and are compatible. Both are gas phase, microanalytical techniques. Gas chromatography is capable of higher separation efficiency than any other current technique, while the mass spectrometer offers detailed structural information and, when corresponding reference spectra are available, can provide conclusive identification of analytes.

The major point of incompatibility between the GC and the MS is the pressure within the active elements of each system. The GC column is normally operated at above atmospheric pressure, while the mass analyzer of the MS must be maintained below 10^{-5} torr. Numerous splitters, separators, and other devices have been developed for overcoming this incompatibility.[45–47] In spite of the progress that has been made in the design of GC-MS interfaces, the link between the two remains a critical stage in the combined operation and a likely source of problems. As a general rule, the connection between the GC and MS should be kept as simple and direct as possible. In line with this principle, there is a current trend toward direct coupling of the GC and MS without separators or splitters. This poses no major problem in the case of capillary column chromatography where the carrier gas flow rate is only 1 to 2 ml/min. It is also commonly done when chemical ionization is used. In this case, the carrier gas (methane) can be used as the CI reagent gas. For the combination of packed columns and electron impact ionization, some type of separator is still normally used. The glass jet separator appears to be preferable for biological samples because it shows the least tendency to cause loss of sample due to decomposition or adsorption. Its major fault is its propensity to become clogged, necessitating instrument shut-down and cleaning.

GC-MS analysis is often more dependent on the proper performance of the gas chromatograph than that of the mass spectrometer. Consequently, before initiating a new GC-MS analysis, it is generally advantageous to check out and optimize the GC conditions on a separate GC unit equipped with a flame ionization detector (FID). The results of this preliminary work will facilitate setting up the mass spectrometer's scan and amplification parameters in order to obtain the best quality mass spectra and make the most efficient use of the GC-MS system. Furthermore, a comparison of the FID chromatogram and the total ion current (TIC) chromatogram provides a valuable assessment of the performance of the two systems. If, for example, the peaks in the TIC chromatogram show more tailing than those in the FID chromatogram, it is likely that there is a problem in the interface, such as cold spots or unswept dead volumes.

The amount of sample required for analysis by GC-MS depends on many factors. As a general guide, most modern GC-MS systems should be able to routinely generate good quality spectra on 10 to 100 ng of compound injected into the gas chromatograph. However, instrument capabilities are being continually improved so that some are now able to generate complete mass spectra on considerably smaller quantities (~ 100 pg). When operated in the selected-ion-monitoring mode, a GC-MS system should be capable of detecting subnanogram quantities of most gas chromatographable compounds.

The selection of gas chromatographic columns is discussed in the chapter on gas chromatography. For GC-MS work it is particularly important to use thermally stable liquid phases. Because of the sensitivity of the mass spectrometer, column bleed is often the primary barrier to lower detection limits. Fortunately, liquid phases are readily available which have excellent thermal stability and cover a wide range of polarities, so that there is little justification for using high-bleed liquid phases such as polyethylene glycols, hydrocarbons, and polyesters.

Packed columns are currently those most widely used because of their capacity, versatility, and general availability. However, glass capillary columns offer attractive advantages: better resolution, lower sample losses due to adsorption and surface-catalyzed decomposition, and elimination of the need for a separator. Glass capillary GC technology has advanced more rapidly in Europe than in North America. Nevertheless, wider recognition in this country of the advantages of a glass capillary column seems inevitable.

Many compounds can be more effectively analyzed by GC-MS after chemical conversion to a derivative. Derivatization is used to increase a compound's volatility or its thermal stability, to improve its GC behavior by substituting lipophilic groups for "active" hydrogens, or to advantageously alter the compound's mass spectrum. For example, the latter two benefits are realized when amphetamine is converted to its trifluoroacetamide (TFA) derivative. Figure 2 compares the EI mass spectra of the drug and its derivative. The most intense peaks in the EI mass spectrum of methamphetamine occur at low mass (M/e = 44, 91, 65), where interference from other compounds is likely. In contrast, the mass spectrum of the TFA derivative shows a base peak at M/e 140, a mass which is more useful for detection in that it is less likely to be masked by ions from other compounds. A recent review[48] contains a detailed discussion of the advantages and potential pitfalls in chemical derivatization and a systematic survey of the many derivatizing agents and techniques in use.

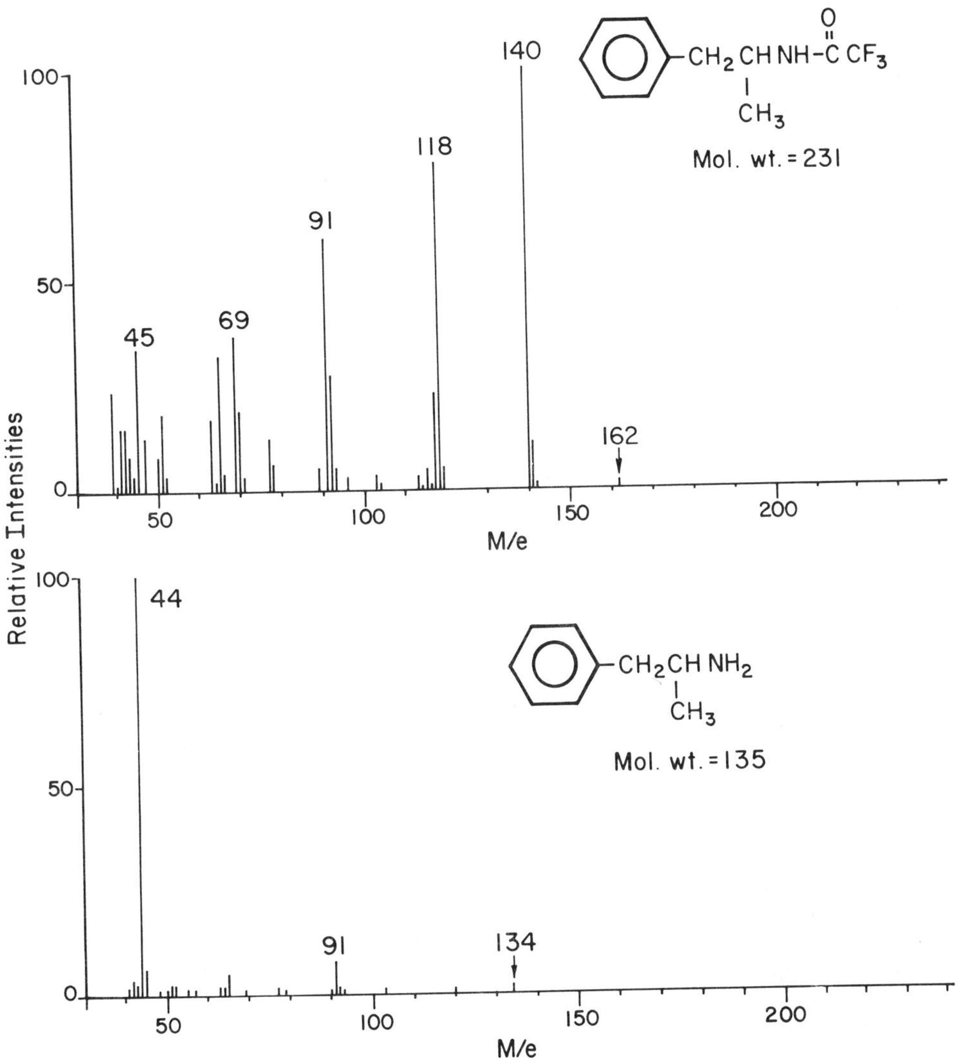
Relative Intensities
100
50
0
45
69
91
118
140
162
50
100
150
200
M/e
CH2 CH NH-C CF3
O
CH3
Mol. wt. = 231
100
50
0
44
91
134
50
100
150
200
M/e
CH2CH NH2
CH3
Mol. wt. = 135

FIGURE 2. Comparison of the EI mass spectra of amphetamine (bottom) and amphetamine trifluoroacetamide (top).

Other Types of Inlets

Many additional types of inlets have been developed to solve specific sampling problems. Those which are likely to be of greatest interest to a toxicology laboratory include devices for direct sampling of gases at atmospheric pressure[49,50] and for adsorption and concentration of organic vapors prior to introduction into the GC-MS.[51] In view of the remarkably successful mating of the gas chromatograph and the mass spectrometer, it is not surprising that there is also tremendous interest in methods for the direct coupling of a liquid chromatograph (LC) to a mass spectrometer. Several approaches to developing LC-MS interfaces have been tried with some success.[52] However, so far all of the methods are limited in the types of solvents that can be handled and are restricted to the analysis of compounds which can be volatilized without decomposition.

Data Handling and Computerization

In the absence of a data system a GC-MS must have two separate recorders, one to record the total ion current signal and another to record mass spectra. The latter must be capable of high-frequency response and a dynamic range of at least 1000. Light beam oscillographic recorders best meet these requirements. Most mass spectrometers now offer mass markers which superimpose a mass scale on the oscillographic recording, thereby greatly facilitating mass identification.

In the manual mode of operation the operator determines when to initiate recording a mass spectral scan. This is normally done by watching the total ion current recording and attempting to initiate a scan just at the instant that the maximum concentration of each component enters the ion source from the GC. There are several problems associated with this technique. It requires careful and continuous attention of the operator for the duration of the GC-MS run. Often it is difficult to anticipate the size of each GC peak so that the mass spectrometer's amplifier gain can be adjusted to give a recording that has a satisfactory intensity. However, the most severe problems are probably the amount of chart paper that is generated in a typical GC-MS analysis and the need to convert the data from the oscillographic recordings into presentations suitable for inclusion in reports. These problems are eliminated when the GC-MS is coupled to a computer.

In the early GC-MS data systems, the computer was used solely for the purpose of acquiring mass spectral data and outputting it in tabular or graphical form. It was soon recognized, however, that the computer can perform many additional functions. These include:

1. Control of the various operating parameters of the GC-MS, i.e., scan rate, mass range, column temperature, carrier gas flow rate, amplifier gain, etc.
2. Data acquisition, including establishment of an optimum signal threshold and peak detection
3. Data processing, such as assignment of masses, normalization of ion intensities, background subtraction, spectra averaging, statistical calculations, and data searching for specific features
4. Data presentation in various tabular and graphical formats using any of several data display devices
5. Computer-aided identification based on matching of spectra against a library of reference spectra or recognition of data patterns associated with specific structural features

Computer-based GC-MS analyses can be grouped into two categories: repetitive scanning or selected ion monitoring. In the former the mass analyzer repetitively scans over the mass range of interest. Scan times are normally 2 to 4 sec so that in a GC-MS analysis consuming 30 min, 400 or more spectra will be entered into the computer. Upon completion of the run, the computer reconstructs a total ionization chromatogram (TIC) by plotting the summation of the ion intensities for each scan vs. scan number. The resulting plot can be displayed on a video screen or drawn by a digital plotter. It should be similar in appearance to a normal gas chromatogram of the same sample. The primary use of the total ionization chromatogram is to indicate which scans contain mass spectral data corresponding to each component of interest. For example, Figure 3 shows the total ionization chromatogram from a GC-MS analysis of an extract of the urine from an emergency room patient intoxicated with an overdose of a drug. Figure 4 is the computer plot of the methane CI mass spectrum (No. 180) corresponding to the major peak in the TIC. The protonated-molecule ion (MH^+) was easily identified in this spectrum on the basis of the very typical intensity pattern for the $M{-}H^+$, MH^+, $MC_2H_5{}^+$, and $MC_3H_5^+$ ions. Further interpretation of the spectrum led to the conclusion that it corresponded to the tricyclic antidepressant drug, amitriptyline. In the same manner, spectra numbers 190, 226, and 243 were plotted, examined, and identified as due to the three major metabolites of amitriptyline.

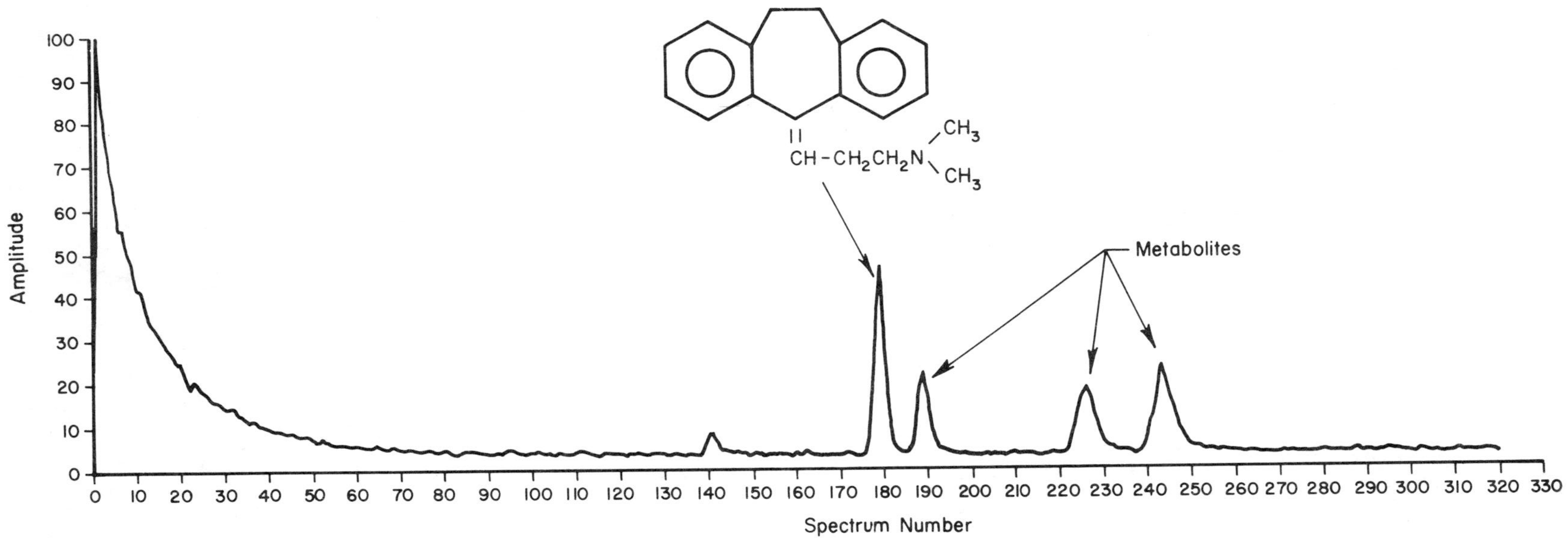
Amplitude
100
90
80
70
60
50
40
30
20
10
0
Spectrum Number
0 10 20 30 40 50 60 70 80 90 100 110 120 130 140 150 160 170 180 190 200 210 220 230 240 250 260 270 280 290 300 310 320 330
Metabolites
CH - CH2CH2N
CH3
CH3

FIGURE 3. Total ion current plot of the GC-MS analysis of a urine extract from an amitriptyline-intoxicated patient. (From Foltz, R. L., *Adv. Mass Spectrom.*, 6, 231, 1975. With permission.)

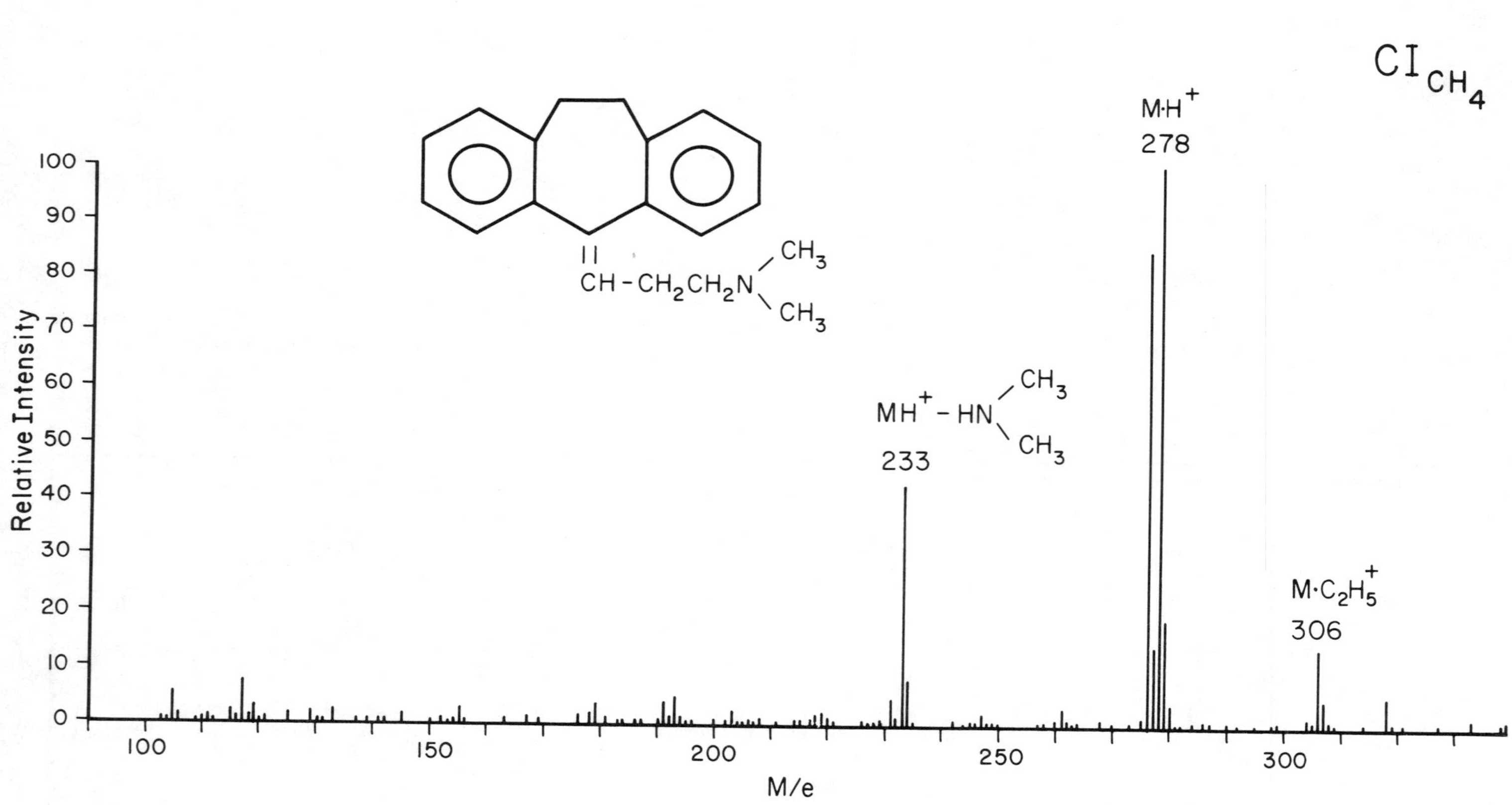
CI CH4
M·H+
278
MH+ - HN CH3 CH3
233
M·C2H5+
306
CH-CH2CH2N CH3 CH3
Relative Intensity
0
10
20
30
40
50
60
70
80
90
100
100
150
200
250
300
M/e

FIGURE 4. Methane CI mass spectrum corresponding to major peak in total ion current plot shown in Figure 3. (From Bryan, F., Foltz, R. L., and Taylor, D. M., *J. Chromatogr. Sci.*, 12, 307, 1974. With permission.)

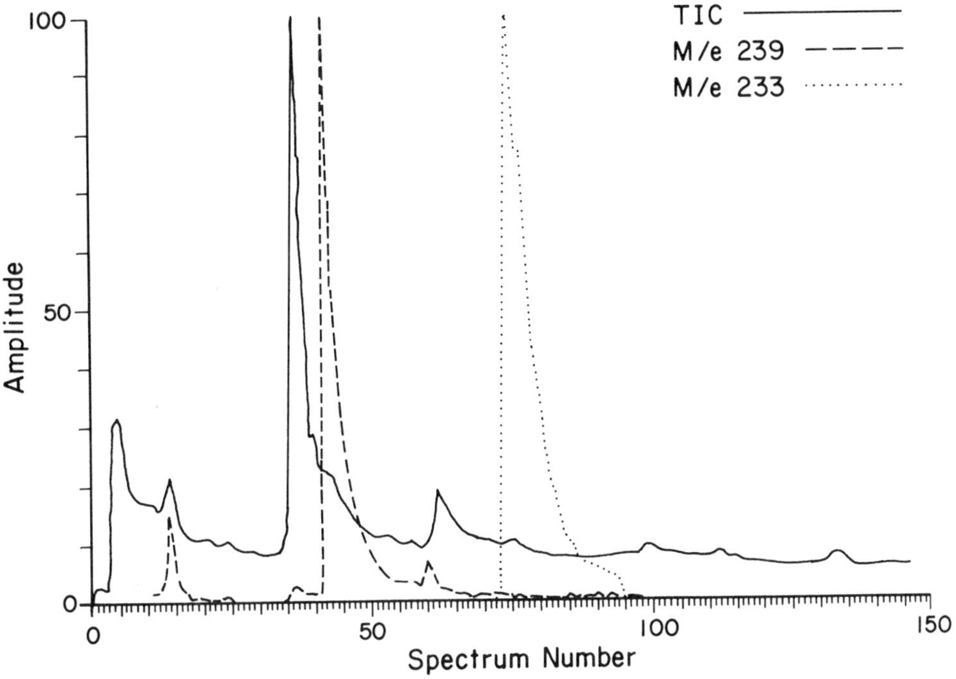
TIC
M/e 239
M/e 233
Amplitude
100
50
0
0
50
100
150
Spectrum Number

FIGURE 5. Computer plots of ion currents from a GC-MS analysis of the extract of blood from a drug overdose patient. Total ion current, solid line; M/e 239 ion current, dash line; and M/e 233 ion current dotted. (From Foltz, R. L., *Adv. Mass Spectrom.*, 6, 231, 1975. With permission.)

A great advantage of this technique is that all of the GC-MS data are stored by the computer and can be rapidly retrieved and examined in many different ways. For example, the technique of plotting the ion current intensity for selected ions is now commonly used as an effective method of uncovering chromatographic peaks which are obscured by other components. Figure 5 consists of superimposed plots of the total ion current (solid line), the M/e 239 ion current (dashed line), and the M/e 233 ion current (dotted line) resulting from the GC-MS analysis of a blood serum extract from a patient (dotted line), resulting from the GC-MS analysis of a blood serum extract from a patient suspected of taking a drug overdose. The major peak in the total ion current plot was readily identified as pentobarbital upon examination of spectrum number 37. The presence of smaller amounts of secobarbital and phenobarbital was indicated by the plots of their protonated-molecule ion intensities (M/e 239 and 233, respectively) Conclusive identification of these drugs was then provided by computer plots of spectra numbers 44 and 75 (Figure 6) after computer subtraction of background.

This technique of plotting the ion current at selected masses vs. time (or spectrum number) has become immensely popular.[53] It enables the mass spectrometer to be used as an extremely selective and sensitive detector. The selected-ion-current plots, referred to as mass chromatograms, can be acquired by having the mass spectrometer continuously monitor only the selected ion masses. The latter method, referred to as selected ion monitoring or mass fragmentography, has the advantages of better quantitative reliability and at least 100 times better sensitivity.[53,54]

Applications

Until a few years ago most laboratories used mass spectrometry mainly for identification of unknown compounds; however, with the emergence of selected ion

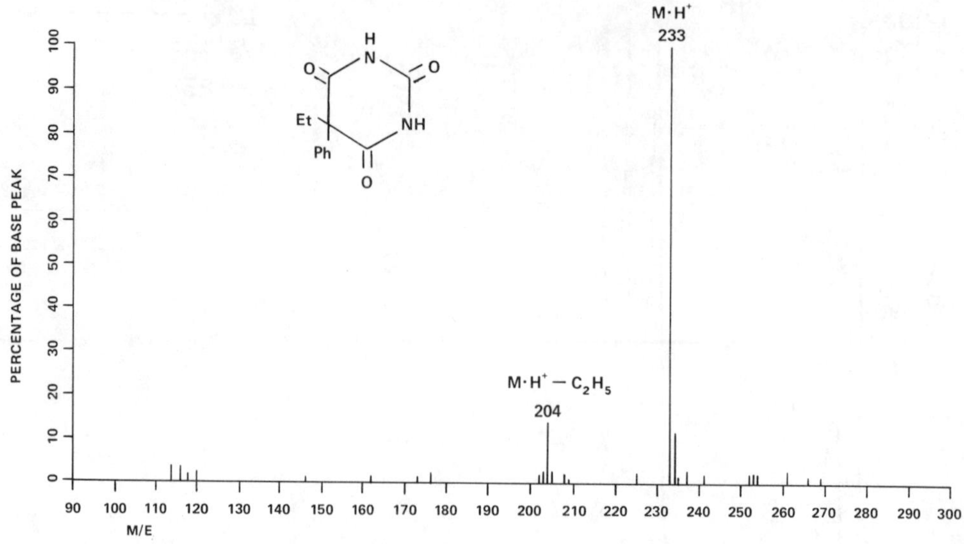
M·H⁺
233
M·H⁺ — C₂H₅
204
PERCENTAGE OF BASE PEAK
M/E
H
N
O
O
Et
NH
Ph
O
0
10
20
30
40
50
60
70
80
90
100
90
100
110
120
130
140
150
160
170
180
190
200
210
220
230
240
250
260
270
280
290
300

FIGURE 6. Computer plot of spectrum number 75 minus background spectrum number 72 in the total ion current plot shown in Figure 5. Based on this methane CI mass spectrum, the compound was identified as phenobarbital.

monitoring and isotope-labeled internal standards, quantitative analysis has assumed an equally important role. The purpose of this section is to briefly describe mass spectrometry techniques that are commonly used to solve toxicological problems. No attempt is made to cover all of the many mass spectrometry techniques that have been developed.

Qualitative Analysis

Molecular Weight Identification

In the identification of an unknown, the first and most useful item of information sought from the mass spectrum is the compound's molecular weight. Often the molecular weight is all that is needed to decide between several alternative structures. Furthermore, most collections of mass spectra have a molecular weight index, so that knowing the unknown's molecular weight is a great aid in locating a matching reference spectrum. Also, once a compound's molecular weight is known, detailed interpretation of its mass spectrum becomes far easier.

Unfortunately, the molecular ion is not always easily identified in electron impact mass spectra. Usually the molecular ion is the highest mass ion in the EI mass spectrum, but it may be too weak to distinguish from background ions. Also, impurities may be present, giving ions at higher mass than the molecular ion. A technique commonly used to identify the molecular ion is to lower the energy of the electron beam. As the energy of the electrons is decreased, the intensities of fragment ions decrease more rapidly than the intensity of the molecular ion. If the electron energy is lowered to within a few electron volts of the compound's ionization potential (8 to 12 eV), frequently the molecular ion will be the only peak observed in the mass spectrum. This technique works well for some compounds but completely fails for others. In essence, it is a method of enhancing the relative intensity of a molecular ion. However, if a compound's molecular ion is not detectable at 70 eV, lowering the electron energy will not make it appear.

Chemical ionization is a more effective technique for identifying a compound's molecular weight. CI mass spectra generally show intense ions in the region of a

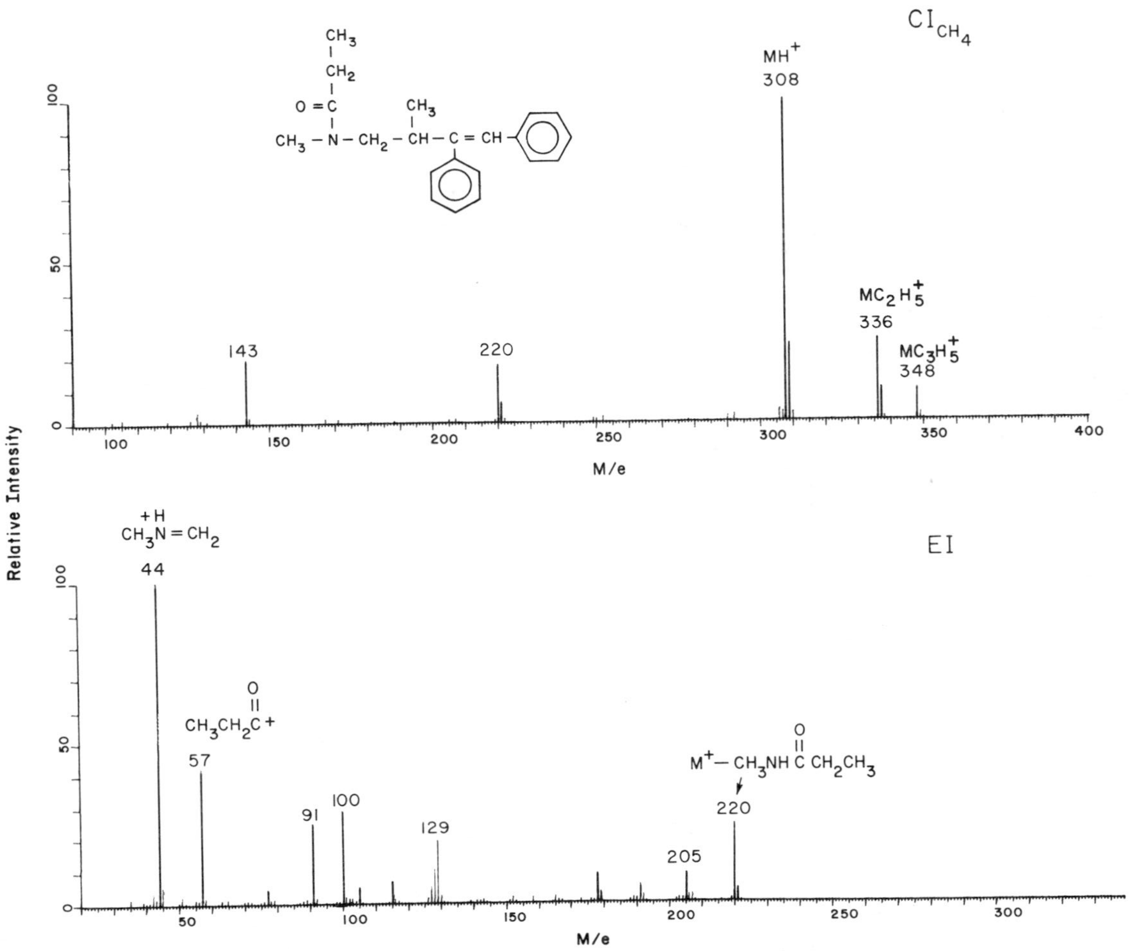
CI CH4
MH+
308
MC2H5+
336
MC3H5+
348
143
220
Relative Intensity
M/e
EI
+H
CH3N=CH2
44
CH3CH2C+
57
91
100
129
205
M+— CH3NH C CH2CH3
220

FIGURE 7. Comparison of the methane CI (top) and EI (bottom) mass spectra of a metabolite of the drug, propoxyphene.

compound's molecular weight. The pattern of these ions is characteristic for each reagent gas and easily recognizable. The EI and methane CI mass spectra of a metabolite of propoxyphene are compared in Figure 7. The molecular ion is undetectable in the EI mass spectrum, while the CI mass spectrum contains an intense peak corresponding to $(M + H)^+$. The weaker peaks at M/e 336 and 348 correspond to attachment of $C_2H_5^+$ and $C_3H_5^+$ to molecules of the drug. The intensity pattern of these peaks is characteristic for methane CI mass spectra and actually helps to identify the compound's molecular weight.

The absence of a molecular ion in the EI mass spectrum (Figure 8) of an unknown metabolite led to a tentative structure identification which subsequently proved incorrect. The compound isolated from the urine of a patient receiving the drug, primidone, was tentatively identified as 2-phenylbutyramide, based primarily on the peak at M/e 163 which appeared to be the molecular ion.[55] However, when the isobutane CI mass spectrum (Figure 9) was subsequently examined, it was immediately apparent that the metabolite had a molecular weight of 206 and, therefore, corresponded to a previously identified metabolite, phenylethyl malondiamide.

Compounds which contain acid-labile substituents may not show intense protonated-molecule ions in their methane or isobutane CI mass spectra. Instead, the most intense peak may result from loss of the protonated substituent. For example, the isobutane CI mass spectrum of cholesterol acetate shows little or no MH^+, but instead has an intense peak corresponding to $MH^+ - HOAc$. In these cases the spectra may not clearly indicate the compound's molecular weight. However, another reagent gas such as ammonia,[28] nitric oxide,[32,33] or trimethylamine[23] can usually be found which will give abundant ions indicative of the compound's molecular weight. Of course, if the compound is too thermally unstable to volatilize intact, neither the EI nor the CI mass spectra will give direct evidence for the molecular weight of the compound.

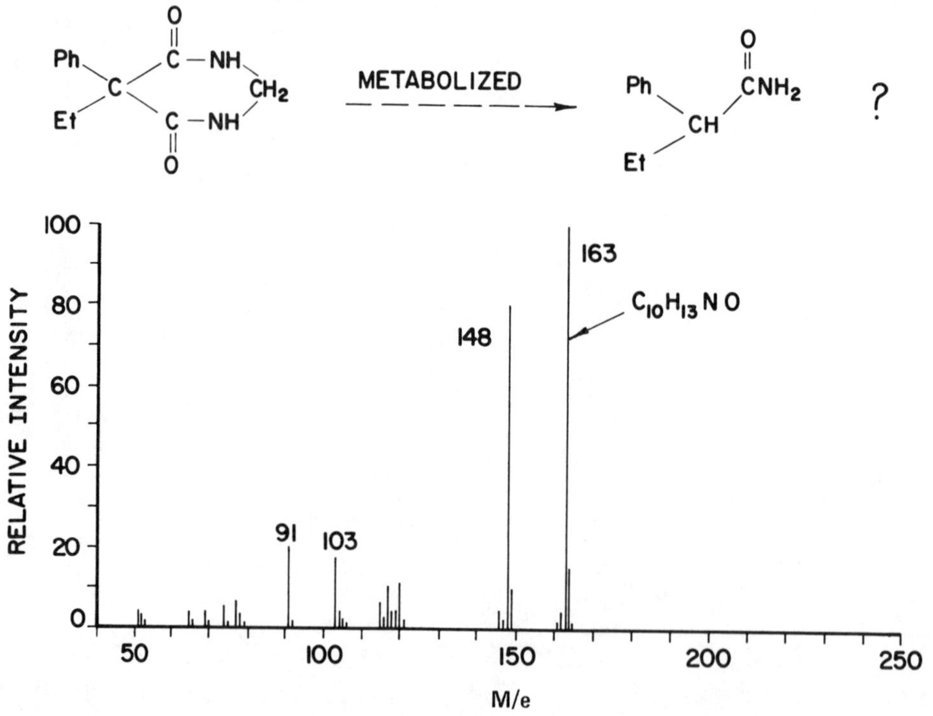
Ph
Et
C
C—NH
CH2
C—NH
O
O
METABOLIZED
Ph
CH
CNH2
Et
?
RELATIVE INTENSITY
100
80
60
40
20
0
163
C10H13NO
148
91
103
50
100
150
200
250
M/e

FIGURE 8. The EI mass spectrum of a metabolite of the drug primidone and its initially proposed structure. (Reprinted with permission from Foltz, R. L., *Chem. Technol.*, 5, 41, 1975. Copyright by the American Chemical Society.)

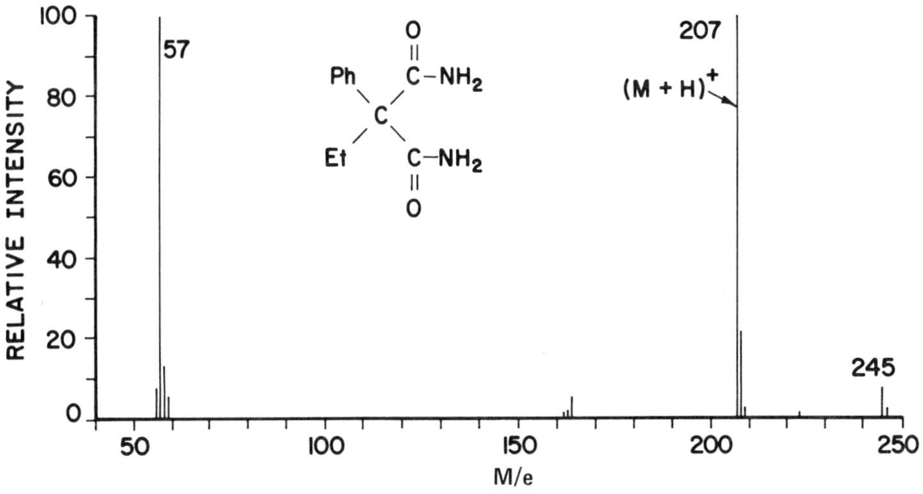
100
80
60
40
20
0
RELATIVE INTENSITY
57
O
Ph
C-NH2
C
Et
C-NH2
O
207
(M + H)+
245
50
100
150
200
250
M/e

FIGURE 9. The isobutane CI mass spectrum and correct structure of a metabolite of primidone. (From Foltz, R. L., *Chem. Technol.*, 5, 41, 1975. With permission.)

Elemental Composition Determination

After a compound's molecular weight has been identified, the next question is often, "What is its elemental composition?" Two mass spectral techniques are commonly used to determine elemental compositions.

1. Accurate mass measurement of the molecular ion
2. Accurate measurement of the relative intensities of the isotope peaks associated with the molecular ion (primarily M + 1 and M + 2)

The first method relies on the facts that atomic weights are not integral numbers (except for carbon) and that every elemental composition has a unique mass if it is expressed with sufficient precision. For example, the precise mass of the molecular ion of amphetamine ($C_9H_{13}N$) is 135.1048. If the mass of this ion can be measured with an accuracy of ±20 ppm or better, it can be distinguished from any other reasonable elemental composition. Accurate mass measurement normally requires a high-resolution mass spectrometer, although computer techniques have been developed to achieve accurate mass measurements with low resolution instruments.[56,57]

The second method for determining elemental compositions is not as generally applicable and reliable but can be useful for the purpose of limiting the number of elemental compositions that should be considered.[3]

Spectrum Matching

In spite of the importance of the molecular ion, it is the fragment ions which contain structural information and endow mass spectra with their diagnostic "fingerprint" patterns. Because of the detail and uniqueness of most EI mass spectra, matching of the spectrum of an unknown with a reference spectrum can constitute a nearly unequivocal identification. In this respect, EI spectra are more useful than CI mass spectra (since EI spectra are normally more complex and, therefore, more characteristic) and are less subject to variation dependent on instrumental conditions.

Mass spectra of unknown compounds are most often identified by simply matching the unknown's spectrum with a known reference spectrum. Spectral matching can be done either manually or by computer. The manual approach is generally used if only a few spectra are to be matched. If large numbers of spectra must be identified, computer matching becomes attractive. Even then, however, the computer matching should be considered only a tentative identification. Conclusive identification requires manual verification and, ultimately, direct comparison with an authentic sample of the compound.

Several problems are associated with spectral matching. First, only a limited number of reference spectra are available. Although several current collections contain more than 40,000 mass spectra, they represent only a small fraction of the number of known organic compounds. Fortunately for toxicologists, the mass spectra of most common toxicants (drugs, pesticides, commercial chemicals, etc.) have been published.[58-60]

A second problem is that of efficiently locating a matching spectrum. Most mass spectral libraries include a molecular weight index. Some collections also include an index ordered according to the M/e value of the most intense peak in each spectrum. These indexes usually make it possible to quickly reduce the collection to a manageable number of possible matches which can be examined in greater detail.

Finally, there is the problem of deciding whether a reference spectrum is sufficiently similar to the spectrum of the unknown to constitute a reasonably certain identification. Mass spectra of the same compound obtained on different instruments or under different operating conditions will show obvious differences in the relative ion intensities. The problem is further complicated by the fact that mass spectra are seldom devoid of ion peaks due to background and sample impurities. The ability to determine which peaks can be ignored and how much variation in relative ion intensities is permissible comes primarily from experience. Nevertheless, the following guidelines may be helpful:

1. Agreement among the relative ion intensities within a small mass range is more important than for those encompassing a wide mass range.
2. The higher mass ions are generally more diagnostic than those occurring at low mass (below about M/e 50).
3. No prominent ions in either spectrum (those with relative intensities above 10%) should be totally missing from the other spectrum, unless they can be attributed to an impurity.

To illustrate these points, Figure 10 compares EI mass spectra of cholesterol obtained on two different mass spectrometers, one a quadrupole and the other a magnetic sector instrument. On initial inspection the spectra appear quite different. The higher mass ions in the spectrum obtained on the quadrupole mass spectrometer have considerably lower relative intensities than the corresponding ions in the spectrum obtained on the magnetic sector instrument. Nevertheless, the patterns of ions within the high mass region are very similar in the two spectra. This can be seen more clearly by expanding those peaks in the quadrupole spectrum by a factor of 5, as shown in the figure. The patterns of peaks in the low mass region of the spectra (below about M/e 100) are not in as good agreement as those at high mass. These differences are primarily a result of impurity and background ions which often contribute to ion intensities in the low mass region.

Two other differences in the spectra in Figure 10, although minor, are worth noting. The intensities of the peak at M/e 368 are significantly different relative to adjacent peaks in the two spectra. This ion is formed by loss of water from the molecular ion. Its relative intensity is strongly influenced by the temperature of the ion source and inlet system. The lower relative intensity of the M/e 368 ion in the quadrupole spectrum suggests that the source temperature was lower in the quadrupole instrument than it was in the magnetic sector instrument. There is also a significant difference in the ratios of the (M + 1)/M ion intensities in the two spectra. The M + 1 ratio is particularly significant in that it is often used to estimate the number of carbon atoms in a compound's elemental composition.[3] For compounds containing only C, H, N, and O, the number of carbon atoms should equal the ratio (M + 1)/M divided by 1.1, the natural abundance of the carbon-13 isotope. For cholesterol ($C_{27}H_{46}O$), the (M + 1)/M ratio should equal 0.30,

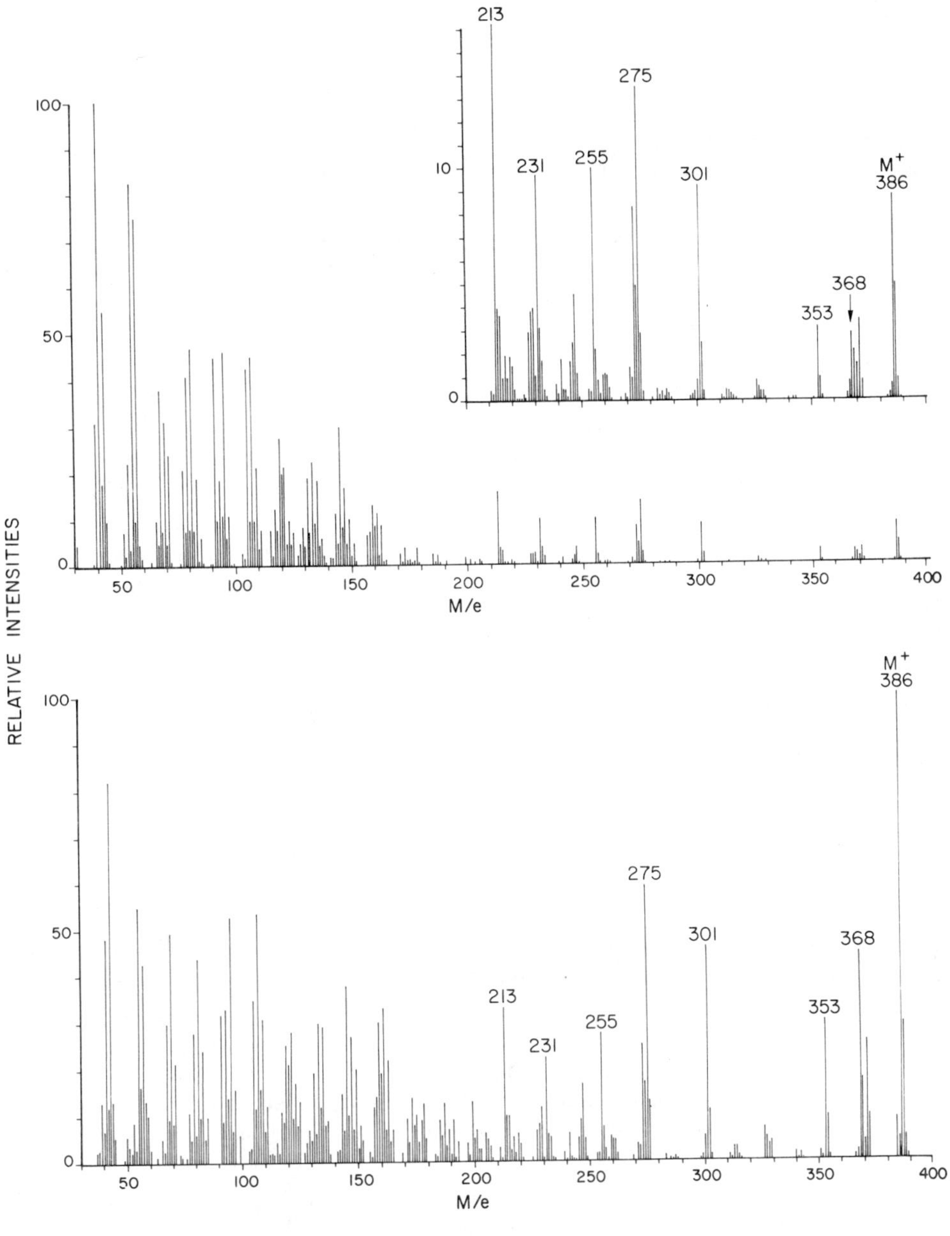
RELATIVE INTENSITIES
M/e
213
231
255
275
301
353
368
M+
386

FIGURE 10. Comparison of the EI mass spectra of cholestrol on two different mass spectrometers. Insert shows expanded relative intensity scale for high mass region of upper spectrum.

which it does in the spectrum obtained on the magnetic sector instrument. However, the (M + 1)/M ratio in the spectrum obtained on the quadrupole instrument is 0.55. This has nothing to do with the type of mass analyzer used, but rather is a consequence of the fact that in this case the quadrupole instrument was equipped with a chemical ionization source. Electron impact mass spectra can be obtained on a chemical ionization source but often show enhanced M + 1 ions due to self-protonation; that is, the sample acts as its own reagent gas and generates ions by ion-molecule reactions.

In spite of the differences noted in these spectra of cholesterol, a reasonably experienced mass spectroscopist would recognize the important matching features and consider the two spectra an acceptable match.

Currently, the most extensive and well-arranged collection of mass spectral data is the Second Edition of *The Eight Peak Index of Mass Spectra* published in 1974.[58] The data in this collection are a condensation of 31,101 mass spectra and are arranged in three separate tables. In Table 1 (Volume 1) the spectra are listed in ascending molecular weight order, subordered on the number of carbon atoms. Table 2 (Volume 2) contains data for spectra also arranged in ascending molecular weight order, but subordered on ascending M/e value of the most abundant ion in each spectrum. In Table 3 (Volumes 3 and 4) the spectra are arranged in ascending M/e value of the most abundant ion in each spectrum and subordered on other M/e values in order of decreasing abundance. Consequently, Tables 1 and 2 require knowledge of the unknown's molecular weight, while Table 3 is useful if the compound's molecular weight is uncertain. A selected list of data relating to drugs is appended to this discussion. *The Atlas of Spectral Data and Physical Constants for Organic Compounds*[59] contains a similar collection of mass spectral data, as well as infrared, nuclear magnetic resonance, and other types of spectral and physical-chemical data. The major limitation of these collections is that the mass spectra are condensed (only the eight most abundant ions), and a conclusive identification of an unknown should include comparison with a complete spectrum. Many collections of complete mass spectra have been published.[60] The most useful of these are

- Stenhagan, E., Abrahamsson, S., and McLafferty, F. W., Eds., *Registry of Mass Spectral Data,* John Wiley and Sons, New York. A magnetic tape version is also available containing 23,879 spectra.
- Markey, S. P., Urban, W. G., and Levine, S. P., Eds., *Mass Spectra of Compounds of Biological Interest,* National Technical Service, U.S. Department of Commerce, Springfield, Virginia. Contains approximately 2,000 mass spectra ordered by molecular weight, as well as an alphabetical index and indexes based on highest peaks every 41 and 50 u.
- *Mass Spectrometry Data Centre Collection,* A.W.R.E., Aldermaston, Reading RG7 4PR, U.K. Contains 7,000 printed spectra submitted by contributors all over the world. Index based on molecular weights.

Four mass spectrometry periodicals publish annual indexes to the mass spectrometry literature which can be very useful in locating mass spectra of specific compounds.[8-11]

All of the previously mentioned collections of mass spectra are limited to data obtained by electron impact. Collections of CI mass spectra are less useful because of the many different reagent gases that can be used and the profound effect that the ion source temperature can have on CI mass spectra. Furthermore, CI mass spectra are more easily predicted than EI mass spectra, so that it is not as important to have a reference spectrum for comparison. Nevertheless, two collections of CI-MS data on drugs have been published.[27,61] A collation of the CI drug data appears in this volume.

A great amount of effort has gone into developing programs which will enable a computer to match mass spectra of unknowns to a computer-stored library of reference spectra.[62] Most computer-matching facilities involving libraries of more than a few thousand spectra utilize relatively large computers because of their larger storage capacity and ability to handle more sophisticated programs. The large computers can be accessed through a teletype terminal via telephone lines. Two computer-based mass spectral matching services are currently in operation. The EPA/NIH Mass Spectral search System offers a number of retrieval and search capabilities in addition to standard spectrum matching. The system is currently operated on the Cyphernetics computer network. The Cornell University Mass Spectral Identification System offers a competitive service with a different approach to computer-based spectral interpretation. As of this writing, both libraries contain more than 40,000 mass spectra. The cost per search ranges from $5 to $15, depending on the specific programs used and the total number of searches conducted.

Most mass spectral data systems also offer the capability of conducting automated spectrum matching. Even though small, dedicated computers are limited in the size of library they can search, it has been demonstrated that a disk-stored library of more than 30,000 spectra can be searched rapidly by a mini-computer using a modified Biemann[63] matching algorithm.[64]

Spectral Interpretation

If an unknown's mass spectrum cannot be matched to a spectrum of a known compound, identification becomes dependent on the analyst's ability to interpret the mass spectrum. If the compound's structure is completely unknown, a detailed interpretation can be extremely difficult. Fortunately, this is seldom the case, as the origin of the sample, its chromatographic behavior, and spectroscopic and physical-chemical data are usually available and provide clues as to the compound's possible structure. An excellent guide to a systematic approach to the interpretation of EI mass spectra has been written by McLafferty,[3] one of the earliest proponents of the application of mass spectrometry to the analysis of a wide range of organic materials.

Considerable effort has been directed toward developing computer programs to assist the analyst in interpreting EI mass spectra.[65–66] The complexity of these programs currently requires that they be maintained on large, central computers.

In addition to the standard EI mass spectrum, other types of mass spectral data can prove invaluable in the identification of an unknown. The complementary nature of EI and CI mass spectra has already been mentioned. The CI mass spectra not only help identify the unknown's molecular weight, but can also provide useful structural information. In general, CI mass spectra are relatively easy to interpret since the prominent fragment ions are almost always formed by simple and logical fragmentation processes.[23] The same cannot be said for EI mass spectra where fragmentation often involves a complex sequence of events.

Quantitative Analysis

In spite of the high cost and complexity of mass spectrometers, they are being used increasingly for quantitative analyses because they offer sensitivity and specificity not achievable by other methods. Often a GC-MS quantitative assay is developed and used as a primary standard for evaluation of less expensive assay methods. Most GC-MS quantitative analyses involve operation of the mass spectrometer in the selected-ion-monitoring mode.[53] Operation in this mode requires either an interactive data system or a hard-wired multiple ion monitor.

Need for Internal Standards

Mass spectrometers tend to be more subject to variations in response than most other analytical tools. Consequently, for quantitative analysis use of an internal standard is a prerequisite for obtaining accurate and reproducible measurements. The normal procedure is to add the internal standard to the sample prior to any extraction, purification, or derivatization. Ideally the internal standard should have identical chemical and physical properties to the compound being analyzed, but still be distinguishable from it. When this is true, any losses occurring during the sample pretreatment will affect the compound being analyzed and the internal standard equally and will not change the ratio of their concentrations.

The types of compounds which have been used as internal standards in GC-MS assays include: isomers,[67] homologues,[68] compounds of similar chemical nature,[69] and isotope-labeled analogues.[70] Of these possibilities, the isotope-labeled analogue comes the closest to having identical chemical and physical properties to the parent compound. In addition to compensating for losses during sample pretreatment, an isotope-labeled internal standard can serve as a "carrier" to prevent losses of small amounts of the sample on the GC column. This technique, involving addition of relatively large amounts (10- to 1,000-fold excess) of the labeled internal standard, has been used to quantitate picomole amounts of prostaglandins in biological samples.[71]

A further advantage of isotope-labeled internal standards is that the isotope-labeled analogue will essentially coelute from the GC column with the sample compound. Therefore, quantitation can be based on measurement of peak heights, which are sometimes easier to measure than peak areas. Actually, deuterium-labeled analogues will normally elute slightly ahead of the unlabeled compound, but the effect on peak shape is negligible.

The major limitation to wider use of isotope-labeled internal standards is the relatively small number of labeled compounds currently available. Often, it is necessary for the analyst to have the isotope-labeled analogue custom-synthesized, usually at high cost. However, the situation is improving, particularly with respect to drugs. The National Institute on Drug Abuse has contracted the synthesis of isotope-labeled analogues of most of the commonly abused drugs.[72] Also, many pharmaceutical companies synthesize isotope-labeled compounds for use in their own research programs and often are willing to supply small amounts to other laboratories. Since the amount of material required for use as an internal standard is very small (a few milligrams is sufficient), a single batch synthesis can easily satisfy the needs of many laboratories. For maximum sensitivity, the isotope-labeled internal standard should have high isotopic purity and a mass increment of three or more atomic mass units.

If isotope-labeled analogues are not available, homologous or isomeric compounds often can be obtained for use as internal standards. In some situations they may actually be preferred.[68]

Operating Conditions

The development of a selected-ion-monitoring assay should include careful consideration and evaluation of the many possible operational parameters. The GC column and operating conditions should be chosen so as to give the shortest possible retention time consistent with adequate resolution of the sample compound and internal standard from potentially interfering materials. Various modes of ionization should be compared. For many compounds, CI has been found to offer better sensitivity and specificity than EI ionization.[73,74] Ammonia CI is particularly well suited for selected-ion-monitoring analysis of compounds containing basic functional groups.[74] Negative ion CI shows promise of being extremely sensitive for certain types of compounds and derivatives.[35] If monitoring one particular ion mass proves unsatisfactory due to interferences, changing the method of ionization will often provide abundant ions at a more satisfactory mass. The number of masses monitored need not be limited to one per compound. If, for example, the mass of the molecular ion and that of a prominent fragment ion are monitored for both the compound of interest and the internal standard, two sets of complementary data will be generated. Contributions by interfering compounds to any of the monitored ion currents will change the molecular ion/fragment ion ratio and will be clearly evident.

Some of these considerations can be illustrated by describing the selected-ion-monitoring procedure developed for quantitation of morphine in urine.[73] Morphine, labeled with three deuteriums on the *N*-methyl group, was used as the internal standard. A carefully measured amount of the internal standard was added to the urine. After adjusting the pH to 8.5, the urine was extracted with chloroform-isopropanol (4:1). The extract was concentrated and treated with *bis*(trimethylsilyl)acetamide in order to convert the morphine to the *bis*(trimethylsilyl)ether derivative which exhibits better gas chromatographic behavior than the parent drug. A portion of the methane CI mass spectra of the *bis*(trimethylsilyl)ethers of morphine and its trideuterated analogue are shown in Figure 11. Any of the three most abundant ions in the morphine spectrum are suitable for selected ion monitoring: M/e 430 $(M + H)^+$, 414 $(M - CH_3)^+$, and 340 $(MH\text{-}HOSiMe_3)^+$. Clearly the ion at M/e 371 would not be suitable, as its formation involves loss of the *N*-methyl group and it is, therefore, formed from both morphine and its deuterated analogue. Figure 12 shows the ion current plots resulting from a selected-ion-monitoring analysis of an extract from an individual who had recently consumed codeine-containing cough medicine. The ion current plots for each morphine/*N*-CD_3 morphine ion pair (340/343, 414/417, and 430/433) are separated for greater clarity. The occurrence of peaks in the M/e 340, 414, and 430 plots coincident with the internal

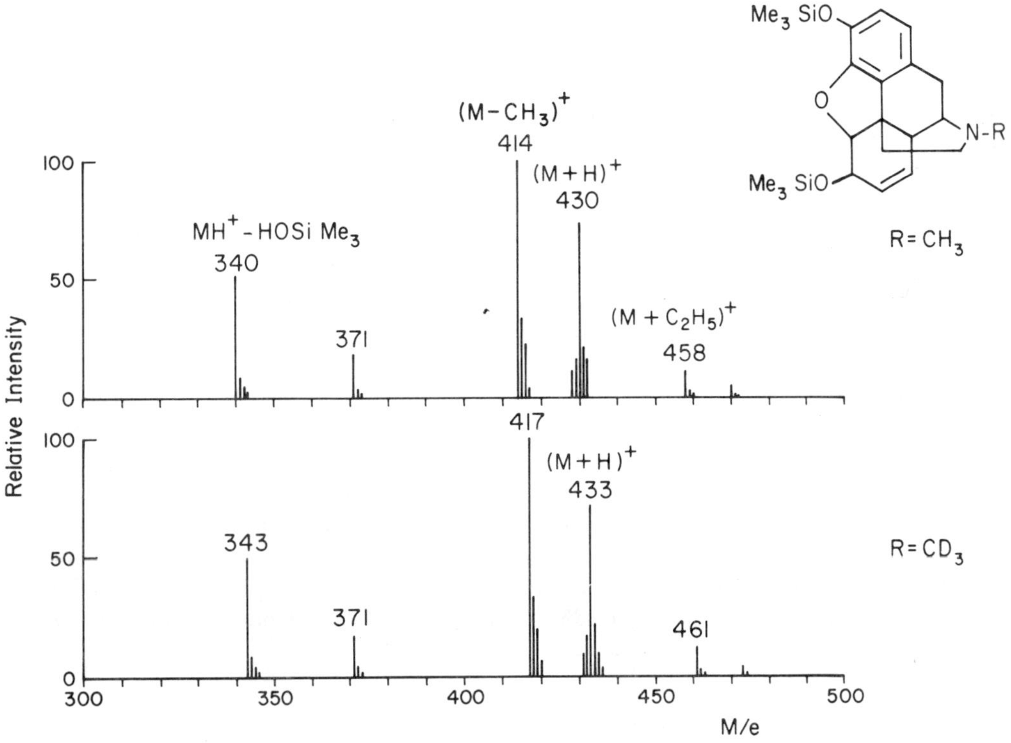

Me3 SiO
O
N-R
Me3 SiO
R= CH3
R= CD3
(M-CH3)+
414
(M+H)+
430
MH+ - HOSi Me3
340
371
(M + C2H5)+
458
417
(M+H)+
433
343
371
461
Relative Intensity
100
50
0
300
350
400
450
500
M/e

FIGURE 11. Methane CI mass spectra of trimethylsilyl derivatives of morphine (above O and *N*-CD_3 morphine (bottom). (From Clarke, P. A. and Foltz, R. L., *Clin. Chem.*, 20, 467, 1974. With permission.)

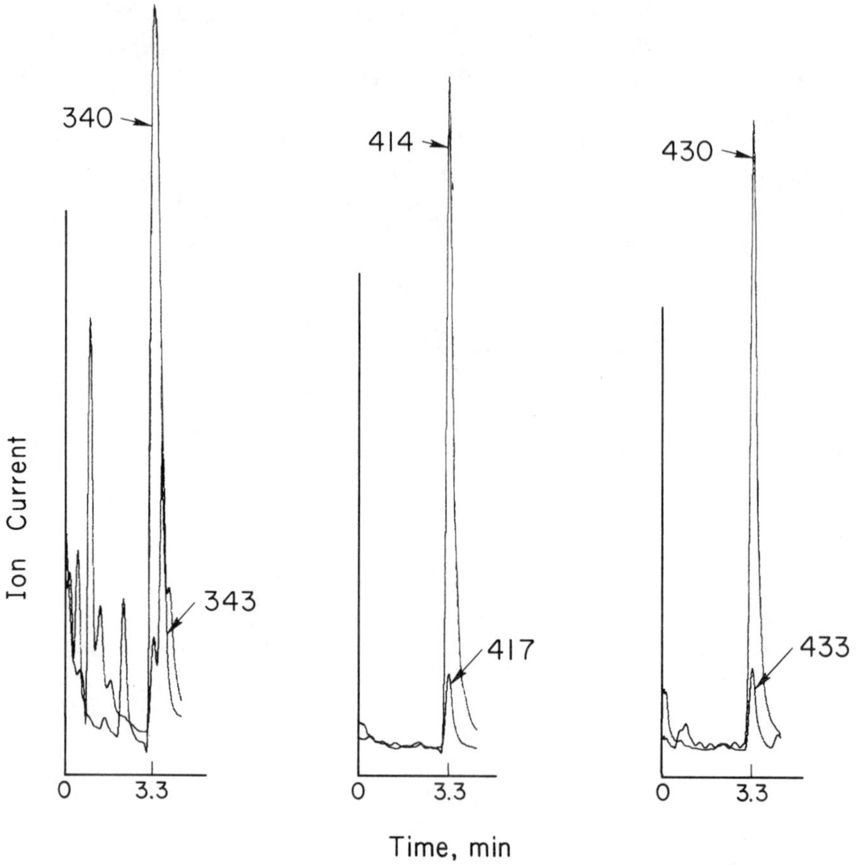
340
343
414
417
430
433
Ion Current
0
3.3
0
3.3
0
3.3
Time, min

FIGURE 12. Ion-current plots for ions monitored in quantitation of morphine in urine. (From Clarke, P. A. and Foltz, R. L., *Clin. Chem.*, 20, 468, 1974. With permission.)

standard peaks in the M/e 343, 417, and 433 plots provides conclusive evidence for morphine in the urine. In spite of the structural similarity of trimethylsilylcodeine and *bis*-(trimethylsilyl) morphine, the former only contributes to the ion current at M/e 343. Based on the 414/417 and 430/433 ion current peak height ratios, the concentration of morphine in the urine was calculated to be 77 ng/ml. It seemed likely in this case that the morphine was solely a metabolic product of the ingested codeine.

Sensitivity and Precision

Numerous articles have appeared in the literature describing quantitation by selected-ion monitoring of biologically important compounds at concentrations below 1 ng/ml.[53] It should not be concluded from these reports that quantitation at such low concentrations is routine. The sensitivity and precision that can be routinely achieved will depend on the extraction, chromatographic, and mass spectral behavior of the compound being analyzed, as well as the capabilities of the GC-MS system and the skill of the analyst. However, as a general guide, quantitation with good precision (10% relative standard deviation) should be fairly easy to achieve down to about the 10-ng/ml level. Below this concentration loss of sample during the pretreatment steps and the GC ionization.[73,74] Ammonia CI is particularly well-suited for selected-ion-monitoring separation can become a severe problem. Also, interferences due to other materials can become significant in spite of the high specificity normally associated with selected ion monitoring. Achievement of adequate quantitative results at very low sample concentrations may require any or all of the following steps:

1. Development of an extraction procedure which couples good sample recovery with substantial purification
2. Silylation of all glassware in contact with the sample to avoid losses due to adsorption
3. Derivatization in order to improve the gas chromatographic behavior of the material being analyzed
4. Evaluation of different modes of ionization to determine which offers the best sensitivity with minimum interferences
5. Use of a stable isotope-labeled analogue to serve as the internal standard and as a "carrier" to minimize adsorption losses
6. Preparation of standard curves on a daily basis
7. Optimization of the instrumental operation conditions, including ion source conditions, integration times, and ion current signal amplification and filtering

Although progress has been made toward automation of GC-MS analyses, most assays still require a considerable amount of manual sample manipulation. The analyst's ability to utilize good laboratory technique in measuring and dispensing standards, as well as extracting, transferring, concentrating, and derivatizing the sample, will have a major effect on the quality of the results.

Instrument Operation

The value of a mass spectrometer to a toxicology laboratory will depend heavily on the expertise of the person responsible for the MS facility and the ability of that person to devote a major portion of his or her time to the many aspects of running such a facility. Too often, management, after deciding to establish a MS facility, will conduct an exhaustive evaluation of instruments, but then hand over responsibility for the facility to a staff member who may have expressed only a casual interest in mass spectrometry, or even to one whose only qualification is that he has time available. In reality, the specific choice of instrument is far less important than finding an analyst with the experience, ability, and interest to make the mass spectrometer a productive component of the laboratory.

Ideally, the person responsible for operating and maintaining the MS system should have at least 2 years' experience in mass spectrometry and a strong background in organic and analytical chemistry. Of course, as with most jobs, a person's overall competence and willingness to learn are more significant than his number of degrees and specific experience. Nevertheless, it is vitally important that the operator have the ability to evaluate all aspects of the instrument's performance and that he do so on a regular basis. The mass spectroscopist must also be willing and able to help prepare samples for analysis, determine the best method for conducting the analysis, and assist in interpretation of the data.

In mass spectrometry, probably more than any of the other common spectroscopic methods, the quality and usefulness of the data are dependent on the operator's skill and experience. For example, the very process of sample introduction usually involves some fractionation (either GC separation or fractional volatilization) and, therefore, some of the spectra recorded during the analysis may reflect a very distorted picture of the true composition of the sample. The operator's ability to distinguish useful data from the useless is a key element in the operation of any mass spectrometry facility.

From the standpoint of optimum performance and minimum down time, it is desirable to have only one operator for each MS system. However, often this is not practical. If maximum sample throughput is important, a single operator simply cannot keep an instrument fully occupied, particularly if he is as involved in the sample preparation and data interpretation as he should be. Furthermore, there are some definite advantages in having the person most knowledgeable regarding each sample perform the MS analysis on that sample. Nevertheless, it should be recognized that the frequency of instrument problems inevitably rises as the number of operators increases. In multi-operator situations one person should be responsible for overseeing the operations, servicing the instruments, monitoring instrument performance, and training new operators.

Individuals submitting samples for mass spectral analysis should be certain that the analyst is provided with all relevant information regarding the sample so that he can decide which inlet and operating conditions should be used. The following information is particularly helpful:

1. What specific information is desired from the analysis?
2. What is known regarding the compound's likely structure? What is the maximum molecular weight anticipated?
3. What is the estimated purity of the sample? How was it isolated? Is there residual solvent present?
4. How volatile is the sample, or more precisely, what temperature will be necessary to volatilize it within the mass spectrometer? In the case of crystalline compounds, the melting point provides a useful clue as to its volatility. Also, a sample's polarity as judged by its solubility and chromatographic behavior is often suggestive of its volatility.
5. Is the sample thermally unstable, or is it sensitive to air or moisture? How should it be stored prior to analysis?

If the sample is to be analyzed by combined GC-MS, suitable GC conditions should be worked out on a separate GC unit and the FID chromatogram submitted with the sample.

Obviously, the purer a sample is the easier it will be to obtain high quality mass spectra suitable for structural identification. However, it may be more efficient with respect to time and cost to attempt a preliminary mass spectral analysis on a crude sample before undertaking a lengthy sample cleanup. Chemical ionization is particularly useful in the analysis of crude samples since it usually enables one to easily identify the molecular weights of the various components of the sample. Often that information is sufficient or at least will indicate whether further sample cleanup is justified.

Maintenance and Troubleshooting

In spite of dramatic improvements in stability and reliability, mass spectrometers remain one of the most attention-requiring of analytical instruments. In addition to having extremely complex electronic circuitry, the mass spectrometer includes key components which can only be operated under high vacuum. If leaks develop in the vacuum system or the ionizer, analyzer, or detector becomes contaminated, the instrument's performance deteriorates. Yet contamination is inevitable since samples are introduced and effectively decomposed within the instrument's vacuum system. When the instrument's performance has degraded to the point that it is no longer acceptable, it must be shut down and the contaminated elements cleaned. The frequency with which cleaning is necessary can be kept to a minimum by using no more sample than is necessary to obtain good-quality spectra. This is particularly true for samples introduced by means of the direct probe. There is a strong tendency by operators not fully accustomed to the sensitivity of a mass spectrometer to load the probe sample holder with hundreds of micrograms of material, when a tenth of a microgram is adequate.

Another common cause of instrument contamination is the presence of cold spots within the ionizer or the GC-MS interface. It is important that the GC-MS interface be uniformly heated to a temperature at least as great as the maximum temperature to which the GC column will be heated. Since the ionizer is under vacuum, it does not have to be maintained at as high a temperature as the GC-MS interface; nevertheless, maintaining it at an elevated temperature (150 to 250°C) will help reduce condensation of contaminants and the need for frequent cleaning. Many laboratories also routinely maintain the analyzer and detector housing at moderately elevated temperatures (100°C) with occasional overnight bakeouts at 250 to 300°C. Optimum temperatures vary with the different instruments, and the operator's manual should be consulted for specific recommendations.

Two additional practices should be adopted in order to reduce the frequency and seriousness of instrument malfunctions: daily monitoring of instrument performance and regular maintenance inspection. These practices are particularly important if more than one operator uses the instrument. To monitor instrument performance, a simple test can be devised requiring only a few minutes at the beginning of each work day. For example, a measured amount of a standard solution can be injected into the GC-MS system under specific operating conditions. The area of the reference material peak in the total ionization chromatogram is measured and compared with previously measured values. The width and shape of the GC peak should also be compared with previous performance checks to detect changes in the GC column behavior.

Considerations in Purchasing a GC-MS System

The prospective buyer of a mass spectrometer system is often bewildered by the assortment of mass spectrometers and ancillary equipment available. Choices range from specialized instruments designed for specific applications to complex, versatile systems requiring considerable skill and experience to operate and maintain. The selection of an instrument should be based on a careful assessment of a laboratory's specific needs. If it will be used primarily as a research tool, or to provide analytical support for various research programs, the benefits of a versatile and high-performance system may be justified. On the other hand, if the primary application will be analysis of large numbers of similar samples, the major considerations should be reliability, ease of operation, and overall cost effectiveness. Unfortunately, these qualities are difficult for the inexperienced user to evaluate. However, there are several sources of useful information which should be explored before deciding upon a specific instrument. If there are other laboratories which have similar analytical needs and they have experience with the GC-MS system of interest, they should be contacted. If feasible, an actual visit to the laboratory will pay dividends in the form of a more candid appraisal of the instrument than would result from a brief telephone inquiry. Most instrument manufacturers maintain applications laboratories where prospective buyers can see an instrument in operation, have their own samples analyzed, and experience hands-on operation of the system. To make most effective use of such opportunities, it is important to take with you samples as nearly representative as possible of the type of sample your laboratory will be asked to analyze. By taking the same samples to several applications laboratories, the performance of each candidate instrument can be directly compared. However, in a comparison of this type it should be recognized that the quality of the results may be more indicative of the skill of the analyst than the capabilities of the instrument. Also, in fairness to application laboratories' analysts, it should be pointed out that they are often expected to set up, perform, and interpret in a few hours a GC-MS analysis of a difficult sample, which under normal circumstances might involve several day's effort. Nevertheless, in spite of their inherent limitations, applications laboratories play a vital role in customer evaluations of competitive systems.

Mass spectrometry technology is undergoing continual evolution so that specific instrument recommendations quickly become dated and of limited value. One of the more significant evolutionary trends is toward manufacture of complete integrated systems. We are already seeing the appearance of systems consisting of a single package containing a gas chromatographic inlet specifically designed for the mass spectrometer, an ion source capable of both electron impact and chemical ionization, and a computer that not only processes and outputs mass spectral data, but also monitors and controls each operation of the system. Ultimately, it should be possible to manufacture an integrated system of this type which will be considerably cheaper than the current total cost for the separate components. However, for the present, most buyers are faced with cost-justifying each optional component. Therefore, the following recommendations are offered as a general guide, recognizing that time and the specific circumstances of your laboratory may alter the wisdom of each suggestion.

The mass spectrometer system should include at least three inlets: a heatable, direct insertion probe; a variable leak inlet; and a gas chromatographic inlet. Of these inlets, only the gas chromatograph constitutes an appreciable additional expenditure. Liquid chromatographic (LC) inlets, when they become commercially available, are likely to be expensive, troublesome, and limited in the types of compounds that they can handle. Before purchasing an LC inlet, be certain that the analytical applications you have in mind cannot be adequately handled by GC-MS or by collection of eluent fractions from an off-line LC unit with subsequent introduction into the mass spectrometer via the direct probe.

The mass spectrometer should be capable of both electron impact and chemical ionization. The two ionization techniques beautifully complement each other. EI is the traditional ionization process and there is a great body of literature relating EI-induced fragmentation to molecular structure. CI is the newer technique, but it has increased rapidly in popularity because it gives simpler spectra which are easier to interpret. It is difficult to envision a toxicology laboratory operation which would not significantly benefit by having both ionization techniques available. Other ionization techniques such as field ionization, field desorption, and photo-ionization are not recommended, either because of their limited applicability or because in their current state of development, they are not suited to routine operation.

The choice of magnetic vs. quadrupole mass spectrometer is not easily made. Both types of analyzers have evolved to where they offer similar capabilities. However, if the instrument will be used primarily as a research tool for the structure elucidation of complex molecules and the determination of fragmentation mechanisms, an instrument with a magnetic analyzer should be favored. On the other hand, if the instrument will be used with a computer for the qualitative and quantitative analysis of mixtures, the quadrupole analyzer offers significant advantages.

The question of whether to buy a data system should be decided based on the anticipated volume of samples, the type of analyses to be performed, and whether or not the additional funds can be obtained. A data system can easily double the cost of a GC-MS. However, at the same time the data system can more than double sample throughput as well as greatly increase the system's capabilities. If the laboratory's current work load does not justify purchasing a data system, it is some comfort to know that one can be added at a future date if the sample load increases and additional funds become available. In fact, a good case can be made for operating a GC-MS system without a data system for an initial period of a year or more, so that the operator becomes familiar with manual operation of the system before being "turned loose" on a computer-based system. A significant danger in operating a computer-based system is that it is more difficult to differentiate good data from bad data. For example, a mass spectrum may consist of ion current signals too weak to be statistically meaningful, but after exhaustive data massaging by the computer (spectrum averaging, background subtraction, normalization, etc.) the outputted spectrum may appear perfectly satisfactory to the inexperienced spectroscopist. This is even more of a problem for the investigator who is not involved in the operation of the instrument. After submitting his sample, he receives a beautifully plotted spectrum. Whether or not the spectrum contains any information relating to the major component of his sample may be difficult to determine.

In spite of these dangers, a data system can be a tremendous asset. It is virtually a necessity if the laboratory's needs include the ability to identify many components in complex mixtures. In these situations the amount of chart paper generated in a single GC-MS run becomes almost unmanageable; whereas with a data system, all of the data are stored on disks or tapes and can be examined easily and rapidly in any of a variety of formats. On the other hand, if the GC-MS system will be used primarily for structure confirmation of purified samples or mixtures containing only a few components, a data system is not necessary. Likewise, if the primary application will be quantitative measurement of drugs in biological samples, a GC-MS system equipped with a hard-wired multiple ion monitor may very well be adequate.

Of necessity, most instrument manufacturers have done a good job keeping pace with the competition. Capabilities and performance characteristics are relatively uniform for instruments within given price brackets. However, the quality of customer-support services does not appear to be as uniform. Since mass spectrometers are complex mechanical-electronic systems requiring frequent servicing, the manufacturer's ability to provide rapid and reliable repair service should be a major consideration in choosing an instrument. Even if your laboratory is capable of performing its own instrument maintenance and repair, it is important to be able to obtain rapid delivery of spare parts. With the current existence of rapid parcel delivery services, there is no excuse for not being able to receive critically needed spare parts within 24 hr after placing a telephone order.

There are additional ways in which an instrument manufacturer can provide valuable customer support. Training courses and users' meetings can be particularly beneficial to laboratories just getting into mass spectrometry. Preparation and circulation of newsletters containing brief descriptions of new GC-MS applications as well as maintenance and servicing tips are functions that can be carried out by the manufacturer and also serve to indicate continuing interest in the customers' needs and problems. The quality and completeness of the owner's manual are important considerations. Many laboratories are capable of doing their own instrument servicing if they have a complete set of schematic diagrams and engineering drawings. Finally, just the basic attitude conveyed by various members of the manufacturer's staff is important. In working with complex instrumentation, it is inevitable that one will encounter frustrating and perplexing problems. A feeling that the manufacturer is eager to provide whatever assistance he can goes a long way in helping relieve the depression that can engulf the analyst during these periods.

REFERENCES

1. **Beynon, J. H., Saunders, R. A., and Williams, A. E.,** *The Mass Spectra of Organic Molecules,* Elsevier, New York, 1968.
2. **Biemann, K.,** Mass Spectrometry, *Organic Applications,* McGraw-Hill, New York, 1962.
3. **McLafferty, F. W.,** *Interpretation of Mass Spectra,* 2nd ed., W. A. Benjamin, Inc., Reading, Massachusetts, 1973.
4. **Budzikiewicz, H., Djerassi, C., and Williams, D. H.,** *Mass Spectrometry of Organic Compounds,* Holden-Day, San Francisco, 1967.
5. **Waller, G. R., Ed.,** *Biochemical Applications of Mass Spectrometry,* Interscience, New York, 1972.

6. **Johnstone, R. A. W.**, Reporter, *Mass Spectrometry, Vol. 3.*, Chemical Society, London, 1975.
7. **Burlingame, A. L., Kimble, B. J., and Derrick, P. J.**, Mass Spectrometry, *Anal. Chem.*, 48, 368R, 1976.
8. **Fenselau, C. C. and Millard, B. J., Eds.**, *Biomedical Mass Spectrometry*, published bi-monthly by Heydon & Sons, London.
9. **Maccoll, A., Ed.**, *Organic Mass Spectrometry*, published monthly by Heydon & Sons, London.
10. **Franzen, J., Quayle, A., and Suec, H. J., Eds.**, *Journal of Mass Spectrometry and Ion Physics*, published monthly by Elsevier, Amsterdam.
11. *Mass Spectrometry Bulletin*, published monthly by the Mass Spectrometry Data Centre, AWRE, Aldermaston, Reading RG7 4PR, U.K.
12. **Milne, G. W. A., Ed.**, *Mass Spectrometry, Techniques and Applications*, Interscience, New York, 1971.
13. **Williams, D. H. and Howe, I.**, *Principles of Organic Mass Spectrometry*, McGraw-Hill, London 1972.
14. **Burlingame, A. A., Ed.**, *Topics in Organic Mass Spectrometry*, John Wiley & Sons, New York, 1970.
15. **West, A. R., Ed.**, *Advances in Mass Spectrometry*, Vol. 6, Applied Sciences Publishers, Barking, Essex, U.K., 1974.
16. **Lawson, A. M. and Draffan, G. H.**, Gas-liquid chromatography-mass spectrometry in biochemistry, pharmacology, and toxicology, *Prog. Med. Chem.*, 12, 1, 1975.
17. **Roboz, J.**, Mass spectrometry in clinical chemistry, *Adv. Clin. Chem.*, 17, 109, 1975.
18. **Jenden, D. J. and Cho, A. K.**, Applications of integrated gas chromatography/mass spectrometry in pharmacology and toxicology, *Annu. Rev. Pharmacol.*, 13, 371, 1972.
19. **Fenselau, C.**, Applications of mass spectrometry to pharmacological problems, in *Methods in Pharmacology*, Vol. 2, Chignell, C. F., Ed., Appleton-Century-Crofts, New York, 1972, chap. 12.
20. **Horning, E. C., Horning, M. G., and Stillwell, R. N.**, Gas-phase analytical methods. Mass spectrometry and GC-MS-COM analytical systems, in *Advances Biomedical Engineering*, Brown, J. H. V. and Dickson, J. R., Eds., Academic Press, New York, 1974.
21. **Field, F. H.**, Chemical ionization mass spectrometry, *Acc. Chem. Res.*, 1, 42, 1968.
22. **Munson, B.**, Chemical ionization mass spectrometry, *Anal. Chem.*, 43(13), 28A, 1971.
23. **Foltz, R. L.**, Structural analysis via chemical ionization mass spectrometry, *Chem. Technol.*, 5, 39, 1976.
24. **Hunt, D. F.**, Reagent gases for chemical ionization mass spectrometry, *Adv. Mass Spectrom.*, 6, 517, 1974.
25. **Jelus, B. L. and Munson, B.**, Reagent gases for GC-MS analysis, *Biomed. Mass Spectrom.*, 1, 96, 1974.
26. **Munson, M. S. B. and Field, F. H.**, Chemical ionization mass spectrometry. General instructions, *J. Am. Chem. Soc.*, 88, 2621, 1966.
27. **Saferstein, R., Chao, J. M., and Manura, J.**, Identification of drugs by CI-MS. Part II. *J. Forensic Sci.*, 19, 463, 1974.
28. **Horton, D., Wander, J. D., and Foltz, R. L.**, Analysis of sugar derivatives by chemical ionization mass spectrometry, *Carbohydr. Res.*, 36, 75, 1975.
29. **Hunt, D. F. and Ryan, J. F.**, Argon-water mixtures as reagents for chemical ionization mass spectrometry, *Anal. Chem.*, 44, 1306, 1972.
30. **Fentiman, A. F., Foltz, R. L., and Kinzer, G. W.**, Identification of noncannabinoid phenols in marihuana smoke condensate using chemical ionization mass spectrometry, *Anal. Chem.*, 45, 580, 1973.
31. **Arsenault, G. P.**, Mixed charge exchange-chemical ionization reactant gases in high pressure mass spectrometry, *J. Am. Chem. Soc.*, 94, 8241, 1972.
32. **Hunt, D. F. and Harvey, T. M.**, Nitric oxide chemical ionization mass spectra of alkanes, *Anal. Chem.*, 47, 1965, 1975.
33. **Jardine, L. and Fenselau, C.**, Charge exchange mass spectra of morphine and tropane alkaloids, *Anal. Chem.*, 47, 730, 1975.
34. **Tannenbaum, H. P., Roberts, J. E., and Dougherty, R. C.**, Negative chemical ionization mass spectrometry – chloride attachment spectra, *Anal. Chem.*, 47, 49, 1975.
35. **Hunt, D. F., Crow, F., and Lambrecht, J.**, Resonance Electron Capture Negative Ion Chemical Ionization Mass Spectrometry, presented at the 24th Annual Conference on Mass Spectrometry and Allied Topics, San Diego, May 9–14, 1976.
36. **Hunt, D. F., Stafford, G. C., Crow, F., and Russell, J. W.**, Instrumentation for Simultaneous Recording of Positive and Negative Ion Chemical Ionization Mass Spectra, presented at the 24th Annual Conference on Mass Spectrometry and Allied Topics, San Diego, May 9–14, 1976.

37. **Beckey, H. D.**, *Field Ionization Mass Spectrometry,* Pergamon Press, Oxford, 1971.
38. **Beckey, H. D. and Schulten, H. R.**, Field desorption mass spectrometry, *Angew. Chem. Int. Ed. Engl.*, 14, 403, 1975.
39. **Horning, E. C., Horning, M. G., Carroll, D. I., Dzidic, I., and Stillwell, R. N.**, New picogram detection systems based on a mass spectrometer with an internal ionization source at atmospheric pressure, *Anal. Chem.*, 45, 936, 1973.
40. **Macfarlane, R. D. and Torgerson, D. F.**, California-252 plasma desorption mass spectrometry, *Science,* 191, 920, 1976.
41. **Cooks, R. G., Beynon, J. H., Caprioli, R. M., and Lester, G. R.**, *Metastable Ions,* Elsevier, New York, 1973.
42. **Desiderio, D. M., Jr.**, Photographic techniques in organic high-resolution mass spectrometry, in *Mass Spectrometry, Techniques and Applications,* Milne, G. W. A., Ed., Interscience, New York, 1971.
43. **Baldwin, M. A. and McLafferty, F. W.**, Direct chemical ionization of relatively involatile samples. Application to underivatized oligopeptides, *Org. Mass Spectrom.*, 7, 1353, 1973.
44. **Down, G. J. and Gwyn, S. A.**, Investigation of direct thin-layer chromatography-mass spectrometry as a drug analysis techniqe, *J. Chromatogr.*, 103, 208, 1975.
45. **McFadden, W., Ed.**, *Techniques of Combined Gas Chromatography/Mass Spectrometry,* Interscience, New York, 1973.
46. **Ryhage, R.**, Integrated gas chromatography mass spectrometry, *Q. Rev. Biophys.*, 6, 311, 1973.
47. **Junk, G. A.**, Gas chromatography-mass spectrometer combinations and their applications, *Int. J. Mass Spectrom. Ion Phys.*, 8, 1, 1972.
48. **Drozd, J.**, Chemical derivatization in gas chromatography, *J. Chromatogr.*, 113, 303, 1975.
49. **Schultz, T. H., Flath, R. A., and Mon, T. R.**, Analysis of orange volatiles with vapor sampling, *J. Agric. Food Chem.*, 19, 1060, 1971.
50. **Westover, L. B., Tou, J. C., and Mark, J. H.**, Novel mass spectrometric sampling device – hollow fiber probe, *Anal. Chem.*, 46, 568, 1974.
51. **Witiak, J. L., Junk, G. A., Calder, G. V., Fritz, J. S., and Svec, H. J.**, A simple fraction collector for gas chromatography. Compatibility with infrared, ultraviolet, nuclear magnetic resonance, and mass spectral identification techniques, *J. Org. Chem.*, 38, 3066, 1973.
52. **McFadden, W. H., Schwartz, H. L., and Evans, S.**, Direct analysis of liquid chromatographic effluents, *J. Chromatogr.*, 122, 389, 1976.
53. **Falkner, F. C., Sweetman, B. J., and Watson, J. T.**, Biomedical applications of selected ion monitoring, *Appl. Spectrosc. Rev.*, 10, 51, 1975.
54. **Middleditch, B. S. and Desiderio, D. M.**, Comparison of selective ion monitoring and repetitive scanning during gas chromatography-mass spectrometry, *Anal. Chem.*, 45, 806, 1973.
55. **Foltz, R. L., Couch, M. W., Greer, M., Scott, K. N., and Williams, C. M.**, Chemical ionization mass spectrometry in the identification of drug metabolites, *Biochem. Med.*, 6, 294, 1973.
56. **Hammar, C.-G., Petterson, G., and Carpenter, P. T.**, Computerized mass fragmentography and peak matching, *Biomed. Mass Spectrom.*, 1, 397, 1974.
57. Finnigan Corp., Sunnyvale, California, 1976.
58. *The Eight Peak Index of Mass Spectra,* compiled by the Mass Spectrometry Data Centre, Her Majesty's Stationary Office, London, U.K., 1974.
59. **Grasselli, J. G., Ed.**, *The Atlas of Spectral Data and Physical Constants for Organic Compounds,* CRC Press, Cleveland, Ohio, 1973.
60. **Middleditch, B. S. and McCloskey, J. A.**, *A Guide to Collections of Mass Spectral Data,* American Society for Mass Spectrometry, May 1974.
61. **Finkle, B. S., Foltz, R. L., and Taylor, D. M.**, A comprehensive GC-MS reference data system for toxicological and biomedical purposes, *J. Chromatogr.*, 12, 304, 1974.
62. **Chapman, J. R.**, *Computers in Mass Spectrometry,* Academic Press, New York, 1976.
63. **Hertz, H. S., Hites, R. A., and Biemann, K.**, Identification of mass spectra by computer-searching a file of known spectra, *Anal. Chem.*, 43, 681, 1971.
64. **Hoyland, J. R.**, Battelle-Columbus Laboratories, Columbus, Ohio, personal communication.
65. **Clerc, T. and Erni, F.**, Identification of organic compounds by computer-aided interpretation of spectra, *Top. Curr. Chem.*, 39, 91, 1973.
66. **Isenhour, T. L., Kowalski, B. R., and Jurs, P. C.**, Applications of pattern recognition to chemistry, *Crit. Rev. Anal. Chem.*, 4, 1, 1974.
67. **Narasimhachari, N.**, The use of isomeric compounds as internal standards for quantitative mass fragmentography of biological samples, *Biochem. Biophys. Res. Commun.*, 56, 36, 1974.
68. **Lee, M. G. and Millard, B. J.**, A comparison of unlabelled and labelled internal standards for quantification by single and multiple ion monitoring, *Biomed. Mass Spectrom.*, 2, 78, 1975.

69. **Kelly, R. W.**, The measurement by gas chromatography-mass spectrometry of oestra-1,3,5-triene-3,15α,16α,17β-tetrol in pregnancy urine, *J. Chromatogr.*, 54, 345, 1971.
70. **Knapp, D. R. and Gaffney, T. E.**, Use of stable isotopes in pharmacology-clinical pharmacology, *Clin. Pharmacol. Ther.*, 13, 308, 1972.
71. **Axen, U., Green, K., Horlin, D., and Samuelsson, B.**, Mass spectrometric determination of picomole amounts of prostaglandins E_2 and $F_{2\alpha}$ using synthetic deuterium labeled carriers, *Biochem. Biophys. Res. Commun.*, 45, 519, 1971.
72. **Foltz, R. L.**, Report on the workshop sponsored by the committee for biological applications at the 23rd Annual Conference of the American Society for Mass Spectrometry, Houston, May 1975, *Biomed. Mass Spectrom.*, 2, 227, 1975.
73. **Clarke, P. A. and Foltz, R. L.**, Quantitative analysis of morphine in urine by gas chromatography-chemical ionization-mass spectrometry using N-CD_3 morphine as an internal standard, *Clin. Chem.*, 20, 465, 1974.
74. **Foltz, R. L., Knowlton, D. A., Lin, D. C. K., and Fentiman, A. F.**, Analysis of Abused Drugs by Selected Ion Monitoring: Quantitative Comparison of Electron Impact and Chemical Ionization, presented at the 2nd Internal Conference on Stable Isotopes, Argonne, Illinois, October, 20–23, 1975.

EIGHT-PEAK INDEX OF EI SPECTRA ARRANGED IN INCREASING ORDER OF MAJOR PEAKS

INTRODUCTION

Catherine E. Costello

The format of the eight-peak index to the collection has been designed to resemble that of the *Eight Peak Index of Mass Spectra*, Table 3, published by the Mass Spectrometry Data Centre, AWRE, Aldermaston, Reading RG7 4PR, UK, since that format is already familiar to many in the field. The spectra are ordered and sub-ordered according to highest, second-highest, etc. mass-to-charge ratio (m/e), the mass-to-charge ratio under consideration at any point of the table being asterisked. A diagram of the arrangement is given below. The names listed in this index are those which appear on the spectra in the collection. Molecular weights (MW) are calculated on the basis of the most abundant isotopes.

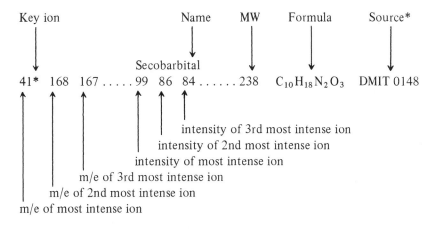

Key ion Name MW Formula Source*

Secobarbital

41* 168 167 . . . . . 99 86 84 . . . . . . 238 $C_{10}H_{18}N_2O_3$ DMIT 0148

intensity of 3rd most intense ion

intensity of 2nd most intense ion

intensity of most intense ion

m/e of 3rd most intense ion

m/e of 2nd most intense ion

m/e of most intense ion

* The sources of these data, as indicated in the Table, are DMIT – MIT Laboratory, BLR – Baylor University, NIHMSC – National Institute of Health, ABB – Abbott Laboratories, CSL – Custom Service Laboratory.

EIGHT-PEAK INDEX OF EI SPECTRA ARRANGED IN INCREASING ORDER OF MAJOR PEAK

Catherine E. Costello

Eight most intense peaks								Relative Intensities								Mol wt	Formula	Drug	Source	
15*	79	94	29	45	33	16	48	99	90	84	25	20	18	18	17	94	$C_2H_6O_2S$	Dimethylsulfone	DMIT	207
44	28*	29	31	43	60	73	41	99	89	82	81	65	60	56	52	357	$C_{19}H_{16}ClNO_4$	Indomethacin	DMIT	192
58	28*	59	221	30	42	0	0	99	22	4	4	3	3	0	0	315	$C_{18}H_{16}ClNS$	Chlorprothixene	DMIT	119
72	28*	165	42	180	178	179	91	99	30	10	7	7	7	6	6	309	$C_{21}H_{27}NO$	Methadone	DMIT	143
82	28*	182	83	77	42	94	105	99	45	44	32	31	29	26	25	303	$C_{17}H_{21}NO_4$	Cocaine	DMIT	146
100	28*	44	72	77	42	105	27	99	35	28	20	17	14	14	12	205	$C_{13}H_{19}NO$	Diethylpropion	DMIT	129
30*	99	28	42	43	41	27	39	99	66	64	62	61	60	31	27	99	C_5H_9NO	Piperidone	DMIT	79
30*	108	107	77	39	51	28	137	99	41	22	13	6	5	5	5	137	$C_8H_{11}NO$	Tyramine	DMIT	196
124	30*	123	153	77	51	125	78	99	82	35	23	12	9	6	4	153	$C_8H_{11}NO_2$	4-(2-Aminoethyl)pyro-catechol	ABB	5
182	30*	167	181	211	151	183	148	99	88	54	49	21	18	13	11	211	$C_{11}H_{17}NO_3$	Mescaline	DMIT	179
31*	49	77	113	82	51	115	29	99	16	11	7	5	5	5	4	148	$C_2H_3Cl_3O$	Chloral hydrate metabolite 1	DMIT	225
36*	57	41	40	44	38	55	39	99	99	83	83	72	67	65	55	157	$C_8H_{15}NO_2$	Oxanamide	DMIT	121
184	40*	169	140	126	183	44	55	99	73	51	42	24	20	17	17	212	$C_{10}H_{16}N_2O_3$	Barbital 1,2 (or 4)-dimethyl derivative	BLR	56
41*	43	56	55	39	84	58	30	99	92	75	50	32	28	27	25	274	$C_{13}H_{26}N_2O_4$	Tybamate (GC)	DMIT	21
41*	43	69	44	39	55	53	70	99	92	46	44	41	38	28	25	278	$C_9H_{15}BrN_2O_3$	Acetylcarbromal	DMIT	155
41*	43	113	96	44	98	39	69	99	52	49	49	49	47	39	36	156	$C_7H_{12}N_2O_2$	Ectylurea	DMIT	120
41*	55	43	69	83	57	67	54	99	94	64	51	34	33	33	26	282	$C_{18}H_{34}O_2$	Oleic acid	DMIT	101
41*	167	168	39	125	97	44	53	99	83	52	38	32	28	23	19	224	$C_{11}H_{16}N_2O_3$	Talbutal	DMIT	163
41*	168	167	43	39	55	97	124	99	79	72	68	56	41	34	29	238	$C_{12}H_{18}N_2O_3$	Secobarbital	DMIT	148
43	41*	184	168	167	97	55	53	99	86	84	58	50	36	35	32	254	$C_{12}H_{18}N_2O_2S$	Thiamylal	NIH MSC	3824
55	41*	343	43	98	39	57	42	99	50	42	41	33	32	25	24	343	$C_{20}H_{25}NO_4$	Naltrexone metabolite 1	DMIT	291
57	41*	55	43	97	69	70	98	99	82	74	62	60	58	48	47	232	$C_{10}H_{20}N_2O_4$	Mebutamate (GC)	DMIT	63
70	41*	43	194	44	42	55	69	99	40	34	31	28	25	25	23	0	—	Frequent urine constituent 2	DMIT	230
149	41*	29	28	57	27	56	104	99	27	25	22	10	10	8	8	278	$C_{16}H_{22}O_4$	Dibutyl phthalate	DMIT	107
154	41*	86	70	44	42	124	98	99	34	16	14	14	12	11	10	0	—	Frequent urine constituent 4	DMIT	246
167	41*	39	124	28	168	29	43	99	67	44	41	39	35	30	24	210	$C_{10}H_{14}N_2O_3$	Aprobarbital	DMIT	173
167	41*	124	80	39	32	28	166	99	89	68	58	54	48	45	41	208	$C_{10}H_{12}N_2O_3$	Allobarbital	DMIT	114
167	41*	168	124	43	97	169	96	99	40	37	32	28	25	18	10	210	$C_{10}H_{14}N_2O_3$	Aprobarbital	NIH MSC	3766
168	41*	124	181	167	141	97	98	99	34	25	20	19	17	16	12	224	$C_{11}H_{16}N_2O_3$	Butalbital	NIH MSC	3817

Eight Most Intense Peaks — E2 (continued)

Eight most intense peaks								Relative Intensities								Mol wt	Formula	Drug	Source	
195	41*	194	53	138	70	137	79	99	97	60	45	39	37	37	27	236	$C_{12}H_{16}N_2O_3$	Allobarbital 1,2-(or 4)-Dimethyl derivative	BLR	0048
196	41*	195	209	138	181	67	43	99	78	52	45	33	25	22	21	252	$C_{13}H_{20}N_2O_3$	Butalbital 1,2 (or 3)-dimethyl derivative	BLR	0060
215	41*	104	77	132	39	128	244	99	49	44	24	20	19	19	19	244	$C_{13}H_{12}N_2O_3$	Alphenal	DMIT	0001
42*	140	41	49	84	112	43	85	99	39	10	8	8	8	7	5	140	$C_6H_{12}N_4$	Methenamine	DMIT	0260
42*	231	45	119	44	74	67	46	99	54	37	32	29	22	20	15	277	$C_9H_7N_7O_2S$	Azathioprine	DMIT	0297
42*	246	232	70	56	143	214	43	99	66	59	49	36	36	28	20	403	$C_{21}H_{26}ClN_3OS$	Perphenazine	DMIT	0136
42*	280	70	143	56	113	72	100	99	99	96	96	95	85	80	79	437	$C_{22}H_{26}F_3N_3OS$	Fluphenazine	DMIT	0134
57	42*	56	43	233	158	91	103	99	42	30	29	24	12	11	11	233	$C_{14}H_{19}NO_2$	Meperidine MTB 1	BLR	0002
57	42*	73	277	103	56	187	43	99	28	28	18	16	15	15	14	277	$C_{15}H_{23}NO_2Si$	Nor-meperidinic acid TMS ester	BLR	0004
59	42*	150	271	44	171	115	128	99	39	31	31	23	19	16	16	271	$C_{18}H_{25}NO$	Methorphan	DMIT	0029
71	42*	43	56	177	77	70	72	99	34	29	29	16	6	6	5	177	$C_{11}H_{15}NO$	Phenmetrazine	DMIT	0039
84	42*	133	162	161	51	119	65	99	27	26	13	10	8	7	7	162	$C_{10}H_{14}N_2$	Nicotine	BLR	0033
123	42*	222	207	179	149	150	166	99	86	79	74	72	61	55	49	222	$C_{10}H_{14}N_4O_2$	1,3-Di-ethyl-9-methylxanthine	BLR	0102
152	42*	98	154	153	28	174	69	99	61	58	42	31	30	27	24	273	$C_{11}H_{12}ClNO_3S$	Chlormezanone	DMIT	0122
212	42*	187	45	70	180	56	98	99	75	68	64	61	58	54	46	427	$C_{24}H_{33}N_3O_2S$	Dixyrazine	DMIT	0139
246	42*	247	377	91	172	47	28	99	18	16	11	9	6	5	5	452	$C_{30}H_{32}N_2O_2$	Diphenoxylate	DMIT	0149
43*	41	184	168	167	97	55	53	99	86	84	58	50	36	35	32	254	$C_{12}H_{18}N_2O_2S$	Thiamylal	NIH MSC	3824
43*	56	45	55	73	29	27	15	99	33	8	7	6	6	6	5	116	$C_5H_8O_3$	Levulinic acid	DMIT	0224
43*	57	41	55	44	81	93	105	99	91	73	46	44	36	36	36	368	$C_{27}H_{44}$	Cholesta-3,5-diene	DMIT	0102
43*	167	237	41	39	124	168	55	99	99	99	65	60	51	41	34	316	$C_{12}H_{17}BrN_2O_3$	5-(2-Bromoallyl)-5-(1-methyl butyl) barbituric acid	DMIT	0256
41	43*	56	55	39	84	58	30	99	92	75	50	32	28	27	25	274	$C_{13}H_{26}N_2O_4$	Tybamate (GC)	DMIT	0021
41	43*	69	44	39	55	53	70	99	92	46	44	41	38	28	25	278	$C_9H_{15}BrN_2O_3$	Acetylcarbromal	DMIT	0155
41	43*	113	96	44	98	39	69	99	52	49	49	49	47	39	36	156	$C_7H_{12}N_2O_2$	Ectylurea	DMIT	0120
44	43*	73	57	60	41	55	69	99	24	24	22	20	17	16	14	284	$C_{18}H_{36}O_2$	Stearic acid	NIH MSC	3820
57	43*	71	41	85	55	56	49	99	89	64	45	42	20	18	15	198	$C_{14}H_{30}$	$C_{14}H_{30}$ Standard	DMIT	0275
57	43*	71	85	41	55	99	56	99	79	73	57	42	32	18	17	310	$C_{22}H_{46}$	$C_{22}H_{46}$ Standard	DMIT	0276
57	43*	71	85	55	41	69	83	99	77	65	44	36	32	28	20	450	$C_{32}H_{66}$	$C_{32}H_{66}$ Standard	DMIT	0277
59	43*	156	121	155	157	158	139	99	71	44	40	37	14	13	10	214	$C_{11}H_{15}ClO_2$	Phenaglycodol	DMIT	0038
73	43*	84	55	69	41	44	85	99	50	37	35	29	28	23	20	145	$C_7H_{15}NO_2$	Emylcamate	DMIT	0060
91	43*	67	81	78	79	106	44	99	79	55	54	50	49	42	41	167	$C_9H_{13}NO_2$	Ethinamate	DMIT	0124

116	43*	85	61	119	71	31	101	99	29	29	19	18	18	15	14	176	$C_6H_8O_6$	Ascorbic acid	ABB	0010
120	43*	138	92	28	42	121	39	99	86	58	55	41	31	20	20	180	$C_9H_8O_4$	Acetylsalicylic acid	DMIT	0109
144	43*	103	44	296	159	146	77	99	59	29	27	26	24	20	20	296	—	Oxacillin MTB 5	DMIT	0159
158	43*	98	41	115	71	159	83	99	65	34	29	28	25	25	24	733	$C_{37}H_{67}NO_{13}$	Erythromycin A	ABB	0004
185	43*	129	41	57	259	157	112	99	61	59	45	44	44	23	17	402	$C_{20}H_{34}O_8$	Acetyl tri-*N*-butyl citrate	DMIT	0314
44*	28	29	31	43	60	73	41	99	89	82	81	65	60	56	52	357	$C_{19}H_{16}ClNO_4$	Indomethacin	DMIT	0192
44*	43	73	57	60	41	55	69	99	24	24	22	20	17	16	14	284	$C_{18}H_{36}O_2$	Stearic acid	NIH MSC	3820
44*	57	43	71	41	55	45	56	99	13	12	6	6	5	4	4	263	$C_{19}H_{21}N$	Nortriptyline	DMIT	0047
44*	69	41	55	43	39	27	71	99	74	72	35	32	28	24	21	236	$C_7H_{13}BrN_2O_2$	Carbromal	DMIT	0108
44*	69	41	208	210	55	43	71	99	78	57	45	44	28	24	22	236	$C_7H_{13}BrN_2O_2$	Carbromal	NIH MSC	3772
44*	91	45	65	42	43	41	51	99	8	6	5	4	3	2	2	135	$C_9H_{13}N$	Amphetamine	NIH MSC	3768
44*	91	65	42	45	120	40	92	99	8	4	3	3	3	2	2	135	$C_9H_{13}N$	Amphetamine	DMIT	0035
44*	100	234	88	105	57	41	91	99	35	24	21	16	16	14	12	325	$C_{21}H_{27}NO_2$	Propoxyphene MTB 3	DMIT	0210
44*	152	28	137	77	65	91	78	99	21	12	8	6	5	5	4	196	$C_{11}H_{18}NO_2$	2,5-Dimethoxyamphetamine	DMIT	0183
44*	166	151	57	43	91	135	209	99	40	12	7	6	5	5	4	209	$C_{12}H_{19}NO_2$	2,5-Dimethoxy-4-methylamphetamine	DMIT	0180
44*	220	100	57	205	91	129	221	99	92	62	42	35	26	20	20	325	$C_{21}H_{27}NO_2$	Propoxyphene MTB 1	DMIT	0070
58	44*	188	130	42	143	59	77	99	21	5	5	4	4	2	2	188	$C_{12}H_{16}N_2$	*N,N*-dimethyltryptamine	DMIT	0181
70	44*	168	41	58	69	125	97	99	80	56	48	40	39	33	32	0	—	Frequent urine constituent 1	DMIT	0229
114	44*	142	365	42	115	263	128	99	90	85	78	46	42	35	34	365	$C_{21}H_{23}N_3OS$	Pericyazine	DMIT	0135
233	44*	304	72	232	235	198	234	99	82	59	58	41	39	37	31	304	$C_{16}H_{17}ClN_2S$	Chlorpromazine MTB 2	DMIT	0212
271	44*	270	214	272	42	43	70	99	35	27	23	19	18	16	14	311	$C_{19}H_{21}NO_3$	Nalorphine	NIH MSC	3801
334	44*	120	77	41	107	55	144	99	44	36	34	33	32	32	32	334	$C_{21}H_{22}N_2O_2$	Strychnine	DMIT	0099
57	45*	56	85	101	125	41	199	99	96	90	90	72	63	59	45	398	$C_{18}H_{39}O_7P$	B-D vacutainer impurity	DMIT	0184
58	45*	43	298	42	41	46	57	99	35	19	11	10	9	8	8	298	$C_{18}H_{22}N_2S$	Trimeprazine	DMIT	0141
146	45*	117	86	288	118	57	232	99	74	30	25	24	19	19	14	320	$C_{16}H_{20}N_2O_5$	*N,N'*-dimethoxymethyl phenobarbital	DMIT	0293
217	45*	70	110	202	270	69	285	99	58	36	33	27	27	24	22	285	$C_{19}H_{27}NO$	Pentazocine	NIH MSC	3803
31	49*	77	113	82	51	115	29	99	16	11	7	5	5	5	4	148	$C_2H_3Cl_3O$	Chloral hydrate MTB 1	DMIT	0225
51*	117	67	69	118	119	101	133	99	75	43	14	8	8	5	2	184	$C_3H_2ClF_5O$	2-Chloro-1,1,2-trifluoroethyl-difluoromethyl ether	ARB	0001
81	53*	330	96	82	332	64	44	99	10	9	6	6	3	3	3	330	$C_{12}H_{11}ClN_2O_5S$	Furosemide	DMIT	0265
160	54*	105	106	159	107	82	39	99	69	55	49	28	27	25	22	0	—	Blood 324, drug 1	DMIT	0234
55*	41	343	43	98	39	57	42	99	50	42	41	33	32	25	24	343	$C_{20}H_{25}NO_4$	Naltrexone MTB 1	DMIT	0291
55*	72	97	118	158	56	41	57	99	81	66	61	58	40	36	34	274	$C_{13}H_{26}N_2O_4$	Tybamate	DMIT	0069
55*	73	413	110	414	84	75	328	99	98	70	29	24	22	19	18	413	$C_{23}H_{31}NO_4Si$	Naltrexone-1 TMS	DMIT	0278
41	55*	43	69	83	57	67	54	99	94	64	51	34	33	33	26	282	$C_{18}H_{34}O_2$	Oleic acid	DMIT	0101
58	55*	28	89	90	118	42	71	99	27	27	24	24	21	20	19	291	$C_{17}H_{25}NO_3$	Cyclopentolate	DMIT	0128
58	55*	97	43	104	158	62	56	99	77	55	53	43	40	39	37	260	$C_{12}H_{24}N_2O_4$	Carisoprodol	DMIT	0065

Eight Most Intense Peaks — E2 (continued)

Eight most intense peaks								Relative Intensities								Mol wt	Formula	Drug	Source	
73	55*	75	44	45	77	542	557	99	24	22	15	14	12	12	10	557	$C_{29}H_{47}NO_4Si_3$	Naltrexone-3 TMS	DMIT	0280
73	55*	75	472	487	110	473	488	99	51	24	24	24	19	19	18	487	$C_{26}H_{41}NO_4SI_2$	Alpha-hydroxy naltrexone-2 TMS	DMIT	0281
73	55*	77	560	559	75	85	44	99	38	31	28	28	26	16	16	559	$C_{29}H_{49}NO_4Si_3$	Alpha-hydroxy naltrexone-3 TMS	DMIT	0282
73	55*	487	486	75	110	45	488	99	46	27	24	21	20	12	12	487	$C_{26}H_{41}NO_4Si_2$	Naltrexone MTB 1-2 TMS	DMIT	0283
83	55*	71	96	114	144	62	56	99	59	44	43	42	34	30	25	218	$C_9H_{18}N_2O_4$	Meprobamate	DMIT	0062
83	55*	98	153	166	41	155	152	99	61	60	52	52	50	26	24	199	$C_{10}H_{17}NO_3$	Methyprylon MTB 1	DMIT	0010
83	55*	114	96	71	144	41	44	99	68	58	56	49	40	32	31	218	$C_9H_{18}N_2O_4$	2-Methyl-2-propyl-1,3-propane diol carbamate	ABB	0011
113	55*	70	42	41	39	85	69	99	81	63	37	20	18	10	8	141	$C_7H_{11}NO_2$	Ethosuximide	DMIT	0205
127	55*	70	41	42	128	69	112	99	60	41	20	18	8	7	7	155	$C_8H_{13}NO_2$	Ethosuximide N-methyl derivative	BLR	0110
341	55*	36	300	342	110	243	256	99	79	58	29	23	23	23	22	341	$C_{20}H_{23}NO_4$	Naltrexone	DMIT	0290
343	55*	110	36	98	302	84	344	99	72	45	41	31	29	23	23	343	$C_{20}H_{25}NO_4$	Alpha-hydroxy naltrexone	DMIT	0292
56*	84	57	203	42	83	77	93	99	64	63	43	30	26	13	11	203	$C_{11}H_{13}N_3O$	Ampyrone	DMIT	0007
56*	84	245	57	203	83	43	42	99	50	48	26	26	25	19	10	245	$C_{13}H_{15}N_3O_2$	4-Acetyl-aminoantipyrine	DMIT	0006
56*	97	231	42	111	77	71	112	99	35	35	12	11	8	7	7	231	$C_{13}H_{17}N_3O$	Aminopyrine	DMIT	0008
43	56*	45	55	73	29	27	15	99	33	8	7	6	6	6	5	116	$C_5H_8O_3$	Levulinic acid	DMIT	0224
57	56*	42	219	43	103	91	158	99	38	32	32	23	13	12	11	219	$C_{13}H_{17}NO_2$	Nor-meperidinic acid methyl ester	BLR	0008
99	56*	72	165	300	228	229	242	99	51	38	30	27	23	23	22	300	$C_{18}H_{21}ClN_2$	Chlorcyclizine	DMIT	0074
99	56*	167	194	266	165	195	207	99	66	60	43	43	37	35	35	266	$C_{18}H_{22}N_2$	Cyclizine	DMIT	0082
57*	41	55	43	97	69	70	98	99	82	74	62	60	58	48	47	232	$C_{10}H_{20}N_2O_4$	Mebutamate (GC)	DMIT	0063
57*	42	56	43	233	158	91	103	99	42	30	29	24	12	11	11	233	$C_{14}H_{19}NO_2$	Meperidine MTB 1	BLR	0002
57*	42	73	277	103	56	187	43	99	28	28	18	16	15	15	14	277	$C_{15}H_{23}NO_2Si$	Nor-meperidinic acid TMS ester	BLR	0004
57*	43	71	41	85	55	56	49	99	89	64	45	42	20	18	15	198	$C_{14}H_{30}$	$C_{14}H_{30}$ standard	DMIT	0275
57*	43	71	85	41	55	99	56	99	79	73	57	42	32	18	17	310	$C_{22}H_{46}$	$C_{22}H_{46}$ standard	DMIT	0276
57*	43	71	85	55	41	69	83	99	77	65	44	36	32	28	20	450	$C_{32}H_{66}$	$C_{32}H_{66}$ standard	DMIT	0277
57*	45	56	85	101	125	41	199	99	96	90	90	72	63	59	45	398	$C_{18}H_{39}O_7P$	B-D vacutainer impurity	DMIT	0184
57*	56	42	219	43	103	91	158	99	38	32	32	23	13	12	11	219	$C_{13}H_{17}NO_2$	Nor-meperidinic acid methyl ester	BLR	0008
57*	73	60	43	55	71	41	69	99	82	80	62	45	40	37	35	284	$C_{18}H_{36}O_2$	Stearic acid	DMIT	0098
57*	233	42	56	43	158	131	160	99	31	26	26	18	18	15	13	233	$C_{14}H_{19}NO_2$	Meperidine MTB 1	NIH MSC	3782
36	57*	41	40	44	38	55	39	99	99	83	83	72	67	65	55	157	$C_8H_{15}NO_2$	Oxanamide	DMIT	0121
43	57*	41	55	44	81	93	105	99	91	73	46	44	36	36	36	368	$C_{27}H_{44}$	Cholesta-3,5-diene	DMIT	0102

44	57*	43	71	41	55	45	56	99	13	12	6	6	5	4	4	263	$C_{19}H_{21}N$	Nortriptyline	DMIT	0047
58	57*	59	29	49	91	42	105	99	4	4	4	4	3	1	1	339	$C_{22}H_{29}NO_2$	Propoxyphene	DMIT	0022
58	57*	59	91	42	77	105	115	99	5	4	4	3	3	3	3	339	$C_{22}H_{29}NO_2$	Propoxyphene	NIH MSC	3813
73	57*	31	43	60	61	44	71	99	94	74	57	52	36	27	22	342	$C_{12}H_{22}O_{11}$	Sucrose	ABB	0003
129	57*	55	70	43	41	71	112	99	74	56	46	43	41	38	33	370	$C_{22}H_{42}O_4$	Dioctyl adipate	DMIT	0106
149	57*	41	43	70	71	167	28	99	47	30	30	28	28	26	23	390	$C_{24}H_{38}O_4$	Dioctyl phthalate	DMIT	0095
149	57*	167	43	71	70	41	55	99	34	32	25	25	19	15	15	390	$C_{24}H_{38}O_4$	Dioctyl phthalate	NIH MSC	3792
205	57*	220	41	206	145	29	81	99	32	21	17	15	12	10	10	220	$C_{15}H_{24}O$	Ionol	DMIT	0220
58*	28	59	221	30	42	0	0	99	22	4	4	3	3	0	0	315	$C_{18}H_{18}ClNS$	Chloprothixene	DMIT	0119
58*	44	188	130	42	143	59	77	99	21	5	5	4	4	2	2	188	$C_{12}H_{16}N_2$	*N,N*-Dimethyltryptamine	DMIT	0181
58*	45	43	298	42	41	46	57	99	35	19	11	10	9	8	8	298	$C_{18}H_{22}N_2S$	Trimeprazine	DMIT	0141
58*	55	28	89	90	118	42	71	99	27	27	24	24	21	20	19	291	$C_{17}H_{25}NO_3$	Cyclopentolate	DMIT	0128
58*	55	97	43	104	158	62	56	99	77	55	53	43	40	39	37	260	$C_{12}H_{24}N_2O_4$	Carisoprodol	DMIT	0065
58*	57	59	29	49	91	42	105	99	4	4	4	4	3	1	1	339	$C_{22}H_{29}NO_2$	Propoxyphene	DMIT	0022
58*	57	59	91	42	77	105	115	99	5	4	4	3	3	3	3	339	$C_{22}H_{29}NO_2$	Propoxyphene	NIH MSC	3813
58*	59	202	91	203	215	218	217	99	5	3	2	2	2	2	1	277	$C_{20}H_{23}N$	Amitriptyline	DMIT	0043
58*	71	44	72	55	41	43	49	99	29	17	14	14	13	13	8	0	—	Blood impurity PK 2	DMIT	0228
58*	71	44	72	55	41	43	49	99	29	17	14	14	13	13	8	0	—	Blood impurity PK 1	DMIT	0227
58*	71	167	72	42	59	202	45	99	61	7	6	4	4	4	3	290	$C_{16}H_{19}ClN_2O$	Carbinoxamine	DMIT	0080
58*	71	180	73	167	182	72	181	99	49	12	9	9	6	5	5	270	$C_{17}H_{22}N_2O$	Doxylamine	DMIT	0085
58*	72	135	71	44	42	107	95	99	78	47	45	16	15	11	10	165	$C_9H_{15}N_3$	Methapyrilene MTB 1	DMIT	0271
58*	73	45	57	43	44	167	165	99	24	8	8	6	6	5	5	333	$C_{17}H_{20}BrNO$	Bromodiphenhydramine	DMIT	0072
58*	73	59	88	45	75	43	56	99	7	6	2	2	2	2	2	237	$C_{13}H_{23}NOSi$	Ephedrine TMS ether	BLR	0108
58*	73	167	165	45	44	168	42	99	24	9	8	7	4	4	4	255	$C_{17}H_{21}NO$	Diphenhydramine	DMIT	0086
58*	77	30	51	49	56	59	42	99	12	6	6	6	6	6	5	165	$C_{10}H_{15}NO$	Pseudoephedrine	DMIT	0214
58*	86	42	334	220	59	87	44	99	77	11	6	5	4	4	4	334	$C_{17}H_{19}CiN_2OS$	3-Hydroxy-chlorpromazine	DMIT	0015
58*	86	289	43	291	85	248	288	99	26	10	6	6	5	5	4	376	$C_{19}H_{21}CiN_2O_2S$	8-Acetoxychlorpromazine	DMIT	0014
58*	86	314	229	185	228	42	85	99	38	30	16	11	10	9	7	314	$C_{18}H_{22}N_2OS$	Methoxypromazine	DMIT	0126
58*	86	318	85	272	320	42	232	99	26	23	12	9	8	6	5	318	$C_{17}H_{19}CiN_2S$	Chlorpromazine	DMIT	0100
58*	86	334	42	85	336	44	243	99	44	27	17	10	10	9	9	334	$C_{17}H_{19}CiN_2OS$	8-Hydroxy-chlorpromazine	DMIT	0016
58*	86	352	85	306	59	266	248	99	31	24	18	10	6	5	5	352	$C_{18}H_{19}F_3N_2S$	Triflupromazine	DMIT	0034
58*	86	376	43	378	42	87	59	99	85	31	15	12	11	10	6	376	$C_{19}H_{21}CiN_2O_2S$	3-Acetoxychlorpromazine	DMIT	0013
58*	91	59	43	56	65	134	39	99	7	5	3	3	3	3	2	149	$C_{10}H_{15}N$	Methamphetamine	NIH MSC	3797
58*	91	59	56	42	65	134	41	99	7	6	5	3	3	2	2	149	$C_{10}H_{15}N$	Methamphetamine	BLR	0107
58*	91	197	72	71	185	184	65	99	93	30	26	25	16	13	8	255	$C_{16}H_{21}N_3$	Tripelennamine	NIH MSC	3826
58*	97	72	71	79	42	78	40	99	68	23	22	14	14	13	10	261	$C_{14}H_{19}N_3S$	Thenyldiamine	DMIT	0157
58*	97	72	71	191	190	78	79	99	71	22	19	11	10	8	8	261	$C_{14}H_{19}N_3S$	Methapyrilene	NIH MSC	3798
58*	97	72	84	71	191	190	261	99	71	28	23	18	14	11	10	261	$C_{14}H_{19}N_3S$	Methapyrilene	DMIT	0054
58*	131	72	71	79	28	42	30	99	18	16	15	11	10	10	8	295	$C_{14}H_{18}CiN_3S$	Chlorothen	DMIT	0055
58*	165	225	359	166	73	199	45	99	11	9	8	7	7	4	3	359	$C_{21}H_{29}NS_2$	Captodiamine	DMIT	0088
58*	180	71	182	44	183	167	181	99	79	70	50	41	39	24	21	270	$C_{17}H_{22}N_2O$	Doxylamine (GC)	DMIT	0241
58*	195	59	72	388	89	315	42	99	18	14	13	7	6	5	5	388	$C_{21}H_{28}N_2O_5$	Trimethobenzamide	DMIT	0316

Eight Most Intense Peaks — E2 (continued)

Eight most intense peaks								Relative Intensities								Mol wt	Formula	Drug	Source	
58*	197	86	42	43	44	85	196	99	16	15	12	10	7	6	6	326	$C_{19}H_{22}N_2OS$	Acetylpromazine	DMIT	0133
58*	204	146	42	44	77	39	51	99	17	5	4	4	3	3	3	284	$C_{12}H_{17}N_2O_4P$	Psilocybin	DMIT	0182
58*	204	146	205	59	42	130	30	99	25	4	3	3	2	2	2	204	$C_{12}H_{16}N_2O$	Psilocin	DMIT	0177
58*	234	44	59	41	42	36	427	99	10	8	5	4	4	3	3	445	$C_{23}H_{27}NO_8$	Narceine	DMIT	0197
58*	246	233	318	86	272	248	232	99	35	26	22	17	13	10	10	334	$C_{17}H_{19}CiN_2OS$	Chlorpromazine MTB 4	DMIT	0097
58*	255	40	72	42	59	71	91	99	9	6	4	3	3	3	3	255	$C_{17}H_{21}NO$	Phenyltoloxamine	DMIT	0049
58*	276	73	41	59	75	77	40	99	12	11	10	9	7	6	6	276	$C_{15}H_{24}N_2OSi$	Psilocin-1 TMS	DMIT	0329
58*	284	86	238	199	85	198	285	99	43	29	23	20	12	11	10	284	$C_{17}H_{20}N_2S$	Promazine	DMIT	0017
58*	290	73	291	348	41	75	77	99	37	33	10	9	7	7	7	348	$C_{18}H_{32}N_2OSi_2$	Psilocin-2 TMS	DMIT	0330
58*	318	86	272	85	320	232	42	99	28	26	12	11	10	9	6	318	$C_{17}H_{19}ClN_2S$	Chlorpromazine	NIH MSC	3776
58*	328	100	42	135	228	229	242	99	13	7	4	4	3	3	3	328	$C_{19}H_{24}N_2OS$	Methotrimeprazine	DMIT	0140
86	58*	30	57	99	87	71	43	99	23	21	19	19	16	14	13	387	$C_{21}H_{23}ClFN_3O$	Flurazepam	DMIT	0323
86	58*	30	87	72	42	56	120	99	49	43	33	20	11	10	9	234	$C_{14}H_{22}N_2O$	Lidocaine	DMIT	0142
86	58*	73	87	319	41	99	245	99	10	5	5	5	4	4	4	319	$C_{18}H_{26}ClN_3$	Chloroquine	NIH MSC	3774
86	58*	99	71	55	100	57	87	99	16	15	9	6	6	5	5	309	$C_{19}H_{35}NO_2$	Dicyclomine	DMIT	0255
121	58*	72	71	78	215	122	77	99	67	13	13	12	11	9	8	286	$C_{16}H_{22}N_4O$	Thonzylamine	DMIT	0031
121	58*	72	79	71	28	78	42	99	88	24	17	17	15	14	11	285	$C_{17}H_{23}N_3O$	Pyrilamine	DMIT	0071
203	58*	205	72	204	167	202	168	99	52	33	25	16	12	10	7	274	$C_{16}H_{19}ClN_2$	Chlorpheniramine	DMIT	0073
203	58*	205	204	72	167	202	168	99	50	32	19	17	15	13	10	274	$C_{16}H_{19}ClN_2$	Chlorpheniramine	NIH MSC	3773
230	58*	231	44	173	105	42	159	99	24	17	11	11	10	8	6	321	$C_{22}H_{27}NO$	Phenazocine	DMIT	0112
235	58*	234	85	280	195	193	35	99	78	64	62	48	29	23	20	280	$C_{19}H_{24}N_2$	Imipramine	DMIT	0066
357	58*	59	102	75	45	373	374	99	97	53	45	37	32	29	25	0	—	Blood 955 drug	DMIT	0327
59*	42	150	271	44	171	115	128	99	39	31	31	23	19	16	16	271	$C_{18}H_{25}NO$	Methorphan	DMIT	0029
59*	43	156	121	155	157	158	139	99	71	44	40	37	14	13	10	214	$C_{11}H_{15}ClO_2$	Phenaglycodol	DMIT	0038
59*	157	156	43	141	41	55	180	99	68	64	37	34	30	26	24	242	$C_{11}H_{18}N_2O_4$	Amobarbital MTB 1	DMIT	0267
58	59*	202	91	203	215	218	217	99	5	3	2	2	2	2	1	277	$C_{20}H_{23}N$	Amitriptyline	DMIT	0043
162	63*	164	98	99	49	62	73	99	64	62	24	24	21	17	15	162	$C_6H_4Cl_2O$	2,4-Dichlorophenol	DMIT	0318
204	63*	146	232	117	143	174	89	99	34	34	28	22	18	18	17	232	$C_{12}H_{12}N_2O_3$	Phenobarbital	NIH MSC	3807
269	64*	205	297	271	43	44	31	99	92	49	45	42	40	38	36	265	$C_7H_8ClN_3O_4S$	Hydrochlorothiazide	DMIT	0266
310	64*	36	28	312	42	43	62	99	57	53	44	43	36	29	25	359	$C_9H_{11}Cl_2N_3O_4S_2$	Methychlothiazide	DMIT	0160
94	66*	39	65	40	95	38	55	99	27	27	22	13	10	10	8	94	C_6H_6O	Phenol	DMIT	0189
207	67*	79	81	141	77	55	91	99	70	48	44	42	42	34	24	236	$C_{12}H_{16}N_2O_3$	Cyclobarbital	DMIT	0169
69*	81	137	41	136	95	44	149	99	54	24	22	15	14	13	13	0	—	Frequent contaminant	DMIT	0232
44	69*	41	55	43	39	27	71	99	74	72	35	32	28	24	21	236	$C_7H_{13}BrN_2O_2$	Carbromal	DMIT	0108
44	69*	41	208	210	55	43	71	99	78	57	45	44	28	24	22	236	$C_7H_{13}BrN_2O_2$	Carbromal	NIH MSC	3772

128	69*	143	73	169	75	159	41	99	98	87	69	52	44	39	35	286	$C_{13}H_{19}ClO_3Si$	2-(4-Chlorophenoxy)-2-methylpropionic acid TMS ester (from clofibrate)	BLR	0023
141	69*	113	45	68	42	41	54	99	98	95	87	60	40	33	30	141	—	Oxacillin MTB 4	DMIT	0150
70*	41	43	194	44	42	55	69	99	40	34	31	28	25	25	23	0	—	Frequent urine constituent 2	DMIT	0230
70*	44	168	41	58	69	125	97	99	80	56	48	40	39	33	32	0	—	Frequent urine constituent 1	DMIT	0229
70*	113	43	42	127	71	44	56	99	98	85	59	56	41	36	27	446	$C_{22}H_{30}N_4O_2S_2$	Thioproperazine	DMIT	0145
70*	154	72	43	125	44	41	55	99	95	39	32	32	31	30	27	0	—	Frequent urine constituent 3	DMIT	0231
71	70*	42	28	57	43	172	25	99	54	49	46	33	31	29	22	247	$C_{15}H_{21}NO_2$	Meperidine	DMIT	0115
71	70*	42	57	233	232	43	44	99	62	40	40	39	27	26	22	233	$C_{14}H_{19}NO_2$	Meperidinic acid methyl ester	BLR	0003
71	70*	42	103	73	291	57	44	99	17	16	14	13	11	11	10	291	$C_{16}H_{25}NO_2Si$	Meperidinic acid TMS ester	BLR	0005
71	70*	42	247	57	44	96	43	99	50	35	33	28	23	23	19	247	$C_{15}H_{21}NO_2$	Meperidine	BLR	0001
71	70*	247	57	42	246	91	103	99	56	38	35	34	32	24	21	247	$C_{15}H_{21}NO_2$	Meperidine	NIH MSC	3784
98	70*	370	126	99	185	244	125	99	13	9	8	7	4	3	3	370	$C_{21}H_{26}N_2S_2$	Thioridazine	DMIT	0004
113	70*	43	42	141	407	71	127	99	74	44	30	30	27	25	23	407	$C_{21}H_{24}F_3N_3S$	Trifluoperazine	DMIT	0131
113	70*	373	141	43	42	44	375	99	89	58	48	44	44	20	20	373	$C_{20}H_{24}ClN_3S$	Prochlorperazine	DMIT	0154
141	70*	55	41	42	69	112	126	99	40	36	10	10	8	6	6	169	$C_9H_{15}NO_2$	Ethosuximide n-ethyl derivative	BLR	0111
198	70*	58	115	71	56	72	269	99	61	58	33	20	11	11	11	327	$C_{19}H_{25}N_3S$	Aminopromazine	DMIT	0138
71*	42	43	56	177	77	70	72	99	34	29	29	16	6	6	5	177	$C_{11}H_{15}NO$	Phenmetrazine	DMIT	0039
71*	70	42	28	57	43	172	25	99	54	49	46	33	31	29	22	247	$C_{15}H_{21}NO_2$	Meperidine	DMIT	0115
71*	70	42	57	233	232	43	44	99	62	40	40	39	27	26	22	233	$C_{14}H_{19}NO_2$	Meperidinic acid methyl ester	BLR	0003
71*	70	42	103	73	291	57	44	99	17	16	14	13	11	11	10	291	$C_{16}H_{25}NO_2Si$	Meperidinic acid TMS ester	BLR	0005
71*	70	42	247	57	44	96	43	99	50	35	33	28	23	23	19	247	$C_{15}H_{21}NO_2$	Meperidine	BLR	0001
71*	70	247	57	42	246	91	103	99	56	38	35	34	32	24	21	247	$C_{15}H_{21}NO_2$	Meperidine	NIH MSC	3784
71*	73	113	61	85	43	31	91	99	81	48	39	37	31	31	30	308	$C_8H_{11}Cl_3O_6$	Alpha Chloralose	DMIT	0328
58	71*	44	72	55	41	43	49	99	29	17	14	14	13	13	8	0	—	Blood impurity PK 2	DMIT	0228
58	71*	44	72	55	41	43	49	99	29	17	14	14	13	13	8	0	—	Blood impurity PK 1	DMIT	0227
58	71*	167	72	42	59	202	45	99	61	7	6	4	4	4	3	290	$C_{16}H_{19}ClN_2O$	Carbinoxamine	DMIT	0080
58	71*	180	73	167	182	72	181	99	49	12	9	9	6	5	5	270	$C_{17}H_{22}N_2O$	Doxylamine	DMIT	0085
73	71*	83	129	256	98	85	97	99	33	24	23	22	20	20	16	256	$C_{16}H_{32}O_2$	Palmitic acid	DMIT	0096
72*	28	165	42	180	178	179	91	99	30	10	7	7	7	6	6	309	$C_{21}H_{27}NO$	Methadone	DMIT	0143
72*	73	198	180	213	284	42	44	99	5	4	3	3	3	2	2	284	$C_{17}H_{20}N_2S$	Promethazine	NIH MSC	3812
72*	73	229	75	230	43	45	153	99	24	11	4	4	3	3	3	301	$C_{18}H_{27}NOSi$	Pronethalol TMS ether	BLR	0109
72*	73	284	198	213	180	199	285	99	3	3	2	2	1	1	0	284	$C_{17}H_{20}N_2S$	Promethazine	DMIT	0018
72*	149	269	255	197	73	254	340	99	16	5	4	4	3	2	2	340	$C_{20}H_{24}N_2OS$	Propiomazine	DMIT	0019
55	72*	97	18	158	56	41	57	99	81	66	61	58	40	36	34	274	$C_{13}H_{26}N_2O_4$	Tybamate	DMIT	0069

Eight Most Intense Peaks — E2 (continued)

Eight most intense peaks								Relative Intensities								Mol wt	Formula	Drug	Source	
58	72*	135	71	44	42	107	95	99	78	47	45	16	15	11	10	165	$C_9H_{15}N_3$	Methapyrilene MTB 1	DMIT	0271
97	72*	55	71	62	110	69	158	99	64	62	42	40	34	34	30	232	$C_{10}H_{20}N_2O_4$	Mebutamate	DMIT	0061
73*	43	84	55	69	41	44	85	99	50	37	35	29	28	23	20	145	$C_7H_{15}NO_2$	Emylcamate	DMIT	0060
73*	55	75	44	45	77	542	557	99	24	22	15	14	12	12	10	557	$C_{29}H_{47}NO_4Si_3$	Naltrexone-3TMS	DMIT	0280
73*	55	75	472	487	110	473	488	99	51	24	24	24	19	19	18	487	$C_{26}H_{41}NO_4Si_2$	Alpha-hydroxy naltrexone-2 TMS	DMIT	0281
73*	55	77	560	559	75	85	44	99	38	31	28	28	26	16	16	559	$C_{29}H_{49}NO_4Si_3$	Alpha-hydroxy naltrexone-3 TMS	DMIT	0282
73*	55	487	486	75	110	45	488	99	46	27	24	21	20	12	12	487	$C_{26}H_{41}NO_4Si_2$	Naltrexone MTB 1-2 TMS	DMIT	0283
73*	57	31	43	60	61	44	71	99	94	74	57	52	36	27	22	342	$C_{12}H_{22}O_{11}$	Sucrose	ABB	0003
73*	71	83	129	256	98	85	97	99	33	24	23	22	20	20	16	256	$C_{16}H_{32}O_2$	Palmitic acid	DMIT	0096
73*	75	44	77	45	47	43	74	99	43	41	34	26	16	14	11	503	$C_{25}H_{41}NO_4Si_3$	Noroxymorphone-3 TMS	DMIT	0287
73*	75	45	44	74	195	258	77	99	24	17	8	8	8	8	6	505	$C_{25}H_{43}NO_4Si_3$	Dihydronoroxymorphone-3 TMS	DMIT	0289
73*	75	45	44	77	431	259	74	99	43	34	26	23	23	15	13	431	$C_{22}H_{33}NO_4Si_2$	Noroxymorphone-2 TMS	DMIT	0286
73*	75	45	274	258	433	74	195	99	24	15	12	11	9	9	8	433	$C_{22}H_{35}NO_4Si_2$	Dihydronoroxymorphone-2 TMS	DMIT	0288
73*	75	129	191	147	74	217	173	99	24	15	15	15	9	8	6	642	$C_{32}H_{66}O_5Si_4$	Prostaglandin F2B TMS ester tri-TMS ether	BLR	0129
73*	75	129	438	478	199	388	419	99	32	20	20	16	14	13	13	509	$C_{27}H_{51}NO_4Si_2$	Prostaglandin A1 methyloxime TMS ester TMS ether	BLR	0121
73*	75	133	355	74	130	426	67	99	27	14	10	8	8	7	6	599	$C_{30}H_{61}NO_5Si_3$	Prostaglandin E1 methyloxime TMS ester di-TMS ether	BLR	0123
73*	75	147	129	191	43	55	117	99	36	19	16	14	12	12	9	644	$C_{32}H_{68}O_5Si_4$	Prostaglandin F1A TMS ester tri TMS ether	BLR	0126
73*	75	147	191	129	74	173	217	99	24	19	17	12	9	7	7	642	$C_{32}H_{66}O_5Si_4$	Prostaglandin F2A TMS ester tri-TMS ether	BLR	0128
73*	75	355	133	426	130	74	225	99	29	19	19	17	13	9	8	599	$C_{30}H_{61}NO_5Si_3$	Prostaglandin 8-iso E1 methyloxime TMS ester TMS ether	BLR	0124
73*	129	128	114	334	349	115	130	99	87	23	21	20	15	13	12	349	$C_{18}H_{31}NO_2Si_2$	Nor-meperidinic acid TMS ester *N*-TMS derivative	BLR	0006
73*	129	305	128	304	276	103	232	99	65	35	33	30	23	22	21	305	$C_{17}H_{27}NO_2Si$	Nor-merperidinic acid ethyl ester *N*-TMS derivative	BLR	0007

73*	147	164	296	253	74	75	149	99	28	14	12	10	7	5	5	296	$C_{14}H_{24}O_3Si_2$	*O*-hydroxyphenylacetic acid trimethylsilyl ether	BLR	0167
73*	191	75	45	74	104	147	167	99	29	17	8	8	8	7	7	516	$C_{25}H_{40}N_2O_4Si_3$	5-(3,4-Dihydroxycyclohexa-1,5-dienyl)-3-methyl-5-phenyl hydantoin TMS Derivative	BLR	0045
73*	193	43	195	210	237	75	252	99	95	84	74	60	40	36	36	252	$C_{13}H_{20}O_3Si$	Propylparaben TMS ether	BLR	0027
73*	193	237	89	162	161	194	177	99	86	67	56	46	42	14	13	252	$C_{13}H_{20}O_3Si$	Phenyl-lactic acid methyl ester trimethylsilyl ether	BLR	0143
73*	217	204	75	147	185	253	45	99	53	37	35	21	18	12	12	676	$C_{29}H_{56}N_2O_{10}Si_3$	Hydroxyamobarbital glucuronide methyl TMS derivative (peak 2)	BLR	0053
73*	225	75	133	353	131	226	74	99	73	37	27	24	15	15	14	597	$C_{30}H_{59}NO_5Si_3$	Prostaglandin E2 methyloxime TMS ester di-TMS ether	BLR	0125
73*	253	75	217	147	185	317	45	99	40	37	23	22	19	16	13	676	$C_{29}H_{56}N_2O_{10}Si_3$	Hydroxyamobarbital glucuronide methyl TMS derivative (peak 1)	BLR	0052
73*	253	217	185	317	75	204	69	99	53	31	28	28	17	17	15	676	$C_{29}H_{56}N_2O_{10}Si_3$	Hydroxypentobarbital glucuronide 1,3-dimethyl deriv methyl ester TMS ether	BLR	0067
73*	265	217	75	204	69	41	147	99	47	25	23	22	21	14	14	688	$C_{30}H_{56}N_2O_{10}Si_3$	Hydroxysecobarbital glucuronide 1,3-di-methyl deriv methyl ester TMS ether	BLR	0077
73*	274	359	75	44	77	45	259	99	55	54	40	36	35	28	26	359	$C_{19}H_{25}NO_4Si$	Noroxymorphone-1 TMS	DMIT	0285
73*	317	169	171	41	318	217	75	99	67	55	18	17	17	16	14	620	$C_{26}H_{45}ClO_9Si_3$	Clofibrate glucuronide methyl ester TMS ether	BLR	0025
73*	317	217	75	147	318	43	79	99	58	20	19	18	15	10	10	688	$C_{32}H_{48}N_2O_9Si_3$	5-(4-hydroxyphenyl)-3-methyl-5-phenylhydantoin glucuronide methyl ester TMS ether	BLR	0044
73*	319	291	348	206	320	45	333	99	97	81	74	61	30	28	25	348	$C_{17}H_{24}N_2O_4Si$	Hydroxyphenobarbital 1,3-dimethyl derivative TMS ether	BLR	0069
73*	323	338	321	143	45	77	154	99	75	35	30	26	23	19	19	338	$C_{19}H_{22}N_2O_2Si$	Methaqualone MTB 4 TMS	DMIT	0312
73*	338	247	323	179	77	75	235	99	69	66	55	49	39	32	28	338	$C_{19}H_{22}N_2O_2Si$	Methaqualone MTB 1 TMS	DMIT	0310
73*	341	43	271	75	147	41	342	99	55	26	23	18	14	12	12	444	$C_{20}H_{40}N_2O_5Si_2$	Dihydroxysecobarbital di-TMS ether 1,3-dimethyl derivative	BLR	0074
73*	341	342	343	45	344	327	91	99	90	64	44	25	19	16	16	342	$C_{18}H_{19}ClN_2OSi$	7-Chloro-1,3-dihydro-5-phenyl-2H-1,4-dibenzodiazepin-2-one TMS deriv	BLR	0113

Eight Most Intense Peaks — E2 (continued)

Eight most intense peaks								Relative Intensities								Mol wt	Formula	Drug	Source	
73*	343	257	256	345	372	283	45	99	77	41	31	31	29	29	27	372	$C_{19}H_{21}ClN_2O_2Si$	7-Cl-1,3-dihydro-3-OH-1-methyl-5-phenyl-2H-1,4-benzodiazephin-2-one TMS ether	BLR	0116
73*	409	75	319	410	381	55	465	99	64	42	27	22	14	14	12	480	$C_{26}H_{48}O_4Si_2$	Prostaglandin A1 TMS ester TMS ether	BLR	0120
73*	429	430	45	431	147	432	75	99	32	32	21	20	14	11	10	430	$C_{21}H_{27}ClN_2O_2Si_2$	7-Cl-3-OH-5-phenyl-1,3-dihydro-2H-1,4-benzodiazepin-2-one di-TMS derivative	BLR	0117
73*	478	75	479	43	132	74	55	99	53	40	22	9	9	8	8	509	$C_{27}H_{51}NO_4Si_2$	Prostaglandin B1 methyloxime TMS ester TMS ether	BLR	0122
73*	483	75	147	191	129	554	367	99	32	31	29	29	28	25	21	644	$C_{32}H_{68}O_5Si_4$	Prostaglandin F1B TMS ester tri-TMS ether	BLR	0127
73*	485	55	470	75	486	44	45	99	44	39	31	23	20	18	18	485	$C_{26}H_{39}NO_4Si_2$	Naltrexone-2 TMS	DMIT	0279
73*	559	55	558	372	560	75	373	99	40	35	32	25	20	20	12	559	$C_{29}H_{49}NO_4Si_3$	Naltrexone MTB 1-3 TMS	DMIT	0284
55	73*	413	110	414	84	75	328	99	98	70	29	24	22	19	18	413	$C_{23}H_{31}NO_4Si$	Naltrexone-1 TMS	DMIT	0278
57	73*	60	43	55	71	41	69	99	82	80	62	45	40	37	35	284	$C_{18}H_{36}O_2$	Stearic acid	DMIT	0098
58	73*	45	57	43	44	167	165	99	24	8	8	6	6	5	5	333	$C_{17}H_{20}BrNO$	Bromodiphenhydramine	DMIT	0072
58	73*	59	88	45	75	43	56	99	7	6	2	2	2	2	2	237	$C_{13}H_{23}NOSi$	Ephedrine TMS ether	BLR	0108
58	73*	167	165	45	44	168	42	99	24	9	8	7	4	4	4	255	$C_{17}H_{21}NO$	Diphenhydramine	DMIT	0086
71	73*	113	61	85	43	31	91	99	81	48	39	37	31	31	30	308	$C_8H_{11}Cl_3O_6$	Alpha chloralose	DMIT	0328
72	73*	198	180	213	284	42	44	99	5	4	3	3	3	2	2	284	$C_{17}H_{20}N_2S$	Promethazine	NIH MSC	3812
72	73*	229	75	230	43	45	153	99	24	11	4	4	3	3	3	301	$C_{18}H_{27}NOSi$	Pronethalol TMS ether	BLR	0109
72	73*	284	198	213	180	199	285	99	3	3	2	2	1	1	0	284	$C_{17}H_{20}N_2S$	Promethazine	DMIT	0018
96	73*	94	42	57	227	142	212	99	95	45	44	41	41	39	26	227	$C_{11}H_{21}NO_2Si$	Scopoline TMS ether	BLR	0037
117	73*	327	75	143	256	118	69	99	50	40	26	18	16	13	12	342	$C_{16}H_{30}N_2O_4Si$	Hydroxypentobarbital 1,3-dimethyl derivative TMS ether	BLR	0083
139	73*	141	111	487	370	75	140	99	35	25	15	15	13	8	8	487	$C_{24}H_{30}ClNO_4Si_2$	1-(4-Chlorobenzoyl)-2-methyl-5-TMSO-indole-3-acetic acid TMS ester	BLR	0020
146	73*	232	334	247	117	246	147	99	59	46	46	33	20	20	19	362	$C_{18}H_{30}N_2O_2Si_2$	Primidone 1,3-di-TMS derivative	BLR	0090

166	73*	149	168	267	74	193	55	99	38	13	12	12	7	7	7	0	—	Blood 324, drug 2	DMIT	0235
169	73*	170	75	249	233	79	171	99	47	20	19	12	11	8	7	338	$C_{16}H_{26}N_2O_4Si$	Hydroxyhexobarbital 3-methyl derivative TMS ether	BLR	0065
179	73*	89	180	223	181	195	74	99	74	19	16	8	4	4	3	238	$C_{12}H_{18}O_3Si$	Mandelic acid methyl ester trimethylsilyl ether	BLR	0181
267	73*	268	269	45	135	193	75	99	99	25	13	12	9	9	8	282	$C_{13}H_{22}O_3Si_2$	Salicylic acid TMS ester TMS ether	BLR	0105
323	73*	338	324	321	45	77	143	99	66	29	24	21	16	16	15	338	$C_{19}H_{22}N_2O_2Si$	Methaqualone MTB 3 TMS	DMIT	0311
344	73*	372	371	346	345	373	374	99	78	48	39	38	31	26	20	372	$C_{19}H_{21}ClN_2O_2Si$	7-Cl-1,3-dihydro-5-(4-hydroxylphenyl)-1-methyl-2H-1,4-benzodiazepin-2-one TMS ether	BLR	0114
354	73*	104	268	325	282	355	77	99	49	42	37	36	26	26	23	354	$C_{19}H_{22}N_2O_3Si$	5-(3-Hydroxyphenyl)-3-methyl-5-phenylhydantoin TMS ether	BLR	0041
74*	87	43	55	143	298	41	57	99	76	33	27	21	21	20	20	298	$C_{19}H_{38}O_2$	Methyl stearate	DMIT	0222
74*	87	270	75	143	43	55	41	99	85	22	19	19	18	17	14	270	$C_{17}H_{34}O_2$	Methyl palmitate	DMIT	0223
73	75*	44	77	45	47	43	74	99	43	41	34	26	16	14	11	503	$C_{25}H_{41}NO_4Si_3$	Noroxymorphone-3 TMS	DMIT	0287
73	75*	45	44	74	195	258	77	99	24	17	8	8	8	8	6	505	$C_{25}H_{43}NO_4Si_3$	Dihydronoroxymor-phone-3 TMS	DMIT	0289
73	75*	45	44	77	431	259	74	99	43	34	26	23	23	15	13	431	$C_{22}H_{33}NO_4Si_2$	Noroxymorphone-2 TMS	DMIT	0286
73	75*	45	274	258	433	74	195	99	24	15	12	11	9	9	8	433	$C_{22}H_{35}NO_4Si_2$	Dihydronoroxymor-phone-2 TMS	DMIT	0288
73	75*	129	191	147	74	217	173	99	24	15	15	15	9	8	6	642	$C_{32}H_{66}O_5Si_4$	Prostaglandin F2B TMS ester tri-TMS ether	BLR	0129
73	75*	129	438	478	199	388	419	99	32	20	20	16	14	13	13	509	$C_{27}H_{51}NO_4Si_2$	Prostaglandin A1 methyloxime TMS ester TMS ether	BLR	0121
73	75*	133	355	74	130	426	67	99	27	14	10	8	8	7	6	599	$C_{30}H_{61}NO_5Si_3$	Prostaglandin E1 methyloxime TMS ester di-TMS ether	BLR	0123
73	75*	147	129	191	43	55	117	99	36	19	16	14	12	12	9	644	$C_{32}H_{68}O_5Si_4$	Prostaglandin F1A TMS ester tri TMS ether	BLR	0126
73	75*	147	191	129	74	173	217	99	24	19	17	12	9	7	7	642	$C_{32}H_{66}O_5Si_4$	Prostaglandin F2A TMS ester tri-TMS ether	BLR	0128
73	75*	355	133	426	130	74	225	99	29	19	19	17	13	9	8	599	$C_{30}H_{61}NO_5Si_3$	Prostaglandin 8-iso E1 methyloxime TMS ester TMS ether	BLR	0124
103	76*	49	50	84	104	51	86	99	57	20	16	14	13	13	10	103	C_7H_5N	Oxacillin MTB 3(C_6H_5CN)	DMIT	0058
77*	205	233	239	267	51	268	75	99	96	82	70	59	57	47	36	286	$C_{15}H_{11}ClN_2O_2$	Oxazepam	DMIT	0194

Eight Most Intense Peaks — E2 (continued)

Eight most intense peaks								Relative Intensities								Mol wt	Formula	Drug	Source	
77*	245	244	105	228	246	247	51	99	97	79	46	42	39	36	33	245	$C_{14}H_{12}ClNO$	2-Methylamino-5-chloro-benzophenone	BLR	0112
58	77*	30	51	49	56	59	42	99	12	6	6	6	6	6	5	165	$C_{10}H_{15}NO$	Pseudoephedrine	DMIT	0214
105	77*	51	76	122	117	50	161	99	78	41	32	29	25	23	23	179	$C_9H_9NO_3$	Hippuric acid	DMIT	0221
105	77*	96	183	51	42	182	28	99	64	56	31	28	26	25	24	431	$C_{22}H_{26}BrNO_3$	Clidinium bromide	DMIT	0175
135	77*	90	92	51	64	136	63	99	37	35	22	10	9	9	8	223	$C_{11}H_{13}NO_4$	2-Hydroxybenzoylgly-cine methyl ester methyl ether	BLR	0031
144	77*	202	43	160	51	44	185	99	85	74	60	41	37	32	32	202	$C_{11}H_{10}N_2O_2$	Oxacillin MTB 2	DMIT	0274
180	77*	104	209	223	252	51	181	99	56	56	56	44	33	27	25	252	$C_{15}H_{12}N_2O_2$	Diphenylhydantoin	DMIT	0168
183	77*	308	184	105	252	93	309	99	67	52	40	18	11	11	8	308	$C_{19}H_{20}N_2O_2$	Phenylbutazone	DMIT	0077
249	77*	235	76	264	247	250	50	99	36	27	23	20	19	17	14	264	$C_{17}H_{16}N_2O$	Ethinazone	DMIT	0252
78*	106	51	104	77	137	50	186	99	89	68	58	45	35	32	21	137	$C_6H_7N_3O$	Isoniazid	DMIT	0237
169	78*	113	171	51	63	44	50	99	57	55	31	20	19	17	16	169	$C_7H_4ClNO_2$	Chlorzoxazone	DMIT	0113
79*	108	107	77	51	50	39	91	99	81	65	64	46	27	25	23	108	C_7H_8O	Benzyl alcohol	DMIT	0190
15	79*	94	29	45	33	16	48	99	90	84	25	20	18	18	17	94	$C_2H_6O_2S$	Dimethylsulfone	DMIT	0207
81*	53	330	96	82	332	64	44	99	10	9	6	6	3	3	3	330	$C_{12}H_{11}ClN_2O_5S$	Furosemide	DMIT	0265
81*	83	67	85	47	48	98	95	99	12	7	7	6	4	4	3	164	$C_3H_4Cl_2F_2O$	2,2-Dichloro-1,1-difluoroe-thyl methyl ether	ABB	0012
81*	91	106	95	79	68	67	78	99	97	84	64	53	48	44	43	167	$C_9H_{13}NO_2$	Ethinamate	NIH MSC	3787
69	81*	137	41	136	95	44	149	99	54	24	22	15	14	13	13	0	—	Frequent contaminant	DMIT	0232
82	81*	44	54	55	83	94	41	99	34	16	14	11	8	5	4	111	$C_5H_9N_3$	Histamine	NIH MSC	3791
136	81*	324	189	55	137	41	42	99	23	20	17	17	15	13	13	324	$C_{20}H_{24}N_2O_2$	Quinidine	DMIT	0264
221	81*	28	79	157	80	27	155	99	56	46	41	29	29	28	27	236	$C_{12}H_{16}N_2O_3$	Hexobarbital	DMIT	0174
221	81*	157	80	79	155	222	77	99	55	32	23	22	17	12	10	236	$C_{12}H_{16}N_2O_3$	Hexobarbital	BLR	0064
221	81*	157	80	155	222	79	156	99	77	63	36	24	19	18	12	235	$C_{12}H_{15}N_2O_3$	Hexobarbital	NIH MSC	3790
235	81*	169	171	79	236	170	91	99	40	26	23	18	13	12	11	250	$C_{13}H_{18}N_2O_3$	Hexobarbital-3-methyl de-rivative	BLR	0062
249	81*	183	250	79	185	184	264	99	35	20	20	15	11	9	9	264	$C_{14}H_{20}N_2O_3$	Hexobarbital 3-ethyl deriv-ative	BLR	0080
82*	28	182	83	77	42	94	105	99	45	44	32	31	29	26	25	303	$C_{17}H_{21}NO_4$	Cocaine	DMIT	0146
82*	81	44	54	55	83	94	41	99	34	16	14	11	8	5	4	111	$C_5H_9N_3$	Histamine	NIH MSC	3791
82*	209	81	104	83	50	173	210	99	31	18	9	8	8	6	6	543	$C_{25}H_{13}D_{27}N_2O_4Si_3$	5-(3,4-Di-D9-TMSO-cy-clohexa-1,5-dienyl)-3-methyl-5-phenyl-1-D9-TMS-hydantoin	BLR	0046

82*	350	280	351	181	43	81	162	99	75	39	19	15	15	13	12	462	$C_{20}H_{22}D_{18}N_2O_5Si_2$	Dihydroxysecobarbital dI-D9-TMS ether 1, 3-dimethyl derivative	BLR	0075
83	82*	96	124	97	42	73	184	99	68	62	33	33	28	17	12	213	$C_{11}H_{23}NOSi$	Tropine TMS ether	BLR	0035
124	82*	83	57	41	94	42	43	99	32	24	23	20	18	16	15	267	$C_{16}H_{29}NO_2$	Anisotropine	DMIT	0270
182	82*	83	105	94	77	96	303	99	95	41	35	33	31	27	25	303	$C_{17}H_{21}NO_4$	Cocaine	NIH MSC	3778
208	82*	123	67	209	42	55	207	99	32	31	17	13	10	9	9	208	$C_9H_{12}N_4O_2$	1,3,7(or 9),8-tetramethylxanthine	BLR	0103
83*	55	71	96	114	144	62	56	99	59	44	43	42	34	30	25	218	$C_9H_{18}N_2O_4$	Meprobamate	DMIT	0062
83*	55	98	153	166	41	155	152	99	61	60	52	52	50	26	24	199	$C_{10}H_{17}NO_3$	Methyprylon MTB 1	DMIT	0010
83*	55	114	96	71	144	41	44	99	68	58	56	49	40	32	31	218	$C_9H_{18}N_2O_4$	2-Methyl-2-propyl-1,3-propane diol carbamate	ABB	0011
83*	82	96	124	97	42	73	184	99	68	62	33	33	28	17	12	213	$C_{11}H_{23}NOSi$	Tropine TMS ether	BLR	0035
83*	84	43	56	55	71	75	62	99	72	63	59	55	45	32	28	218	$C_9H_{18}N_2O_4$	Meprobamate	NIH MSC	3795
81	83*	67	85	47	48	98	95	99	12	7	7	6	4	4	3	164	$C_3H_4Cl_2F_2O$	2,2-Dichloro-1,1-difluoroethyl methyl ether	ABB	0012
113	83*	40	56	67	97	28	39	99	36	31	12	11	9	8	4	113	$C_3H_3N_3O_2$	2-Nitro-imidazole	ABB	0006
124	83*	82	73	94	96	125	42	99	18	17	14	12	10	10	9	361	$C_{20}H_{31}NO_3Si$	Atropine TMS ether	BLR	0034
124	83*	82	94	289	42	96	125	99	24	22	20	17	13	12	10	289	$C_{17}H_{23}NO_3$	Hyoscyamine (atropine)	DMIT	0158
140	83*	82	124	96	97	167	125	99	88	59	41	24	24	22	21	307	$C_{21}H_{25}NO$	Benztropine	DMIT	0050
84*	42	133	162	161	51	119	65	99	27	26	13	10	8	7	7	162	$C_{10}H_{14}N_2$	Nicotine	BLR	0033
84*	85	56	28	91	36	55	30	99	9	6	5	5	4	4	3	233	$C_{14}H_{19}NO_2$	Methylphenidate	NIH MSC	3816
84*	91	55	77	85	182	83	65	99	14	5	5	5	5	3	2	265	$C_{17}H_{19}N_3$	Antazoline	DMIT	0020
84*	91	55	150	41	56	83	85	99	18	8	7	4	4	4	4	233	$C_{14}H_{19}NO_2$	Methylphenidate	DMIT	0051
84*	133	42	162	161	28	39	51	99	26	20	19	17	17	9	7	162	$C_{10}H_{14}N_2$	Nicotine	DMIT	0056
84*	198	245	112	97	70	197	96	99	86	76	61	59	57	51	48	356	$C_{20}H_{24}N_2S_2$	Thioridazine MTB 3	DMIT	0291
84*	296	42	85	180	55	221	41	99	7	6	5	3	3	3	2	296	$C_{18}H_{20}N_2S$	Pyrathiazine	DMIT	0092
56*	84*	57	203	42	83	77	93	99	64	63	43	30	26	13	11	203	$C_{11}H_{13}N_3O$	Ampyrone	DMIT	0007
56	84*	245	57	203	83	43	42	99	50	48	26	26	25	19	10	245	$C_{13}H_{15}N_3O_2$	4-Acetyl-aminoantipyrine	DMT	0006
83	84*	43	56	55	71	75	62	99	72	63	59	55	45	32	28	218	$C_9H_{18}N_2O_4$	Meprobamate	NIH MSC	3795
85	84*	183	107	77	55	56	184	99	74	70	43	22	18	18	11	267	$C_{18}H_{21}NO$	Azacyclonol	DMIT	0045
205	84*	96	166	248	122	246	81	99	74	68	31	22	22	22	19	248	$C_{17}H_{20}D_5N$	Phencyclidine (D5-phenyl)	DMIT	0319
85*	84	183	107	77	55	56	184	99	74	70	43	22	18	18	11	267	$C_{18}H_{21}NO$	Azacyclonol	DMIT	0045
84	85*	56	28	91	36	55	30	99	9	6	5	5	4	4	3	233	$C_{14}H_{19}NO_2$	Methylphenidate	NIH MSC	3816
86*	58	30	57	99	87	71	43	99	23	21	19	19	16	14	13	387	$C_{21}H_{23}ClFN_3O$	Flurazepam	DMIT	0323
86*	58	30	87	72	42	56	120	99	49	43	33	20	11	10	9	234	$C_{14}H_{22}N_2O$	Lidocaine	DMIT	0142
86*	58	73	87	319	41	99	245	99	10	5	5	5	4	4	4	319	$C_{18}H_{26}ClN_3$	Chloroquine	NIH MSC	3774
86*	58	99	71	55	100	57	87	99	16	15	9	6	6	5	5	309	$C_{19}H_{35}NO_2$	Dicyclomine	DMIT	0255
86*	99	91	144	58	56	41	87	99	15	13	11	9	8	6	6	289	$C_{18}H_{27}NO_2$	Caramiphen	DMIT	0164
86*	99	120	58	30	65	71	87	99	25	16	10	8	7	6	5	236	$C_{13}H_{20}N_2O_2$	Procaine	DMIT	0202
86*	99	120	58	87	65	92	71	99	24	16	7	7	6	6	5	236	$C_{13}H_{20}N_2O_2$	Procaine	NIH MSC	3811
86*	126	259	58	112	99	400	74	99	31	15	14	12	11	11	7	399	$C_{23}H_{30}ClN_3O$	Quinacrine	NIH MSC	3815

Eight Most Intense Peaks — E2 (continued)

Eight most intense peaks								Relative Intensities								Mol wt	Formula	Drug	Source	
86*	298	83	58	87	30	85	180	99	10	6	5	5	4	4	3	298	$C_{18}H_{22}N_2S$	Diethazine	DMIT	0111
58	86*	42	334	220	59	87	44	99	77	11	6	5	4	4	4	334	$C_{17}H_{19}ClN_2OS$	3-Hydroxy-chlorpromazine	DMIT	0015
58	86*	289	43	291	85	248	288	99	26	10	6	6	5	5	4	376	$C_{19}H_{21}ClN_2O_2S$	8-Acetoxychlorpromazine	DMIT	0014
58	86*	314	229	185	228	42	85	99	38	30	16	11	10	9	7	314	$C_{18}H_{22}N_2OS$	Methoxypromazine	DMIT	0126
58	86*	318	85	272	320	42	232	99	26	23	12	9	8	6	5	318	$C_{17}H_{19}ClN_2S$	Chlorpromazine	DMIT	0100
58	86*	334	42	85	336	44	243	99	44	27	17	10	10	9	9	334	$C_{17}H_{19}ClN_2OS$	8-Hydroxy-chlorpromazine	DMIT	0016
58	86*	352	85	306	59	266	248	99	31	24	18	10	6	5	5	352	$C_{18}H_{19}F_3N_2S$	Triflupromazine	DMIT	0034
58	86*	376	43	378	42	87	59	99	85	31	15	12	11	10	6	376	$C_{19}H_{21}ClN_2O_2S$	3-Acetoxychlorpromazine	DMIT	0013
181	86*	72	182	85	152	108	99	99	66	61	24	12	11	9	8	419	$C_{21}H_{26}BrNO_3$	Methantheline bromide	DMIT	0053
74	87*	43	55	143	298	41	57	99	76	33	27	21	21	20	20	298	$C_{19}H_{38}O_2$	Methyl stearate	DMIT	0222
74	87*	270	75	143	43	55	41	99	85	22	19	19	18	17	14	270	$C_{17}H_{34}O_2$	Methyl palmitate	DMIT	0223
89*	250	235	73	219	146	118	90	99	38	32	18	17	14	13	10	250	$C_{13}H_{18}O_3Si$	*o*-Hydroxycinnamic acid methyl ester trimethylsilyl ether	BLR	0175
117	90*	89	63	28	39	118	116	99	37	25	15	11	9	9	8	117	C_8H_7N	Indole	DMIT	0187
135	90*	77	92	136	134	51	64	99	24	19	13	8	5	4	4	223	$C_{11}H_{13}NO_4$	2-Hydroxybenzoylglycine methyl ester methyl ether	BLR	0106
135	90*	134	136	164	105	77	91	99	48	11	8	6	4	4	2	223	$C_{11}H_{13}NO_4$	Salicyluric acid methyl ester methyl ether	BLR	0153
91*	43	67	81	78	79	106	44	99	79	55	54	50	49	42	41	167	$C_9H_{13}NO_2$	Ethinamate	DMIT	0124
91*	106	28	44	78	51	177	79	99	82	69	55	47	39	26	25	298	$C_{16}H_{18}N_4O_2$	Nialamide	DMIT	0166
91*	106	127	110	57	92	65	104	99	31	31	17	15	13	12	12	231	$C_{12}H_{13}N_3O_2$	Isocarboxazid	DMIT	0041
91*	110	79	231	272	298	41	77	99	90	80	78	77	76	72	64	298	$C_{20}H_{26}O_2$	Norethindrone	DMIT	0262
91*	135	119	163	120	44	78	92	99	24	22	17	14	12	10	9	163	$C_{10}H_{13}NO$	Primaclone MTB 2	DMIT	0251
91*	150	59	92	151	119	118	105	99	61	10	8	6	3	2	1	150	$C_9H_{10}O_2$	Phenylacetic acid methyl ester	BLR	0139
91*	150	65	92	59	39	107	108	99	65	20	14	12	11	9	7	150	$C_9H_{10}O_2$	Methylphenidate MTB 1	DMIT	0253
91*	159	82	68	158	42	65	118	99	96	90	54	51	26	19	16	159	$C_{11}H_{13}N$	*N*-Methyl-*n*-2-propynyl-benzylamine	ABB	0009
44	91*	45	65	42	43	41	51	99	8	6	5	4	3	2	2	135	$C_9H_{13}N$	Amphetamine	NIH MSC	3768
44	91*	65	42	45	120	40	92	99	8	4	3	3	3	2	2	135	$C_9H_{13}N$	Amphetamine	DMIT	0035
58	91*	59	43	56	65	134	39	99	7	5	3	3	3	3	2	149	$C_{10}H_{15}N$	Methamphetamine	NIH MSC	3797
58	91*	59	56	42	65	134	41	99	7	6	5	3	3	2	2	149	$C_{10}H_{15}N$	Methamphetamine	BLR	0107
58	91*	197	72	71	185	184	65	99	93	30	26	25	16	13	8	255	$C_{16}H_{21}N_3$	Tripelennamine	NIH MSC	3826
81	91*	106	95	79	68	67	78	99	97	84	64	53	48	44	43	167	$C_9H_{13}NO_2$	Ethinamate	NIH MSC	3787
84	91*	55	77	85	182	83	65	99	14	5	5	5	5	3	2	265	$C_{17}H_{19}N_3$	Antazoline	DMIT	0020
84	91*	55	150	41	56	83	85	99	18	8	7	4	4	4	4	233	$C_{14}H_{19}NO_2$	Methylphenidate	DMIT	0051

105	91*	104	77	133	92	51	65	99	76	48	32	24	16	11	11	136	$C_8H_{12}N_2$	Phenelzine	DMIT	0036
118	91*	178	44	92	65	90	119	99	51	30	25	20	10	10	8	178	$C_9H_{10}N_2O_2$	Phenacetyl urea	ABB	0002
188	91*	105	119	65	189	104	78	99	70	52	25	14	12	11	7	188	$C_{11}H_{12}N_2O$	1-(*p*-Tolyl)-3-methyl-pyrazol-5-one	ABB	0008
200	91*	84	28	242	243	115	129	99	66	47	46	36	31	30	29	243	$C_{17}H_{25}N$	Phencyclidine	DMIT	0144
200	91*	243	84	242	186	166	201	99	48	26	25	23	21	18	17	243	$C_{17}H_{25}N$	Phencyclidine	NIH MSC	3806
92*	120	39	138	64	63	65	38	99	91	46	45	43	30	23	19	138	$C_7H_6O_3$	Salicylic acid	DMIT	0090
92*	156	108	65	162	119	174	189	99	81	64	50	46	37	30	16	253	$C_{10}H_{11}N_3O_3S$	Sulfamethoxazole	DMIT	0302
93	92*	39	41	77	91	27	79	99	29	23	23	22	21	21	17	136	$C_{10}H_{16}$	Pinene (from turpentine)	DMIT	0226
120	92*	152	121	65	64	93	63	99	54	49	29	18	12	12	10	152	$C_8H_8O_3$	Methyl salicylate	NIH MSC	3799
93*	92	39	41	77	91	27	79	99	29	23	23	22	21	21	17	136	$C_{10}H_{16}$	Pinene (from turpentine)	DMIT	0226
93*	109	199	77	119	162	55	324	99	78	34	31	27	25	24	21	324	$C_{19}H_{20}N_2O_3$	*p*-Hydroxyphenylbutazone	DMIT	0118
94*	66	39	65	40	95	38	55	99	27	27	22	13	10	10	8	94	C_6H_6O	Phenol	DMIT	0189
94*	138	42	108	136	154	97	137	99	94	57	55	38	36	32	27	303	$C_{17}H_{21}NO_4$	Scopolamine	NIH MSC	3818
94*	138	42	108	136	154	103	137	99	82	62	57	37	35	28	28	303	$C_{17}H_{21}NO_4$	Scopolamine	DMIT	0263
138	94*	73	108	42	154	137	136	99	69	46	33	33	31	27	23	375	$C_{20}H_{29}NO_4Si$	Scopolamine TMS ether	BLR	0036
95*	115	196	41	67	43	96	209	99	69	63	47	42	31	28	22	252	$C_{13}H_{20}N_2O_3$	Butalbital 2,4(or 4,6)-dimethyl derivative	BLR	0061
126	95*	41	81	39	69	108	83	99	96	79	70	48	43	38	36	152	$C_{10}H_{16}O$	Camphor	DMIT	0188
180	95*	68	53	123	96	181	94	99	51	36	13	12	8	8	5	180	$C_7H_8N_4O_2$	Theophylline	DMIT	0201
208	95*	193	67	180	123	73	43	99	45	25	17	17	16	14	12	208	$C_9H_{12}N_4O_2$	Theophylline 7-ethyl derivative	BLR	0096
222	95*	194	166	207	179	123	67	99	53	39	38	31	28	20	18	222	$C_{10}H_{14}N_4O_2$	1,7-Di-ethyl-3-methylxanthine	BLR	0099
96*	73	94	42	57	227	142	212	99	95	45	44	41	41	39	26	227	$C_{11}H_{21}NO_2Si$	Scopoline TMS ether	BLR	0037
188	96*	77	56	28	39	55	105	99	63	63	48	42	26	26	24	188	$C_{11}H_{12}N_2O$	Antipyrine	DMIT	0030
188	96*	77	56	105	55	187	189	99	59	44	28	19	15	13	13	188	$C_{11}H_{12}N_2O$	Antipyrine	NIH MSC	3769
287	96*	215	286	288	243	229	230	99	61	47	44	29	25	22	20	287	$C_{21}H_{21}N$	Cyproheptadine	DMIT	0208
97*	72	55	71	62	110	69	158	99	64	62	42	40	34	34	30	232	$C_{10}H_{20}N_2O_4$	Mebutamate	DMIT	0061
97*	98	55	82	199	198	180	296	99	94	93	44	36	33	31	31	296	$C_{18}H_{20}N_2S_1$	Methdilazine	DMIT	0003
97*	107	204	191	44	78	79	189	99	56	27	27	16	10	8	8	247	$C_{13}H_{17}N_3S$	Methapyrilene MTB 1	DMIT	0233
97*	165	164	206	84	249	135	136	99	87	76	64	45	44	35	32	249	$C_{15}H_{23}NS$	1-(1-Thiophenyl cyclohexyl) piperidine	DMIT	0325
56	97*	231	42	111	77	71	112	99	35	35	12	11	8	7	7	231	$C_{13}H_{17}N_3O$	Aminopyrine	DMIT	0008
58	97*	72	71	79	42	78	40	99	68	23	22	14	14	13	10	261	$C_{14}H_{19}N_3S$	Thenyldiamine	DMIT	0157
58	97*	72	71	191	190	78	79	99	71	22	19	11	10	8	8	261	$C_{14}H_{19}N_3S$	Methapyrilene	NIH MSC	3798
58	97*	72	84	71	191	190	261	99	71	28	23	18	14	11	10	261	$C_{14}H_{19}N_3S$	Methapyrilene	DMIT	0054
98*	70	370	126	99	185	244	125	99	13	9	8	7	4	3	3	370	$C_{21}H_{26}N_2S_2$	Thioridazine	DMIT	0004
98*	99	218	85	131	219	84	69	99	7	6	4	3	3	2	2	287	$C_{19}H_{29}NO$	Cycrimine	DMIT	0081
98*	176	147	42	119	51	41	91	99	28	11	10	6	6	5	4	176	$C_{10}H_{12}N_2O$	Cotinine (from nicotine)	DMIT	0206
97	98*	55	82	199	198	180	296	99	94	93	44	36	33	31	31	296	$C_{18}H_{20}N_2S_1$	Methdilazine	DMIT	0003
99*	56	72	165	300	228	229	242	99	51	38	30	27	23	23	22	300	$C_{18}H_{21}ClN_2$	Chlorcyclizine	DMIT	0074

Eight Most Intense Peaks — E2 (continued)

Eight most intense peaks								Relative Intensities								Mol wt	Formula	Drug	Source	
99*	56	167	194	266	165	195	207	99	66	60	43	43	37	35	35	266	$C_{18}H_{22}N_2$	Cyclizine	DMIT	0082
99*	114	98	167	70	165	96	168	99	40	31	27	15	13	10	8	281	$C_{19}H_{23}NO$	Diphenylpyraline	DMIT	0083
30	99*	28	42	43	41	27	39	99	66	64	62	61	60	31	27	99	C_5H_9NO	Piperidone	DMIT	0079
86	99*	91	144	58	56	41	87	99	15	13	11	9	8	6	6	289	$C_{18}H_{27}NO_2$	Caramiphen	DMIT	0164
86	99*	120	58	30	65	71	87	99	25	16	10	8	7	6	5	236	$C_{13}H_{20}N_2O_2$	Procaine	DMIT	0202
86	99*	120	58	87	65	92	71	99	24	16	7	7	6	6	5	236	$C_{13}H_{20}N_2O_2$	Procaine	NIH MSC	3811
98	99*	218	85	131	219	84	69	99	7	6	4	3	3	2	2	287	$C_{19}H_{29}NO$	Cycrimine	DMIT	0081
100*	28	44	72	77	42	105	27	99	35	28	20	17	14	14	12	205	$C_{13}H_{19}NO$	Diethylpropion	DMIT	0129
100*	101	72	197	312	84	179	212	99	7	3	3	3	2	2	2	312	$C_{19}H_{24}N_2S$	Ethopropazine	DMIT	0033
44	100*	234	88	105	57	41	91	99	35	24	21	16	16	14	12	325	$C_{21}H_{27}NO_2$	Propoxyphene MTB 3	DMIT	0210
100	101*	72	197	312	84	179	212	99	7	3	3	3	2	2	2	312	$C_{19}H_{24}N_2S$	Ethopropazine	DMIT	0033
103*	76	49	50	84	104	51	86	99	57	20	16	14	13	13	10	103	C_7H_5N	Oxacillin MTB 3 (C_6H_5CN)	DMIT	0058
104*	164	91	105	133	103	165	78	99	41	29	23	8	6	4	3	164	$C_{10}H_{12}O_2$	β-Phenylpropionic acid methyl ester	BLR	0160
104*	175	44	77	43	41	55	56	99	99	41	31	29	27	18	15	204	$C_{11}H_{12}N_2O_2$	Mephenytoin MTB 1	DMIT	0245
104*	189	103	51	77	78	105	39	99	25	17	13	13	13	12	5	189	$C_{11}H_{11}NO_2$	Phensuximide	DMIT	0012
180	104*	266	77	237	57	209	71	99	60	48	40	38	37	25	22	266	$C_{16}H_{14}N_2O_2$	Diphenylhydantoin 3-methyl derivative	BLR	0038
189	104*	77	190	51	105	103	132	99	85	22	12	9	8	5	5	218	$C_{12}H_{14}N_2O_2$	Mephenytoin	DMIT	0005
105*	77	51	76	122	117	50	161	99	78	41	32	29	25	23	23	179	$C_9H_9NO_3$	Hippuric acid	DMIT	0221
105*	77	96	183	51	42	182	28	99	64	56	31	28	26	25	24	431	$C_{22}H_{26}BrNO_3$	Clidinium bromide	DMIT	0175
105*	91	104	77	133	92	51	65	99	76	48	32	24	16	11	11	136	$C_8H_{12}N_2$	Phenelzine	DMIT	0036
105*	134	193	77	162	106	79	147	99	35	10	9	8	7	4	3	193	$C_{10}H_{11}NO_3$	Hippuric acid methyl ester	BLR	0159
189	105*	165	201	166	285	390	190	99	83	20	20	17	17	17	14	390	$C_{25}H_{27}ClN_2$	Meclizine	DMIT	0087
204	105*	104	133	77	72	205	78	99	94	71	23	19	13	13	12	204	$C_{11}H_{12}N_2O_2$	Ethotoin	DMIT	0068
106*	137	78	136	79	105	107	138	99	76	33	18	12	7	6	5	137	$C_7H_7NO_2$	Pyridine-3-carboxylic acid methyl ester	BLR	0140
78	106*	51	104	77	137	50	186	99	89	68	58	45	35	32	21	137	$C_6H_7N_3O$	Isoniazid	DMIT	0237
91	106*	28	44	78	51	177	79	99	82	69	55	47	39	26	25	298	$C_{16}H_{18}N_4O_2$	Nialamide	DMIT	0166
91	106*	127	110	57	92	65	104	99	31	31	17	15	13	12	12	231	$C_{12}H_{13}N_3O_2$	Isocarboxazid	DMIT	0041
107*	108	77	51	79	39	53	50	99	92	24	19	18	17	15	13	108	C_7H_8O	*p*-Cresol	DMIT	0151
107*	108	78	79	51	80	77	106	99	52	42	18	12	11	11	10	214	$C_{13}H_{14}N_2O$	Phenyramidol	DMIT	0048
107*	149	57	78	108	79	77	72	99	78	70	60	56	42	37	37	261	$C_{16}H_{23}NO_2$	Ethoheptazine	DMIT	0057
97	107*	204	191	44	78	79	189	99	56	27	27	16	10	8	8	247	$C_{13}H_{17}N_3S$	Methapyrilene MTB 1	DMIT	0233
108	107*	91	182	109	77	65	79	99	23	22	18	12	12	10	9	182	$C_{10}H_{14}O_3$	Mephenesin	DMIT	0002
176	107*	90	89	77	79	105	70	99	96	56	54	45	41	33	33	176	$C_9H_8N_2O_2$	2-Imino-5-phenyl-4-oxazolidinone	ABB	0007

108*	107	91	182	109	77	65	79	99	23	22	18	12	12	10	9	182	$C_{10}H_{14}O_3$	Mephenesin	DMIT	0002
108*	109	43	80	137	179	53	81	99	90	66	29	28	25	24	24	179	$C_{10}H_{13}NO_2$	Acetophenetidin	BLR	0009
108*	109	44	179	137	28	80	81	99	83	51	51	43	26	16	15	179	$C_{10}H_{13}NO_2$	Acetophenetidin	DMIT	0117
108*	123	165	43	52	80	53	122	99	57	49	37	14	14	10	9	151	$C_9H_{11}O_2$	Acetaminophen methyl ether	BLR	0010
30	108*	107	77	39	51	28	137	99	41	22	13	6	5	5	5	137	$C_8H_{11}NO$	Tyramine	DMIT	0196
79	108*	107	77	51	50	39	91	99	81	65	64	46	27	25	23	108	C_7H_8O	Benzyl alcohol	DMIT	0190
107	108*	77	51	79	39	53	50	99	92	24	19	18	17	15	13	108	C_7H_8O	*p*-Cresol	DMIT	0151
107	108*	78	79	51	80	77	106	99	52	42	18	12	11	11	10	214	$C_{13}H_{14}N_2O$	Phenyramidol	DMIT	0048
109*	151	43	79	80	53	108	110	99	26	26	19	12	10	10	8	151	$C_8H_9NO_2$	Acetaminophen	DMIT	0116
109*	151	43	80	81	108	86	57	99	30	22	13	11	10	9	8	151	$C_8H_9NO_2$	Acetaminophen	NIH MSC	3765
93	109*	199	77	119	162	55	324	99	78	34	31	27	25	24	21	324	$C_{19}H_{20}N_2O_3$	*p*-Hydroxyphenylbutazone	DMIT	0118
108	109*	43	80	137	179	53	81	99	90	66	29	28	25	24	24	179	$C_{10}H_{13}NO_2$	Acetophenetidin	BLR	0009
108	109*	44	179	137	28	80	81	99	83	51	51	43	26	16	15	179	$C_{10}H_{13}NO_2$	Acetophenetidin	DMIT	0117
121	109*	151	65	93	43	39	271	99	83	59	29	21	20	15	12	271	$C_{15}H_{13}NO_4$	Phenetsal	DMIT	0156
180	109*	55	67	82	137	181	108	99	25	23	21	20	9	8	6	180	$C_7H_8N_4O_2$	Theobromine	DMIT	0191
194	109*	55	67	82	15	18	42	99	59	37	22	17	14	11	10	194	$C_8H_{10}N_4O_2$	Caffeine	DMIT	0059
194	109*	55	67	82	43	42	41	99	55	33	25	18	15	13	10	194	$C_8H_{10}N_4O_2$	Caffeine	BLR	0095
91	110*	79	231	272	298	41	77	99	90	80	78	77	76	72	64	298	$C_{20}H_{26}O_2$	Norethindrone	DMIT	0262
139	111*	141	75	170	50	113	172	99	38	35	26	23	15	12	10	170	$C_8H_7ClO_2$	4-Chlorobenzoic acid methyl ester (from indomethacin)	BLR	0015
310	111*	112	58	199	212	96	41	99	92	84	79	65	64	61	55	310	$C_{19}H_{22}N_2S$	Mepazine	DMIT	0104
184	112*	42	55	170	183	212	58	99	21	19	19	17	13	13	12	240	$C_{12}H_{20}N_2O_3$	Butethal 1,3-dimethyl derivative	BLR	0058
113*	55	70	42	41	39	85	69	99	81	63	37	20	18	10	8	141	$C_7H_{11}NO_2$	Ethosuximide	DMIT	0205
113*	70	43	42	141	407	71	127	99	74	44	30	30	27	25	23	407	$C_{21}H_{24}F_3N_3S$	Trifluoperazine	DMIT	0131
113*	70	373	141	43	42	44	375	99	89	58	48	44	44	20	20	373	$C_{20}H_{24}ClN_3S$	Prochlorperazine	DMIT	0154
113*	83	40	56	67	97	28	39	99	36	31	12	11	9	8	4	113	$C_3H_3N_3O_2$	2-Nitro-imidazone	ABB	0006
70	113*	43	42	127	71	44	56	99	98	85	59	56	41	36	27	446	$C_{22}H_{30}N_4O_2S_2$	Thioproperazine	DMIT	0145
399	113*	70	141	43	72	400	71	99	97	79	72	34	28	28	22	399	$C_{22}H_{29}N_3S_2$	Torecane	DMIT	0094
114*	44	142	365	42	115	263	128	99	90	85	78	46	42	35	34	365	$C_{21}H_{23}N_3OS$	Pericyazine	DMIT	0135
99	114*	98	167	70	165	96	168	99	40	31	27	15	13	10	8	281	$C_{19}H_{23}NO$	Diphenylpyraline	DMIT	0083
115*	117	89	53	109	51	91	39	99	29	24	20	15	13	12	8	144	C_7H_9ClO	Ethchlorvynol	NIH MSC	3808
115*	208	193	130	179	117	91	129	99	74	42	38	21	21	20	16	208	$C_{16}H_{16}$	Propoxyphene MTB 2-pk 2	DMIT	0204
115*	208	193	130	179	117	91	129	99	74	42	38	21	21	20	16	208	$C_{16}H_{16}$	Propoxyphene MTB 2-pk 1	DMIT	0203
95	115*	196	41	67	43	96	209	99	69	63	47	42	31	28	22	252	$C_{13}H_{20}N_2O_3$	Butalbital 2,4(or 4,6)-dimethyl derivative	BLR	0061
116*	43	85	61	119	71	31	101	99	29	29	19	18	18	15	14	176	$C_6H_8O_6$	Ascorbic acid	ABB	0010
117	116*	145	90	89	63	51	39	99	90	30	27	26	22	17	15	145	$C_{10}H_{11}N$	Phenobarbital MTB 1	DMIT	0249

Eight Most Intense Peaks — E2 (continued)

Eight most intense peaks								Relative Intensities								Mol wt	Formula	Drug	Source	
117*	73	327	75	143	256	118	69	99	50	40	26	18	16	13	12	342	$C_{16}H_{30}N_2O_4Si$	Hydroxypentobarbital 1,3-dimethyl derivative TMS ether	BLR	0083
117*	90	89	63	28	39	118	116	99	37	25	15	11	9	9	8	117	C_8H_7N	Indole	DMIT	0187
117*	116	145	90	89	63	51	39	99	90	30	27	26	22	17	15	145	$C_{10}H_{11}N$	Phenobarbital MTB 1	DMIT	0249
117*	146	43	91	32	115	39	116	99	54	48	47	38	34	30	29	218	$C_{12}H_{14}N_2O_2$	Primaclone	DMIT	0161
117*	189	132	115	91	160	77	39	99	88	55	51	47	44	30	28	217	$C_{13}H_{15}NO_2$	Glutethimide	DMIT	0028
51	117*	67	69	118	119	101	133	99	75	43	14	8	8	5	2	184	$C_3H_2ClF_5O$	2-Chloro-1,1,2-trifluoroethyl-difluoromethyl ether	ABB	0001
115	117*	89	53	109	51	91	39	99	29	24	20	15	13	12	8	144	C_7H_9ClO	Ethchlorvynol	NIH MSC	3808
118	117*	103	189	78	77	58	119	99	22	16	16	11	11	10	10	189	$C_{11}H_{11}NO_2$	Methsuximide MTB 1	DMIT	0269
189	117*	132	160	115	91	190	103	99	60	54	31	17	14	12	9	217	$C_{13}H_{15}NO_2$	Glutethimide	NIH MSC	3786
204	117*	232	118	161	77	115	146	99	29	22	20	17	16	16	16	232	$C_{12}H_{12}N_2O_3$	Phenobarbital	DMIT	0162
218	117*	146	118	219	28	161	39	99	20	18	16	13	12	9	9	246	$C_{13}H_{14}N_2O_3$	Mephobarbital	DMIT	0171
118*	91	178	44	92	65	90	119	99	51	30	25	20	10	10	8	178	$C_9H_{10}N_2O_2$	Phenacetyl urea	ABB	0002
118*	117	103	189	78	77	58	119	99	22	16	16	11	11	10	10	189	$C_{11}H_{11}NO_2$	Methsuximide MTB 1	DMIT	0269
118*	243	104	77	129	231	130	128	99	99	52	43	39	35	34	32	272	$C_{15}H_{16}N_2O_3$	Alphenal 1,3-dimethyl derivative	BLR	0049
218	118*	146	117	103	77	219	91	99	32	31	19	18	14	13	12	246	$C_{13}H_{14}N_2O_3$	Mephobarbital	BLR	0071
119*	151	120	92	152	91	93	65	99	91	22	16	7	2	2	1	151	$C_8H_9NO_2$	*o*-Aminobenzoic acid methyl ester	BLR	0138
120*	43	138	92	28	42	121	39	99	86	58	55	41	31	20	20	180	$C_9H_8O_4$	Acetylsalicylic acid	DMIT	0109
120*	92	152	121	65	64	93	63	99	54	49	29	18	12	12	10	152	$C_8H_8O_3$	Methyl salicylate	NIH MSC	3799
120*	138	43	92	64	65	63	42	99	59	48	34	10	9	8	6	180	$C_9H_8O_4$	Acetyl Salicyclic acid	NIH MSC	3763
120*	151	92	121	152	65	93	122	99	86	11	6	6	1	1	0	151	$C_8H_9NO_2$	*p*-Aminobenzoic acid methyl ester	BLR	0137
120*	152	43	121	92	63	65	64	99	81	42	41	39	12	11	9	194	$C_{10}H_{10}O_4$	Acetyl salicyclic acid methyl ester	BLR	0029
92	120*	39	138	64	63	65	38	99	91	46	45	43	30	23	19	138	$C_7H_6O_3$	Salicylic acid	DMIT	0090
195	120*	43	210	135	75	73	92	99	52	46	33	23	20	15	15	252	$C_{12}H_{16}O_4Si$	Acetyl salicylic acid TMS ester	BLR	0104
121*	58	72	71	78	215	122	77	99	67	13	13	12	11	9	8	286	$C_{16}H_{22}N_4O$	Thonzylamine	DMIT	0031
121*	58	72	79	71	28	78	42	99	88	24	17	17	15	14	11	285	$C_{17}H_{23}N_3O$	Pyrilamine	DMIT	0071
121*	109	151	65	93	43	39	271	99	83	59	29	21	20	15	12	271	$C_{15}H_{13}NO_4$	Phenetsal	DMIT	0156
121*	152	93	65	39	122	63	153	99	36	21	16	10	8	4	3	152	$C_8H_8O_3$	Methyl *p*-hydroxybenzoate	DMIT	0185
121*	180	91	181	122	148	107	93	99	87	19	9	8	6	3	2	180	$C_{10}H_{12}O_3$	*o*-Methoxyphenylacetic acid methyl ester	BLR	0163

121*	180	122	181	0	0	0	0	99	49	7	4	0	0	0	0	180	$C_{10}H_{12}O_3$	*p*-Methoxyphenylacetic acid methyl ester	BLR	0130
121*	194	134	163	122	195	135	108	99	39	19	15	8	3	2	0	194	$C_{11}H_{14}O_3$	*p*-Methoxyphenylpropionic acid methyl ester	BLR	0158
180	121*	181	122	59	91	148	182	99	74	10	7	2	2	1	1	180	$C_{10}H_{12}O_3$	*m*-Methoxyphenylacetic acid methyl ester	BLR	0131
256	121*	185	224	257	65	43	41	99	53	52	17	14	12	11	8	285	$C_{13}H_{19}NO_4S$	Probenecid	NIH MSC	3810
292	121*	263	293	249	77	92	215	99	57	55	38	37	30	27	25	292	$C_{19}H_{16}O_3$	Warfarin MTB 1	DMIT	0326
123*	42	222	207	179	149	150	166	99	86	79	74	72	61	55	49	222	$C_{10}H_{14}N_4O_2$	1,3-Di-ethyl-9-methylxan-thine	BLR	0102
108	123*	165	43	52	80	53	122	99	57	49	37	14	14	10	9	151	$C_9H_{11}O_2$	Acetaminophen methyl ether	BLR	0010
141	123*	42	41	151	55	169	96	99	93	57	48	35	32	32	29	401	$C_{21}H_{24}ClN_3OS$	Pipamazine	DMIT	0137
124*	30	123	153	77	51	125	78	99	82	35	23	12	9	6	4	153	$C_8H_{11}NO_2$	4-(2-Aminoethyl) pyro-ca-techol	ABB	0005
124*	82	83	57	41	94	42	43	99	32	24	23	20	18	16	15	267	$C_{16}H_{29}N_2$	Anisotropine	DMIT	0270
124*	83	82	73	94	96	125	42	99	18	17	14	12	10	10	9	361	$C_{20}H_{31}NO_3Si$	Atropine TMS ether	BLR	0034
124*	83	82	94	289	42	96	125	99	24	22	20	17	13	12	10	289	$C_{17}H_{23}NO_3$	Hyoscyamine (atropine)	DMIT	0158
124*	223	109	77	122	123	52	95	99	69	40	20	15	14	12	12	223	$C_{11}H_{13}NO_4$	Mephenoxalone	DMIT	0123
126*	95	41	81	39	69	108	83	99	96	79	70	48	43	38	36	152	$C_{10}H_{16}O$	Camphor	DMIT	0188
86	126*	259	58	112	99	400	74	99	31	15	14	12	11	11	7	399	$C_{23}H_{30}ClN_3O$	Quinacrine	NIH MSC	3815
127*	55	70	41	42	128	69	112	99	60	41	20	18	8	7	7	155	$C_8H_{13}NO_2$	Ethosuximide *N*-methyl de-rivative	BLR	0110
128*	69	143	73	169	75	159	41	99	98	87	69	52	44	39	35	286	$C_{13}H_{19}ClO_3Si$	2-(4-Clorophenoxy)-2-methylpropionic acid TMS ester (from clofibrate)	BLR	0023
128*	130	169	41	129	228	75	69	99	32	14	13	11	8	6	5	228	$C_{11}H_{13}ClO_3$	2-(4-Chlorophenoxy)-2-methylpropionic acid methyl ester (from clofi-brate)	BLR	0024
128*	130	169	87	41	129	242	59	99	32	21	14	13	11	8	8	242	$C_{12}H_{15}ClO_3$	Clofibrate	BLR	0022
129*	57	55	70	43	41	71	112	99	74	56	46	43	41	38	33	370	$C_{22}H_{42}O_4$	Dioctyl adipate	DMIT	0106
129*	157	128	130	187	158	156	188	99	21	15	11	8	2	1	0	187	$C_{11}H_9NO_2$	Quinoline-2-carboxylic acid methyl ester	BLR	0161
129*	158	130	143	115	128	91	77	99	97	80	68	66	45	31	24	158	$C_{12}H_{14}$	1-Phenyl cyclohexene	DMIT	0127
73	129*	128	114	334	349	115	130	99	87	23	21	20	15	13	12	349	$C_{18}H_{31}NO_2Si_2$	Nor-meperidinic acid TMS ester *N*-TMS derivative	BLR	0006
73	129*	305	128	304	276	103	232	99	65	35	33	30	23	22	21	305	$C_{17}H_{27}NO_2Si$	Nor-meperidinic acid etheyl ester *N*-TMS derivative	BLR	0007
130*	131	291	147	232	292	73	79	99	11	8	1	1	1	1	1	261	$C_{14}H_{19}NO_2Si$	Indoleacetic acid methyl es-ter trimethylsilyl ether	BLR	0169
130*	189	131	79	190	52	103	51	99	57	10	9	7	3	2	0	189	$C_{11}H_{11}NO_2$	3-Indoleacetic acid methyl ester	BLR	0145

Eight Most Intense Peaks — E2 (continued)

Eight most intense peaks								Relative Intensities								Mol wt	Formula	Drug	Source	
130*	203	131	204	144	145	205	117	99	47	9	6	3	3	0	0	201	$C_{12}H_{11}NO_2$	Indolepropionic acid methyl ester	BLR	0157
130*	217	143	131	186	218	144	117	99	43	21	9	6	5	4	0	217	$C_{13}H_{15}NO_2$	3-Indolebutyric acid methyl ester	BLR	0151
128	130*	169	41	129	228	75	69	99	32	14	13	11	8	6	5	228	$C_{11}H_{13}ClO_3$	2-(4-Chlorophenoxy)-2-methylpropionic acid methyl ester (from clofibrate)	BLR	0024
128	130*	169	87	41	129	242	59	99	32	21	14	13	11	8	8	242	$C_{12}H_{15}ClO_3$	Clofibrate	BLR	0022
185	130*	55	56	41	87	57	144	99	86	32	29	28	27	26	25	258	$C_{14}H_{26}O_4$	Dibutyl adipate	DMIT	0186
131*	162	79	103	161	52	132	163	99	75	35	32	19	13	11	8	162	$C_{10}H_{10}O_2$	Cinnamic acid methyl ester	BLR	0144
131*	327	73	143	75	132	328	169	99	31	28	19	16	7	5	4	342	$C_{16}H_{30}N_2O_4Si$	Hydroxyamobarbital 1,3-dimethyl derivative TMS ether	BLR	0079
58	131*	72	71	79	28	42	30	99	18	16	15	11	10	10	8	295	$C_{14}H_{18}ClN_3S$	Chlorothen	DMIT	0055
130	131*	291	147	232	292	73	79	99	11	8	1	1	1	1	1	261	$C_{14}H_{19}NO_2Si$	Indoleacetic acid methyl ester trimethylsilyl ether	BLR	0169
255	131*	254	125	257	256	57	58	99	70	53	38	35	31	15	13	325	$C_{19}H_{20}ClN_3$	Clemizole	DMIT	0037
132*	133	56	115	30	28	77	51	99	82	61	59	43	43	34	31	133	$C_9H_{11}N$	Tranylcypromine	DMIT	0110
84	133*	42	162	161	28	39	51	99	26	20	19	17	17	9	7	162	$C_{10}H_{14}N_2$	Nicotine	DMIT	0056
132	133*	56	115	30	28	77	51	99	82	61	59	43	43	34	31	133	$C_9H_{11}N$	Tranylcypromine	DMIT	0110
444	133*	330	358	445	387	69	105	99	38	33	32	28	26	25	17	472	$C_{18}H_{15}F_7N_2O_5$	Hydroxyphenobarbital 1,3-dimethyl HFB derivative	BLR	0070
105	134*	193	77	162	106	79	147	99	35	10	9	8	7	4	3	193	$C_{10}H_{11}NO_3$	Hippuric acid methyl ester	BLR	0159
135*	77	90	92	51	64	136	63	99	37	35	22	10	9	9	8	223	$C_{11}H_{13}NO_4$	2-Hydroxybenzoylglycine methyl ester methyl ether	BLR	0031
135*	90	77	92	136	134	51	64	99	24	19	13	8	5	4	4	223	$C_{11}H_{13}NO_4$	2-Hydroxybenzoylglycine methyl ester methyl ether	BLR	0106
135*	90	134	136	164	105	77	91	99	48	11	8	6	4	4	2	223	$C_{11}H_{13}NO_4$	Salicyluric acid methyl ester methyl ether	BLR	0153
135*	152	77	92	64	63	136	107	99	56	25	24	15	11	10	9	194	$C_{11}H_{14}O_3$	Propylparaben methyl ether	BLR	0026
135*	166	133	77	105	137	134	136	99	49	46	16	13	12	10	8	166	$C_9H_{10}O_3$	*o*-Methoxybenzoic acid methyl ester	BLR	0135
135*	166	136	167	77	107	92	137	99	58	7	5	4	4	1	0	166	$C_9H_{10}O_3$	*p*-Methoxybenzoic acid methyl ester	BLR	0133
135*	197	73	136	329	417	43	193	99	33	29	16	13	13	7	7	0	—	BHA-type antioxidant	DMIT	0248
91	135*	119	163	120	44	78	92	99	24	22	17	14	12	10	9	163	$C_{10}H_{13}NO$	Primaclone MTB 2	DMIT	0251

136	135*	52	28	137	109	29	18	99	11	8	8	7	6	6	4	136	$C_5H_4N_4O$	Allopurinol	DMIT	0296
164	135*	136	149	97	84	163	91	99	67	56	37	21	20	18	17	164	$C_{10}H_{12}S$	1-Thiophenyl cyclohexene	DMIT	0324
166	135*	107	167	136	77	108	59	99	89	14	10	8	4	3	2	166	$C_9H_{10}O_3$	*m*-Methoxybenzoic acid methyl ester	BLR	0134
136*	81	324	189	55	137	41	42	99	23	20	17	17	15	13	13	324	$C_{20}H_{24}N_2O_2$	Quinidine	DMIT	0264
136*	135	52	28	137	109	29	18	99	11	8	8	7	6	6	4	136	$C_5H_4N_4O$	Allopurinol	DMIT	0296
222	136*	166	194	150	123	207	67	99	61	33	24	21	18	14	13	222	$C_{10}H_{14}N_4O_2$	1,3-Di-ethyl-7-methylxan-thine	BLR	0100
106	137*	78	136	79	105	107	138	99	76	33	18	12	7	6	5	137	$C_7H_7NO_2$	Pyridine-3-carboxylic acid methyl ester	BLR	0140
138*	94	73	108	42	154	137	136	99	69	46	33	33	31	27	23	375	$C_{20}H_{29}NO_4Si$	Scopolamine TMS ether	BLR	0036
94	138*	42	108	136	154	97	137	99	94	57	55	38	36	32	27	303	$C_{17}H_{21}NO_4$	Scopolamine	NIH MSC	3818
94	138*	42	108	136	154	103	137	99	82	62	57	37	35	28	28	303	$C_{17}H_{21}NO_4$	Scopolamine	DMIT	0263
120	138*	43	92	64	65	63	42	99	59	48	34	10	9	8	6	180	$C_9H_8O_4$	Acetyl salicyclic acid	NIH MSC	3763
195	138*	41	53	194	80	110	58	99	93	76	50	48	43	27	26	236	$C_{12}H_{16}N_2O_3$	Allobarbital 1,3-dimethyl derivative	BLR	0047
139*	73	141	111	487	370	75	140	99	35	25	15	15	13	8	8	487	$C_{24}H_{30}ClNO_4Si_2$	1-(4-Chlorobenzoyl)-2-methyl-5-TMSO indole-3-acetic acid TMS ester	BLR	0020
139*	111	141	75	170	50	113	172	99	38	35	26	23	15	12	10	170	$C_8H_7ClO_2$	4-Chlorobenzoic acid methyl ester (from indo-methacin)	BLR	0015
139*	141	73	111	312	429	140	75	99	32	19	19	17	15	8	8	429	$C_{22}H_{24}ClNO_4Si$	Indomethacin TMS ester	BLR	0014
139*	141	111	371	140	75	113	158	99	32	23	16	8	7	7	6	371	$C_{20}H_{18}ClNO_4$	Indomethacin methyl ester	BLR	0013
139*	141	429	111	73	140	431	113	99	32	17	17	12	8	7	6	429	$C_{22}H_{24}ClNO_4Si$	1-4-(Chlorobenzoyl)-2-methyl-5-tmso-indole-3-acetic acid methyl ester	BLR	0019
213	139*	75	169	111	215	141	77	99	70	62	46	39	37	23	20	228	$C_{10}H_{13}ClO_2Si$	4-Chlorobenzoic acid TMS ester (from indomethacin)	BLR	0016
140*	83	82	124	96	97	167	125	99	88	59	41	24	24	22	21	307	$C_{21}H_{25}NO$	Benztropine	DMIT	0050
42	140*	41	49	84	112	43	85	99	39	10	8	8	8	7	5	140	$C_6H_{12}N_4$	Methenamine	DMIT	0260
155	140*	83	98	55	56	57	97	99	76	53	46	33	8	7	7	183	$C_{10}H_{17}NO_2$	Methyprylon	DMIT	0009
141*	69	113	45	68	42	41	54	99	98	95	87	60	40	33	30	141	—	Oxacillin MTB 4	DMIT	0150
141*	70	55	41	42	69	112	126	99	40	36	10	10	8	6	6	169	$C_9H_{15}NO_2$	Ethosuximide *N*-ethyl de-rivative	BLR	0111
141*	123	42	41	151	55	169	96	99	93	57	48	35	32	32	29	401	$C_{21}H_{24}ClN_3OS$	Pipamazine	DMIT	0137
141*	156	55	41	28	27	29	142	99	74	45	43	35	30	26	23	212	$C_{10}H_{16}N_2O_3$	Butethal	DMIT	0172
139	141*	73	111	312	429	140	75	99	32	19	19	17	15	8	8	429	$C_{22}H_{24}ClNO_4Si$	Indomethacin TMS ester	BLR	0014
139	141*	111	371	140	75	113	158	99	32	23	16	8	7	7	6	371	$C_{20}H_{18}ClNO_4$	Indomethacin methyl ester	BLR	0013
139	141*	429	111	73	140	431	113	99	32	17	17	12	8	7	6	429	$C_{22}H_{24}ClNO_4Si$	1-(4-Chlorobenzoyl)-2-methyl-5-TMSO-indole-3-acetic acid methyl ester	BLR	0019

Eight Most Intense Peaks — E2 (continued)

Eight most intense peaks								Relative Intensities								Mol wt	Formula	Drug	Source	
156	141*	41	43	155	157	98	169	99	57	11	9	9	8	7	5	198	$C_9H_{14}N_2O_3$	Probarbital	DMIT	0052
156	141*	41	157	55	98	142	43	99	74	14	14	13	10	9	8	240	$C_{12}H_{20}N_2O_3$	Hexethal	NIH MSC	3762
156	141*	43	41	157	55	69	71	99	70	32	22	20	13	10	10	226	$C_{11}H_{18}N_2O_3$	Pentobarbital	NIH MSC	3804
156	141*	43	41	157	55	71	39	99	55	30	22	22	10	7	7	226	$C_{11}H_{18}N_2O_3$	Pentobarbital	DMIT	0027
156	141*	98	155	55	112	41	83	99	89	20	20	18	18	16	11	184	$C_8H_{12}N_2O_3$	Barbital	NIH MSC	3770
156	141*	157	41	43	142	197	55	99	53	19	15	11	10	9	9	226	$C_{11}H_{18}N_2O_3$	Amobarbital	DMIT	0025
156	141*	157	41	55	43	142	197	99	67	29	16	15	12	12	10	226	$C_{11}H_{18}N_2O_3$	Amobarbital	NIH MSC	3767
156	141*	157	69	45	41	197	43	99	58	43	37	33	23	21	15	242	$C_{11}H_{18}N_2O_4$	Pentobarbital MTB 1	DMIT	0011
207	141*	81	67	79	208	80	77	99	28	18	17	14	13	10	7	236	$C_{12}H_{16}N_2O_3$	Cyclobarbital	NIH MSC	3780
221	141*	81	79	222	41	67	93	99	22	16	14	14	9	9	8	250	$C_{13}H_{18}N_2O_3$	Heptabarbital	NIH MSC	3788
221	141*	222	81	79	41	93	95	99	18	14	12	10	7	7	6	250	$C_{13}H_{18}N_2O_3$	Heptabarbital	DMIT	0026
254	143*	411	70	255	42	380	157	99	55	51	37	35	30	27	24	411	$C_{23}H_{29}N_3O_2S$	Acetophenazine	DMIT	0091
268	143*	425	55	70	41	40	269	99	61	58	54	50	48	47	41	425	$C_{24}H_{31}N_3O_2S$	Carphenazine	MIT	0093
144*	43	103	44	296	159	146	77	99	59	29	27	26	24	20	20	296	—	Oxacillin MTB 5	DMIT	0159
144*	77	202	43	160	51	44	185	99	85	74	60	41	37	32	32	202	$C_{11}H_{10}N_2O_2$	Oxacillin MTB 2	DMI	0274
146*	45	117	86	288	118	57	232	99	74	30	25	24	19	19	14	320	$C_{16}H_{20}N_2O_5$	*N,N'*-dimethoxymethyl phenobarbital	DMIT	0293
146*	73	232	334	247	117	246	147	99	59	46	46	33	20	20	19	362	$C_{18}H_{30}N_2O_2Si_2$	Primidone 1,3-di-TMS derivative	BLR	0090
146*	161	103	160	104	44	91	77	99	76	69	48	30	28	28	22	161	$C_{10}H_{11}NO$	Oxacillin MTB 1	DMIT	0273
146*	190	117	118	189	161	103	115	99	90	59	42	34	31	22	16	218	$C_{12}H_{14}N_2O_2$	Primaclone	NIH MSC	3809
146*	190	117	189	161	118	91	103	99	76	55	30	26	25	18	17	218	$C_{12}H_{14}N_2O_2$	Primaclone	BLR	0078
146*	247	117	45	232	118	275	230	99	69	54	48	33	32	30	24	290	$C_{15}H_{18}N_2O_4$	*N*-methyl, *N'*-methoxymethyl phenobarbital	DMIT	0294
146*	260	118	117	103	261	91	232	99	99	38	34	26	24	17	16	288	$C_{16}H_{20}N_2O_3$	Phenobarbital 1,3-diethyl derivative	BLR	0068
117	146*	43	91	32	115	39	116	99	54	48	47	38	34	30	29	218	$C_{12}H_{14}N_2O_2$	Primaclone	DMIT	0161
232	146*	117	175	118	233	103	188	99	44	41	41	40	22	18	18	260	$C_{14}H_{16}N_2O_3$	Phenobarbital 1,3-dimethyl derivative	BLR	0084
232	146*	118	175	117	233	103	188	99	29	27	26	21	15	10	10	260	$C_{14}H_{16}N_2O_3$	*N,N*-dimethyl phenobarbital	DMIT	0295
235	146*	118	117	77	236	178	103	99	27	27	26	19	14	13	12	263	$C_{14}H_{13}D_3N_2O_3$	*N*-methyl, *N*-d3-methyl phenobarbital	DMIT	0322
246	146*	117	118	247	175	103	77	99	46	33	28	20	20	15	14	274	$C_{15}H_{18}N_2O_3$	Mephobarbital 3-ethyl derivative	BLR	0072
147*	165	201	167	166	105	203	117	99	75	74	62	48	47	37	35	432	$C_{28}H_{33}ClN_2$	Buclizine	DMIT	0076
147*	189	73	74	148	75	190	146	99	62	29	17	16	14	11	10	204	$C_7H_{20}N_2OSi_2$	Urea d-TMS derivative	BLR	0032

73	147*	164	296	253	74	75	149	99	28	14	12	10	7	5	5	296	$C_{14}H_{24}O_3Si_2$	*o*-Hydroxyphenylacetic acid trimethylsilyl ether	BLR	0167
163	148*	91	103	44	117	120	164	99	77	41	26	19	15	13	11	206	$C_{11}H_{14}N_2O_2$	Primaclone MTB 1	DMIT	0250
149*	41	29	28	57	27	56	104	99	27	25	22	10	10	8	8	278	$C_{16}H_{22}O_4$	Dibutyl phthalate	DMIT	0107
149*	57	41	43	70	71	167	28	99	47	30	30	28	28	26	23	390	$C_{24}H_{38}O_4$	Diocytl phthalate	DMIT	0095
149*	57	167	43	71	70	41	55	99	34	32	25	25	19	15	15	390	$C_{24}H_{38}O_4$	Diocytl phthalate	NIH MSC	3792
149*	150	41	57	56	76	104	205	99	13	10	7	6	5	5	5	336	$C_{18}H_{24}O_6$	Butyl carbobutoxymethyl phthalate	NIH MSC	3771
149*	150	41	223	57	205	56	104	99	22	12	10	9	9	8	8	278	$C_{16}H_{22}O_4$	Di-*N*-butyl phthalate	NIH MSC	3783
149*	263	57	77	104	41	133	76	99	29	21	15	14	14	11	9	322	$C_{18}H_{26}O_5$	Butyl butoxyethyl phthalate	DMIT	0243
72	149*	269	255	197	73	254	340	99	16	5	4	4	3	2	2	340	$C_{20}H_{24}N_2OS$	Propiomazine	DMIT	0019
107	149*	57	78	108	79	77	72	99	78	70	60	56	42	37	37	261	$C_{16}H_{23}NO_2$	Ethoheptazine	DMIT	0057
91	150*	59	92	151	119	118	105	99	61	10	8	6	3	2	1	150	$C_9H_{10}O_2$	Phenylacetic acid methyl ester	BLR	0139
91	150*	65	92	59	39	107	108	99	65	20	14	12	11	9	7	150	$C_9H_{10}O_2$	Methylphenidate MTB 1	DMIT	0253
149	150*	41	57	56	76	104	205	99	13	10	7	6	5	5	5	336	$C_{18}H_{24}O_6$	Butyl carbobutoxymethyl phthalate	NIH MSC	3771
149	150*	41	223	57	205	56	104	99	22	12	10	9	9	8	8	278	$C_{16}H_{22}O_4$	Di-*N*-butyl phthalate	NIH MSC	3783
222	150*	194	179	166	207	109	43	99	77	29	25	22	22	21	19	222	$C_{10}H_{14}N_4O_2$	3,7-Diethyl-1-methylxanthine	BLR	0098
235	150*	165	236	79	137	250	164	99	25	24	21	18	16	14	12	250	$C_{13}H_{18}N_2O_3$	Hexobarbital-2(or 4)-methyl derivative	BLR	0063
151*	210	152	211	107	195	59	153	99	74	8	7	1	1	0	0	210	$C_{11}H_{14}O_4$	3,4-Dimethoxyphenylacetic acid methyl ester	BLR	0136
151*	224	164	152	225	149	165	193	99	68	12	8	7	6	3	2	224	$C_{12}H_{16}O_4$	3,4-Dimethoxyphenylpropionic acid methyl ester	BLR	0152
109	151*	43	79	80	53	108	110	99	26	26	19	12	10	10	8	151	$C_8H_9NO_2$	Acetaminophen	DMIT	0116
109	151*	43	80	81	108	86	57	99	30	22	13	11	10	9	8	151	$C_8H_9NO_2$	Acetaminophen	NIH MSC	3765
119	151*	120	92	152	91	93	65	99	91	22	16	7	2	2	1	151	$C_8H_9NO_2$	*o*-Aminobenzoic acid methyl ester	BLR	0138
120	151*	92	121	152	65	93	122	99	86	11	6	6	1	1	0	151	$C_8H_9NO_2$	*p*-Aminobenzoic acid methyl ester	BLR	0137
222	151*	121	122	123	105	77	93	99	37	23	22	21	20	17	16	254	$C_{13}H_6D_8N_2O_3$	*N*-d3-methyl phenobarbital (D5-ethyl)	DMIT	0321
239	151*	121	179	122	123	105	77	99	36	26	24	19	19	16	15	271	$C_{14}H_5D_{11}N_2O_3$	*N,N*-d3-methyl phenobarbital (D5-ethyl)	DMIT	0320
152*	42	98	154	153	28	174	69	99	61	58	42	31	30	27	24	273	$C_{11}H_{12}ClNO_3S$	Chlormezanone	DMIT	0122
44	152*	28	137	77	65	91	78	99	21	12	8	6	5	5	4	196	$C_{11}H_{18}NO_2$	2,5-Dimethoxyamphetamine	DMIT	0183
120	152*	43	121	92	63	65	64	99	81	42	41	39	12	11	9	194	$C_{10}H_{10}O_4$	Acetylsalicylic acid methyl ester	BLR	0029
121	152*	93	65	39	122	63	153	99	36	21	16	10	8	4	3	152	$C_8H_8O_3$	Methyl *p*-hydroxybenzoate	DMIT	0185

Eight Most Intense Peaks — E2 (continued)

Eight most intense peaks								Relative Intensities								Mol wt	Formula	Drug	Source	
135	152*	77	92	64	63	136	107	99	56	25	24	15	11	10	9	194	$C_{11}H_{14}O_3$	Propylparaben methyl ether	BLR	0026
154*	41	86	70	44	42	124	98	99	34	16	14	14	12	11	10	0	—	F requent urine constituent 4	DMIT	0246
70	154*	72	43	125	44	41	55	99	95	39	32	32	31	30	27	0	—	Frequent urine constituent 3	DMIT	0231
155*	140	83	98	55	56	57	97	99	76	53	46	33	8	7	7	183	$C_{10}H_{17}NO_2$	Methyprylon	DMIT	0009
170	155*	41	169	55	112	39	83	99	76	30	24	22	21	17	16	198	$C_9H_{14}N_2O_3$	Metharbital	DMIT	0130
156*	141	41	43	155	157	98	169	99	57	11	9	9	8	7	5	198	$C_9H_{14}N_2O_3$	Probarbital	DMIT	0052
156*	141	41	157	55	98	142	43	99	74	14	14	13	10	9	8	240	$C_{12}H_{20}N_2O_3$	Hexethal	NIH MSC	3762
156*	141	43	41	157	55	69	71	99	70	32	22	20	13	10	10	226	$C_{11}H_{18}N_2O_3$	Pentobarbital	NIH MSC	3804
156*	141	43	41	157	55	71	39	99	55	30	22	22	10	7	7	226	$C_{11}H_{18}N_2O_3$	Pentobarbital	DMIT	0027
156*	141	98	155	55	112	41	83	99	89	20	20	18	18	16	11	184	$C_8H_{12}N_2O_3$	Barbital	NIH MSC	3770
156*	141	157	41	43	142	197	55	99	53	19	15	11	10	9	9	226	$C_{11}H_{18}N_2O_3$	Amobarbital	DMIT	0025
156*	141	157	41	55	43	142	197	99	67	29	16	15	12	12	10	226	$C_{11}H_{18}N_2O_3$	Amobarbital	NIH MSC	3767
156*	141	157	69	45	41	197	43	99	58	43	37	33	23	21	15	242	$C_{11}H_{18}N_2O_4$	Pentobarbital MTB 1	DMIT	0011
92	156*	108	65	162	119	174	189	99	81	64	50	46	37	30	16	253	$C_{10}H_{11}N_3O_3S$	Sulfamethoxazole	DMIT	0302
141	156*	55	41	28	27	29	142	99	74	45	43	35	30	26	23	212	$C_{10}H_{16}N_2O_3$	Butethal	DMIT	0172
59	157*	156	43	141	41	55	180	99	68	64	37	34	30	26	24	242	$C_{11}H_{18}N_2O_4$	Amobarbital MTB 1	DMIT	0267
129	157*	128	130	187	158	156	188	99	21	15	11	8	2	1	0	187	$C_{11}H_9NO_2$	Quinoline-2-carboxylic acid methyl ester	BLR	0161
158*	43	98	41	115	71	159	83	99	65	34	29	28	25	25	24	733	$C_{37}H_{67}NO_{13}$	Erythromycin A	ABB	0004
129	158*	130	143	115	128	91	77	99	97	80	68	66	45	31	24	158	$C_{12}H_{14}$	1-Phenyl cyclohexene	DMIT	0127
91	159*	82	68	158	42	65	118	99	96	90	54	51	26	19	16	159	$C_{11}H_{13}N$	*N*-methyl-*N*-2-propynyl-benzylamine	ABB	0009
160*	54	105	106	159	107	82	39	99	69	55	49	28	27	25	22	0	—	Blood 324, drug 1	DMIT	0234
160*	219	161	220	145	74	69	83	99	64	11	8	2	1	0	0	219	$C_{12}H_{13}NO_3$	5-Methoxyindoleacetic acid methyl ester	BLR	0155
160*	235	266	77	251	146	69	76	99	93	89	73	71	59	50	47	266	$C_{16}H_{14}N_2O_2$	Methaqualone MTB 1	DMIT	0215
160*	266	235	77	251	146	247	76	99	57	51	44	34	28	27	23	266	$C_{16}H_{14}N_2O_2$	Methaqualone MTB 1	DMIT	0305
146	161*	103	160	104	44	91	77	99	76	69	48	30	28	28	22	161	$C_{10}H_{11}NO$	Oxacillin MTB 1	DMIT	0273
162*	63	164	98	99	49	62	73	99	64	62	24	24	21	17	15	162	$C_6H_4Cl_2O$	2,4-Dichlorophenol	DMIT	0318
131	162*	79	103	161	52	132	163	99	75	35	32	19	13	11	8	162	$C_{10}H_{10}O_2$	Cinnamic acid methyl ester	BLR	0144
285	162*	42	28	44	31	215	70	99	63	58	55	38	34	34	28	285	$C_{17}H_{19}NO_3$	Morphine	DMIT	0103
285	162*	215	42	286	124	284	174	99	38	31	21	20	19	18	16	285	$C_{17}H_{19}NO_3$	Morphine	NIH MSC	3800
299	162*	229	42	214	300	124	188	99	51	39	34	24	21	21	20	299	$C_{18}H_{21}NO_3$	Codeine	DMIT	0042
299	162*	229	124	300	214	298	42	99	38	28	22	20	15	15	14	299	$C_{18}H_{21}NO_3$	Codeine	NIH MSC	3779

163*	148	91	103	44	117	120	164	99	77	41	26	19	15	13	11	206	$C_{11}H_{14}N_2O_2$	Primaclone MTB 1	DMIT	0250
164*	135	136	149	97	84	163	91	99	67	56	37	21	20	18	17	164	$C_{10}H_{12}S$	1-Thiophenyl cyclohexene	DMIT	0324
104	164*	91	105	133	103	165	78	99	41	29	23	8	6	4	3	164	$C_{10}H_{12}O_2$	β-Phenylpropionic acid methyl ester	BLR	0160
250	164*	222	194	235	251	207	150	99	45	34	31	21	13	12	10	250	$C_{12}H_{18}N_4O_2$	8-Methylxanthine 1,3,7-triethyl derivative	BLR	0101
165*	194	147	77	121	176	118	51	99	76	48	47	38	34	32	30	194	$C_{10}H_{10}O_4$	Meconin	DMIT	0213
58	165*	255	359	166	73	199	45	99	11	9	8	7	7	4	3	359	$C_{21}H_{29}NS_2$	Captodiamine	DMIT	0088
97	165*	164	206	84	249	135	136	99	87	76	64	45	44	35	32	249	$C_{15}H_{23}NS$	1-(1-Thiophenyl cyclohexyl) piperidine	DMIT	0325
147	165*	201	167	166	105	203	117	99	75	74	62	48	47	37	35	432	$C_{28}H_{33}ClN_2$	Buclizine	DMIT	0076
167	165*	117	119	83	169	130	132	99	91	90	89	58	54	43	42	200	C_2HCl_5	Pentachloroethane	NIH MSC	3802
193	165*	121	120	194	93	65	92	99	67	57	21	12	11	8	8	358	$C_{19}H_{18}O_7$	Sal ethyl carbonate	DMIT	0064
196	165*	197	181	166	59	94	121	99	51	9	6	4	3	1	1	196	$C_{10}H_{12}O_4$	3,4-Dimethoxybenzoic acid methyl ester	BLR	0156
408	165*	409	243	167	241	166	410	99	59	33	30	23	20	16	12	566	$C_{32}H_{24}Br_2$	Alpha, alpha-dibromo-alpha, alpha, alpha, alpha-tetraphenylxylene	DMIT	0331
166*	73	149	168	267	74	193	55	99	38	13	12	12	7	7	7	0	—	Blood 324, drug 2	DMIT	0235
166*	135	107	167	136	77	108	59	99	89	14	10	8	4	3	2	166	$C_9H_{10}O_3$	*m*-Methoxybenzoic acid methyl ester	BLR	0134
44	166*	151	57	43	91	135	209	99	40	12	7	6	5	5	4	209	$C_{12}H_{19}NO_2$	2,5-Dimethoxy-4-methylamphetamine	DMIT	0180
135	166*	133	77	105	137	134	136	99	49	46	16	13	12	10	8	166	$C_9H_{10}O_3$	*o*-Methoxybenzoic acid methyl ester	BLR	0135
135	166*	136	167	77	107	92	137	99	58	7	5	4	4	1	0	166	$C_9H_{10}O_3$	*p*-Methoxybenzoic acid methyl ester	BLR	0133
167*	41	39	124	28	168	29	43	99	67	44	41	39	35	30	24	210	$C_{10}H_{14}N_2O_3$	Aprobarbital	DMIT	0173
167*	41	124	80	39	32	28	166	99	89	68	58	54	48	45	41	208	$C_{10}H_{12}N_2O_3$	Allobarbital	DMIT	0114
167*	41	168	124	43	97	169	96	99	40	37	32	28	25	18	10	210	$C_{10}H_{14}N_2O_3$	Aprobarbital	NIH MSC	3766
167*	165	117	119	83	169	130	132	99	91	90	89	58	54	43	42	200	C_2HCl_5	Pentachloroethane	NIH MSC	3802
167*	168	41	97	124	195	153	53	99	74	43	27	24	17	15	14	224	$C_{11}H_{16}N_2O_3$	Talbutal	NIH MSC	3822
167*	211	181	180	151	45	168	108	99	36	32	27	25	24	23	21	0	—	Blood 315 drug	DMIT	0238
41	167*	168	39	125	97	44	53	99	83	52	38	32	28	23	19	224	$C_{11}H_{16}N_2O_3$	Talbutal	DMIT	0163
43	167*	237	41	39	124	168	55	99	99	99	65	60	51	41	34	316	$C_{12}H_{17}BrN_2O_3$	5-(2-Bromoallyl)-5-(1 methyl butyl) barbituric acid	DMIT	0256
168	167*	41	43	124	97	169	195	99	85	37	33	21	19	18	16	238	$C_{12}H_{18}N_2O_3$	Secobarbital	NIH MSC	3819
168	167*	41	45	70	169	43	69	99	71	67	55	47	41	37	33	254	$C_{12}H_{18}N_2O_4$	Secobarbital MTB 1	DMIT	0268
168	167*	41	181	124	97	169	141	99	66	38	17	15	11	11	10	224	$C_{11}H_{16}N_2O_3$	Butalbital	DMI	0067
168	167*	43	97	169	41	124	153	99	81	26	25	16	16	14	13	246	$C_7H_7BrN_2O_3$	5-(2-Bromoallyl) barbituric acid	DMIT	0257

Eight Most Intense Peaks — E2 (continued)

Eight most intense peaks								Relative Intensities								Mol wt	Formula	Drug	Source	
182	167*	183	168	44	106	79	77	99	47	40	12	12	11	9	9	199	$C_{13}H_{13}NO$	Doxylamine MTB 1	DMIT	0239
199	167*	198	166	99	154	69	77	99	52	21	18	8	7	5	4	199	$C_{12}H_9NS$	Phenothiazine	DMIT	0147
168*	41	124	181	167	141	97	98	99	34	25	20	19	17	16	12	224	$C_{11}H_{16}N_2O_3$	Butalbital	NIH MSC	3817
168*	167	41	43	124	97	169	195	99	85	37	33	21	19	18	16	238	$C_{12}H_{18}N_2O_3$	Secobarbital	NIH MSC	3819
168*	167	41	45	70	169	43	69	99	71	67	55	47	41	37	33	254	$C_{12}H_{18}N_2O_4$	Secobarbital MTB 1	DMIT	0268
168*	167	41	181	124	97	169	141	99	66	38	17	15	11	11	10	224	$C_{11}H_{16}N_2O_3$	Butalbital	DMIT	0067
168*	167	43	97	169	41	124	153	99	81	26	25	16	16	14	13	246	$C_7H_7BrN_2O_3$	5-(2-Bromoallyl) barbituric acid	DMIT	0257
168*	170	113	78	63	169	43	76	99	33	18	16	14	10	9	9	168	$C_7H_5ClN_2O$	Zoxazolamine	NIH MSC	3781
41	168*	167	43	39	55	97	124	99	79	72	68	56	41	34	29	238	$C_{12}H_{18}N_2O_3$	Secobarbital	DMIT	0148
167	168*	41	97	124	195	153	53	99	74	43	27	24	17	15	14	224	$C_{11}H_{16}N_2O_3$	Talbutal	NIH MSC	3822
169*	73	170	75	249	233	79	171	99	47	20	19	12	11	8	7	338	$C_{16}H_{26}N_2O_4Si$	Hydroxyhexobarbital 3-methyl derivative TMS ether	BLR	0065
169*	78	113	171	51	63	44	50	99	57	55	31	20	19	17	16	169	$C_7H_4ClNO_2$	Chlorzoxazone	DMIT	0113
169*	184	41	43	183	69	112	55	99	84	27	18	14	14	13	12	254	$C_{13}H_{22}N_2O_3$	Pentobarbital 1,3-dimethyl derivative	BLR	0082
169*	184	41	112	185	69	183	55	99	90	24	15	9	9	8	8	240	$C_{12}H_{20}N_2O_3$	Butabarbital 1,3-dimethyl derivative	BLR	0057
169*	184	140	185	126	226	55	41	99	80	25	23	16	15	14	13	254	$C_{13}H_{22}N_2O_3$	Amobarbital 1,2 ester(or 4)-dimethyl derivative	BLR	0051
169*	184	183	126	112	83	41	40	99	99	35	26	21	15	14	13	212	$C_{10}H_{16}N_2O_3$	Barbital 1,3-dimethyl derivative	BLR	0055
169*	184	185	170	55	112	183	69	99	86	17	15	12	12	10	9	254	$C_{13}H_{22}N_2O_3$	Amobarbital 1,3-dimethyl derivative	BLR	0050
169*	194	43	195	409	71	69	170	99	65	48	26	16	13	11	11	692	$C_{22}H_{22}F_{14}N_2O_7$	Dihydroxysecobarbital 1,3-dimethyl derivative di-HFB derivative	BLR	0076
184	169*	69	223	185	41	183	224	99	76	74	42	41	25	19	12	478	$C_{18}H_{21}F_7N_2O_5$	Hydroxypentobarbital 1,3-dimethyl derivative HFB derivative	BLR	0066
170*	155	41	169	55	112	39	83	99	76	30	24	22	21	17	16	198	$C_9H_{14}N_2O_3$	Metharbital	DMIT	0130
168	170*	113	78	63	169	43	76	99	33	18	16	14	10	9	9	168	$C_7H_5ClN_2O$	Zoxazolamine	NIH MSC	3781
172*	187	42	84	129	144	44	91	99	66	65	54	48	39	34	32	261	$C_{16}H_{23}NO_2$	Alphaprodine	DMIT	0153
172*	187	84	57	42	188	44	43	99	56	36	30	28	22	21	20	261	$C_{16}H_{23}NO_2$	Alphaprodine	NIH MSC	3764

174*	233	175	159	131	130	234	158	99	36	14	13	9	5	5	5	233	$C_{13}H_{15}NO_3$	2-Methyl-5-methoxyindole-3-acetic acid methyl ester (from indomethacin)	BLR	0017
174*	291	73	175	159	131	75	158	99	17	13	13	7	6	5	5	291	$C_{15}H_{21}NO_3Si$	2-Methyl-5-methoxyindole-3-acetic acid TMS ester (from indomethacin)	BLR	0021
174*	382	161	187	159	175	383	41	99	60	31	25	22	19	18	15	382	$C_{27}H_{42}O$	3,5-Cholestadiene-7-one	DMIT	0254
104	175*	44	77	43	41	55	56	99	99	41	31	29	27	18	15	204	$C_{11}H_{12}N_2O_2$	Mephenytoin MTB 1	DMIT	0245
176*	107	90	89	77	79	105	70	99	96	56	54	45	41	33	33	176	$C_9H_8N_2O_2$	2-imino-5-phenyl-4-oxazolidinone	ABB	0007
98	176*	147	42	119	51	41	91	99	28	11	10	6	6	5	4	176	$C_{10}H_{12}N_2O$	Cotinine (from nicotine)	DMIT	0206
336	176*	334	338	88	337	175	168	99	76	51	48	38	16	12	12	334	$C_{14}H_8Br_2$	9,10-Dibromoanthracene	DMIT	0332
179*	73	89	180	223	181	195	74	99	74	19	16	8	4	4	3	238	$C_{12}H_{18}O_3Si$	Mandelic acid methyl ester TMS ether	BLR	0181
179*	238	163	180	223	73	239	181	99	55	15	15	15	12	10	4	238	$C_{12}H_{18}O_3Si$	*p*-Hydroxyphenylacetic acid methyl ester TMS ether	BLR	0141
179*	252	180	253	192	73	131	89	99	60	15	11	10	9	9	8	252	$C_{13}H_{20}O_3Si$	*p*-Hydroxyphenylpropionic acid methyl ester TMS ether	BLR	0174
179*	340	267	73	341	180	342	268	99	84	31	24	22	14	7	6	340	$C_{16}H_{28}O_4Si_2$	3,4-Dihydroxyphenylpropionic acid methyl ester trimethylsilyl ether	BLR	0177
209	179*	210	135	59	89	161	193	99	24	16	14	13	12	10	9	224	$C_{11}H_{16}O_3Si$	Salicylic acid methyl ester TMS ether	BLR	0028
209	179*	210	193	89	161	177	211	99	14	14	6	5	4	4	4	224	$C_{11}H_{16}O_3Si$	*o*-Hydroxybenzoic acid methyl ester TMS ether	BLR	0150
326	179*	267	327	73	180	328	268	99	84	27	25	16	11	9	5	326	$C_{15}H_{26}O_4Si_2$	3,4-Dihydroxyphenyl acetic acid methyl ester TMS ether	BLR	0183
180*	77	104	209	223	252	51	181	99	56	56	56	44	33	27	25	252	$C_{15}H_{12}N_2O_2$	Diphenylhydantoin	DMIT	0168
180*	95	68	53	123	96	181	94	99	51	36	13	12	8	8	5	180	$C_7H_8N_4O_2$	Theophylline	DMIT	0201
180*	104	266	77	237	57	209	71	99	60	48	40	38	37	25	22	266	$C_{16}H_{14}N_2O_2$	Diphenylhydantoin 3-methyl derivative	BLR	0038
180*	109	55	67	82	137	181	108	99	25	23	21	20	9	8	6	180	$C_7H_8N_4O_2$	Theobromine	DMIT	0191
180*	121	181	122	59	91	148	182	99	74	10	7	2	2	1	1	180	$C_{10}H_{12}O_3$	*m*-Methoxyphenylacetic acid methyl ester	BLR	0131
180*	181	77	152	182	51	44	90	99	25	11	9	8	8	6	5	181	$C_{13}H_{11}N$	Doxylamine MTB 2	DMIT	0240
180*	209	104	223	252	77	181	165	99	52	51	48	43	36	26	25	252	$C_{15}H_{12}N_2O_2$	Diphenylhydantoin	NIH MSC	3785
180*	209	280	77	181	104	251	165	99	58	36	33	32	28	24	22	280	$C_{17}H_{16}N_2O_2$	5,5-Diphenylhydantoin 3-ethyl derivative	BLR	0092
180*	266	237	104	209	77	189	181	99	85	52	49	40	29	28	25	266	$C_{16}H_{14}N_2O_2$	Diphenylhydantoin MTB 1	DMIT	0259

Eight Most Intense Peaks — E2 (continued)

Eight most intense peaks								Relative Intensities								Mol wt	Formula	Drug	Source	
180*	308	279	208	77	251	165	104	99	80	78	61	49	26	26	20	308	$C_{19}H_{20}N_2O_2$	5,5-Diphenylhydantoin 1,3-diethyl derivative	BLR	0093
58	180*	71	182	44	183	167	181	99	79	70	50	41	39	24	21	270	$C_{17}H_{22}N_2O$	Doxylamine (GC)	DMIT	0241
121	180*	91	181	122	148	107	93	99	87	19	9	8	6	3	2	180	$C_{10}H_{12}O_3$	*o*-Methoxyphenylacetic acid methyl ester	BLR	0163
121	180*	122	181	0	0	0	0	99	49	7	4	0	0	0	0	180	$C_{10}H_{12}O_3$	*p*-Methoxyphenylacetic acid methyl ester	BLR	0130
208	180*	109	67	42	137	55	179	99	30	29	16	14	13	12	11	208	$C_9H_{12}N_4O_2$	Theobromine 1-ethyl derivative	BLR	0097
296	180*	267	210	104	77	134	297	99	52	42	33	31	23	21	20	296	$C_{17}H_{16}N_2O_3$	3-Hydroxydiphenylhydantoin methyl ether 3-methyl derivative	BLR	0040
181*	86	72	182	85	152	108	99	99	66	61	24	12	11	9	8	419	$C_{21}H_{26}BrNO_3$	Methantheline bromide	DMIT	0053
181*	223	166	73	43	45	208	75	99	70	68	58	54	25	22	17	223	$C_{11}H_{17}NO_2Si$	Acetaminophen TMS ether	BLR	0011
180	181*	77	152	182	51	44	90	99	25	11	9	8	8	6	5	181	$C_{13}H_{11}N$	Doxylamine MTB 2	DMIT	0240
182*	30	167	181	211	151	183	148	99	88	54	49	21	18	13	11	211	$C_{11}H_{17}NO_3$	Mescaline	DMIT	0179
182*	82	83	105	94	77	96	303	99	95	41	35	33	31	27	25	303	$C_{17}H_{21}NO_4$	Cocaine	NIH MSC	3778
182*	167	183	168	44	106	79	77	99	47	40	12	12	11	9	9	199	$C_{13}H_{13}NO$	Doxlyamine MTB 1	DMIT	0239
183*	77	308	184	105	252	93	309	99	67	52	40	18	11	11	8	308	$C_{19}H_{20}N_2O_2$	Phenylbutazone	DMIT	0077
184*	40	169	140	126	183	44	55	99	73	51	42	24	20	17	17	212	$c_{10}H_{16}N_2O_3$	Barbital 1,2(or 4)-dimethyl derivative	BLR	0056
184*	112	42	55	170	183	212	58	99	21	19	19	17	13	13	12	240	$C_{12}H_{20}N_2O_3$	Butethal 1,3-dimethyl derivative	BLR	0058
184*	169	69	223	185	41	183	224	99	76	74	42	41	25	19	12	478	$C_{18}H_{21}F_7N_2O_5$	Hydroxypentobarbital 1,3-dimethyl derivative HFB derivative	BLR	0066
184*	185	92	65	108	186	66	183	99	83	60	40	26	11	9	9	249	$C_{11}H_{11}N_3O_2S$	Sulphapyridine	NIH MSC	3821
169	184*	41	43	183	69	112	55	99	84	27	18	14	14	13	12	254	$C_{13}H_{22}N_2O_3$	Pentobarbital 1,3-dimethyl derivative	BLR	0082
169	184*	41	112	185	69	183	55	99	90	24	15	9	9	8	8	240	$C_{12}H_{20}N_2O_3$	Butabarbital 1,3-dimethyl derivative	BLR	0057
169	184*	140	185	126	226	55	41	99	80	25	23	16	15	14	13	254	$C_{13}H_{22}N_2O_3$	Amobarbital 1,2(or 4)-dimethyl derivative	BLR	0051
169	184*	183	126	112	83	41	40	99	99	35	26	21	15	14	13	212	$C_{10}H_{16}N_2O_3$	Barbital 1,3-dimethyl derivative	BLR	0055
169	184*	185	170	55	112	183	69	99	86	17	15	12	12	10	9	254	$C_{13}H_{22}N_2O_3$	Amobarbital 1,3-dimethyl derivative	BLR	0050

185*	43	129	41	57	259	157	112	99	61	59	45	44	44	23	17	402	$C_{20}H_{34}O_8$	Acetyl tri-*N*-butyl citrate	DMI	0314
185*	130	55	56	41	87	57	144	99	86	32	29	28	27	26	25	258	$C_{14}H_{26}O_4$	Dibutyl adipate	DMIT	0186
185*	214	184	257	109	104	77	82	99	73	73	63	62	59	47	43	177	$C_{11}H_7D_5N_2$	Diphenylhydantoin (D5-phenyl)	DMIT	0315
184	185*	92	65	108	186	66	183	99	83	60	40	26	11	9	9	249	$C_{11}H_{11}N_3O_2S$	Sulphapyridine	NIH MSC	3821
246	185*	445	70	125	154	213	87	99	93	81	73	65	59	55	50	445	$C_{23}H_{28}ClN_3O_2S$	Thiopropazate	DMIT	0046
172	187*	42	84	129	144	44	91	99	66	65	54	48	39	34	32	261	$C_{16}H_{23}NO_2$	Alphaprodine	DMIT	0153
172	187*	84	57	42	188	44	43	99	56	36	30	28	22	21	20	261	$C_{16}H_{23}NO_2$	Alphaprodine	NIH MSC	3764
188*	91	105	119	65	189	104	78	99	70	52	25	14	12	11	7	188	$C_{11}H_{12}N_2O$	1-(*p*-Tolyl)-3-methyl-pyrazol-5-one	ABB	0008
188*	96	77	56	28	39	55	105	99	63	63	48	42	26	26	24	188	$C_{11}H_{12}N_2O$	Antipyrine	DMIT	0030
188*	96	77	56	105	55	187	189	99	59	44	28	19	15	13	13	188	$C_{11}H_{12}N_2O$	Antipyrine	NIH MSC	3769
203	188*	232	117	204	40	115	70	99	53	23	16	13	11	10	9	260	$C_{14}H_{16}N_2O_3$	Phenobarbital 1,2(or 4)-dimethyl derivative	BLR	0085
247	188*	160	63	215	43	106	216	99	59	38	31	30	25	20	20	247	$C_{14}H_{17}NO_3$	1,2-Dimethyl-5-methoxyindole-3-acetic acid methyl ester (from indomethacin)	BLR	0018
189*	104	77	190	51	105	103	132	99	85	22	12	9	8	5	5	218	$C_{12}H_{14}N_2O_2$	Mephenytoin	DMIT	0005
189*	105	165	201	166	285	390	190	99	83	20	20	17	17	17	14	390	$C_{25}H_{27}ClN_2$	Meclizine	DMIT	0087
189*	117	132	160	115	91	190	103	99	60	54	31	17	14	12	9	217	$C_{13}H_{15}NO_2$	Glutethimide	NIH MSC	3786
104	189*	103	51	77	78	105	39	99	25	17	13	13	13	12	5	189	$C_{11}H_{11}NO_2$	Phensuximide	DMIT	0012
117	189*	132	115	91	160	77	39	99	88	55	51	47	44	30	28	217	$C_{13}H_{15}NO_2$	Glutethimide	DMIT	0028
130	189*	131	79	190	52	103	51	99	57	10	9	7	3	2	0	189	$C_{11}H_{11}NO_2$	3-Indoleacetic acid methyl ester	BLR	0145
147	189*	73	74	148	75	190	146	99	62	29	17	16	14	11	10	204	$C_7H_{20}N_2OSi_2$	Urea di-TMS derivative	BLR	0032
146	190*	117	118	189	161	103	115	99	90	59	42	34	31	22	16	218	$C_{12}H_{14}N_2O_2$	Primaclone	NIH MSC	3809
146	190*	117	189	161	118	91	103	99	76	55	30	26	25	18	17	218	$C_{12}H_{14}N_2O_2$	Primaclone	BLR	0078
73	191*	75	45	74	104	147	167	99	29	17	8	8	8	7	7	516	$C_{25}H_{40}N_2O_4Si_3$	5-(3,4-Dihydroxycyclohexa-1,5-dienyl)-3-methyl-5-phenyl hydantoin TMS deriv	BLR	0045
222	191*	223	207	79	147	164	190	99	23	13	8	6	2	2	1	222	$C_{12}H_{14}O_4$	3,4-Dimethoxycinnamic acid methyl ester	BLR	0154
192*	224	193	73	164	179	208	191	99	24	17	14	10	7	7	6	225	$C_{11}H_{17}NO_3Si$	3-Hydroxyanthranilic acid methyl ester trimethylsilyl ether	BLR	0147
193	192*	194	191	165	180	95	167	99	27	17	16	15	14	8	8	236	$C_{15}H_{12}N_2O$	Carbamazepine	DMIT	0258
193*	165	121	120	194	93	65	92	99	67	57	21	12	11	8	8	358	$C_{19}H_{18}O_7$	Sal ethyl carbonate	DMIT	0064
193*	192	194	191	165	180	95	167	99	27	17	16	15	14	8	8	236	$C_{15}H_{12}N_2O$	Carbamazephine	DMIT	0258
193*	312	194	313	73	314	195	281	99	61	15	15	14	5	4	3	312	$C_{14}H_{24}O_4Si_2$	3,4-Dihydroxybenzoic acid methyl ester trimethylsilyl ether	BLR	0185

Eight Most Intense Peaks — E2 (continued)

Eight most intense peaks								Relative Intensities								Mol wt	Formula	Drug	Source	
73	193*	43	195	210	237	75	252	99	95	84	74	60	40	36	36	252	$C_{13}H_{20}O_3Si$	Propylparaben TMS ether	BLR	0027
73	193*	237	89	162	161	194	177	99	86	67	56	46	42	14	13	252	$C_{13}H_{20}O_3Si$	Phenyl-lactic acid methyl ester trimethylsilyl ether	BLR	0143
297	193*	298	299	73	312	194	79	99	39	24	9	8	6	5	4	312	$C_{14}H_{24}O_4Si_2$	2,3-Dihydroxybenzoic acid methyl ester trimethylsilyl ether	BLR	0187
324	193*	206	73	325	75	194	45	99	77	66	58	29	24	15	13	339	$C_{15}H_{25}NO_4Si_2$	2-Hydroxybenzoylglycine TMS ester TMS ether	BLR	0030
194*	109	55	67	82	15	18	42	99	59	37	22	17	14	11	10	194	$C_8H_{10}N_4O_2$	Caffeine	DMIT	0059
194*	109	55	67	82	43	42	41	99	55	33	25	18	15	13	10	194	$C_8H_{10}N_4O_2$	Caffeine	BLR	0095
121	194*	134	163	122	195	135	108	99	39	19	15	8	3	2	0	194	$C_{11}H_{14}O_3$	*p*-Methoxyphenylpropionic acid methyl ester	BLR	0158
165	194*	147	77	121	176	118	51	99	76	48	47	38	34	32	30	194	$C_{10}H_{10}O_4$	Meconin	DMIT	0213
169	194*	43	195	409	71	69	170	99	65	48	26	16	13	11	11	692	$C_{22}H_{22}F_{14}N_2O_7$	Dihydroxysecobarbital di-HFB derivative 1,3-dimethyl derivative	BLR	0076
195*	41	194	53	138	70	137	79	99	97	60	45	39	37	37	27	236	$C_{12}H_{16}N_2O_3$	Allobarbital 1,2(or 4)-dimethyl derivative	BLR	0048
195*	120	43	210	135	75	73	92	99	52	46	33	23	20	15	15	252	$C_{12}H_{16}O_4Si$	Acetyl salicylic acid TMS ester	BLR	0104
195*	138	41	53	194	80	110	58	99	93	76	50	48	43	27	26	236	$C_{12}H_{16}N_2O_3$	Allobarbital 1,3-dimethyl derivative	BLR	0047
195*	196	41	138	53	111	181	58	99	77	67	48	35	30	27	25	238	$C_{12}H_{18}N_2O_3$	Aprobarbital 1,3-dimethyl derivative	BLR	0054
195*	196	168	153	343	331	136	121	99	81	77	48	38	37	15	11	343	$C_{19}H_{21}NO_5$	Trimethobenzamide MTB 1	DMIT	0317
195*	196	197	69	41	138	169	223	99	97	32	26	23	20	13	13	490	$C_{19}H_{21}F_7N_2O_5$	Hydroxysecobarbital HFB derivative 1,3-dimethyl derivative	BLR	0073
195*	235	234	193	208	266	194	84	99	93	77	69	69	63	58	56	266	$C_{18}H_{22}N_2$	Desipramine	DMIT	0044
58	195*	59	72	388	89	315	42	99	18	14	13	7	6	5	5	388	$C_{21}H_{28}N_2O_5$	Trimethobenzamide	DMIT	0316
196	195*	41	138	181	111	209	169	99	66	48	20	19	15	14	13	252	$C_{13}H_{20}N_2O_3$	Butalbital 1,3-dimethyl derivative	BLR	0059
238	195*	223	163	239	191	179	146	99	39	28	21	19	18	15	14	238	$C_{12}H_{18}O_3Si$	*m*-Hydroxyphenylacetic acid methyl ester trimethylsilyl ether	BLR	0148
365	195*	221	28	366	31	29	212	99	57	57	44	40	36	34	33	578	$C_{32}H_{38}N_2O_8$	Deserpidine	DMIT	0176

196*	41	195	209	138	181	67	43	99	78	52	45	33	25	22	21	252	$C_{13}H_{20}N_2O_3$	Butalbital 1,2(or 4)-dimethyl derivative	BLR	0060
196*	165	197	181	166	59	94	121	99	51	9	6	4	3	1	1	196	$C_{10}H_{12}O_4$	3,4-Dimethoxybenzoic acid methyl ester	BLR	0156
196*	195	41	138	181	111	209	169	99	66	48	20	19	15	14	13	252	$C_{13}H_{20}N_2O_3$	Butalbital 1,3-dimethyl derivative	BLR	0059
195	196*	41	138	53	111	181	58	99	77	67	48	35	30	27	25	238	$C_{12}H_{18}N_2O_3$	Aprobarbital 1,3-dimethyl derivative	BLR	0054
195	196*	168	153	343	331	136	121	99	81	77	48	38	37	15	11	343	$C_{19}H_{21}NO_5$	Trimethobenzamide MTB 1	DMIT	0317
195	196*	197	69	41	138	169	223	99	97	32	26	23	20	13	13	490	$C_{19}H_{21}F_7N_2O_5$	Hydroxysecobarbital HFB derivative 1,3-dimethyl derivative	BLR	0073
58	197*	86	42	43	44	85	196	99	16	15	12	10	7	6	6	326	$C_{19}H_{22}N_2OS$	Acetylpromazine	DMIT	0133
135	197*	73	136	329	417	43	193	99	33	29	16	13	13	7	7	0	—	BHA-type antioxidant	DMIT	0248
198*	70	58	115	71	56	72	269	99	61	58	33	20	11	11	11	327	$C_{19}H_{25}N_3S$	Aminopromazine	DMIT	0138
198*	277	197	186	199	154	91	171	99	47	20	17	15	14	11	10	277	$C_{13}H_{11}NO_2S_2$	Thioridazine MTB 2	DMIT	0218
84	198*	245	112	97	70	197	96	99	86	76	61	59	57	51	48	356	$C_{20}H_{23}N_2S_2$	Thioridazine MTB 3	DMIT	0219
233	198*	235	201	234	232	154	197	99	62	38	26	23	15	14	11	233	$C_{12}H_8ClNS$	Chlorpromazine MTB 1	DMIT	0211
245	198*	230	186	246	197	185	199	99	46	31	23	14	13	9	8	245	$C_{13}H_{11}NS_2$	Thioridazine MTB 1	DMIT	0217
199*	167	198	166	99	154	69	77	99	52	21	18	8	7	5	4	199	$C_{12}H_9NS$	Phenothiazine	DMIT	0147
200*	91	84	28	242	243	115	129	99	66	47	46	36	31	30	29	243	$C_{17}H_{25}N$	Phencyclidine	DMIT	0144
200*	91	243	84	242	186	166	201	99	48	26	25	23	21	18	17	243	$C_{17}H_{25}N$	Phencyclidine	NIH MSC	3806
201*	203	45	166	165	299	56	132	99	34	24	22	21	20	17	16	374	$C_{21}H_{27}ClN_2O_2$	Hydroxyzine	DMIT	0078
203*	58	205	72	204	167	202	168	99	52	33	25	16	12	10	7	274	$C_{16}H_{19}ClN_2$	Chlorpheniramine	DMIT	0073
203*	58	205	204	72	167	202	168	99	50	32	19	17	15	13	10	274	$C_{16}H_{19}ClN_2$	Chlorpheniramine	NIH MSC	3773
203*	188	232	117	204	40	115	70	99	53	23	16	13	11	10	9	260	$C_{14}H_{16}N_2O_3$	Phenobarbital 1,2(or 4)-dimethyl derivative	BLR	0085
203	232	132	175	204	233	130	118	99	65	36	32	16	11	11	10	216	$C_{13}H_{16}N_2O$	Aminoglutethimide	DMIT	0165
130	203*	131	204	144	145	205	117	99	47	9	6	3	3	0	0	201	$C_{12}H_{11}NO_2$	Indolepropionic acid methyl ester	BLR	0157
201	203*	45	166	165	299	56	132	99	34	24	22	21	20	17	16	374	$C_{21}H_{27}ClN_2O_2$	Hydroxyzine	DMIT	0078
204*	63	146	232	117	143	174	89	99	34	34	28	22	18	18	17	232	$C_{12}H_{12}N_2O_3$	Phenobarbital	NIH MSC	3807
204*	105	104	133	77	72	205	78	99	94	71	23	19	13	13	12	204	$C_{11}H_{12}N_2O_2$	Ethotoin	DMIT	0068
204*	117	232	118	161	77	115	146	99	29	22	20	17	16	16	16	232	$C_{12}H_{12}N_2O_3$	Phenobarbital	DMIT	0162
58	204*	146	42	44	77	39	51	99	17	5	4	4	3	3	3	284	$C_{12}H_{17}N_2O_4P$	Psilocybin	DMIT	0182
58	204*	146	205	59	42	130	30	99	25	4	3	3	2	2	2	204	$C_{12}H_{16}N_2O$	Psilocin	DMIT	0177
205*	57	220	41	206	145	29	81	99	32	21	17	15	12	10	10	220	$C_{15}H_{24}O$	Ionol	DMIT	0220
205*	84	96	166	248	122	246	81	99	74	68	31	22	22	22	19	248	$C_{17}H_{20}D_5N$	Phencyclidine (D5-PH)	DMIT	0319
205*	240	125	91	242	186	84	129	99	54	27	25	24	19	17	16	311	$C_{20}H_{22}ClN$	Pyrrobutamine	DMIT	0040
77	205*	233	239	267	51	268	75	99	96	82	70	59	57	47	36	286	$C_{15}H_{11}ClN_2O_2$	Oxazepam	DMIT	0194
220	205*	221	28	147	77	178	42	99	24	18	16	9	9	8	8	413	$C_{22}H_{23}NO_7$	Noscapine	DMIT	0198

Eight Most Intense Peaks — E2 (continued)

Eight most intense peaks								Relative Intensities								Mol wt	Formula	Drug	Source	
207*	67	79	81	141	77	55	91	99	70	48	44	42	42	34	24	236	$C_{12}H_{16}N_2O_3$	Cyclobarbital	DMIT	0169
207*	141	81	67	79	208	80	77	99	28	18	17	14	13	10	7	236	$C_{12}H_{16}N_2O_3$	Cyclobarbital	NIH MSC	3780
242	207*	244	208	165	243	270	269	99	90	25	18	17	16	15	12	270	$C_{16}H_{15}ClN_2$	Medazepam	DMIT	0242
208*	82	123	67	209	42	55	207	99	32	31	17	13	10	9	9	208	$C_9H_{12}N_4O_2$	1,3.7(or 9),8-tetramethylxanthine	BLR	0103
208*	95	193	67	180	123	73	43	99	45	25	17	17	16	14	12	208	$C_9H_{12}N_4O_2$	Theophylline 7-ethyl derivative	BLR	0096
208*	180	109	67	42	137	55	179	99	30	29	16	14	13	12	11	208	$C_9H_{12}N_4O_2$	Theobromine 1-ethyl derivative	BLR	0097
115	208*	193	130	179	117	91	129	99	74	42	38	21	21	20	16	208	$C_{16}H_{16}$	Propoxyphene MTB 2-pk 2	DMIT	0204
115	208*	193	130	179	117	91	129	99	74	42	38	21	21	20	16	208	$C_{16}H_{16}$	Propoxyphene MTB 2-pk 1	DMIT	0203
209	208*	278	207	193	194	84	200	99	91	58	36	22	19	17	17	278	$C_{19}H_{22}N_2$	Triprolidine	DMIT	0075
279	208*	77	308	280	104	149	180	99	62	45	41	28	21	19	19	308	$C_{19}H_{20}N_2O_2$	5,5-Diphenylhydantoin 2,3-diethyl derivative	BLR	0094
209*	179	210	135	59	89	161	193	99	24	16	14	13	12	10	9	224	$C_{11}H_{16}O_3Si$	Salicylic acid methyl ester TMS ether	BLP	0028
209*	179	210	193	89	161	177	211	99	14	14	6	5	4	4	4	224	$C_{11}H_{16}O_3Si$	*o*-Hydroxybenzoic acid methyl estertrimethylsilyl ether	BLR	0150
209*	208	278	207	193	194	84	200	99	91	58	36	22	19	17	17	278	$C_{19}H_{22}N_2$	Triprolidine	DMIT	0075
209*	210	73	211	253	89	225	135	99	17	16	4	4	3	2	2	268	$C_{13}H_{20}O_4Si$	*p*-Methoxymandelic acid methyl ester trimethylsilyl ether	BLR	0165
209*	224	177	89	210	225	149	193	99	89	37	21	16	16	12	11	224	$C_{11}H_{16}O_3Si$	*m*-Hydroxybenzoic acid methyl ester trimethylsilyl ether	BLR	0142
209*	224	225	193	210	135	73	211	99	96	17	17	15	5	5	5	224	$C_{11}H_{16}O_3Si$	*p*-Hydroxybenzoic acid methyl ester trimethylsilyl ether	BLR	0149
82	209*	81	104	83	50	173	210	99	31	18	9	8	8	6	6	543	$C_{25}H_{13}D_{27}N_2O_4Si_3$	5-(3,4-di-d9-TMSO-cyclohexa-1,5-dienyl)-3-methyl-5-phenyl-1-d9-TMS-hydantoin	BLR	0046
180	209*	104	223	252	77	181	165	99	52	51	48	43	36	26	25	252	$C_{15}H_{12}N_2O_2$	Diphenylhydantoin	NIH MSC	3785
180	209*	280	77	181	104	251	165	99	58	36	33	32	28	24	22	280	$C_{17}H_{16}N_2O_2$	5,5-Diphenylhydantoin 3-ethyl derivative	BLR	0092
151	210*	152	211	107	195	59	153	99	74	8	7	1	1	0	0	210	$C_{11}H_{14}O_4$	3,4-Dimethoxyphenylacetic acid methyl ester	BLR	0136

209	210*	73	211	253	89	225	135	99	17	16	4	4	3	2	2	268	$C_{13}H_2O_4Si$	*p*-Methoxymandelic acid methyl ester trimethylsilyl ether	BLR	0165
167	211*	181	180	151	45	168	108	99	36	32	27	25	24	23	21	0	—	Blood 315 drug	DMIT	0238
212*	42	187	45	70	180	56	98	99	75	68	64	61	58	54	46	427	$C_{24}H_{33}N_3O_2S$	Dixyrazine	DMIT	0139
213*	139	75	169	111	215	141	77	99	70	62	46	39	37	23	20	228	$C_{10}H_{13}ClO_2Si$	4-Chlorobenzoic acid TMS ester (from indomethacin)	BLR	0016
185	214*	184	257	109	104	77	82	99	73	73	63	62	59	47	43	177	$C_{11}H_7D_5N_2$	Diphenylhydantoin (D5-phenyl)	DMIT	0315
215*	41	104	77	132	39	128	244	99	49	44	24	20	19	19	19	244	$C_{13}H_{12}N_2O_3$	Alphenal	DMIT	0001
217*	45	70	110	202	270	69	285	99	58	36	33	27	27	24	22	285	$C_{19}H_{27}NO$	Pentazocine	NIH MSC	3803
217*	218	132	146	117	103	246	118	99	26	25	24	22	18	18	13	274	$C_{15}H_{18}N_2O_3$	Mephobarbital 4-ethyl derivative	BLR	0081
73	217*	204	75	147	185	253	45	99	53	37	35	21	18	12	12	676	$C_{29}H_{56}N_2O_{10}Si_3$	Hydroxyamobarbital glucuronide methyl TMS derivative (peak 2)	BLR	0053
130	217*	143	131	186	218	144	117	99	43	21	9	6	5	4	0	217	$C_{13}H_{15}NO_2$	3-Indolebutyric acid methyl ester	BLR	0151
218*	117	146	118	219	28	161	39	99	20	18	16	13	12	9	9	246	$C_{13}H_{14}N_2O_3$	Mephobarbital	DMIT	0171
218*	118	146	117	103	77	219	91	99	32	31	19	18	14	13	12	246	$C_{13}H_{14}N_2O_3$	Mephobarbital	BLR	0071
217	218*	132	146	117	103	246	118	99	26	25	24	22	18	18	13	274	$C_{15}H_{18}N_2O_3$	Mephobarbital 4-ethyl derivative	BLR	0081
277	218*	278	219	79	279	93	220	99	89	21	18	10	6	4	4	249	$C_{14}H_{19}NO_3$	5-Hydroxyindoleacetic acid methyl ester trimethylsilyl ether	BLR	0170
160	219*	161	220	145	74	69	83	99	64	11	8	2	1	0	0	219	$C_{12}H_{13}NO_3$	5-Methoxyindoleacetic acid methyl ester	BLR	0155
338	219*	339	73	220	77	340	75	99	98	29	26	17	16	11	5	338	$C_{16}H_{26}O_4Si_2$	3,4-Dihydroxycinnamic acid methyl ester trimethylsilyl ether	BLR	0179
220*	205	221	28	147	77	178	42	99	24	18	16	9	9	8	8	413	$C_{22}H_{23}NO_7$	Noscapine	DMIT	0198
44	220*	100	57	205	91	129	221	99	92	62	42	35	26	20	20	325	$C_{21}H_{27}NO_2$	Propoxyphene MTB 1	DMIT	0070
235	220*	73	75	204	145	236	130	99	59	49	37	37	21	17	13	350	$C_{17}H_{30}N_2O_2Si_2$	Ethylphenylmalondiamide di-TMS derivative (from primidone)	BLR	0091
221*	81	28	79	157	80	27	155	99	56	46	41	29	29	28	27	236	$C_{12}H_{16}N_2O_3$	Hexobarbital	DMIT	0174
221*	81	157	80	79	155	222	77	99	55	32	23	22	17	12	10	236	$C_{12}H_{16}N_2O_3$	Hexobarbital	BLR	0064
221*	81	157	80	155	222	79	156	99	77	63	36	24	19	18	12	235	$C_{12}H_{15}N_2O_3$	Hexobarbital	NIH MSC	3790
221*	141	81	79	222	41	67	93	99	22	16	14	14	9	9	8	250	$C_{13}H_{18}N_2O_3$	Heptabarbital	NIH MSC	3788
221*	141	222	81	79	41	93	95	99	18	14	12	10	7	7	6	250	$C_{13}H_{18}N_2O_3$	Heptabarbital	DMIT	0026
323	221*	181	222	223	207	324	72	99	64	41	36	32	29	22	21	323	$C_{20}H_{25}N_3O$	Lysergide	DMIT	0178

Eight Most Intense Peaks — E2 (continued)

Eight most intense peaks								Relative Intensities								Mol wt	Formula	Drug	Source	
222*	95	194	166	207	179	123	67	99	53	39	38	31	28	20	18	222	$C_{10}H_{14}N_4O_2$	17-Di-ethyl-3-methylxanthine	BLR	0099
222*	136	166	194	150	123	207	67	99	61	33	24	21	18	14	13	222	$C_{10}H_{14}N_4O_2$	1,3 Di-ethyl-7-methylxanthine	BR	0100
222*	150	194	179	166	207	109	43	99	77	29	25	22	22	21	19	222	$C_{10}H_{14}N_4O_2$	3, 7 Di-ethyl-1-methylxanthine	BLR	0098
222*	151	121	122	123	105	77	93	99	37	23	22	21	20	17	16	254	$C_{13}H_6D_8N_2O_3$	*N*-d3-methyl phenobarbital (D5-ethyl)	DMIT	0321
222	191	223	207	79	147	164	190	99	23	13	8	6	2	2	1	222	$C_{12}H_{14}O_4$	3,4-Dimethoxycinnamic acid methyl ester	BLR	0154
124	223*	109	77	122	123	52	95	99	69	40	20	15	14	12	12	223	$C_{11}H_{13}NO_4$	Mephenoxalone	DMIT	0123
181	223*	166	73	43	45	208	75	99	70	68	58	54	25	22	17	223	$C_{11}H_{17}NO_2Si$	Acetaminophen TMS ether	BLR	0011
224	223*	41	225	125	209	43	109	99	80	18	16	15	15	14	13	294	$C_{16}H_{26}N_2O_3$	Secobarbital 1,3-di-ethyl derivative	BLR	0089
317	223*	75	147	217	43	318	73	99	57	45	43	43	31	28	26	557	$C_{24}H_{43}NO_8Si_3$	Acetaminophen glucuronide methyl ester TMS ether	BLR	0012
224*	223	41	225	125	209	43	109	99	80	18	16	15	15	14	13	294	$C_{16}H_{26}N_2O_3$	Secobarbital 1,3-di-ethyl derivative	BLR	0089
224*	239	254	193	225	240	255	223	99	64	60	17	16	11	11	10	254	$C_{12}H_{18}O_4Si$	3-Methoxy-4-hydroxybenzoic acid methyl ester trimethylsilyl ether	BLR	0172
151	224*	164	152	225	149	165	193	99	68	12	8	7	6	3	2	224	$C_{12}H_{16}O_4$	3,4-Dimethoxyphenylpropionic acid methyl ester	BLR	0152
192	224*	193	73	164	179	208	191	99	24	17	14	10	7	7	6	225	$C_{11}H_{17}NO_3Si$	3-Hydroxyanthranilic acid methyl ester trimethysilyl ether	BLR	0147
209	224*	177	89	210	225	149	193	99	89	37	21	16	16	12	11	224	$C_{11}H_{16}O_3Si$	*m*-Hydroxybenzoic acid methyl ester trimethylsilyl ether	BLR	0142
209	224*	225	193	210	135	73	211	99	96	17	17	15	5	5	5	224	$C_{11}H_{16}O_3Si$	*p*-Hydroxybenzoic acid methyl ester trimethylsilyl ether	BLR	0149
73	225*	75	133	353	131	226	74	99	73	37	26	24	15	15	14	597	$C_{30}H_{59}NO_5Si_3$	Prostaglandin E2 methyloxime TMS ester di-TMS ether	BLR	0125
227*	310	174	284	147	160	173	199	99	40	39	36	30	29	28	27	310	$C_{21}H_{26}O_2$	Mestranol	DMIT	0261
230*	58	231	44	173	105	42	159	99	24	17	11	11	10	8	6	321	$C_{22}H_{27}NO$	Phenazocine	DMIT	0112

230*	231	58	44	105	42	173	91	99	17	12	7	7	6	6	5	321	$C_{22}H_{27}NO$	Phenazocine	NIH MSC	3805
231*	314	271	258	43	193	41	246	99	79	44	42	41	31	31	22	314	$C_{21}H_{30}O_2$	Tetrahydrocannabinol-delta-8	DMIT	0132
42	231*	45	119	44	74	67	46	99	54	37	32	29	22	20	15	277	$C_9H_7N_7O_2S$	Azathioprine	DMIT	0297
230	231*	58	44	105	42	173	91	99	17	12	7	7	6	6	5	321	$C_{22}H_{27}NO$	Phenazocine	NIH MSC	3805
232*	146	117	175	118	233	103	188	99	44	41	41	40	22	18	18	260	$C_{14}H_{16}N_2O_3$	Phenobarbital 1,3-dimethyl derivative	BLR	0084
232*	146	118	175	117	233	103	188	99	29	27	26	21	15	10	10	260	$C_{14}H_{16}N_2O_3$	*N,n*-dimethyl phenobarbital	DMIT	0295
203	232*	132	175	204	233	130	118	99	65	36	32	16	11	11	10	216	$C_{13}H_{16}N_2O$	Aminoglutethimide	DMIT	0165
233*	44	304	72	232	235	198	234	99	82	59	58	41	39	37	31	304	$C_{16}H_{17}ClN_2S$	Chloropromazine MTB 2	DMIT	0212
233*	198	235	201	234	232	154	197	99	62	38	26	23	15	14	11	233	$C_{12}H_8ClNS$	Chlorpromazine MTB 1	DMIT	0211
57	233*	42	56	43	158	131	160	99	31	26	26	18	18	15	13	233	$C_{14}H_{19}NO_2$	Meperidine MTB 1	NIH MSC	3782
174	233*	175	159	131	130	234	158	99	36	14	13	9	5	5	5	233	$C_{13}H_{15}NO_3$	2-Methyl-5-methoxyindole-3-acetic acid ME ester (from indomethacin)	BLR	0017
58	234*	44	59	41	42	36	427	99	10	8	5	4	4	3	3	445	$C_{23}H_{27}NO_8$	Narceine	DMIT	0197
236	234*	238	155	157	50	75	49	99	48	47	40	39	26	26	17	234	$C_6H_4Br_2$	*p*-Dibromobenzene	DMIT	0333
235*	58	234	85	280	195	193	35	99	78	64	62	48	29	23	20	280	$C_{19}H_{24}N_2$	Imipramine	DMIT	0066
235*	81	169	171	79	236	170	91	99	40	26	23	18	13	12	11	250	$C_{13}H_{18}N_2O_3$	Hexobarbital-3-methyl derivative	BLR	0062
235*	146	118	117	77	236	178	103	99	27	27	26	19	14	13	12	263	$C_{14}H_{13}D_3N_2O_3$	*N*-Methyl, *N*-d3-methyl phenobarbital	DMIT	0322
235*	150	165	236	79	137	250	164	99	25	24	21	18	16	14	12	250	$C_{13}H_{18}N_2O_3$	Hexobarbital-2(or 4)-methyl derivative	BLR	0063
235*	220	73	75	204	145	236	130	99	59	49	37	37	21	17	13	350	$C_{17}H_{30}N_2O_2Si_2$	Ethylphenylmalondiamide di-TMS derivative (from primidone)	BLR	0091
235*	237	165	212	246	75	176	36	99	65	40	23	16	15	15	15	352	$C_{14}H_9Cl_5$	DDT	DMIT	0167
235*	250	91	233	65	236	40	76	99	40	37	29	22	20	18	17	250	$C_{16}H_{14}N_2O$	Methaqualone	DMIT	0193
160	235*	266	77	251	146	69	76	99	93	89	73	71	59	50	47	266	$C_{16}H_{14}N_2O_2$	Methaqualone MTB 1	DMIT	0215
195	235*	234	193	208	266	194	84	99	93	77	69	69	63	58	56	266	$C_{18}H_{22}N_2$	Desipramine	DMIT	0044
250	235*	251	219	203	236	73	252	99	35	20	15	12	6	5	5	250	$C_{13}H_{18}O_3Si$	*p*-Hydroxycinnamic acid methyl ester trimethylsilyl ether	BLR	0146
236*	234	238	155	157	50	75	49	99	48	47	40	39	26	26	17	234	$C_6H_4Br_2$	*p*-Dibromobenzene	DMIT	0333
235	237*	165	212	246	75	176	36	99	65	40	23	16	15	15	15	352	$C_{14}H_9Cl_5$	DDT	DMIT	0167
252	237*	253	221	238	177	209	149	99	34	12	6	4	3	2	1	252	$C_{13}H_{16}O_5$	3,4,5-Trimethoxycinnamic acid methyl ester	BLR	0173
238*	195	223	163	239	191	179	146	99	39	28	21	19	18	15	14	238	$C_{12}H_{18}O_3Si$	*m*-Hydroxyphenylacetic acid methyl ester trimethylsilyl ether	BLR	0148

Eight Most Intense Peaks — E2 (continued)

Eight most intense peaks								Relative Intensities								Mol wt	Formula	Drug	Source	
179	238*	163	180	223	73	239	181	99	55	15	15	15	12	10	4	238	$C_{12}H_{18}O_3Si$	*p*-Hydroxyphenylacetic acid methyl ester trimethylsilyl ether	BLR	0141
268	238*	209	253	269	179	239	210	99	51	41	36	20	15	9	6	268	$C_{13}H_{20}O_4Si$	3-Methoxy-4-hydroxyphenylacetic acid methyl ester trimethylsilyl ether	BLR	0171
239*	151	121	179	122	123	105	77	99	36	26	24	19	19	16	15	271	$C_{14}H_5D_{11}N_2O_3$	*N,N*-d3-Methyl phenobarbital (D5-ethyl)	DMIT	0320
239*	240	73	298	241	283	89	165	99	18	12	6	5	4	2	2	270	$C_{14}H_{22}O_5$	3,4-Dimethoxymandelic acid methyl ester trimethylsilyl ether	BLR	0166
224	239*	254	193	225	240	255	223	99	64	60	17	16	11	11	10	254	$C_{12}H_{18}O_4Si$	3-Methoxy-4-hydroxybenzoic acid methyl ester trimethylsilyl ether	BLR	0172
205	240*	125	91	242	186	84	129	99	54	27	25	24	19	17	16	311	$C_{20}H_{22}ClN$	Pyprobutamine	DMIT	0040
239	240*	73	298	241	283	89	165	99	18	12	6	5	4	2	2	270	$C_{14}H_{22}O_5$	3,4-Dimethoxymandelic acid methyl ester trimethylsilyl ether	BLR	0166
242*	207	244	208	165	243	270	269	99	90	25	18	17	16	15	12	270	$C_{16}H_{15}ClN_2$	Medazepam	DMIT	0242
242*	270	269	241	243	271	244	103	99	88	78	67	48	39	34	30	270	$C_{15}H_{11}ClN_2O$	Diazepam MTB 1	BLR	0115
118	243*	104	77	129	231	130	128	99	99	52	43	39	35	34	32	272	$C_{15}H_{16}N_2O_3$	Alphenal 1,3-dimethyl derivative	BLR	0049
245*	198	230	186	246	197	185	199	99	46	31	23	14	13	9	8	245	$C_{13}H_{11}NS_2$	Thioridazine MTB 1	DMIT	0217
77	245*	244	105	228	246	247	51	99	97	79	46	42	39	36	33	245	$C_{14}H_{12}ClNO$	2-Methylamino-5-chlorobenzophenone	BLR	0112
276	245*	261	123	275	277	229	201	99	38	22	19	16	16	9	9	276	$C_{13}H_{16}N_4O_3$	Trimethoprim MTB 3	DMT	0300
246*	42	247	377	91	172	47	28	99	18	16	11	9	6	5	5	452	$C_{30}H_{32}N_2O_2$	Diphenoxylate	DMIT	0149
246*	146	117	118	247	175	103	77	99	46	33	28	20	20	15	14	274	$C_{15}H_{18}N_2O_3$	Mephobarbital 3-ethyl derivative	BLR	0072
246*	185	445	70	125	154	213	87	99	93	81	73	65	59	55	50	445	$C_{23}H_{28}ClN_3O_2S$	Thiopropazate	DMIT	0046
246*	247	42	120	218	172	106	91	99	18	12	8	7	7	6	6	352	$C_{22}H_{28}N_2O_2$	Anileridine	DMIT	0032
246*	366	106	247	133	260	234	367	99	53	24	19	18	17	16	15	366	$C_{23}H_{30}N_2O_2$	Piminodine	DMIT	0125
42	246*	232	70	56	143	214	43	99	66	59	49	36	36	28	20	403	$C_{21}H_{26}ClN_3OS$	Perphenazine	DMIT	0136
58	246*	233	318	86	272	248	232	99	35	26	22	17	13	10	10	334	$C_{17}H_{19}ClN_2OS$	Chlorpromazine MTB 4	DMIT	0097
247*	188	160	63	215	43	106	216	99	59	38	31	30	25	20	20	247	$C_{14}H_{17}NO_3$	1,2-Dimethyl-5-methoxyindole-3-acetic acid methyl ester (from indomethacin)	BLR	0018

247*	248	249	250	219	212	106	211	99	50	36	15	12	10	5	5	248	$C_{12}H_{13}ClN_4$	Pyrimethamine	NIH MSC	3814
146	247*	117	45	232	118	275	230	99	69	54	48	33	32	30	24	290	$C_{15}H_{18}N_2O_4$	*N*-Methyl, *N′*-methoxymethyl phenobarbital	DMIT	0294
246	247*	42	120	218	172	106	91	99	18	12	8	7	7	6	6	352	$C_{22}H_{28}N_2O_2$	Anileridine	DMIT	0032
249	247*	58	72	167	248	18	250	99	95	71	30	23	20	15	15	318	$C_{16}H_{19}BrN_2$	Parabromdylamine	DMIT	0084
247	248*	249	250	219	212	106	211	99	50	36	15	12	10	5	5	248	$C_{12}H_{13}ClN_4$	Pyrimethamine	NIH MSC	3814
249*	77	235	76	264	247	250	50	99	36	27	23	20	19	17	14	264	$C_{17}H_{16}N_2O$	Ethinazone	DMIT	0252
249*	81	183	250	79	185	184	264	99	35	20	20	15	11	9	9	264	$C_{14}H_{20}N_2O_3$	Hexobarbital 3-ethyl derivative	BLR	0080
249*	247	58	72	167	248	168	250	99	95	71	30	23	20	15	15	318	$C_{16}H_{19}BrN_2$	Parabromdylamine	DMIT	0084
250*	164	222	194	235	251	207	150	99	45	34	31	21	13	12	10	250	$C_{12}H_{18}N_4O_2$	8-Methylxanthine 1,3,7-triethyl derivative	BLR	0101
250*	235	251	219	203	236	73	252	99	35	20	15	12	6	5	5	250	$C_{13}H_{18}O_3Si$	*p*-Hydroxycinnamic acid methyl ester trimethylsilyl ether	BLR	0146
89	250*	235	73	219	146	118	90	99	38	32	18	17	14	13	10	250	$C_{13}H_{18}O_3Si$	*o*-Hydroxycinnamic acid methyl ester trimethylsilyl ether	BLR	0175
235	250*	91	233	65	236	40	76	99	40	37	29	22	20	18	17	250	$C_{16}H_{14}N_2O$	Methaqualone	DMIT	0193
280	250*	281	265	251	282	249	73	99	66	20	20	12	5	4	3	280	$C_{14}H_{20}O_4Si$	3-Methoxy-4-hydroxycinnamic acid methyl ester trimethylsilyl ether	BLR	0168
251*	266	77	249	252	76	51	91	99	45	30	30	21	18	14	13	266	$C_{16}H_{14}N_2O_2$	Methaqualone MTB 2	DMIT	0216
251*	266	91	249	132	92	63	252	99	82	50	44	20	16	15	15	266	$C_{16}H_{14}N_2O_2$	Methaqualone MTB 5 (GC)	DMIT	0309
251*	266	249	77	143	252	76	267	99	50	39	25	23	17	13	9	266	$C_{16}H_{14}N_2O_2$	Methaqualone MTB 4	DMIT	0307
251*	266	249	77	252	148	83	76	99	44	34	19	17	11	11	9	266	$C_{16}H_{14}N_2O_2$	Methaqualone MTB 3	DMIT	0306
251*	266	249	91	252	65	132	267	99	66	46	30	17	15	15	12	266	$C_{16}H_{14}N_2O_2$	Methaqualone MTB 5	DMIT	0308
251*	280	72	134	77	265	208	175	99	62	61	46	37	28	21	20	280	$C_{17}H_{16}N_2O_2$	Diphenylhydantoin 1,3-dimethyl derivative	BLR	0039
252*	237	253	221	238	177	209	149	99	34	12	6	4	3	2	1	252	$C_{13}H_{16}O_5$	3,4,5-Trimethoxycinnamic acid methyl ester	BLR	0173
179	252*	180	253	192	73	131	89	99	60	15	11	10	9	9	8	252	$C_{13}H_{20}O_3Si$	*p*-Hydroxyphenylpropionic acid methyl ester trimethylsilyl ether	BLR	0174
253	252*	43	254	104	235	57	251	99	90	21	17	15	13	9	8	253	$C_{12}H_{11}N_7$	Triamterene	NIH MSC	3825
253*	252	43	254	104	235	57	251	99	90	21	17	15	13	9	8	253	$C_{12}H_{11}N_7$	Triamterene	NIH MSC	3825
73	253*	75	217	147	185	317	45	99	40	37	23	22	19	16	13	676	$C_{29}H_{56}N_2O_{10}Si_3$	Hydroxyamobarbital glucuronide methyl TMS derivative (peak 1)	BLR	0052
73	253*	217	185	317	75	204	69	99	53	31	28	28	17	17	15	676	$C_{29}H_{56}N_2O_{10}Si_3$	Hydroxypentobarbital glucuronide 1,3-dimethyl deriv methyl ester TMS ether	BLR	0067

Eight Most Intense Peaks — E2 (continued)

Eight most intense peaks								Relative Intensities								Mol wt	Formula	Drug	Source	
254*	143	411	70	255	42	380	157	99	55	51	37	35	30	27	24	411	$C_{23}H_{29}N_3O_2S$	Acetophenazine	DMIT	0091
255*	131	254	125	257	256	57	58	99	70	53	38	35	31	15	13	325	$C_{19}H_{20}ClN_3$	Clemizole	DMIT	0037
58	255*	40	72	42	59	71	91	99	9	6	4	3	3	3	3	255	$C_{17}H_{21}NO$	Phenyltoloxamine	DMIT	0049
256*	121	185	224	257	65	43	41	99	53	52	17	14	12	11	8	285	$C_{13}H_{19}NO_4S$	Probenecid	NIH MSC	3810
256*	283	255	284	257	258	285	165	99	75	60	59	48	38	34	31	284	$C_{16}H_{13}ClN_2O$	Diazepam	DMIT	0024
256*	283	284	255	257	258	285	222	99	87	69	42	41	35	23	18	284	$C_{16}H_{13}ClN_2O$	Diazepam	NIH MSC	3827
256*	283	284	285	255	257	258	286	99	86	76	40	38	38	36	30	284	$C_{16}H_{13}ClN_2O$	Diazepam	BLR	0119
290	259*	275	291	243	123	200	43	99	30	29	19	16	14	8	8	290	$C_{14}H_{18}N_4O_3$	Trimethoprim	DMIT	0303
290	259*	306	275	243	123	43	81	99	44	36	33	26	26	19	16	306	$C_{14}H_{18}N_4O_4$	Trimethoprim MTB 2	DMIT	0299
260*	261	42	202	57	217	203	218	99	77	34	25	25	20	19	18	261	$C_{19}H_{19}N$	Phenindamine	DMIT	0152
146	260*	118	117	103	261	91	232	99	99	38	34	26	24	17	16	288	$C_{16}H_{20}N_2O_3$	Phenobarbital 1,3-diethyl derivative	BLR	0068
261*	290	233	148	262	133	176	260	99	80	55	44	26	10	8	8	290	$C_{15}H_{18}N_2O_4$	Hydroxyphenobarbital 1,3-dimethyl derivative methyl ether	BLR	0086
260	261*	42	202	57	217	203	218	99	77	34	25	25	20	19	18	261	$C_{19}H_{19}N$	Phenindamine	DMIT	0152
276	261*	123	277	245	201	243	81	99	53	20	15	14	13	12	10	276	$C_{13}H_{16}N_4O_3$	Trimethoprim MTB 4	DMIT	0301
149	263*	57	77	104	41	133	76	99	29	21	15	14	14	11	9	322	$C_{18}H_{26}O_5$	Butyl butoxyethyl phthalate	DMIT	0243
73	265*	217	75	204	69	41	147	99	47	25	23	22	21	14	14	688	$C_{30}H_{56}N_2O_{10}Si_3$	Hydroxysecobarbital glucuronide 1,3-di-methyl deriv methyl ester TMS ether	BLR	0077
160	266*	235	77	251	146	247	76	99	57	51	44	34	28	27	23	266	$C_{16}H_{14}N_2O_2$	Methaqualone MTB 1	DMIT	0305
180	266*	237	104	209	77	189	181	99	85	52	49	40	29	28	25	266	$C_{16}H_{14}N_2O_2$	Diphenlhydantoin MTB 1	DMIT	0259
251	266*	77	249	252	76	51	91	99	45	30	30	21	18	14	13	266	$C_{16}H_{14}N_2O_2$	Methaqualone MTB 2	DMIT	0216
251	266*	91	249	132	92	63	252	99	82	50	44	20	16	15	15	266	$C_{16}H_{14}N_2O_2$	Methaqualone MTB 5 (GC)	DMIT	0309
251	266*	249	77	143	252	76	267	99	50	39	25	23	17	13	9	266	$C_{16}H_{14}N_2O_2$	Methaqualone MTB 4	DMIT	0307
251	266*	249	77	252	148	83	76	99	44	34	19	17	11	11	9	266	$C_{16}H_{14}N_2O_2$	Methaqualone MTB 3	DMIT	0306
251	266*	249	91	252	65	132	267	99	66	46	30	17	15	15	12	266	$C_{16}H_{14}N_2O_2$	Methaqualone MTB 5	DMIT	0308
267*	73	268	269	45	135	193	75	99	99	25	13	12	9	9	8	282	$C_{13}H_{22}O_3Si_2$	Salicylic acid TMS ester TMS ether	BLR	0105
267*	268	269	73	311	75	283	89	99	22	9	7	4	2	2	1	326	$C_{15}H_{26}O_4Si_2$	*p*-Hydroxymandelic acid methyl ester trimethylsilyl ether	BLR	0182
296	267*	210	219	134	180	77	297	99	65	60	60	38	31	24	21	296	$C_{17}H_{16}N_2O_3$	*p*-Hydroxydiphenylhydantoin methyl ether 3-methyl derivative	BLR	0042
268*	143	425	55	70	41	40	269	99	61	58	54	50	48	47	41	425	$C_{24}H_{31}N_3O_2S$	Carphenazine	DMIT	0093

268*	238	209	253	269	179	239	210	99	51	41	36	20	15	9	6	268	$C_{13}H_{20}O_4Si$	3-Methoxy-4-hydroxyphenylacetic acid methyl ester trimethylsilyl ether	BLR	0171
268*	269	285	270	77	233	284	91	99	71	50	42	34	30	29	26	285	$C_{15}H_{12}ClN_3O$	Chlordiazephoxide MTB 1	DMIT	0244
267	268*	269	73	311	75	283	89	99	22	9	7	4	2	2	1	326	$C_{15}H_{26}O_4Si_2$	*p*-Hydroxymandelic acid methyl ester trimethylsilyl ether	BLR	0182
295	268*	297	97	57	270	62	64	99	70	44	43	35	30	27	25	295	$C_7H_6ClN_3O_4S_2$	Chlorthiazide	NIH MSC	3775
269*	64	205	297	271	43	44	31	99	92	49	45	42	40	38	36	265	$C_7H_8ClN_3O_4S$	Hydrochlorothiazide	DMIT	0266
268	269*	285	270	77	233	284	91	99	71	50	42	34	30	29	26	285	$C_{15}H_{12}ClN_3O$	Chlordiazepoxide MTB 1	DMIT	0244
242	270*	269	241	243	271	244	103	99	88	78	67	48	39	34	30	270	$C_{15}H_{11}ClN_2O$	Diazepam MTB 1	BLR	0115
271*	44	270	214	272	42	43	70	99	35	27	23	19	18	16	14	311	$C_{19}H_{21}NO_3$	Nalorphine	NIH MSC	3801
73	274*	359	75	44	77	45	259	99	55	54	40	36	35	28	26	359	$C_{19}H_{25}NO_4Si$	Noroxymorphone-1 TMS	DMIT	0285
386	275*	107	43	95	55	81	368	99	73	68	63	63	58	57	56	386	$C_{27}H_{46}O$	Cholesterol	DMIT	0105
276*	245	261	123	275	277	229	201	99	38	22	19	16	16	9	9	276	$C_{13}H_{16}N_4O_3$	Trimethoprim MTB 3	DMIT	0300
276*	261	123	277	245	201	243	81	99	53	20	15	14	13	12	10	276	$C_{13}H_{16}N_4O_3$	Trimethoprim MTB 4	DMIT	0301
276*	277	262	105	220	278	200	205	99	84	48	26	23	22	18	13	277	$C_{20}H_{23}N$	Methadone MTB 1	DMIT	0236
58	276*	73	41	59	75	77	40	99	12	11	10	9	7	6	6	276	$C_{15}H_{24}N_2OSi$	Psilocin-1 TMS	DMIT	0329
277*	218	278	219	79	279	93	220	99	89	21	18	10	6	4	4	249	$C_{14}H_{19}NO_3$	5-Hydroxyindoleacetic acid methyl ester trimethylsilyl ether	BLR	0170
198	277*	197	186	199	154	91	171	99	47	20	17	15	14	11	10	277	$C_{13}H_{11}NO_2S_2$	Thioridazine MTB 2	DMIT	0218
276	277*	262	105	220	278	200	205	99	84	48	26	23	22	18	13	277	$C_{20}H_{23}N$	Methadone MTB 1	DMIT	0236
354	277*	325	104	268	73	269	355	99	85	85	65	65	60	30	25	354	$C_{19}H_{22}N_2O_3Si$	5-(4-Hydroxyphenyl)-3-methyl-5-phenylhydantoin TMS ether	BLR	0043
279*	208	77	308	280	104	149	180	99	62	45	41	28	21	19	19	308	$C_{19}H_{20}N_2O_2$	5,5-Diphenylhydantoin 2,3-diethyl derivative	BLR	0094
280*	250	281	265	251	282	249	73	99	66	20	20	12	5	4	3	280	$C_{14}H_{20}O_4Si$	3-Methoxy-4-hydroxycinnamic acid methyl ester trimethylsilyl ether	BLR	0168
42	280*	70	143	56	113	72	100	99	99	96	96	95	85	80	79	437	$C_{22}H_{26}F_3N_3OS$	Fluphenazine	DMIT	0134
251	280*	72	134	77	265	208	175	99	62	61	46	37	28	21	20	280	$C_{17}H_{16}N_2O_2$	Diphenylhydantoin 1,3-dimethyl derivative	BLR	0039
310	280*	295	311	281	73	75	79	99	88	21	15	11	8	4	4	310	$C_{15}H_{22}O_5Si$	3,5-Dimethoxy-4-hydroxycinnamic acid methyl ester trimethylsilyl ether	BLR	0164
282*	283	28	284	299	241	247	77	99	42	34	34	25	19	15	13	299	$C_{16}H_{14}ClN_3O$	Chlordiazepoxide	NIH MSC	3794
282*	283	284	285	247	219	220	241	99	82	50	29	25	21	20	17	299	$C_{16}H_{14}ClN_3O$	Chlordiazepoxide (GC)	DMIT	0209
282*	283	284	285	299	77	241	91	99	72	40	24	22	18	15	12	299	$C_{16}H_{14}ClN_3O$	Chlordiazepoxide	BLR	0118
282*	283	284	299	285	77	56	41	99	71	43	34	23	22	19	18	299	$C_{16}H_{14}ClN_3O$	Chlordiazepoxide	DMIT	0023
283	282*	256	176	157	43	41	57	99	62	54	44	43	42	40	37	283	$C_{19}H_{25}NO$	Levallorphan	NIH MSC	3793

Eight Most Intense Peaks — E2 (continued)

Eight most intense peaks								Relative Intensities								Mol wt	Formula	Drug	Source	
283*	282	256	176	157	43	41	57	99	62	54	44	43	42	40	37	283	$C_{19}H_{25}NO$	Levallorphan	NIH MSC	3793
256	283*	255	284	257	258	285	165	99	75	60	59	48	38	34	31	284	$C_{16}H_{13}ClN_2O$	Diazepam	DMIT	0024
256	283*	284	255	257	258	285	222	99	87	69	42	41	35	23	18	284	$C_{16}H_{13}ClN_2O$	Diazepam	NIH MSC	3827
256	283*	284	285	255	257	258	286	99	86	76	40	38	38	36	30	284	$C_{16}H_{13}ClN_2O$	Diazepam	BLR	0119
282	283*	28	284	299	241	247	77	99	42	34	34	25	19	15	13	299	$C_{16}H_{14}ClN_3O$	Chlordiazepoxide	NIH MSC	3794
282	283*	284	285	247	219	220	241	99	82	50	29	25	21	20	17	299	$C_{16}H_{14}ClN_3O$	Chlordiazepoxide (GC)	DMIT	0209
282	283*	284	285	299	77	241	91	99	72	40	24	22	18	15	12	299	$C_{16}H_{14}ClN_3O$	Chlordiazepoxide	BLR	0118
282	283*	284	299	285	77	56	41	99	71	43	34	23	22	19	18	299	$C_{16}H_{14}ClN_3O$	Chlordiazepoxide	DMIT	0023
58	284*	86	238	199	85	198	285	99	43	29	23	20	12	11	10	284	$C_{17}H_{20}N_2S$	Promazine	DMIT	0017
285*	162	42	28	44	31	215	70	99	63	58	55	38	34	34	28	285	$C_{17}H_{19}NO_3$	Morphine	DMIT	0103
285*	162	215	42	286	124	284	174	99	38	31	21	20	19	18	16	285	$C_{17}H_{19}NO_3$	Morphine	NIH MSC	3800
287*	96	215	286	288	243	229	230	99	61	47	44	29	25	22	20	287	$C_{21}H_{21}N$	Cyproheptadine	DMIT	0208
289*	290	306	259	243	275	257	228	99	44	25	16	9	9	8	8	306	$C_{14}H_{18}N_4O_4$	Trimethoprim MTB 1	DMIT	0298
290*	259	275	291	243	123	200	43	99	30	29	19	16	14	8	8	290	$C_{14}H_{18}N_4O_3$	Trimethoprim	DMIT	0303
290*	259	306	275	243	123	43	81	99	44	36	33	26	26	19	16	306	$C_{14}H_{18}N_4O_4$	Trimethoprim MTB 2	DMIT	0299
58	290*	73	291	348	41	75	77	99	37	33	10	9	7	7	7	348	$C_{18}H_{32}N_2OSi_2$	Psilocin-2 TMS	DMIT	0330
261	290*	233	148	262	133	176	260	99	80	55	44	26	10	8	8	290	$C_{15}H_{18}N_2O_4$	Hydroxyphenobarbital 1,3-dimethyl derivative methyl ether	BLR	0086
289	290*	306	259	243	275	257	228	99	44	25	16	9	8	8	8	306	$C_{14}H_{18}N_4O_4$	Trimethoprim MTB 1	DMIT	0298
174	291*	73	175	159	131	75	158	99	17	13	13	7	6	5	5	291	$C_{15}H_{21}NO_3Si$	2-Methyl-5-methoxyindole-3-acetic acid TMS ester(from indomethacin)	BLR	0021
292*	121	263	293	249	77	9	215	99	57	55	38	37	30	27	25	292	$C_{19}H_{16}O_3$	Warfarin MTB 1	DMIT	0326
295*	268	297	97	57	270	62	64	99	70	44	43	35	30	27	25	295	$C_7H_6ClN_3O_4S_2$	Chlorthiazide	NIH MSC	3775
296*	180	267	210	104	77	134	297	99	52	42	33	31	23	21	20	296	$C_{17}H_{16}N_2O_3$	3-Hydroxydiphenylhydantoin methyl ether 3-methyl derivative	BLR	0040
296*	267	210	219	134	180	77	297	99	65	60	60	38	31	24	21	296	$C_{17}H_{16}N_2O_3$	*p*-Hydroxydiphenylhydantoin methyl ether 3-methyl derivative	BLR	0042
84	296*	42	85	180	55	212	41	99	7	6	5	3	3	3	2	296	$C_{18}H_{20}N_2S$	Pyrathiazine	DMIT	0092
311	296*	42	44	255	312	310	253	99	51	42	32	30	22	20	15	311	$C_{19}H_{21}NO_3$	Thebaine	DMIT	0195
297*	193	298	299	73	312	194	79	99	39	24	9	8	6	5	4	312	$C_{14}H_{24}O_4Si_2$	2,3-Dihydroxybenzoic acid methyl ester trimethylsilyl ether	BLR	0187

297*	298	73	299	341	75	267	356	99	27	14	11	6	4	4	4	356	$C_{16}H_{28}O_5Si_2$	3-Methoxy-4-hydroxymandelic acid methyl ester trimethylsilyl ether	BLR	0176
297*	298	299	281	73	312	77	265	99	24	10	4	3	3	2	2	312	$C_{14}H_{24}O_4Si_2$	2,6-Dihydroxybenzoic acid methyl ester trimethylsilyl ether	BLR	0186
297*	298	299	312	281	267	73	313	99	23	8	5	3	1	1	1	312	$C_{14}H_{24}O_4Si_2$	2,4-Dihydroxybenzoic acid methyl ester trimethylsilyl ether	BLR	0184
297*	298	312	299	267	79	281	73	99	20	12	9	4	3	2	2	312	$C_{14}H_{24}O_4Si_2$	2,5-Dihydroxybenzoic acid methyl ester trimethylsilylether	BLR	0132
312	297*	313	265	314	298	281	299	99	31	24	12	9	6	3	3	312	$C_{14}H_{24}O_4Si_2$	3,5-Dihydroxybenzoic acid methyl ester trimethylsilyl ether	BLR	0188
298*	594	58	593	299	609	595	564	99	37	26	21	17	16	15	8	694	$C_{38}H_{44}Cl2N_2O_6$	Tubocurarine chloride	DMIT	0170
86	298*	83	58	87	30	85	180	99	10	6	5	5	4	4	3	298	$C_{18}H_{22}N_2S$	Diethazine	DMIT	0111
297	298*	73	299	341	75	267	356	99	27	14	11	6	4	4	4	356	$C_{16}H_{28}O_5Si_2$	3-Methoxy-4-hydroxymandelic acid methyl ester trimethylsilyl ether	BLR	0176
297	298*	299	281	73	312	77	265	99	24	10	4	3	3	2	2	312	$C_{14}H_{24}O_4Si_2$	2,6-Dihydroxybenzoic acid methyl ester trimethylsilyl ether	BLR	0186
297	298*	299	312	281	267	73	313	99	23	8	5	3	1	1	1	312	$C_{14}H_{24}O_4Si_2$	2,4-Dihydroxybenzoic acid methyl ester trimethylsilyl ether	BLR	0184
297	298*	312	299	267	79	281	73	99	20	12	9	4	3	2	2	312	$C_{14}H_{24}O_4Si_2$	2,5-Dihydroxybenzoic acid methyl ester trimethylsilylether	BLR	0132
299*	162	229	42	214	300	124	188	99	51	39	34	24	21	21	20	299	$C_{18}H_{21}NO_3$	Codeine	DMIT	0042
299*	162	229	124	300	214	298	42	99	38	28	22	20	15	15	14	299	$C_{18}H_{21}NO_3$	Codeine	NIH MSC	3779
299*	300	330	43	217	231	41	193	99	21	16	9	9	8	8	8	330	$C_{21}H_{30}O_3$	Tetrahydrocannabinol MTB 1	DMIT	0304
314	299*	231	271	43	41	315	243	99	80	55	41	36	35	24	22	314	$C_{21}H_{30}O_2$	Tetrahydrocannabinol	DMIT	0089
314	299*	231	271	243	258	41	43	99	95	80	50	32	30	28	28	314	$C_{21}H_{30}O_2$	Tetrahydrocannabinol	NIH MSC	3823
299	300*	330	43	217	231	41	193	99	21	16	9	9	9	8	8	330	$C_{21}H_{30}O_3$	Tetrahydrocannabinol MTB 1	DMIT	0304
332	303*	275	304	190	276	134	333	99	81	64	49	21	20	17	17	332	$C_{18}H_{24}N_2O_4$	Hydroxyphenobarbital 1,3-di-ethyl derivative ethyl ether	BLR	0087
180	308*	279	208	77	251	165	104	99	80	78	61	49	26	26	20	308	$C_{19}H_{20}N_2O_2$	5,5-Diphenylhydantoin 1,3-diethyl derivative	BLR	0093

Eight Most Intense Peaks — E2 (continued)

Eight most intense peaks								Relative Intensities								Mol wt	Formula	Drug	Source	
310*	64	36	28	312	42	43	62	99	57	53	44	43	36	29	25	359	$C_8H_{11}Cl_2N_3O_4S_2$	Methychlothiazide	DMIT	0160
310*	111	112	58	199	212	96	41	99	92	84	79	65	64	61	55	310	$C_{19}H_{22}N_2S$	Mepazine	DMIT	0104
310*	280	295	311	281	73	75	79	99	88	21	15	11	8	4	4	310	$C_{15}H_{22}O_5Si$	3,5-Dimethoxy-4-hydroxy-cinnamic acid methyl ester trimethylsilyl ether	BLR	0164
227	310*	174	284	147	160	173	199	99	40	39	36	30	29	28	27	310	$C_{21}H_{26}O_2$	Mestranol	DMIT	0261
311*	296	42	44	255	312	310	253	99	51	42	32	30	22	20	15	311	$C_{19}H_{21}NO_3$	Thebaine	DMIT	0195
312*	297	313	265	314	298	281	299	99	31	24	12	9	6	3	3	312	$C_{14}H_{24}O_4Si_2$	3,5-Dihydroxybenzoic acid methyl ester trimethylsilyl ether	BLR	0188
312*	371	43	297	399	281	298	313	99	53	50	44	36	34	24	22	399	$C_{22}H_{25}NO_6$	Colchicine	DMIT	0272
193	312*	194	313	73	314	195	281	99	61	15	15	14	5	4	3	312	$C_{14}H_{24}O_4Si_2$	3,4-Dihydroxybenzoic acid methyl ester trimethylsilyl ether	BLR	0185
314*	299	231	271	43	41	315	243	99	80	55	41	36	35	24	22	314	$C_{21}H_{30}O_2$	Tetrahydrocannabinol	DMIT	0089
314*	299	231	271	243	258	41	43	99	95	80	50	32	30	28	28	314	$C_{21}H_{30}O_2$	Tetrahydrocannabinol	NIH MSC	3823
231	314*	271	258	43	193	41	246	99	79	44	42	41	31	31	22	314	$C_{21}H_{30}O_2$	Tetrahydrocannabinol-delta-8	DMIT	0132
315*	393	451	316	44	91	253	197	99	26	24	18	18	16	15	15	0	—	OV-17 column bleed	DMIT	0247
317*	223	75	147	217	43	318	73	99	57	45	43	43	31	28	26	557	$C_{24}H_{43}NO_8Si_3$	Acetaminophen glucuronide methyl ester TMS ether	BLR	0012
73	317*	169	171	41	318	217	75	99	67	55	18	17	17	16	14	620	$C_{26}H_{45}ClO_9Si_3$	Clofibrate glucuronide methyl ester TMS ether	BLR	0025
73	317*	217	75	147	318	43	79	99	58	20	19	18	15	10	10	688	$C_{32}H_{48}N_2O_9Si_3$	5-(4-Hydroxyphenyl)-3-methyl-5-phenylhydantoin glucuronide methyl ester TMS ether	BLR	0044
58	318*	86	272	85	320	232	42	99	28	26	12	11	10	9	6	318	$C_{17}H_{19}ClN_2S$	Chlorpromazine	NIH MSC	3776
73	319*	291	348	206	320	45	333	99	97	81	74	61	30	28	25	348	$C_{17}H_{24}N_2O_4Si$	Hydroxyphenobarbital 1,3-dimethyl derivative TMS ether	BLR	0069
323*	73	338	324	321	45	77	143	99	66	29	24	21	16	16	15	338	$C_{19}H_{22}N_2O_2Si$	Methaqualone MTB 3 TMS	DMIT	0311
323*	221	181	222	223	207	324	72	99	64	41	36	32	29	22	21	323	$C_{20}H_{25}N_3O$	Lysergide	DMIT	0178
323*	338	73	91	321	65	324	45	99	64	55	53	37	30	27	25	338	$C_{19}H_{22}N_2O_2Si$	Methaqualone MTB 5 TMS	DMIT	0313

323*	338	206	324	339	308	325	89	99	48	40	25	24	15	10	9	266	$C_{13}H_{18}O_4Si$	*p*-Hydroxyphenylpyruvic acid methyl ester trimethylsilyl ether	BLR	0178
73	323*	338	321	143	45	77	154	99	75	35	30	26	23	19	19	338	$C_{19}H_{22}N_2O_2Si$	Methaqualone MTB 4 TMS	DMIT	0312
324*	193	206	73	325	75	194	45	99	77	66	58	29	24	15	13	339	$C_{15}H_{25}NO_4Si_2$	2-Hydroxybenzoylglycine TMS ester TMS ether	BLR	0030
324*	338	339	308	293	325	220	340	99	93	75	27	22	22	18	16	339	$C_{20}H_{21}NO_4$	Papaverine	DMIT	0200
326*	179	267	327	73	180	328	268	99	84	27	25	16	11	9	5	326	$C_{15}H_{26}O_4Si_2$	3,4-Dihydroxyphenylacetic acid methyl ester trimethylsilyl ether	BLR	0183
326*	327	194	253	311	267	73	328	99	24	15	15	14	9	8	8	326	$C_{15}H_{26}O_4Si_2$	2,5-Dihydroxyphenylacetic acid methyl ester trimethylsilyl ether	BLR	0180
327*	369	268	43	310	215	42	204	99	65	54	49	47	30	28	25	369	$C_{21}H_{23}NO_5$	Heroin	NIH MSC	3789
131	327*	73	143	75	132	328	169	99	31	28	19	16	7	5	4	342	$C_{16}H_{30}N_2O_4Si$	Hydroxyamobarbital 1,3-dimethyl derivative TMS ether	BLR	0079
326	327*	194	253	311	267	73	328	99	24	15	15	14	9	8	8	326	$C_{15}H_{26}O_4SI_2$	2,5-Dihydroxyphenyl acetic acid methyl ester trimeththylsilyl ether	BLR	0180
58	328*	100	42	135	228	229	242	99	13	7	4	4	3	3	3	328	$C_{19}H_{24}N_2OS$	Methotrimeprazine	DMIT	0140
332*	303	275	304	190	276	134	333	99	81	64	49	21	20	17	17	332	$C_{18}H_{24}N_2O_4$	Hydroxyphenobarbital 1,3-di-ethyl derivative ethyl ether	BLR	0087
332*	333	304	335	176	303	334	162	99	85	66	48	40	38	34	31	353	$C_{20}H_{19}NO_5$	Chelidonine	DMIT	0199
332	333*	304	335	176	303	334	162	99	85	66	48	40	38	34	31	353	$C_{20}H_{19}NO_5$	Chelidonine	DMIT	0199
334*	44	120	77	41	107	55	144	99	44	36	34	33	32	32	32	334	$C_{21}H_{22}N_2O_2$	Strychnine	DMIT	0099
336*	176	334	338	88	337	175	168	99	76	51	48	38	16	12	12	334	$C_{14}H_8Br_2$	9,10-Dibromoanthracene	DMIT	0332
338*	219	339	73	220	77	340	75	99	98	29	26	17	16	11	5	338	$C_{16}H_{26}O_4Si_2$	3,4-Dihydroxycinnamic acid methyl ester trimethylsilyl ether	BLR	0179
73	338*	247	323	179	77	75	235	99	69	66	55	49	39	32	28	338	$C_{19}H_{22}N_2O_2Si$	Methaqualone MTB 1 TMS	DMIT	0310
323	338*	73	91	321	65	324	45	99	64	55	53	37	30	27	25	338	$C_{19}H_{22}N_2O_2Si$	Methaqualone MTB 5 TMS	DMIT	0313
323	338*	206	324	339	308	325	89	99	48	40	25	24	15	10	9	266	$C_{13}H_{18}O_4Si$	*p*-Hydroxyphenylpyruvic acid methyl ester trimethylsilyl ether	BLR	0178
324	338*	339	308	293	325	220	340	99	93	75	27	22	22	18	16	339	$C_{20}H_{21}NO_4$	Papaverine	DMIT	0200
179	340*	267	73	341	180	342	268	99	84	31	24	22	14	7	6	340	$C_{16}H_{28}O_4Si_2$	3,4-Dihydroxyphenylpropionic acid methyl ester trimethylsilyl ether	BLR	0177
341*	55	36	300	342	110	243	256	99	79	58	29	23	23	23	22	341	$C_{20}H_{23}NO_4$	Naltrexone	DMIT	0290

Eight Most Intense Peaks — E2 (continued)

Eight most intense peaks								Relative Intensities								Mol wt	Formula	Drug	Source	
73	341*	43	271	75	147	41	342	99	55	26	23	18	14	12	12	444	$C_{20}H_{40}N_2O_5Si_2$	Dihydroxysecobarbital di-TMS ether 1,3-dimethyl derivative	BLR	0074
73	341*	342	343	45	344	327	91	99	90	64	44	25	19	16	16	342	$C_{18}H_{19}ClN_2OSi$	7-Chloro-1,3-dihydro-5-phenyl-2H-1,4-dibenzodiazepin-2-one TMS deriv	BLR	0113
343*	55	110	36	98	302	84	344	99	72	45	41	31	29	23	23	343	$C_{20}H_{25}NO_4$	Alpha-hydroxy naltrexone	DMIT	0291
73	343*	257	256	345	372	283	45	99	77	41	31	31	29	29	27	372	$C_{19}H_{21}ClN_2O_2Si$	7-Cl-1,3-dihydro-3-hydroxy-1-methyl-5-phenyl-2H-1,4-benzodiazepin-2-one TMS ether	BLR	0116
344*	73	372	371	346	345	373	374	99	78	48	39	38	31	26	20	372	$C_{19}H_{21}ClN_2O_2Si$	7-Cl-1,3-dihydro-5-(4-hydroxyphenyl)-1-methyl-2H-1,4-benzodiazepin-2-one TMS ether	BLR	0114
376	347*	319	348	377	361	73	192	99	84	44	31	24	22	20	14	376	$C_{19}H_{28}N_2O_4Si$	Hydroxyphenobarbital 1,3-di-ethyl derivative TMS ether	BLR	0088
82	350*	280	351	181	43	81	162	99	75	39	19	15	15	13	12	462	$C_{20}H_{22}D_{18}N_2O_5Si_2$	Dihydroxysecobarbital di-d9-TMS ether 1,3-di-methyl derivative	BLR	0075
354*	73	104	268	325	282	355	77	99	49	42	37	36	26	26	23	354	$C_{19}H_{22}N_2O_3Si$	5-(3-Hydroxyphenyl)-3-methyl-5-phenylhydantoin TMS ether	BLR	0041
354*	277	325	104	268	73	269	355	99	85	85	65	65	60	30	25	354	$C_{19}H_{22}N_2O_3Si$	5-(4-Hydroxyphenyl)-3-methyl-5-phenylhydantoin TMS ether	BLR	0043
355*	356	357	414	358	399	237	415	99	32	14	9	3	3	3	3	414	$C_{18}H_{34}O_5Si_3$	3,4-Dihydroxymandelic acid methyl ester trimethylsilyl ether	BLR	0162
355	356*	357	414	358	399	237	415	99	32	14	9	3	3	3	3	414	$C_{18}H_{34}O_5Si_3$	3,4-Dihydroxymandelic acid methyl ester trimethylsilyl ether	BLR	0162
357*	58	59	102	75	45	373	374	99	97	53	45	37	32	29	25	0	—	Blood 955 drug	DMIT	0327
365*	195	221	28	366	31	29	212	99	57	57	44	40	36	34	33	578	$C_{32}H_{38}N_2O_8$	Deserpidine	DMIT	0176
246	366*	106	247	133	260	234	367	99	53	24	19	18	17	16	15	366	$C_{23}H_{30}N_2O_2$	Piminodine	DMIT	0125
368*	386	275	149	353	147	145	247	99	68	61	58	55	53	51	51	386	$C_{27}H_{46}O$	Cholesterol	NIH MSC	3777

327	369*	268	43	310	215	42	204	99	65	54	49	47	30	28	25	369	$C_{21}H_{23}NO_5$	Heroin	NIH MSC	3789
312	371*	43	297	399	281	298	313	99	53	50	44	36	34	24	22	399	$C_{22}H_{25}NO_6$	Colchicine	DMIT	0272
376*	347	319	348	377	361	73	192	99	84	44	31	24	22	20	14	376	$C_{19}H_{28}N_2O_4Si$	Hydroxyphenobarbital 1,3-di-ethyl derivative TMS ether	BLR	0088
174	382*	161	187	159	175	383	41	99	60	31	25	22	19	18	15	382	$C_{27}H_{42}O$	3,5-Cholestadiene-7-one	DMIT	0254
386*	275	107	43	95	55	81	368	99	73	68	63	63	58	57	56	386	$C_{27}H_{46}O$	Cholesterol	DMIT	0105
368	386*	275	149	353	147	145	247	99	68	61	58	55	53	51	51	386	$C_{27}H_{46}O$	Cholesterol	NIH MSC	3777
315	393*	451	316	44	91	253	197	99	26	24	18	18	16	15	15	0	—	OV-17 column bleed	DMIT	0247
399*	113	70	141	43	72	400	71	99	97	79	72	34	28	28	22	399	$C_{22}H_{29}N_3S_2$	Torecane	DMIT	0094
408*	165	409	243	167	241	166	410	99	59	33	30	23	20	16	12	566	$C_{32}H_{24}Br_2$	Alpha,alpha-dibromo-alpha, alpha, alpha, alpha-tetraphenylxylene	DMIT	0331
73	409*	75	319	410	381	55	465	99	64	42	27	22	14	14	12	480	$C_{26}H_{48}O_4Si_2$	Prostaglandin A1 TMS ester TMS ether	BLR	0120
73	429*	430	45	431	147	432	75	99	32	32	21	20	14	11	10	430	$C_{21}H_{27}ClN_2O_2Si_2$	7-Cl-3-hydroxy-5-phenyl-1,3-dihydro-2H-1,4-benzodiazepin-2-one di-TMS derivative	BLR	0117
444*	133	330	358	445	387	69	105	99	38	33	32	28	26	25	17	472	$C_{18}H_{15}F_7N_2O_5$	Hydroxyphenobarbital 1,3-dimethyl derivative HFB derivative	BLR	0070
73	478*	75	479	43	132	74	55	99	53	40	22	9	9	8	8	509	$C_{27}H_{51}NO_4Si_2$	Prostaglandin B1 methyloxime TMS esther TMS ether	BLR	0122
73	483*	75	147	191	129	554	367	99	32	31	29	29	28	25	21	644	$C_{32}H_{68}O_5Si_4$	Prostaglandin F1B TMS ester tri-TMS ether	BLR	0127
73	485*	55	470	75	486	44	45	99	44	39	31	23	20	18	18	485	$C_{26}H_{39}NO_4Si_2$	Naltrexone-2 TMS	DMIT	0279
73	559*	55	558	372	560	75	373	99	40	35	32	25	20	20	12	559	$C_{29}H_{49}NO_4Si_3$	Naltrexone MTB 1-3 TMS	DMIT	0284
298	594*	58	593	299	609	595	564	99	37	26	21	17	16	15	8	694	$C_{38}H_{44}Cl_2N_2O_6$	Tubocurarine chloride	DMIT	0170

MOLECULAR WEIGHT INDEX

Michael Caplis

mol wt	Substance	CI Isobutane (relative abundance) MH^+	CI Isobutane (relative abundance) Major peaks	CI CH_4 (relative abundance) MH^+	CI CH_4 (relative abundance) Major peaks	EI Base peak
60	Ethylenediamine	61(100)				
75	Glycine	76(100)				
75	Thioacetamide	76(100)	75(80)			
76	Thiourea	77(100)				
85	Piperidine	86(100)	85(20); 84(18)			
88	Putrescine	89(100)	72(60); 73(30)			
89	Sarcosine	90(100)				
90	Oxalic acid	91(100)				
	Phenol					94
	Dimethylsulphone					79
	Piperidone					30
102	Pentanediamine	103(100)	86(39)			103
103	Oxacillin mtb.					
104	Malonic acid	105(100)				107
108	p-Cresol					
108	p-Phenylenediamine			109(83)	108(108); 107(1)	79
108	Benzyl alcohol					
109	Aminophenol	110(100)	109(40)	110(100)	109(73); 93(11); 92(8)	
109	3-Pyridine methanol	110(100)	92(14)	110(100)	92(45)	
110	Hydroquinone	111(100)	110(50)			
110	Resorcinol	111(100)	110(100)	111(100)	110(9)	
111	Histamine	112(100)				82
113	2-Nitro-imadazole					113

MOLECULAR WEIGHT INDEX (continued)

mol wt	Substance	CI: Isobutane (relative abundance) MH⁺	CI: Isobutane (relative abundance) Major peaks	CI: CH_4 (relative abundance) MH⁺	CI: CH_4 (relative abundance) Major peaks	EI Base peak
116	Dimethylgloyxime	117(100)	116(16)			
116	Levolinic acid					43
116	Fumaric acid	117(100)	73(25); 99(10)			
117	Indole					117
118	Benzimdazole	119(100)	118(20)			
120	Dithiooxamide	121(100)	120(22)			
121	Phenethylamine	122(100)	121(35); 105(15)			
121	Tromethamine	122(100)	104(15)			
122	Benzoic acid	123(100)	105(35)			
122	Nicotinamide	123(100)		123(100)	106(3)	
124	Diaminophenol	125(100)	124(35)			
126	Ethchlorvynol (dehyd. prod.)			127(100)		
126	Phloroglucinol	127(100)	126(25)			
126	Pyrogallol	127(100)	126(40)			
127	Coniine	128(100)	126(24)			
128	Barbituric acid	129(100)				
128	Isobarbituric acid	129(100)				
132	Paraldehyde					45
133	Tranylcypromine	134(100)	132(25); 133(20)	134(25)	117(100); 91(88); 133(52); 119(26)	132
135	Acetanilide			136(100)	94(14)	
135	Amphetamine	136(100)		136(100)	119(65); 91(9); 134(5)	44
136	Allopurinol			137(100)	136(11)	136
136	Pinene					93

136	Phenelzine	137(100)	135(34); 122(22); 105(12)			105
137	*o*-Aminobenzoic acid	138(100)	137(35); 120(35)			
137	β-hydroxy-β-phenylehylamine	138(100)	120(50)			
137	Isoniazide	138(100)	137(10)	138(100)	108(22)	78
137	Phenacetin mtb.			138(100)	121(4)	
137	Pyridine-3-carboxcylic acid methyl ester					106
137	Salicylamide	138(100)	137(11)	138(100)	121(20); 95(4)	
137	Salicylaldoxine	138(100)	137(22); 120(15); 91(10)			
137	Tyramine	138(100)	121(50); 108(30); 107(10)			30
138	*p*-Hydroxybenzoic acid	139(100)	121(35)			
138	Pentylenetrazole	139(100)		139(100)	96(8)	
138	Salicylic acid	139(100)		139(100)	121(90); 95(8)	93
140	Dimethyldihydroresorcinol	141(100)				
140	Methenamine	141(100)	112(45); 140(35)			42
141	Cyclopentamine				143(100); 140(4) 126(23); 111(20)	
141	Ethosuximide	142(100)		142(100)	114(9); 113(6)	113
141	Meparifynol carbamate			142(<1)	99(100)	
	Oxacillin mtb.					141
141	Tropine	142(60)	124(100); 140(50); 141(20)			
143	Trimethadione	144(100)		144(100)	100(10); 128(2)	
144	Ethchlorvynol	145(–%)	127(100); 129(34); 109(10)	145(18)	107(100); 109(1); 115(86); 143(45)	115
				145(8)	127(100); 115(1); 129(33); 117(33); 143(25); 109(16)	
145	Emylcamate			146(10)	102(100); 101(8)	73
146	Phenobarbital mtb.					117
145	Quinolinol			146(100)	145(7)	

MOLECULAR WEIGHT INDEX (continued)

mol wt	Substance	CI: Isobutane (relative abundance) MH⁺	CI: Isobutane (relative abundance) Major peaks	CI: CH_4 (relative abundance) MH⁺	CI: CH_4 (relative abundance) Major peaks	EI Base peak
146	Coumarin			147(100)		
146	*p*-Dichlorobenzene			147(100)	149(50); 175(5); 177(2)	
147	Isatin	148(100)				
148	Chloral hydrate mtb.					31
149	*o*-Acetotoluide			150(100)	149(5); 107(5) 106(4)	
149	*p*-Dimethylaminobenzaldehyde	150(100)	149(15)			
149	Methamphetamine	150(100)		150(36)	119(100); 148 120(10)	158
149	p-Methylamphetamine	150(100)	133(49)			
149	Phentermine	150(100)				
149	Phenylpropylmethylamine	150(100)		150(90)	119(100); 91(63); 148(55); 134(8)	
150	*p*-Aminoacetanilide			151(100)	150(14); 108(3); 109(2)	
150	Methylphenidate mtb.			151(24)	91(100); 119(30)	91
150	Pheniprazine			151(100)	91(98); 119(93); 145(26); 149(11)	
150	Phenylacetic acid methyl ester					91
150	Tartaric acid	151(100)	105(33); 123(15)			
150	Thymol			151(62)	109(100); 149(4); 135(37); 137(35)	
151	Acetaminophen	152(100)		152(100)	109(11); 134(2)	109
151	Acetaminophen methyl ester					108

151	Amantadine	152(100)	135(48); 151(32) 150(15)			
151	*p*-Hydroxyamphetamine	152(100)	135(95)	152(42)	135(100); 107 150(3)	
151	Methylsalicylamide	152(100)	151(73); 121(25); 120(13)			
151	Norpseudoephedrine			152(45)	134(100)	
151	*o*-Nitrobenzaldehyde	152(100)				
151	Phenylpropanolamine	152(100)	134(50); 107(25)	152(5)	134(100); 117 150(2)	
151	*p*-Aminobenzoic acid methyl ester					120
151	*o*-Aminobenzoic acid methyl ester					119
152	Mandelic acid	153(−%)	135(100); 107(65)			
152	Methylparaben			153(100)	109(11); 121(10)	
152	Methyl p-hydroxybenzoate					121
152	Methylsalicylate	153(100)	152(35); 120(35); 121(20)	153(100)	121(15)	120
152	Camphor					126
152	3-Methylsalicylate	153(80)	135(100); 134(82); 152(70); 105(42); 106(34); 78(25)			
152	Vanillin	153(100)	152(15)	153(100)	125(12)	
153	4-(2-Aminoethyl)pyro-catecol					124
153	*p*-Aminosalicylic acid	154(100)	153(28); 135(13); 136(12)	154(100)	136(20); 110(3)	
153	Hydroxytyramine	154(100)	137(24); 124(24)			
154	Chloroacetophenone	155(100)	157(35); 139(20); 154(17)			
154	Gentisic acid	155(100)				
155	Bemegride			156(100)	128(3)	
155	Propylhexedrine			156(100)	154(39); 140(9); 125(4)	

MOLECULAR WEIGHT INDEX (continued)

		CI				EI
		Isobutane (relative abundance)		CH_4 (relative abundance)		
mol wt	Substance	MH^+	Major peaks	MH^+	Major peaks	Base peak
155	Ethosuximide *n*-methyl derivative					127
156	Ectylurea	157(100)	114(10)	157(73)	97(100); 114(17)	41
157	Oxanamide	158(100)	131(10)	158(68)	140(100); 141(4); 113(24); 97(17)	36
157	Paramethadione			158(100)		
157	Violuric acid	158(100)	142(33)			
158	Phencyclidine contaminant			159(100)	157(23); 119(16)	
158	1-Phenyl cyclohexene					129
159	Pagyline	160(100)	159(20)			
159	N-Methyl-N-2-propynylbenzyl-amine					91
160	Hydralazine	161(100)		161(100)	160(37); 132(20)	
160	Tolazoline	161(100)		161(100)	159(12)	
161	Oxacillin mtb.					146
162	Nicotine	163(100)		163(100)	161(27)	84
162	2,4-Dichlorophenol					162
162	Cinnamic acid methyl ester					131
163	*m*-Acetylcysteine	164(100)				
163	Mephentermine	164(100)		164(45)	72(100); 133(10) 162(10)	
163	Primaclone mtb.					91
163	*N,N'*-Dimethylamphetamine			164(100)	119(10)	
164	1-Thiophenyl cyclohexene					164
164	Chloralhydrate			165(22)	147(100); 149(9) 111(75); 113(49)	82

164	β-Phenylpropionic acid; methyl ester					104
164	Eugenol			165(100)	137(10); 150(8); 105(2)	
164	2,2-Dichloro-1,1-difluoroethyl methyl ether					81
165	Benzocaine	166(100)		166(100)	138(15); 120(15); 122(8); 94(5)	120
165	Methapyrilene mtb.					58
165	Ephedrine	166(100)	148(50)	166(29)	148(100); 164(8)	
165	Pseudoephedrine					58
165	Ethylaminbenzoate			166(100)	138(22); 120(20); 122(14)	
165	4-Hydroxphenylisopropyl-methylamine	166(100)	135(10)			
165	Isoephedrine	166(100)	107(42); 148(15)			
165	*m*-Methoxyamphetamine	166(100)	149(10); 138(10)			
165	*p*-Methoxyamphetamine	166(100)	149(45); 122(15)			
166	Atrolactic acid	167(100)	149(58); 107(37); 166(30)			
166	*p*-Methoxybenzoic acid methyl ester					135
166	Ethionamide	167(100)	166(25)			
166	*o*-Methoxybenzoic acid methyl ester					135
166	Phthalic acid	167(25)	167(25); 123(12); 149(100)			
166	m-Methoxybenzoic acid methyl ester					166
166	Vanillal			167(100)	139(25); 111(6)	
167	Ethinamate	168(–%)	125(100); 107(66); 163(13)	168(0.5)	107(100); 125(10); 166(1)	91
167	*d*-4-Hydroxynorephedrine	168(50)	150(100); 123(17)			
167	2-Mercaptobenzothiazole			169(100)		

MOLECULAR WEIGHT INDEX (continued)

mol wt	Substance	CI Isobutane (relative abundance) MH⁺	CI Isobutane Major peaks	CI CH_4 (relative abundance) MH⁺	CI CH_4 Major peaks	EI Base peak
167	*m*-Nitrobenzoic acid	168(100)	138(45)			
167	Orthocaine	168(100)	136(10)			
167	Phenylephrine	168(100)	150(33); 123(10)	168(15)	150(100); 166(5)	
168	Cyclopal	169(100)	235(98)			
168	*o*-Dinitrobenzene	168(100)	152(15)			
168	Zoxazolamine			169(100)	133(12)	168
169	Chlorzoxazone	170(100)	172(35); 118(20); 136(10)	170(100)		169
169	Diphenylamine	170(100)	169(45)			
169	Norepinephrine				105(100); 132(50); 151(29)	
169	Ethosuximide n-ethyl derivative					141
169	Piperidione			170(100)	152(86); 141(1) 168(7)	
169	Pyridoxine	170(100)	152(15)			
170	4-Chlorobenzoic acid methyl ester					139
171	Metronidazole	172(100)				
172	5-Ethyl-2-thiobarbital	173(100)				
172	Sulfanilamide			173(100)	156(67); 93(3)	
173	1-Nitroso-2-naphthol	158(100)	174(40); 129(10)			
173	Quinaldic acid	174(100)				
174	Methyltryptamine	175(100)	158(50); 131(40)			
176	Ascorbic acid	177(100)				116
176	2-Imino-5-phenyl-4-oxazolidinone					176
176	Cotinine (nicotine mtb.)			177(100)	98(23)	98

177	Phenmetrazine	178(100)		178(100)	100(18); 176(10)	71
177	Diphenylhydantoin (D5-ph)				91(8); 119(5)	185
178	Anthracene	179(100)	178(14)			
178	Nikethamide	179(100)		179(100)	106(9); 177(8)	
178	Ninhydrin	161(100)	177(20); 132(15)			
178	Phenacemide	179(100)	136(40)	179(100)	136(74); 118(9); 91(7)	
178	Phenacetyl urea					118
178	Acridine	180(100)	179(24); 181(14)			
179	Hippuric acid	180(100)	105(54); 135(31); 134(21); 162(15)			105
179	Methylenedioxy-amphetamine	180(100)	136(10)	180(100)	163(49); 136(14)	
179	Methoxyphenamine			180(74)	149(100); 178(19); 121(13)	
179	Acetophenenetidin					108
179	*n*-Methylephedrine	180(50)	72(100); 162(20)			
179	Napthoquinoline	180(100)				
179	Phenacetin	180(100)	179(12)	180(100)	152(11)	
180	Acetylsalicylic acid (aspirin)	181(17)	121(100); 139(20); 163(18); 138(17); 180(10)	181(<1)	121(100); 139(68); 163(25)	120
180	1,7-Dimethylxanthine			181(100)		
180	Frutose	181(–%)	163(100); 145(85); 127(25)			
180	Galactose	181(–%)	163(100); 145(85); 127(25)			
180	Glucose	181(–%)	163(100); 145(85); 127(25)			
180	Maltose	163(100)	145(85); 127(40)			
180	Mannose	181(–%)	163(100); 145(85); 127(25)			
180	*o*-Phentroline	181(100)				
180	Propylparaben			181(90)	139(100); 121(60); 95(18); 167(10)	

MOLECULAR WEIGHT INDEX (continued)

		CI				EI
		Isobutane (relative abundance)		CH_4 (relative abundance)		
mol wt	Substance	MH^+	Major peaks	MH^+	Major peaks	Base peak
180	Theobromine	181(100)	180(10)	181(100)	138(2)	180
180	Theophylline	181(100)	180(12)	181(100)	124(5); 95(2)	180
180	p-Methoxyphenylacetic acid methyl ester	181(100)	180(15)			121
180	*o*-Methoxyphenylacetic acid methyl ester					121
180	*m*-Methoxyphenylacetic acid methyl ester					180
181	Styramate			182(1)	121(100); 164(43); 107(4)	
181	Doxylamine					180
182	Mannitol	183(100)	165(20); 147(10); 129(10)	183(47)	129(100); 111(61); 99(55); 103(33) 165(21)	
182	Mephenesin	183(100)	147(30); 109(17); 165(15)	183(21)	147(100); 135(79); 121(65); 165(48) 109(43)	108
182	Methyprylon					155
182	Sorbitol	183(100)	165(20); 147(10); 129(10)			
183	Adrenaline	184(25)	166(100)	184(14)	166(100); 164(5); 182(3); 123(3)	
183	Chlorphentermine	184(100)	186(30)			
183	Methyprylon	184(100)		184(100)	166(60); 155(13); 182(6)	
183	Saccharin	184(100)		184(14)		

184	Apronalide			185(100)	142(51); 97(43) 125(29)	
184	Barbital	185(100)		185(100)	156(5); 142(2) 141(1)	156
184	2-Chloro-1,1,2-trifluoro-ethyl difluoromethyl ether					51
184	Benzidine	185(100)	184(60)			
184	2,4-Dinitrophenol	185(100)	155(40)			
185	Ecgonine (cocaine mtb.)	186(100)	124(30: 141(25); 168(21); 185(17)	186(100)	168(42); 124(3)	
187	Quinoline-2-carboxylic acid methyl ester					129
188	Antipyrine	189(100)		189(100)	217(30); 190(12)	188
188	Dimethyltryptamine	189(100)				58
188	1-(p-tolyl)-3-methyl-pyrazol-5-one					188
189	*o*-Chlorobenzalmalonitrile	189(100)	191(33)			
189	Methsuximide mtb.					118
189	*a*-Phenylglutaramide			190(100)	102(15); 104(5)	
189	3-Indolacetic acid methyl ester					130
189	Phensuximide			190(100)	104(5)	104
190	Terpin			191(<1)	137(100); 96(3)	
191	Chloropropamide mtb.			192(100)	175(10)	
191	Phendimetrazine	192(100)		192(100)	105(31); 190(22); 114(13)	
192	Nicotine			193(100)	114(11); 175(9)	
192	1-Piperidine-cyclohexane-carbonitrile (PCC)	193(–%)	166(100); 192(10); 149(10); 124(5)			
193	Butamben			194(100)	138(92); 120(35); 166(35); 94(13)	
193	Hippuric acid methyl ester					
194	Butylparaben			195(73)	139(100); 121(25); 167(19); 95(10)	

MOLECULAR WEIGHT INDEX (continued)

mol wt	Substance	CI Isobutane (relative abundance) MH^+	CI Isobutane (relative abundance) Major peaks	CI CH_4 (relative abundance) MH^+	CI CH_4 (relative abundance) Major peaks	EI Base peak
194	Meconin					165
194	Caffeine	195(100)		195(100)	138(2)	194
194	Propylparaben methyl ester					135
194	Acetylsalicylic acid methyl ester					120
194	*p*-Methoxyphenylpropionic acid methyl ester					121
195	2-5-Dimethoxyamphetamine	196(100)	179(25); 152(25)			
195	3,5-Dimethoxyamphetamine			196(15)	91(100); 179(6); 152(30)	
195	*p*-Xenylcarbimide	196(40)	170(100); 195(30); 169(30)			
196	Cantharidin	197(100)				
196	2,5-Dimethoxyamphetamine					47
196	3,4-Dimethyoxybenzoic acid methyl ester					196
197	Dopa	198(100)				
197	Methyprylon mtb.			198(100)	128(22); 170(15)	
197	Metanephrine	198(100)	180(27)			
198	2,4-Dinitrophenylhydrazine	199(100)				
198	Glyceryl guaiacolate	199(100)	198(60); 124(50); 125(30)	199(75)	125(100); 151(60); 163(45); 181(25); 137(20)	
198	Guanethidine			199(79)	140(100); 197(29); 114(16); 182(13)	

198	Metharbital	199(100)		199(100)	170(4)	170
198	Probarbital	199(100)				156
198	Xanthydrol	181(100)	197(25)			
199	Chlorophenylalanine	200(100)	202(35); 154(30)			
199	Methyprylon mtb.			200(14)	182(100); 154(36)	83
199	Doxylamine mtb.			200(100)	182(55); 171(4); 182(2)	182
199	Phenothiazine	200(45)	199(100)	200(100)	199(57)	199
200	Tetrahydrazoline	201(100)				
200	Thiobarbituric acid	201(–%)	145(100)			
200	Pentachloroethane					167
201	Indolepropionic acid methyl ester					130
202	Oxacillin mtb.					144
203	Aminoantipyrine	204(100)	203(50); 84(42)	204(100)	203(20); 94(5)	
				204(100)	120(13); 201(5)	
203	Ampyrone					56
203	Methsuximide	204(100)		204(100)		
				204(100)	118(40); 205(15); 232(10)	
204	Psilocin					58
204	Bufotenine	205(100)	198(11)			58
204	Mephenytoin mtb.					104
204	Ethotoin			205(100)	127(8)	204
204	Thozalinone			205(100)	136(53)	
205	Diethylpropion	206(100)	204(30); 100(28); 135(15)	206(100)	100(51); 204(11); 190(6)	100
206	*l*-Buprofen			207(100)	160(98); 188(10); 190(9)	
206	Lidocaine mtb.			207(100)	169(43); 205(22)	
206	Pivalylbenzhydrazine			207(100)	91(50); 106(10); 205(4)	
206	Primaclone mtb.					163

MOLECULAR WEIGHT INDEX (continued)

mol wt	Substance	CI Isobutane (relative abundance) MH⁺	CI Isobutane (relative abundance) Major peaks	CI CH_4 (relative abundance) MH⁺	CI CH_4 (relative abundance) Major peaks	EI Base peak
206	Primidone mtb.			207(14)	162(100); 163(20); 190(5)	
208	Allobarbital	209(100)	169(11)	209(100)	141(21); 169(6)	167
208	Neostigmine	209(100)				
208	Pilocarpine	209(100)	208(15)	209(100)	95(4)	
208	Theobromine 1-ethyl derivative					208
208	Propoxyphene decomp. prod.			209(4)	131(100); 159(2); 105(6); 208(3)	105
208	Theophylline 7-ethyl derivative					208
208	1,3,7(or 9),8-Tetramethyl-xanthine					208
209	2,5-Dimethoxy-4-methyl-amphetamine	210(100)				44
209	Hydroxyphenamate			210(<1)	131(100); 149(3); 192(20)	
210	3,4-Dimethoxyphenyl-acetic acid methyl ester					151
210	Aprobarbital	211(100)	171(13)	211(100)	169(76)	167
210	Naphazoline	211(100)	210(20); 209(18)	211(100)	209(11)	
211	1,3-Diphenylguanidine			212(100)	94(37); 211(32); 119(22); 195(20)	
211	Isoproterenol	212(50)	194(100)			
211	Mescaline	212(100)		212(100)	195(90); 182(14); 168(10); 151(8)	182
211	Methoxamine	212(100)	194(55); 152(20); 168(18); 167(15)	212(12)	194(100); 168(9); 210(2)	
212	Barbital dimethyl deriv.			213(100)	184(8); 211(3)	169

212	Butabarbital	213(100)		213(97)	157(100); 156(2); 185(10)	
212	Butethal	213(100)		213(100)	156(10)	141
213	Phenazopyridine	214(100)				
214	Methylaurate			215(100)	213(20); 181(11); 183(8)	
214	Phenaglycodol	215(–%)	197(100); 155(45); 199(35); 157(18); 198(15); 156(15)	215(2)	197(100); 199(3); 103(16); 121(4)	59
214	Phenylsalicylate	215(100)		215(67)	121(100); 149(20)	
214	Phenyramidol	215(65)	107(100); 95(23); 108(13); 197(10)	215(59)	197(100); 107(30); 108(22); 137(5)	107
215	Atrazine			216(73)	180(99); 181(11); 215(6)	
216	Diethyltryptamine	217(100)				
216	Primidone mtb.			217(100)	146(22); 161(16); 189(13)	
216	Aminoglutethimide					203
217	Glutethimide	218(100)		218(100)	190(9); 189(6)	117
217	3-Indolebutyric acid methyl ester					130
218	2-Methyl-2-propyl-1,3-propane diol carbamate					
218	Mephenytoin	219(100)	189(10)			189
218	Methetoin			219(80)	141(17); 189(11); 162(4)	
218	Meprobamate	219(100)	158(61)	219(5)	158(100); 115(40);	83
218	Primidone	219(100)	164(13); 181(12)	219(80)	162(100); 190(2) 163(12)	
218	Primaclone					117
218	Sulfosalicylic acid	219(100)	218(65); 201(25); 200(20)			

MOLECULAR WEIGHT INDEX (continued)

mol wt	Substance	CI Isobutane (relative abundance) MH$^+$	CI Isobutane Major peaks	CI CH_4 (relative abundance) MH$^+$	CI CH_4 Major peaks	EI Base peak
219	Nor-meperidinic acid methyl ester					57
219	5-Methoxyindoleacetic acid methyl ester					160
220	Ionol					205
220	Doxepin mtb.			221(100)	220(22)	
220	Furildioxime	221(100)	205(25); 203(10)			
220	Prilocaine	221(100)		221(100)	164(7); 219(4); 136(2)	
220	Rompun			221(100)	90(20); 205(10)	
221	Metaxalone	221(100)				
222	Acetazolamide	223(100)	207(18)	223(100)	181(27)	
222	Bromoisovalum			223(100)	182(84); 180(86) 143(59); 100(26)	
222	Diethyl phthalate			223(<1)	177(100); 149(2) 178(12)	
222	3,7-Diethyl-1-methylxanthine					222
222	1,7-Diethyl-3-methylxanthine					222
222	1,3-Diethyl-7-methylxanthine					222
222	1,3-Diethyl-9-methylxanthine					123
222	3,4-Dimethoxycinnamic acid methyl ester					222
223	2-Hydroxybenzoyl glycine methyl ester					135
223	Salicyluric acid methyl ester methyl ether					135

223	Mephenoxalone	224(100)				124
223	Vanillic acid diethylamide			224(100)	110(22); 151(9)	
224	Butalbital					168
224	Allylbarbital	225(100)	185(17)	225(100)	168(10); 185(7)	
224	Talbutal	225(100)	185(22)	225(100)	169(40)	71
224	Vinbarbital	225(100)		225(90)	157(100)	
224	3,4-Dimethoxyphenylpropionic acid methyl ester					151
225	Furazolidone	226(100)				
226	Amobarbital	227(100)		227(100)	157(12); 156(11)	156
226	Metyrapone	227(100)	226(36); 120(12)			
226	Pentobarbital	227(100)		227(80)	157(100); 185(1); 156(8)	156
227	α-Benzoin oxime	210(100)	107(100); 104(90); 228(50)			
228	2-(4-Chlorophenoxy)-2-methylpropionic acid methyl ester (from chlofibrate)					128
228	Picric acid	200(100)	230(70)			
229	1-(1-Phenylcyclo-hexyl)-pyrrolidine	230(100)	229(65); 228(40); 159(22); 186(13)			
229	1-(1-Phenylcyclo-pentyl)-piperadine	230(100)	229(75); 145(43); 228(40); 220(12); 187(11); 201(9)			
231	Aminopyrine			232(100)	231(40); 113(20); 230(16)	56
231	Dipyrone			232(6)	218(100); 217(2); 120(24)	
231	Metazocine					231
231	Fenfluramine	232(100)	72(68); 212(25)			
231	Isocarboxazid			232(100)	91(89); 119(12); 154(9)	91

MOLECULAR WEIGHT INDEX (continued)

		CI				EI
		Isobutane (relative abundance)		CH_4 (relative abundance)		
mol wt	Substance	MH^+	Major peaks	MH^+	Major peaks	Base peak
232	Aminoglutethimide			233(100)	205(19); 94(12)	
232	Chlorprothixene mtb.			233(100)	197(60); 232(45)	
232	Mebutamate	233(100)	172(75)	233(6)	111(100); 172(7); 69(18)	97
232	Mebutamate (GC)					57
232	Phenobarbital	233(100)		233(100)	204(6)	204
233	Chlorpromazine mtb.			234(100)	233(98); 198(50)	233
233	Meperidine acid methyl ester					71
233	Dimethyl meperidine			234(100)	160(48); 232(16); 188(10)	
233	Meperidine mtb.					57
233	Glutethimide mtb.			234(100)	188(30); 232(25)	
233	Methylphenidate	234(100)	84(70); 151(10)	234(98)	91(100); 151(56); 112(30); 119(25)	84
233	2-Methyl-5-methoxyindole-3-acetic acid methyl ester (from indomethacin)					174
234	Lidocaine	235(100)	86(13)	235(100)	233(15); 132(2); 148(2)	86
234	*p*-Dibromobenzene					236
234	Sparteine			235(42)	233(100); 98(21); 137(8)	
235	Hexobarbital					
235	Azapetine	236(100)	234(25)			
235	Procaine amide	236(100)	99(18); 136(12)			

236	Allobarbital dimethyl derivative			236(100)	169(29); 195(16); 197(9)	195
236	Butethamine			237(33)	100(100); 120(8); 135(11); 164(10)	
236	Carbamazepine	237(100)	194(14)	237(100)	238(20); 193(15); 192(14)	193
236	Carbromal	237(100)	239(98)	237(53)	194(100); 196(9); 114(84); 157(70)	44
236	Cyclobarbital	237(100)	157(10)	237(60)	157(100); 207(1); 235(7)	207
236	Hexobarbital	237(100)		237(26)	157(100)	221
236	Procaine	237(100)	100(16); 99(12)	237(12)	100(100); 120(4); 164(9); 235(5)	86
				237(100)	100(80); 120(18); 86(12); 221(8); 164(5)	
236	Ketamine			238(100)	240(25); 220(12); 209(8)	
238	Nitrofurantoin	239(100)	238(20)			
238	Secobarbital	239(100)	199(36)	239(51)	169(100); 197(1); 168(11)	41
238	Aprobarbital 1,3-dimethyl derivative					195
239	Benzphetamine	240(100)	148(85)	240(1)	148(100); 119(9) 91(6); 238(2)	
240	Alizarin	241(100)				
240	Butabarbital dimethyl derivative			241(100)	185(66); 169(16); 213(10); 239(5)	169
240	Diphenylcarbazone	243(100)	241(78); 226(50); 228(45); 150(40); 152(35)			
240	Butethal 1,3-dimethyl derivative					184
240	Pheniramine	241(100)		241(29)	196(100); 239(2)	

MOLECULAR WEIGHT INDEX (continued)

mol wt	Substance	CI: Isobutane (relative abundance) MH⁺	CI: Isobutane (relative abundance) Major peaks	CI: CH_4 (relative abundance) MH⁺	CI: CH_4 (relative abundance) Major peaks	EI Base peak
240	Hexethal					156
241	Mefenamic acid			242(100)	224(48); 241(32) 225(8)	
241	Methocarbamol	242(100)	199(100); 118(83); 124(33); 224(10)	242(10)	118(100); 163(40); 199(27); 224(10)	
241	Sulfaguanidine			242(1)	215(100); 173(5); 156(42)	
242	Pentobarbital mtb.					156
242	Clofibrate	243(100)	115(55); 245(30); 128(13); 169(12)	243(100)	115(78); 169(27); 129(10)	128, 59
242	Amobarbital mtb.					
242	Diphenylcarbazide	152(100)	94(20); 243(15)			
242	5-Ethyl-5-(3-hydroxy-1-methyl-butyl)-barbituric acid	243(40)	225(100); 157(23)			
242	Thiopental	243(100)	227(25)	243(100)	173(38); 227(20); 157(10); 201(10)	172
243	Phencyclidine	244(100)	242(63); 159(32)	244(43)	159(100); 243(9); 242(45); 200(23)	200
243	Norlevorphanol					243
244	Alphenal	245(100)	113(43); 205(14)	245(100)		215
244	Dianisidine	245(100)				
244	Xylometazoline	245(100)		245(100)	244(23); 243(23); 229(14)	
245	Thioridazine mtb.					245
245	Chlorphenesin carbamate	246(55)	203(100); 205(30); 185(20); 167(20); 248(19); 187(7); 169(7)			

245	4-Acetyl-aminoantipyrin					56
245	Flurazepam mtb.			246(100)	245(60); 181(20); 210(18)	
245	2-Methylamino-5-chlorobenzophenone					77
246	Carbocaine			247(27)	98(100); 96(6)	
246	5-(2-Bromoallyl)barbituric acid					168
246	Mephobarbital	247(100)		247(100)	218(13); 245(3)	218
246	Mepivacaine			247(77)	98(100); 245(20)	
247	Methapyrilene mtb.					97
247	Meperidine	248(100)		248(100)	246(20); 174(5); 202(3)	71
247	1,2-Dimethyl-5-methoxyindole-3-acetic acid methyl ester (from indomethacin)					247
247	Ketobemidone					70
248	Phencyclidine (D5-PH)					205
248	Pyrimethamine					247
248	Piridocaine			249(100)	112(88); 120(45); 110(18)	
249	Cinchophen			250(100)	232(5); 206(2)	
249	2,6-Dimethyl-4-isopropylbenzaldehyde thiosemicarbazone	250(100)	174(18); 158(15)			
249	Sulfapyridine	250(100)	185(20)			184
249	1-(1-Thiophenyl cyclohexyl) piperidine					97
250	Heptabarbital			251(42)	157(100); 95(28); 221(11); 249(6)	221
250	8-Methyxanthine 1,3,7-tri-ethyl derivative					250
250	Hexobarbital methyl derivative			251(60)	171(100); 235(1); 249(7)	235

MOLECULAR WEIGHT INDEX (continued)

		CI				EI
		Isobutane (relative abundance)		CH_4 (relative abundance)		
mol wt	Substance	MH^+	Major peaks	MH^+	Major peaks	Base peak
250	Methaqualone	251(100)	250(30); 235(20)	251(100)	118(40); 146(20)	235
250	Sulfadiazine	251(100)	186(60)	251(100)	156(16); 96(5)	
251	Propoxyphene mtb.			252(64)	143(100); 250(6); 221(24)	
252	Butalbital dimethyl derivative					
252	Benzopyrene	253(100)	252(59)			
252	Cyclizine mtb.	·		253(10)	167(100); 195(7)	
252	Diphenyhydantoin	253(100)		253(100)	175(25); 210(7); 225(4)	180
252	3,4,5-Trimethoxycinnamic acid methyl ester					252
253	Sulfamethoxazole	254(100)				92
253	Triamterene			254(100)		253
254	Dyphylline	255(100)	254(40)			
254	N-D3-Methyl-phenobarbital (D5-Et)					222
254	Pentobarbital dimethyl derivative			255(100)	185(45); 169(7); 253(4)	169
254	Tetra-methyldiaminodiphenyl-methane	255(100)	134(40)			
254	Thiamylal	255(100)		255(84)	185(100); 213(3)	43
254	Secobarbital mtb.					168
254	Amobarbital dimethyl derivative					169
255	Cotarnine	256(–%)	204(100); 118(40)			
255	Diphenhydramine	256(100)	167(25)	256(52)	167(100); 209(7) 254(5)	58

255	Phenyltoloxamine	256(100)		256(100)	254(60)	58
255	Sulfathiazole			256(100)	156(71); 101(3)	
255	Tripelenamine			256(51)	211(100); 254(4); 185(27); 239(13)	58
256	Ethoxazene	257(100)	256(45); 138(15); 133(15)			
256	Palmitic acid	257(100)	256(90); 239(85)	257(100)	255(30); 239(22) 237(22)	73
257	1-Dromoran	258(100)	256(50); 257(25)			
257	Levorphanol	258(100)	256(40); 257(35); 151(20)			
258	Dibutyl adipate					185
259	Propranolol	260(100)	72(60)			
260	Carisoprodol	261(100)	176(22); 158(21); 200(19); 218(11)	261(8); 158(100)	176(100); 158(39); 200(32); 97(23); 176(99); 200(50); 261(15)	58
260	Cyclophosphamide	261(100)	211(30); 225(25); 213(11); 227(10)			
260	Methaphenilene			261(100)	97(64); 202(47); 216(28); 259(24)	
260	Oxymetazocine	261(100)				
260	Phenobarbital dimethyl			261(100)	232(16)	232
261	Alphaprodine	262(26)	188(100); 187(50); 172(35)			172
261	Ethoheptazine	248(100)		262(100)	260(32); 216(12); 188(5)	107
261	Methapyrilene	262(100)	166(27); 191(13)	262(23)	217(100); 260(16); 191(10); 245(5)	58
261	Phenindamine	262(100)		262(100)	261(58); 260(45); 219(23); 247(7)	260
261	Piperocaine			262(100)	140(40); 112(24); 260(21); 246(12)	

MOLECULAR WEIGHT INDEX (continued)

mol wt	Substance	CI Isobutane (relative abundance) MH⁺	CI Isobutane Major peaks	CI CH_4 (relative abundance) MH⁺	CI CH_4 Major peaks	EI Base peak
261	Thenyldiamine			262(18)	217(100); 260(12); 191(6); 245(6)	58
262	Methohexital	263(100)		263(100)	183(52); 211(18); 221(15); 155(6)	
263	N-Methyl, N-D3-methyl-phenobarbital					235
263	Nortriptyline	264(100)		264(100)	233(45); 262(40)	44
263	Protriptyline	264(100)				
263	Dimethylthiambutene					248
263	Hydroxypethidine					71
264	Ethinazene					249
264	Hexobarbital 3-ethyl derivative					249
264	p-Dimethylaminobenzyl-rhodamine	265(100)	177(15)			
264	Methaqualone mtb.			265(100)	235(10); 118(6)	
264	Picrolonic acid	233(100)	249(65); 234(45); 232(15); 265(10)			
264	Sulfamerazine			265(100)	156(21); 110(4)	
264	Tetracaine	265(100)	194(14)	265(32)	176(100); 263(14); 220(5)	
264	Thiamine			265(3)	144(100); 126(99); 158(38)	
265	Antazoline			266(100)	84(100); 196(22); 85(15)	84
265	Doxepin mtb.			266(100)	221(25); 235(19); 264(12)	
265	Intracaine			266(58)	100(100); 86(22); 149(13)	

265	Propoxyphene mtb.			266(100)	264(98); 143(52); 221(16)	
266	Cyclizine			267(5)	167(100); 195(1) 189(11)	99
266	Desipramine	267(100)	222(21)	267(100)	230(38); 208(29); 195(29); 222(16)	195
266	Methaqualone mtb.			267(100)	265(14)	160
266	Methaqualone mtb.			267(100)	118(6); 134(4)	251
266	Methaqualone mtb.			267(20)	249(100); 118(1); 146(8)	160
266	Methaqualone mtb.					251
266	Methaqualone mtb.					251
266	Methaqualone mtb.					251
266	Methaqualone mtb. (GC)					251
266	Diphenylhydantoin					180
266	Diphenylhydantoin 3-methyl derivative					180
266	Pentachlorophenol	267(100)	269(67); 265(67); 271(20)			
266	Secobarbital dimethyl derivative			267(73)	197(100); 225(1); 265(5)	
267	Apomorphine	268(100)	267(22); 266(20)			266
267	Azacyclonol			268(49)	250(100); 190(6); 266(15); 183(5)	85
267	Anistropine					124
267	Pipradrol			268(21)	250(100); 190(1) 266(7)	
267	Sulfisoxazole	268(100)				
268	Diethylstillbestrol	269(100)	268(35)			
268	Diphenylhydantoin mtb.			269(100)	175(73); 191(10);	
268	Lysergic acid			269(100)	268(52); 224(44); 194(33)	

MOLECULAR WEIGHT INDEX (continued)

		CI				EI
		Isobutane (relative abundance)		CH_4 (relative abundance)		
mol wt	Substance	MH^+	Major peaks	MH^+	Major peaks	Base peak
269	Orphenadrine	270(80)	181(100); 224(11); 198(10)	270(7)	181(100); 182(1); 209(14)	
270	2-Chloroprocaine			271(15)	100(100); 154(5); 271(3)	
270	Medazepam					242
270	Diazepam mtb.			271(100)	242(5); 243(4)	242
270	Doxylamine	184(100)	271(50); 182(40)	271(33)	182(100); 269(1); 90(15); 210(14)	
270	Doxylamine (GC)					58
270	Mecloqualone	271(90)	235(100); 273(32); 118(10)			58
270	Methyl palmitate			271(100)	269(87); 237(12); 239(10)	74
270	Sulfamethizole	271(100)				
270	Tolbutamide	271(100)	172(100); 198(25)			
271	*d*-Methorpan	272(100)	270(64); 271(21)	271(100)	270(51)	59
271	Phenetsal					121
271	Normorphine			272(41)	254(100); 271(16); 270(5)	
271	N,N-D3-Methyl-phenobarbital (05-ET)					239
272	Alphenal dimethyl derivative			273(100)	255(10)	118, 243
272	Dimethisoquin			273(40)	271(100); 228(80); 202(20)	
272	Quinalizarin	273(100)	257(90); 241(80); 272(40); 256(30); 240(30)			

273	4-Bromo-2,5-dimethoxy-amphet- amine	274(100)	276(98)	274(25) 230(19); 232(18)	257(100); 259(91);	
273	Chlormezanone			274(40)	154(100); 156(3); 121(12)	152
273	Chlorpromazine mtb.			274(78)	273(100); 238(5)	
274	Chlorpheniramine	275(100)	277(30); 230(22); 203(10)	275(39)	230(100); 232(29); 239(21); 273(18)	203
274	Mephobarbital ethyl derivative					246
274	Tybamate	275(100)	232(40); 158(30);			55
274	Tybamate (GC)		176(30); 214(20)			41
275	Homatropine	276(10)	124(100); 107(14)	276(2)	124(100); 105(1); 135(6); 258(3)	
274	Tybamate (GC)					41
275	Physostigmine			276(100)	275(75); 219(55); 274(25); 162(5)	
275	Proheptazine					57
275	Physostigmine salicylate	276(100)	139(38); 219(18)			
276	Trimethoprim mtb.					276
276	Chlorpropamide	277(–%)	192(100); 194(35)			
276	Trimethoprim mtb.					276
276	Cyclandelate	277(30)	125(100); 107(20); 135(18)			
277	Thiordazine mtb.					198
277	Amitriptyline	278(100)		278(100)	276(70); 233(23);	58
277	Azathioprine					42
277	Methadone mtb.	278(98)	277(100)	278(100)	279(22)	276
277	Ethylmethylthiambutene					262
278	Acetylcarbromal	279(100)	281(98)	279(59)	237(100); 239(93); 149(47); 151(46)	41
278	Amydricaine			279(46)	157(100); 277(52); 256(30)	

MOLECULAR WEIGHT INDEX (continued)

mol wt	Substance	CI: Isobutane (relative abundance) MH⁺	CI: Isobutane (relative abundance) Major peaks	CI: CH_4 (relative abundance) MH⁺	CI: CH_4 (relative abundance) Major peaks	EI Base peak
278	Dibutyl phthalate			279(2)	149(100); 205(2); 177(8)	149
278	Sulfamethazine			279(100)	156(8); 214(7); 124(2)	
278	Sulfisomidine	279(100)	214(40)	279(100)	124(6); 215(5); 156(3)	
278	Triprolidine			279(48)	208(100); 277(2); 236(7)	209
279	Amitriptyline mtb.			280(10)	262(100); 231(3); 260(12)	
279	Doxepine	280(100)		280(100)	221(21); 278(20)	
280	Diphenylhydantoin dimethyl derivative			281(100)	203(23)	251
280	Imipramine	281(100)		281(89)	235(100); 195(6); 208(59)	235
280	5,5-Diphenylhydantoin 5-ethyl derivative					180
280	Sulfameter	281(100)	216(10)			
280	Sulfamethoxypyridazine			281(100)	156(3); 216(3); 126(2); 188(2)	
281	Alverine	282(100)	176(30)			
281	Diphenylpyraline			282(15)	165(100); 280(12); 109(13)	99
281	Chloralbetaine					82
281	Flufenamic acid			282(83)	262(100); 264(47); 281(31)	

282	Oleic acid	283(100)	265(25); 264(15)			41
282	Salicylic acid bis-TMS derivative			283(17)	193(100); 267(89)	
283	Levallorphan	284(100)	282(50); 283(25)	284(100)	283(43); 282(29); 266(14); 256(8)	283
284	Diazepam	285(100)	287(30)	285(100)	162(6)	256
		285(100)				
284	Mazindol	285(100)	287(33); 267(20); 255(20); 269(10); 257(7)			
284	Promazine	285(100)	284(15)	285(100)	284(96); 212(12); 199(7); 240(6)	58
				285(100)	86(35); 200(15); 226(10); 254(2)	
284	Psilocybin					58
284	Promethazine	285(100)	284(27)	285(100)	198(93); 283(58); 284(54); 240(34)	72
284	Stearic acid	285(100)	284(70); 267(45); 283(26)	285(100)	283(19); 265(13); 267(8)	57
284	Sulfachloropyridazine	285(100)	220(60); 287(40); 251(20)			
285	Dihydrocodeinone mtb.			286(100)	257(20); 284(15)	
285	Chlordiazepoxide mtb.					268
285	Hydromorphone	286(100)				
285	Isothipendyl			286(37)	241(100); 200(34); 284(21); 214(17)	
285	Morphine	286(20)	268(100)	286(32)	268(100); 285(20)	285
285	Pentazocine	286(100)	284(50); 285(40)	286(78)	217(100); 290(96) 284(41); 200(20)	217
285	Probenecid	286(100)		286(100)	256(23); 268(4); 185(3); 284(3)	256
285	Prothipendyl			286(100)	241(66); 285(65); 284(7); 269(3)	
285	Pyrilamine	286(100)	241(10)	286(24)	121(100); 117(30); 241(22)	121

MOLECULAR WEIGHT INDEX (continued)

mol wt	Substance	CI: Isobutane (relative abundance) MH⁺	CI: Isobutane (relative abundance) Major peaks	CI: CH_4 (relative abundance) MH⁺	CI: CH_4 (relative abundance) Major peaks	EI Base peak
286	Methallenestrill	287(58)	199(100); 269(25); 229(18)			
286	Oxazepam	287(100)	269(99); 271(40); 289(35); 288(20); 270(20)	287(40)	279(100); 271(35); 257(5)	77
286	Promethazine			287(55)	213(100); 230(25); 200(15)	
286	Thonzylamine			287(29)	121(100); 242(14); 285(14); 216(8)	121
287	Cyproheptadine	288(100)	287(30)			287
287	Dihydromorphine	288(100)	287(18)			287
287	Allylprodine					172
287	Procyclidine			288(100)	286(50); 204(36); 210(8); 270(3)	
287	Cycrimine					98
288	Flurazepam mtb.			289(100)	193(18); 253(4)	
288	Lindane			289(<1)	219(100); 217(72); 221(37); 181(21)	
288	Phenobarbital 1,3-diethyl derivative					146, 260
288	Testosterone			289(100)	271(22); 272(5)	
289	Atropine	290(17)	124(100)	290(2)	124(100); 288(1); 140(1)	124
289	Benzoylecgonine	290(100)	124(95); 168(30); 123(20); 122(20); 105(10)	290(100)	168(81); 124(11); 196(8); 272(3)	

289	Dyclonine			290(15)	98(100); 288(11); 205(3); 234(2)	
289	Atropine	290(17)	124(100)	290(2)	124(100); 288(1); 140(1)	
289	Benzoylecgonine	290(100)	124(95); 168(30); 123(20); 122(20); 105(10)	290(100)	168(81); 124(11); 196(8); 272(3)	
289	Caramiphen					86
289	Dyclonine			290(15)	98(100); 288(11); 205(3); 234(2)	
289	Hyoscyamine	290(15)	124(100); 237(20)			124
290	Carbinoxamine	291(100)	293(30)	291(73)	202(100); 204(35); 289(18); 255(17)	58
290	N-Methyl-*N′*-methoxymethyl phenobarbital					146
290	Trimethoprim					290
290	Hydroxyphenobarbital 1,3-dimethyl derivative methyl ether					261
291	Cyclopentolate					58
291	Thiambutene					276
291	Amitriptyline mtb.			294(23)	276(100); 292(29); 231(20)	
292	Warfarin mtb.					292
294	Cinchonidine	295(100)	296(20)			
294	Cinchonine	295(100)	277(10); 136(10)	295(100)	136(58); 277(15); 279(6)	
294	Secobarbital 1,3-diethyl derivative					224
294	Proparacaine			295(47)	100(100); 99(76); 178(19)	
294	Propoxycaine			295(4)	100(100); 99(30); 178(24)	
295	Chlorothen			296(10)	113(100); 129(23); 251(17)	58

MOLECULAR WEIGHT INDEX (continued)

mol wt	Substance	CI Isobutane (relative abundance) MH⁺	CI Isobutane (relative abundance) Major peaks	CI CH_4 (relative abundance) MH⁺	CI CH_4 (relative abundance) Major peaks	EI Base peak
295	Normethadone					58
295	Chlorothiazide			296(100)	279(5); 295(4); 281(2)	295
296	Disulfiram	297(–%)	150(100); 116(80); 118(60); 149(27); 117(15)			
296	Oxacillin mtb.					144
296	3-Hydroxydiphenylhydantoin methyl ether 3-methyl derivative					296
296	p-Hydroxydiphenyl hydantoin methyl ether 3-methyl derivative					296
296	Ethynylestrodiol	297(50)	279(100); 296(32); 280(21); 295(19); 213(12)			
296	Pyrathiazine					84
296	Meclofenamic acid			296(100)	278(75)	
296	Methdilazine	297(100)				97
297	Hydrochlorothiazide	286(100)	298(70); 300(55); 288(45)	298(100)	286(99); 269(72); 271(31)	269
298	Diethazine			299(14)	100(100); 86(34); 298(100)	86
298	Norethindrone					91
298	Methyl stearate			299(100)	297(76); 265(8); 267(5)	74
298	Nialamide			299(100)	192(30); 91(29); 138(18)	91

298	Phenacaine	299(10)	262(18)			
298	Trimeprazine	299(100)	200(55); 298(50); 240(42); 239(20); 199(20)			58
299	Chlorodiazepoxide	300(33)	284(100); 286(33); 283(30); 285(30); 302(10)	300(100)	299(38); 264(24); 286(6)	282
299	Chlordiazepoxide (GC)					282
299	Codeine	300(22)	282(100)	300(86)	282(100); 299(35); 298(21)	299
299	Metopon					299
299	Dihydrocodeinone	300(100)		300(100)	257(34);298(21)	299
299	Nylidrin	300(100)	282(25)	300(49)	176(100); 282(47); 206(22)	
300	Chlorcyclizine			301(51)	201(100); 265(56); 299(36); 203(35)	99
300	Promazine mtb.			301(–%)	300(100)	
300	Promethazine mtb.			301(100)		
300	Sulfaquinoxaline			301(10)	146(100); 237(78); 210(22)	
301	Dihydrocodeine	302(100)	284(25); 301(20)	302(100)	105(25); 151(10); 184(10)	301
301	Isoxsuprine	302(100)	178(16); 284(12)			
301	Oxymorphone	302(100)				69
301	Pentazocine mtb.			302(100)	217(55); 284(43); 300(33); 230(22)	
301	Pentazocine mtb.			302(35)	284(100); 217(30); 230(28); 300(24)	
301	Morphine-N-oxide					285
301	Trihexyphenidyl	302(100)	300(12)			
302	Butallylonal			303(68)	167(100); 247(93); 249(92); 223(59)	
302	Ethacrynic acid	303(100)	305(65); 285(10)			

MOLECULAR WEIGHT INDEX (continued)

		CI				EI
		Isobutane (relative abundance)		CH_4 (relative abundance)		
mol wt	Substance	MH^+	Major peaks	MH^+	Major peaks	Base peak
302	Naproxen trimethyl silyl ether			303(30)	185(100); 287(20); 213(10); 231(5)	
303	Cocaine	304(100)	182(33)	304(20)	182(100); 272(2);	82
303	Hyoscine			304(4)	138(100); 139(9)	
303	Scopolamine	304(16)	138(100)	304(4)	138(100); 156(6)	94
305	Chlorpromazine mtb.			306(21)	290(100); 292(33); 254(16)	233
305	Iodochlorhydroxyquinoline			306(100)	180(54); 179(28); 129(17)	
306	Trimethoprim mtb.					289
306	Butacaine	307(100)	263(20)	307(100)	263(70); 120(54); 142(47); 170(29)	
306	Trimethoprim					290
307	Benztropine			308(5)	97(100); 124(45); 167(41); 140(13)	140
307	Propoxyphene mtb.			308(100)	143(20); 220(18); 306(4)	
308	Alpha chloralose					71
308	Amfonelic acid			309(100)	264(10); 291(5)	
308	Phenylbutazone	309(100)	190(50)	309(100)	120(15); 183(8)	183
308	Warfarin	309(100)	251(15); 291(10)			
308	5,5-Diphenylhydantoin diethyl derivative					180, 279
309	Acetyl sulfisoxazole	310(100)	155(55); 268(10)			
309	Dicyclomine	310(100)	100(55); 86(48); 99(35)			86

309	Isomethadone					53
309	Methadone	310(100)		310(100)	265(60); 308(28)	295
309	Mepazine					310
310	Cannabinol	311(100)				
310	Mestranol					227
310	Ibogaine	311(100)				136
310	Sulfadimethoxine			311(100)	156(90); 184(20); 140(20); 246(10)	
311	Adiphenine	312(100)	100(65); 86(20)			
311	Biperiden	312(100)				
311	Pyrrobutamine					205
311	Thebaine					311
311	Nalorphine					271
311	Methadol					72
312	Dydrogesterone	313(100)	312(72); 268(33); 207(15)			
312	Ethopropazine			313(5)	100(100); 114(53); 198(6); 311(3)	100
313	Ethylmorphine	314(11)	296(100)			313
313	Flurazepam mtb.			314(100)	313(10); 218(6); 278(2)	
314	Cannabidiol	315(100)		315(100)	193(50); 313(44); 135(35); 231(17)	231
314	Methoxypremazine					58
314	Tetrahydrocannabinol	315(100)		315(100)	313(55); 299(15); 135(7); 193(5)	314
314	Tetrahydrocannabinol-delta-8					231
315	Chlorprothixene	316(100)	318(30)	316(100)	257(20); 314(10)	58
315	Codeine-N-oxide					299
315	Oxycodone	316(100)		316(100)	298(6)	315
316	5-(2-Bromoallyl)-5-(1-methyl butyl) barbituric acid					43
318	Brompheniramine	319(100)	321(98); 166(15)			

MOLECULAR WEIGHT INDEX (continued)

		CI				EI
		Isobutane (relative abundance)		CH_4 (relative abundance)		
mol wt	Substance	MH^+	Major peaks	MH^+	Major peaks	Base peak
318	Chlorpromazine	319(100)	318(32); 234(30)	319(100)	319(4); 246(5); 248(4)	58
318	Parabromdylamine					249
318	Phenolphthalein	319(100)		319(100)	255(30)	
319	Chlorequine					86
320	Chlorpromazine mtb.			321(100)	305(25); 304(22); 234(10); 233(9)	
320	N,N′-Dimethoxymethyl phenobarbital					146
321	Phenazocine	322(100)	230(80)	322(18)	230(100); 320(5); 306(3)	230
322	Chloramphenicol	323(100)	325(65); 305(34); 307(18); 289(18); 291(10); 327(10)	232(29)	305(100); 307(68); 275(18)	
322	Butyl butoxyethyl phthalate					149
323	Lysergic acid diethylamide	324(100)		324(100)	323(46); 281(29); 322(17)	
323	Lysergide					323
323	Lysergic acid-*N*-methylpropylamide	324(100)	323(13)			
323	Piperidolate	324(37)	112(100); 111(40)			
324	Oxyphenbutazone	325(100)	324(70); 199(35)	325(99)	120(7); 309(4); 190(4)	93
324	p-Hydroxyphenylbutazone					93
324	Phenylbutazone-alcohol mtb.			94(100)	325(35); 185(30); 185(30); 234(20); 122(20); 115(18)	

324	Diampromide					162
324	Quinidine	325(100)	136(18)	325(100)	307(12); 136(8); 295(4); 323(4)	136
324	Quinine	325(100)	136(20)	325(100)	136(37; 307(10); 323(7)	
325	Clemizole			326(100)	256(80); 324(28); 255(25)	255
325	Propoxyphene mtb.					44
325	Propoxyphene mtb.					77
325	Mephenzolate bromide					97
325	Ergonovine	326(100)	268(35); 308(12); 325(12)			
326	Acetylpromazine			327(1)	61(100); 85(98); 145(55)	58
				327(100)	86(40); 242(3); 268(2)	
327	Dimenoxadol					57
327	Aminopromazine			238(11)	117(100); 129(20); 115(16)	198
327	Benactyzine	328(100)	310(74); 183(20)			
327	O[6] -Monoacetylmorphine	328(100)	268(55); 327(30)			
327	Naloxone	328(100)	327(35)			
328	Methotrimeprazine					58
329	Cinnamoylocaine	330(100)	182(40); 329(15)			
329	Ecgonine bis-TMS deriv.			330(100)	314(80); 240(47)	
330	Fursemide					81
330	Tetrahydrocannabinol mtb.			331(48)	313(100); 330(47); 329(25); 190(9)	299
332	Flurescein	333(100)				
332	Hydroxyphenobarbital 1,3-diethyl derivative ethyl ether					332
332	Tetraphenylethylene	333(100)	389(100)			
333	Bromodiphenhydramine	334(100)	245(100); 247(98) 336(90)	334(9)	245(100); 247(98); 164(9); 332(5)	58

MOLECULAR WEIGHT INDEX (continued)

mol wt	Substance	CI: Isobutane (relative abundance) MH⁺	CI: Isobutane (relative abundance) Major peaks	CI: CH_4 (relaitve abundance) MH⁺	CI: CH_4 (relaitve abundance) Major peaks	EI Base peak
333	Carbetapentane			334(30)	113(100); 129(2); 144(18); 100(15)	
334	3-Hydroxy-chlorpromazine					58
334	8-Hydroxy-chlorpromazine					58
334	Chlorpromazine mtb.			335(100)	319(20); 318(17); 233(13); 234(12)	58
336	9,10-Dibromoanthracene					336
334	Strychnine	335(100)	334(25)	335(100)	334(30); 333(22)	334
336	Bishydroxycoumarin	337(100)	163(32)			
336	Butylcarbobutoxymethyl-phthalate			337(<1)	149(100); 263(24); 205(10); 177(9)	149
336	Diphenylbenzidine	337(100)	336(95)			
336	Fentanyl	337(100)	245(25)			245
338	Chlorthalidone	323(100)	321(70); 339(50)			
339	Bromothen			340(26)	72(100); 297(30); 295(30)	
339	Methantheline			340(3)	86(100); 100(88); 109(61)	
339	Methylergonovine	340(100)	322(26); 339(15)			
339	Papaverine	340(100)		340(100)	151(3)	
				340(100)	341(13); 339(7)	
339	Propoxyphene			340(16)	131(100); 266(28); 338(19)	58
340	Propiomazine					72
341	Acetylcodeine	342(15)	282(100)	342(32)	282(100); 341(16); 340(13)	
341	Naltrexone					341

341	Thebacon					341
342	Lactose	343(–%)	163(100); 145(85); 127(25)			
342	Sucrose	343(–%)	163(100); 145(85); 127(25)			73
343	Naltrexone Metab.					55
343	Dibucaine	344(100)		344(100)	100(14); 342(12); 271(4)	
343	Alpha-hydroxy naltrexone					343
343	Trimethobenzamide mtb.					195
346	Morpheridine					246
347	Phenomorphan					256
349	Dipipanone					112
350	Strychnine-*N*-oxide	351(20)	335(100); 333(25)			
351	7-Iodo-8-hydroxyquinoline-5-sulfonic acid	272(100)				
352	Anileridine	353(100)	351(25); 212(24); 246(14); 234(10)			246
352	Allobarbital bis-TMS derivative			353(100)	281(90); 209(85); 337(25)	
352	Griseofulvin	353(100)	355(35); 352(30)			
352	Triflupromazine	353(100)		353(72)	352(100); 280(5); 282(4)	58
352	DDT					235
353	Acenocoumarin	354(100)	324(15); 221(10); 163(10)			
353	Chelidonine					332
353	Methysergide	354(100)	353(15)			
353	Acetylmethadol					72
354	Yohimbine	355(100)	354(75); 353(40); 337(10); 352(10)	355(100)	354(43); 337(24); 323(10)	
356	Thioridazine					84
357	Indomethacine	358(100)	357(44); 360(38)	358(100)	139(21); 141(7)	44
				358(–%)	314(100); 313(40); 315(20); 316(20)	

MOLECULAR WEIGHT INDEX (continued)

		CI				EI
		Isobutane (relative abundance)		CH_4 (relative abundance)		
mol wt	Substance	MH^+	Major peaks	MH^+	Major peaks	Base peak
358	Sal ethyl carbonate					193
358	Prednisone			359(100)	341(57); 299(18); 329(15)	
359	Captodiamine					58
359	Methylclothiazide					310
360	Prednisolone			361(16)	61(100); 343(54); 325(49)	
361	Furethidine					246
362	Hydrocortisone			363(46)	61(100); 345(54); 303(43)	
365	Pericyazine					114
366	Piminodine	367(100)	234(12)			246
367	Dioxyline	368(100)				
368	Cholesta-3,5-diene					43
369	Diacetylmorphine (heroin)	370(33)	310(100); 268(14)	370(40)	310(100); 368(30); 268(20)	
370	Dioctyladipate			371(15)	129(100); 147(24); 157(11); 259(7)	129
370	Thioridazine	371(100)	370(77)	371(100)	370(84); 98(81); 126(38)	98
371	Indomethacin methyl ester					139
371	Codeine TMS derivative			372(64)	282(100); 313(38); 356(15)	
371	Dihydrodiacetylmorphine	372(100)	312(15); 330(12)			
373	Prochlorperazine	374(100)	376(30); 373(27); 234(20)	374(7)	99(100); 373(5); 113(4)	113
374	Hydroxyzine	375(100)	377(30)			201
375	Benzylmorphine					284

376	Riboflavin			377(1)	243(100); 244(15); 362(13)	
376	3-Acetoxychlorpromazine					58
376	8-Acetoxychlorpromazine					58
378	Doxapram	379(100)		379(100)	100(13); 101(2)	
379	Trichlormethazide	380(95)	382(100); 384(33)			
380	Phenylbutazone trimethylsilyl ether	381(100)	309(85); 35(5)			
382	3,5-Cholestadiene-7-one					174
386	Cholesterol	387(–%)	369(100); 212(67); 385(32)	387(7)	369(100); 385(36); 367(27); 353(15)	368, 386
386	Clonitazene					86
386	Mesoridazine			387(–%)	371(100); 370(4); 369(15); 372(13)	
387	Flurazepam	388(100)	86(44); 390(36)	388(100)	100(43); 99(27) 315(13); 386(13)	86
388	Trimethobenzamide	389(100)				58
390	Di-*n*-octyl phthalate	391(100)	149(15); 279(10)	391(20)	149(100); 113(4); 167(32); 279(11)	149
390	Meclizine	391(100)	393(30); 390(18); 189(18); 389(11); 243(10)	391(39)	201(100); 189(40); 203(34)	189
392	Moramide					100
394	Brucine	395(100)		395(100)	394(52); 380(17)	
394	Triamcinalone			395(16)	347(100); 327(7); 375(54)	
398	Tributoxyethyl phosphate			399(100)	299(68); 199(47)	
398	Pholcodine					114
399	Torecane					399
399	Quinacrine					86
399	Colchicine	400(100)	399(25); 372(20); 386(15); 371(15)	400(100)	312(3)	312
399	Epinephrine tris-TMS derivative			400(20)	310(100); 384(9); 355(80)	

MOLECULAR WEIGHT INDEX (continued)

		CI				EI
		Isobutane (relative abundance)		CH_4 (relative abundance)		
mol wt	Substance	MH^+	Major peaks	MH^+	Major peaks	Base peak
399	Thiethylperazine	400(100)	401(95); 402(25)			
401	Pipamazine					141
402	Dehydrocholic acid	403(100)	285(60)			
402	Acetyl tri-N-butyl citrate					185
403	Perhenazine	404(100)	406(30); 234(19)			42
404	Hexachlorophene	405(50)	407(100); 409(80); 411(40)			
404	Sulfinpyrazone	405(35)	279(100); 278(55); 280(20)			
407	Trifluoperazine	408(100)	407(30); 278(13); 302(100); 279(10)	408(100)	407(67); 388(52); 406(39)	112
408	γ-Tetrahydropyranyl ether of phenylbutazone alcohol mtb.			325(100)	85(48); 101(25); 120(10); 409(2)	
411	Etorphine					411
411	Acetophenazine	412(55)	117(100); 99(80)			254
413	Noscapine	414(–%)	220(100); 221(15)			220
416	Lasix, methyl ester, mono trimethyl silyl ether	417(25)	81(100); 113(55); 97(45)	417(30)	81(100); 73(25); 114(20)	
416	Methylprednisolone			417(43)	339(100); 339(3); 400(27)	
419	Methantheline bromide					181
421	Bendroflumethiazide	422(100)	320(38)			
425	Carphenazine					268
427	Dixyrazine					212
429	Morphine bis-TMS derivative			430(74)	414(100); 340(5); 371(18)	

431	Clidinium bromide					105
432	Buclizine					147
437	Fluphenazine					42
439	Dipicrylamine	440(100)	439(18)			
439	Polythiazide	440(90)	300(100); 302(45); 442(40); 266(40); 406(30); 361(30)			
442	Decafluorotriphenyl phosphine			443(100)	365(80); 275(10); 217(5)	
442	Methacycline	443(100)	198(38); 425(10); 400(10)			
444	Doxycycline hyclate	445(100)	444(12); 427(10)			
444	Tetracycline	445(40)	427(100); 428(30); 426(28); 257(25)			
445	Thiopropazate					246
445	Narceine					58
446	Thioproperazine					70
452	Diphenoxylate			453(100)	246(8); 451(7)	246
460	Oxytetracycline	461(100)	443(30); 198(30); 460(16)			
467	Oxethazine			468(100)	145(14); 277(12); 466(10)	
478	Chlortetracycline	479(100)	481(40); 461(20)			
478	Rhodamine	479(–%)	443(100); 398(20)			
480	Emetine	481(100)	479(60); 480(48)			
530	Alpha tocopherol succinate			531(92)	101(100); 265(5); 165(31)	
566	Alpha, alpha-dibromo-alpha, alpha, alpha, alpha-tetraphenyl-xylene					408
578	Deserpidine					365
602	Picrotoxin	293(100)	585(80); 603(70)	609(92);		

MOLECULAR WEIGHT INDEX (continued)

		CI				EI
		Isobutane (relative abundance)		CH_4 (relative abundance)		
mol wt	Substance	MH^+	Major peaks	MH^+	Major peaks	Base peak
608	Reserpine			609(100)	213(100); 607(24); 195(24); 213(70); 195(25); 129(20)	
694	Tubocurarine chloride			695(–%)	595(100); 596(43); 609(25)	298
733	Erythromycin A					158

* Relative abundance

CI DATA—ALPHABETICAL INDEX*

Michael E. Caplis and Larry S. Eichmeier

		CI			
		Isobutane		Methane	
Mol wt	Substance	MH+ (relative abundance)	Major peaks	MH+ (relative abundance)	Major peaks
353	Acenocoumarin	354(100)	324(15); 221(10); 163(10)		
151	Acetaminophen	152(100)		152(100)	109(11); 134(2)
135	Acetanilide			136(100)	94(14)
222	Acetazolamide	223(100)	207(18)	223(100)	181(27)
411	Acetophenazine	412(55)	117(100); 99(80)		
149	*o*-Acetotoluide			150(100)	149(5); 107(5); 106(4)
278	Acetylcarbromal	279(100)	281(98)	279(59)	237(100); 239(98); 149(47); 151(46)
341	Acetylcodeine	342(15)	282(100)	342(32)	282(100); 341(16); 340(13)
163	*m*-Acetylcysteine	164(100)			
326	Acetylpromazine			327(1)	61(100); 85(98); 145(55)
				327(100)	86(40); 242(3); 268(2)
180	Acetylsalicylic acid (aspirin)	181(17)	121(100); 139(20); 163(18); 138(17); 180(10)	181(<1)	121(100); 139(20); 163(25)
309	Acetyl sulfisoxazole	310(100)	155(55); 268(10)		
179	Acridine	180(100)	179(24); 181(14)		
311	Adiphenine	312(100)	100(65); 86(20)		
183	Adrenaline	184(25)	166(100)	184(14)	166(100); 164; 182(3); 123(3)
240	Alizarin	241(100)			91(6); 238(2)
208	Allobarbital	209(100)	169(11)	209(100)	141(21); 169(6)
352	Allobarbital bis-TMS deriv.			353(100)	281(90); 209(85); 337(25)
236	Allobarbital dimethyl deriv.			236(100)	169(29); 195(16); 297(9)
136	Allopurinol			137(100)	136(11)
224	Allylbarbital	225(100)	185(17)	225(100)	168(10); 185(7)
530	Alpha tocopherol succinate			531(92)	101(100); 265(5); 165(31)
261	Alphaprodine	262(26)	188(100); 187(50); 172(35)		
244	Alphenal	245(100)	113(43); 205(14)	245(100)	
272	Alphenal dimethyl deriv.			273(100)	255(10)
281	Alverine	282(100)	176(30)		
151	Amantadine	152(100)	135(48); 151(32); 150(15)		
308	Amfonelic acid			309(100)	264(10); 291(5)
150	*p*-Aminoacetanilide			151(100)	150(14); 108(3); 109(2)
203	Aminoantipyrine	204(100)	203(50); 84(42)	204(100)	203(20); 94(5); 120(13); 201(5)
137	*o*-Aminobenzoic acid	138(100)	137(35); 120(35)		
232	Aminoglutethimide			233(100)	205(19); 94(1)
109	Aminophenol	110(100)	109(40)	110(100)	109(73); 93(1); 92(8)
327	Aminopromazine			238(11)	117(100); 129; 115(16)
231	Aminopyrine			232(100)	231(40); 113(20); 230(16)
153	p-Aminosalicylic acid	154(100)	153(28); 135(13); 136(12)	154(100)	136(20); 110(3)
277	Amitriptyline	278(100)		278(100)	276(70); 233(23)
279	Amitriptyline mtb.			280(10)	262(100); 231(3); 260(12)
293	Amitriptyline mtb.			294(23)	276(100); 292(29); 231(20)
226	Amobarbital	227(100)		227(100)	157(12); 156(11)
135	Amphetamine	136(100)		136(100)	119(65); 91(9); 134(5)
278	Amydricaine			279(46)	157(100); 277(52); 256(30)

* Data from Brian S. Finkle, Center for Human Toxicology, University of Utah; Norman C. Lair, Suburban Hospital, Bethesda, Maryland; Jack D. Henion, Assistant Professor of Toxicology, New York State College of Veterinary Medicine; Richard Saferstein, Forensic Science Bureau, New Jersey State Police Laboratory.

CI DATA—ALPHABETICAL INDEX (continued)

		CI			
		Isobutane		Methane	
Mol wt	Substance	MH⁺ (relative abundance)	Major peaks	MH⁺ (relative abundance)	Major peaks
352	Anileridine	353(100)	351(25); 212(24); 246(14); 234(10)		
265	Antazoline			266(100)	84(100); 196(22); 85(15)
178	Anthracene	179(100)	178(14)		
188	Antipyrine	189(100)		189(100)	217(30); 190(12)
210	Aprobarbital	211(100)	171(13)	211(100)	169(76)
267	Apomorphine	268(100)	267(22); 266(20)		
184	Apronalide			185(100)	142(51); 97(43); 125(29)
176	Ascorbic acid	177(100)			
215	Atrazine			216(73)	180(99); 181(11); 215(6)
166	Atrolactic acid	167(100)	149(58); 107(37); 166(30)		
289	Atropine	290(17)	124(100)	290(2)	124(100); 288(1); 140(1)
267	Azacyclonol			268(49)	250(100); 190(6); 266(15); 183(5)
235	Azapetine	236(100)	234(25)		
277	Azathioprine				
184	Barbital	185(100)		185(100)	156(5); 142(2); 141(1)
212	Barbital dimethyl deriv.			213(100)	184(8); 211(3)
128	Barbituric acid	129(100)			
155	Bemegride			156(100)	128(3)
327	Benactyzine	328(100)	310(74); 183(20)		
421	Bendroflumethiazide	422(100)	320(38)		
184	Benzidine	185(100)	184(60)		
118	Benzimdazole	119(100)	118(20)		
165	Benzocaine	166(100)		166(100)	138(15); 120(11); 122(8); 94(5)
122	Benzoic acid	123(100)	105(35)		
227	α-Benzoin oxime	210(100)	107(100); 104(90); 228(50)		
252	Benzopyrene	253(100)	252(59)		
289	Benzoylecgonine	290(100)	124(95); 168(30); 123(20); 122(20); 105(10)	290(100)	168(81); 124(11); 196(8); 272(3)
239	Benzphetamine	240(100)	148(85)	240(1)	148(100); 119(9)
307	Benztropine			308(5)	97(100); 124(45); 167(41); 140(13)
311	Biperiden	312(100)			
336	Bishydroxycoumarin	337(100)	163(32)		
273	4-Bromo-2,5-dimethoxyamphetamine	274(100)	276(98)	274(25)	257(100); 259(9); 230(19); 232(18)
333	Bromodiphenhydramine	334(100)	245(100); 247(98); 336(90)	334(9)	245(100); 247(98); 164(9); 332(5)
222	Bromoisovalum			223(100)	182(84); 180(83); 143(59); 100(26)
339	Bromothen			340(26)	72(100); 297(30); 295(30)
318	Brompheniramine	319(100)	321(98); 166(15)		
394	Brucine	395(100)		395(100)	394(52); 380(17)
204	Bufotenine	205(100)	198(11)		
206	I-Buprofen			207(100)	160(98); 188(10); 190(9)
212	Butabarbital	213(100)		213(97)	157(100); 156(2); 185(10)
240	Butabarbital dimethyl deriv.			241(100)	185(66); 169(16); 213(10); 239(5)
306	Butacaine	307(100)	263(20)	307(100)	263(70); 120(54); 142(47); 170(29)
302	Butallylonal			303(68)	167(100); 247(93); 249(92); 223(59)
193	Butamben			194(100)	138(92); 120(35); 166(35); 94(13)

CI DATA—ALPHABETICAL INDEX (continued)

		CI			
		Isobutane		Methane	
Mol wt	Substance	MH⁺ (relative abundance)	Major peaks	MH⁺ (relative abundance)	Major peaks
212	Butethal	213(100)		213(100)	156(10)
236	Butethamine			237(33)	100(100); 120(8); 135(11); 164(10)
336	Butylcarbobutoxymethylphthalate			337(<1)	149(100); 263(24); 205(10); 177(9)
194	Butylparaben			195(73)	139(100); 121(25); 167(19); 95(10)
194	Caffeine	195(100)		195(100)	138(2)
314	Cannabidiol	315(100)		315(100)	193(50); 313(44); 135(35); 231(17)
310	Cannabinol	311(100)			
196	Cantharidin	197(100)			
236	Carbamazepine	237(100)	194(14)	237(100)	238(20); 193(15); 192(14)
333	Carbetapentane			334(30)	113(100); 129(2); 144(18); 100(15)
290	Carbinoxamine	291(100)	293(30)	291(73)	202(100); 204(37); 289(18); 255(17)
246	Carbocaine			247(27)	98(100); 96(6)
236	Carbromal	237(100)	239(98)	237(53)	194(100); 196(9); 114(84); 157(70)
260	Carisoprodol	261(100)	176(22); 158(21); 200(19); 218(11)	261(8)	176(100); 158(33); 200(32); 97(23)
				158(100)	176(99); 200(50); 261(15)
273	Chlormezanone			274(40)	154(100); 156(3); 121(12)
154	Chloroacetophenone	155(100)	157(35); 139(20); 154(17)		
164	Chloralhydrate			165(22)	147(100); 149(97); 111(75); 113(49)
322	Chloramphenicol	323(100)	325(65); 305(34); 307(18); 289(18); 291(10); 327(10)		
300	Chlorcyclizine			301(51)	201(100); 265(56); 299(36); 203(35)
299	Chlorodiazepoxide	300(33)	284(100); 286(33); 283(30); 285(30); 302(10)	300(100)	299(38); 264(24); 286(6)
189	*o*-Chlorobenzalmalonitrile	189(100)	191(33)		
245	Chlorphenesin carbamate	246(55)	203(100); 205(30); 185(20); 167(20); 248(19); 187(7); 169(7)		
183	Chlorphentermine	184(100)	186(30)		
199	Chlorophenylalanine	200(100)	202(35); 154(30)		
270	2-Chloroprocaine			271(15)	100(100); 154(5); 271(3)
318	Chlorpromazine	319(100)	318(32); 234(30)	319(100)	319(4); 246(5); 248(4)
233	Chlorpromazine mtb.			234(100)	233(98); 198(50)
373	Chlorpromazine mtb.			274(78)	273(100); 238(5)
320	Chlorpromazine mtb.			321(100)	305(25); 304(22); 234(10); 233(9)
305	Chlorpromazine mtb.			306(21)	290(100); 292(33); 254(16)
334	Chlorpromazine mtb.			335(100)	319(20); 318(17); 233(13); 234(12)
276	Chlorpropamide	277(-)	192(100);194(35)		
191	Chlorpropamide mtb.			192(100)	175(10)
295	Chlorothen			296(10)	113(100); 129(23); 251(17)
338	Chlorthalidone	323(100)	321(70); 339(50)		
295	Chlorothiazide			296(100)	279(5); 295(4); 281(2)
274	Chlorpheniramine	275(100)	277(30); 230(22); 203(10)	275(39)	230(100); 232(2); 239(21); 273(18)

CI DATA—ALPHABETICAL INDEX (continued)

		CI			
		Isobutane		Methane	
Mol wt	Substance	MH⁺ (relative abundance)	Major peaks	MH⁺ (relative abundance)	Major peaks
315	Chlorprothixene	316(100)	318(30)	316(100)	257(20); 314(10); 259(7)
232	Chlorprothixene mtb.			233(100)	197(60); 232(45)
478	Chlortetracycline	479(100)	481(40); 461(20)		
169	Chlorzoxazone	170(100)	172(35); 118(20); 136(10)	170(100)	
386	Cholesterol	387(-)	369(100); 212(67); 385(32)	387(7)	369(100); 385(30); 367(27); 353(15)
294	Cinchonidine	295(100)	296(20)		
294	Cinchonine	295(100)	277(10); 136(10)	295(100)	136(58); 277(15);279(6)
249	Cinchophen			250(100)	232(5); 206(2)
329	Cinnamoylcoccaine	330(100)	182(40); 329(15)		
325	Clemizole			326(100)	256(80); 324(28); 255(25)
242	Clofibrate	243(100)	115(55); 245(30); 128(13; 169(12)	243(100)	115(78); 169(27); 129(10)
303	Cocaine	304(100)	182(33)	304(20)	182(100); 272(2)
299	Codeine	300(22)	282(100)	300(86)	282(100); 299(35); 298(21)
371	Codeine TMS deriv.			372(64)	282(100); 313( ); 356(15)
399	Colchicine	400(100)	399(25); 372(20); 386(15); 371(15)	400(100)	312(3)
127	Coniine	128(100)	126(24)		
255	Cotarnine	256(-%)	204(100); 118(40)		
176	Cotinine (nicotine mtb.)			177(100)	98(23)
146	Coumarin			147(100)	
276	Cyclandelate	277(30)	125(100); 107(20); 135(18)		
266	Cyclizine			267(5)	167(100); 195(1); 189(11)
252	Cyclizine mtb.			253(10)	167(100); 195(7)
236	Cyclobarbital	237(100)	157(10)	237(60)	157(100); 207(1); 235(7)
168	Cyclopal	169(100)	235(98)		
141	Cyclopentamine				143(100); 140(40);126(23); 111(20)
260	Cyclophosphamide	261(100)	211(30); 225(25); 213(11); 227(10)		
287	Cyproheptadine	288(100)	287(30)		
442	Decafluorotriphenyl phosphine			443(100)	365(80); 275(10); 217(5)
402	Dehydrocholic acid	403(100)	285(60)		
266	Desipramine	267(100)	222(21)	267(100)	230(38); 208(29); 195(29); 195(29); 222(16)
369	Diacetylmorphine (heroin)	370(33)	310(100); 268(14)	370(40)	319(100); 368(30); 268(20)
124	Diaminophenol	125(100)	124(35)		
244	Dianisidine	245(100)			
284	Diazepam	285(100)	287(30)	285(100)	162(6)
270	Diazepam mtb.			271(100)	242(5); 243(4)
343	Dibucaine	344(100)		344(100)	100(14); 342(12); 271(4)
278	Dibutyl phthalate			279(2)	149(100); 205(20); 177(8)
146	p-Dichlorobenzene			147(100)	149(50); 175(5); 177(2)
309	Dicyclomine	310(100)100(55); 86(48); 99(35)			
298	Diethazine			299(14)	100(100); 86(34); 298(100)
222	Diethyl phthalate			223(<1)	177(100); 149(2); 178(12)
205	Diethylpropion	206(100)	204(30); 100(28); 135(15)	206(100)	100(51); 204(11); 190(6)
268	Diethylstilbestrol	269(100)	268(35)		
216	Diethyltryptamine	217(100)			

CI DATA—ALPHABETICAL INDEX (continued)

		CI			
		Isobutane		Methane	
Mol wt	Substance	MH⁺ (relative abundance)	Major peaks	MH⁺ (relative abundance)	Major peaks
301	Dihydrocodeine	302(100)	284(25); 301(20)	302(100)	105(25); 151(10);184(10)
299	Dihydrocodeinone	300(100)		300(100)	257(34); 298(21)
285	Dihydrocodeinone mtb.			286(100)	257(20); 284(15)
371	Dihydrodiacetylmorphine	372(100)	312(15) 330(12)		
287	Dihydromorphine	288(100)	287(18)		
272	Dimethisoquin			273(40)	271(100); 228(80); 202(20)
195	2-5-Dimethoxyamphetamine	196(100)	179(25); 152(25)		
195	3,5-Dimethoxyamphetamine			196(15)	91(100); 179(62); 152(30)
209	2,5-Dimethoxy-4-methyl-amphetamine	210(200)			
149	p-Dimethylaminobenzaldehyde	150(100)	149(15)		
264	p-Dimethylaminobenzylrhodamine	265(100)	177(15)		
163	N,N′-Dimethylamphetamine			164(100)	119(10)
140	Dimethyldihydroresorcinol	141(100)			
116	Dimethylgloyxime	117(100)	116(16)		
249	2,6-Dimethyl-4-isopropylbenzaldehyde thiosemicarbazone	250(100)	174(18); 174(18); 158(15)		
233	Dimethylmeperidine			234(100)	160(48); 232(16); 188(10)
188	Dimethyltryptamine	189(100)			
180	1,7-Dimethylxanthine			181(100)	
168	o-Dinitrobenzene	168(100)	152(15)		
184	2,4-Dinitrophenol	185(100)	155(40)		
198	2,4-Dinitrophenylhydrazine	199(100)			
300	Dioctyladipate			371(15)	129(100); 147(24); 157(11); 259( )
390	Di-n-octyl phthalate	391(100)	149(15); 279(10)	391(20)	149(100); 113(4); 167(32); 279( )
367	Dioxyline	368(100)			
452	Diphenoxylate			453(100)	246(8); 451(7)
169	Diphenylamine	170(100)	169(45)		
336	Diphenylbenzidine	337(100)	336(95)		
242	Diphenylcarbazide	152(100)	94(20); 243(15)		
240	Diphenylcarbazone	243(100)	241(78); 226(50); 228(45); 150(40); 152(35)		
211	1,3-Diphenylguanidine			212(100)	94(37); 211(32); 119(22); 195(20)
252	Diphenylhydantoin	253(100)		253(100)	175(25); 210( ); 225(4)
268	Diphenylhydantoin mtb.			269(100)	175(73); 191(10); 241(4)
280	Diphenylhydantoin dimethyl deriv.			281(100)	203(23)
255	Diphenhydramine	256(100)	167(25)	256(52)	167(100); 209(7); 254(5)
281	Diphenylpyraline			282(15)	165(100); 280(12); 109(13)
439	Dipicrylamine	440(100)	439(18)		
349	Dipyrome			232(6)	218(100); 217(2); 120(24)
231	Disulfiram	297(-)	150(100); 116(80); 118(60); 149(27); 117(15)		
296	Dithiooxamide	121(100)	120(22)		
120	Dopa	198(100)			
427	Doxapram	379(100)		379(100)	100(13); 101(2)
197	Doxepine	280(100)		280(100)	221(21); 278(20)
378	Doxepin mtb.			266(100)	221(25); 235(19); 264(12)
220	Doxepin mtb.			221(100)	220(22)
444	Doxycycline hyclate	445(100)	444(12); 427(10)		
270	Doxylamine	184(100)	271(50; 182(40)	271(33)	182(100); 269(1); 90(15); 210(14)
237	l-Dormoran	258(100)	256(50); 257(25)		

CI DATA—ALPHABETICAL INDEX (continued)

		CI			
		Isobutane		Methane	
Mol wt	Substance	MH⁺ (relative abundance)	Major peaks	MH⁺ (relative abundance)	Major peaks
289	Dyclonine			290(15)	98(100); 288(11); 205(3); 234(2)
312	Dydrogesterone	313(100)	312(72); 268(33); 207(15)		
254	Dyphylline	255(100)	254(40)		
329	Ecgonine bis-TMS deriv.			330(100)	314(80); 240(47)
185	Ecgonine (cocaine mtb.)	186(100)	124(30); 141(25); 168(21); 185(17)		
156	Ectylurea	157(100)	114(10)	157(73)	97(100); 114(17)
480	Emetine	481(100)	479(60); 480(48)		
145	Emylcamate			146(10)	102(100); 101(80)
165	Ephedrine	166(100)	148(50)	166(29)	148(100); 164(8)
399	Epinephrine tris-TMS deriv.			400(20)	310(100); 384(90); 355(80)
325	Ergonovine	326(100)	268(35); 308(12); 325(12)		
302	Ethacrynic acid	303(100)	305(65); 285(10)		
144	Ethchlorvynol	145(-)	127(100); 129(34); 109(10)	145(8)	107(100); 109(10); 115(86); 143(45)
				145(8)	127(100); 115(10); 129(33); 117(33); 143(25); 109(16)
126	Ethchlorvynol (dehyd. prod.)			127(100)	
167	Ethinamate	168(-)	125(100); 107(66); 163(13)	168(0.5)	107(100; 125(10); 166(1)
166	Ethionamide	167(100)	166(25)		
261	Ethoheptazine	248(100)		262(100)	260(32); 216(12); 188(5)
312	Ethopropazine			313(5)	100(100); 114(53); 198(6); 311(3)
141	Ethosuximide	142(100)		142(100)	114(9); 113(6)
204	Ethotoin			205(100)	127(8)
256	Ethoxazene	257(100)	256(45); 138(15); 133(15)		
165	Ethylaminobenzoate			166(100)	138(22); 120(22); 122(14)
60	Ethylenediamine	61(100)			
242	5-Ethyl-5-(3-hydroxyl-methylbutyl)-barbituric acid	243(40)	225(100); 157(23)		
313	Ethylmorphine	314(11)	296(100)		
172	5-Ethyl-2-thiobarbital	173(100)			
296	Ethynylestrodiol	297(50)	279(100); 296(32); 280(21); 295(19); 213(12)		
164	Eugenol			165(100)	137(10); 150(8); 105(2)
231	Fenfluramine	232(100)	72(68); 212(25)		
336	Fentanyl	337(100)	245(25)		
281	Flufenamic acid			282(83)	262(100); 264(47); 281(31)
387	Flurazepam	388(100)	86(44); 390(36)	388(100)	100(43); 99(27); 315(13); 386(13)
245	Flurazepam mtb.			246(100)	245(60); 181(20); 210(18)
288	Flurazepam mtb.			289(100)	193(18); 253(4)
313	Flurazepam mtb.			314(100)	313(10); 218(6); 278(2)
332	Flurescein	333(100)			
180	Frutose	181(-)	163(100); 145(85); 127(25)		
116	Fumaric acid	117(100)	73(25); 99(10)		
225	Furazolidone	226(100)			
361	Furildioxime	221(100)	205(25); 203(10)		
220	Galactose	181(-)	163(100); 145(85); 127(25)		
154	Gentisic acid	155(100)			
180	Glucose	181(-)	163(100); 145(85); 127(25)		
217	Glutethimide	218(100)		218(100)	190(9); 189(6)
233	Glutethimide mtb.			234(100)	188(30); 232(25)

CI DATA—ALPHABETICAL INDEX (continued)

Mol wt	Substance	CI Isobutane MH⁺ (relative abundance)	CI Isobutane Major peaks	CI Methane MH⁺ (relative abundance)	CI Methane Major peaks
198	Glyceryl Guaiacolate	199(100)	198(60); 124(50); 125(30)	199(75)	125(100); 151( ); 163(45); 181( ); 137(20)
75	Glycine	76(100)			
352	Griseofulvin	353(100)	355(35); 352(30)		
198	Guanethidine			199(79)	140(100); 197(27); 114(16); 182(13)
250	Heptabarbital			251(42)	157(100); 95(28); 221(11); 249(6)
404	Hexachlorophene	405(50)	407(100); 409(80); 411(40)		
236	Hexobarbital	237(100)		237(26)	157(100)
250	Hexobarbital methyl deriv.			251(60)	171(100); 235(16); 249(7)
179	Hippuric acid	180(100)	105(54); 135(31); 134(21); 162(15)		
111	Histamine	112(100)			
	Homatropine	276(10)	124(100); 107(14)	276(2)	124(100); 105(1); 135(6); 258(3)
160	Hydralazine	161(100)		161(100)	160(37); 132(20)
297	Hydrochlorothiazide	286(100)	298(70); 300(55); 288(45)	298(100)	286(99); 269(72); 271(31)
362	Hydrocortisone			363(46)	61(100); 345(54); 303(43)
285	Hydromorphone	286(100)			
110	Hydroquinone	111(100)	110(50)		
151	*p*-Hydroxyamphetamine	152(100)	135(95)	152(42)	135(100); 107(5); 150(3)
138	p-Hydroxybenzoic acid	139(100)	121(35)		
167	d-4-Hydroxynorephedrine	168(50)	150(100); 123(17)		
209	Hydroxyphenamate			210(<1)	131(100); 149(3); 192(20)
137	β-Hydroxy-β-phenylethylamine	138(100)	120(50)		
165	4-Hydroxyphenylisopropylmethylamine	166(100)	135(10)		
153	Hydroxytyramine	154(100)	137(24); 124(24)		
374	Hydroxyzine	375(100)	377(30)		
303	Hyoscine			304(4)	138(100); 139(9)
289	Hyoscyamine	290(15)	124(100); 237(20)		
310	Ibogaine	311(100)			
280	Imipramine	281(100)		281(89)	235(100); 195(6); 208(59)
352	Indomethacine	358(100)	357(44); 360(38)	358(100)	139(21); 141(7)
				358(-)	314(100); 313(40); 315(20); 316(20)
265	Intracaine			266(58)	100(100); 86(2); 149(13)
305	Iodochlorhydroxyquinoline			306(100)	180(54); 179(28); 129(17)
351	7-Iodo-8-hydroxyquinoline-5-sulfonic acid	272(100)			
147	Isatin	148(100)			
128	Isobarbituric acid	129(100)			
231	Isocarboxazid			232(100)	91(89); 119(12); 154(9)
165	Isoephedrine	166(100)	107(42); 148(15)		
137	Isoniazide	138(100)	137(10)	138(100)	108(22)
211	Isoproterenol	212(50)	194(100)		
285	Isothipendyl			286(37)	241(100); 200(34); 284(21); 214(17)
301	Isoxsuprine	302(100)	178(16); 284(12)		
237	Ketamine			238(100)	240(25); 220(1); 209(8)
342	Lactose	343(-)	163(100); 145(85); 127(25)		
416	Lasix, methyl ester, mono trimethyl silyl ether	417(25)	81(100); 113(55); 97(45)	417(30)	81(100); 73(25); 114(20)
283	Levallorphan	284(100)	282(50); 283(25)	284(100)	283(43); 282(29); 266(14); 257(8)

CI DATA—ALPHABETICAL INDEX (continued)

		CI			
		Isobutane		Methane	
Mol wt	Substance	MH⁺ (relative abundance)	Major peaks	MH⁺ (relative abundance)	Major peaks
257	Levorphanol	258(100)	256(40); 257(35); 151(20)		
234	Lidocaine	235(100)	86(13)	235(100)	233(15); 132(2); 148(2)
206	Lidocaine mtb.			207(100)	169(43); 205
288	Lindane			289(<1)	219(100); 217(72); 221(37); 181(21)
268	Lysergic acid			269(100)	268(52); 224(44); 194(33)
323	Lysergic acid diethylamide	324(100)		324(100)	323(46); 281(29); 322(17)
323	Lysergic acid-N-methylpropylamide	324(100)	323(13)		
104	Malonic acid	105(100)			
180	Maltose	163(100)	145(85); 127(40)		
152	Mandelic acid	153(-)	135(100); 107(65)		
182	Mannitol	183(100)	165(20); 147(10); 129(10)	183(47)	129(100); 111(61); 99(55); 103(33); 165(21)
180	Mannose	181(-)	163(100); 145(85); 127(25)		
284	Mazindol	285(100)	287(33); 267(20); 255(20); 269(10); 257(7)		
232	Mebutamate	233(100)	172(75)	233(6)	111(100); 172(7); 69(18)
390	Meclizine	391(100)	393(30); 390(18); 189(18); 389(11); 243(10)	391(39)	201(100); 189(40); 203(34)
296	Meclofenamic acid			296(100)	278(75)
270	Mecloqualon	271(90)	235(100); 273(32); 118(10)		
241	Mefenamic acid			242(100)	224(48); 241(32); 225(8)
141	Meparfynol carbamate			142(<1)	99(100)
247	Meperidine	.248(100)		248(100)	246(20); 174(5); 202(3)
182	Mephenesin	183(100)	147(30); 109(17); 183(21)	147(100); 135(79); 121(65); 165(48); 109(43)	
223	Mephenoxalone	224(100)			
163	Mephentermine	164(100)		164(45)	72(100); 133(10); 162(10)
218	Mephenytoin	219(100)	189(10)		
246	Mephobarbital	247(100)		247(100)	218(13); 245(3)
246	Mepivacaine			247(77)	98(100); 245(20)
218	Meprobamate	219(100)	158(61)	219(5)	158(100); 115(4)
167	2-Mercaptobenzothiazole			169(100)	
386	Mesoridazine			387(-)	371(100); 370(4); 369(15); 372(13)
211	Mescaline	212(100)		212(100)	195(90); 182(14); 168(10); 151(8)
197	Metanephrine	198(100)	180(27)		
221	Metaxalone	221(100)			
442	Methacycline	443(100)	198(38); 425(10); 400(10)		
309	Methadone	310(100)		310(100)	265(60); 308(28)
277	Methadone mtb.	278(98)	277(100)	278(100)	279(22)
149	Methamphetamine	150(100)		150(36)	119(100); 148(33); 120(10)
339	Methantheline			340(3)	86(100); 100(88); 109(61)
260	Methaphenilene			261(100)	97(64); 202(47); 216(28); 259(24)
261	Methapyrilene	262(100)	166(27); 191(13)	262(23)	217(100); 260(16); 191(10); 245(5)

CI DATA—ALPHABETICAL INDEX (continued)

		CI			
		Isobutane		Methane	
Mol wt	Substance	MH⁺ (relative abundance)	Major peaks	MH⁺ (relative abundance)	Major peaks
250	Methaqualone	251(100)	250(30); 235(20); 251(100)	118(40); 146(20)	
266	Methaqualone mtb.			267(100)	265(14)
266	Methaqualone mtb.			267(100)	118(6); 134(4)
266	Methaqualone mtb.			267(20)	249(100); 118(1); 146(8)
264	Methaqualone mtb.			265(100)	235(10); 118(6)
198	Metharbital	199(100)		199(100)	170(4)
140	Methenamine	141(100)	112(45); 140(35)		
286	Methallenestril	287(58)	199(100); 269(25); 229(18)		
218	Methetoin			219(80)	141(17); 189(11); 162(4)
296	Methdilazine	297(100)			
241	Methocarbamol	242(100)	199(100); 118(83); 124(33); 224(10)	242(10)	118(100); 163(40); 199(27); 224(1)
271	d-Methorpan	272(100)	270(64); 271(21)	271(100)	270(51)
211	Methoxamine	212(100)	194(55); 152(20); 168(18); 167(15)	212(12)	194(100); 168(9); 210(2)
165	m-Methoxyamphetamine	166(100)	149(10); 138(10)		
165	p-Methoxyamphetamine	166(100)	149(45); 122(15)		
179	Methoxyphenamine			180(74)	149(100); 178(19); 121(13)
203	Methsuximide	204(100)		204(100)	
214	Methylaurate			215(100)	213(20); 181(11); 183(8)
152	Methylparaben			153(100)	109(11); 121(10)
233	Methylphenidate	234(100)	84(70); 151(10)	234(98)	91(100); 151(56); 112(30); 119(25)
152	3-Methylsalicylate	153(80)	135(100); 134(82); 152(70); 105(42); 106(34); 78(25)		
262	Methohexital	263(100)		263(100)	183(52); 211(18); 221(15); 155(6)
149	p-Methylamphetamine	150(100)	133(49)		
179	Methylenedioxyamphetamine	180(100)	136(10)	180(100)	163(49); 136(14)
179	n-Methylephedrine	180(50)	72(100); 162(10)		
339	Methylergonovine	340(100)	322(26); 339(15)		
270	Methyl palmitate			271(100)	269(87); 237(12); 239(10)
150	Methylphenidate mtb.			151(24)	92(100); 119(30)
416	Methylprednisolone			417(43)	339(100); 339(3); 400(27)
183	Methylprylone	184(100)		184(100)	166(60); 155(13); 182(6)
151	Methylsalicylamide	152(100)	151(73); 121(25); 120(13)		
152	Methylsalicylate	153(100)	152(35); 120(35); 121(20)	153(100)	121(15)
353	Methysergide	354(100)	353(15)		
298	Methyl stearate			299(100)	297(57); 265(8); 267(5)
174	Methyltryptamine	175(100)	158(50); 131(40)		
197	Methyprylon mtb.			198(100)	128(22); 170(15)
199	Methyprylon mtb.			200(14)	182(100); 154(36)
				200(100)	182(55); 171(4); 182(2)
171	Metronidazole	172(100)			
226	Metyrapone	227(100)	226(36); 120(12)		
327	0⁶-Monoacetylmorphine	328(100)	268(55);327(30)		
285	Morphine	286(20)	268(100)	286(32)	268(100); 285(20)
429	Morphine bis-TMS deriv.			430(74)	414(100); 340(5); 371(18)
327	Naloxone	328(100)	327(35)		
210	Naphazoline	211(100)	210(20); 209(18)	211(100)	209(11)
302	Naproxen trimethyl silyl ether			303(30)	185(100); 287(20); 213(10); 231(5)
179	Napthoquinoline	180(100)			

CI DATA—ALPHABETICAL INDEX (continued)

		CI			
		Isobutane		Methane	
Mol wt	Substance	MH⁺ (relative abundance)	Major peaks	MH⁺ (relative abundance)	Major peaks
208	Neostigmine	209(100)			
298	Nialamide			299(100)	192(30); 92(29); 138(18)
122	Nicotinamide	123(100)		123(100)	106(3)
162	Nicotine	163(100)		163(100)	161(27)
192	Nicotine mtb.			193(100)	114(11); 175(9)
178	Nikethamide	179(100)		179(100)	106(9); 177(8)
178	Ninhydrin	161(100)	177(20); 132(15)		
151	o-Nitrobenzaldehyde	152(100)			
167	m-Nitrobenzoic acid	168(100)	138(45)		
238	Nitrofurantoin	239(100)	238(20)		
173	1-Nitroso-2-naphthol	158(100)	174(40); 129(10)		
169	Norepinephrine				105(100); 132(5); 151(29)
271	Normorphine			272(41)	254(100); 271(16); 270(5)
151	Norpseudoephedrine			152(45)	134(100)
263	Nortriptyline	264(100)		264(100)	233(45); 262(40)
413	Noscapine	414(-)	220(100); 221(15)		
299	Nylidrin	300(100)	282(25)	300(49)	176(100); 282(47); 206(22)
282	Oleic acid	283(100)	265(25); 264(15)		
269	Orphenadrine	270(80)	181(100); 224(11); 198(10)	270(7)	181(100); 182(1); 209(14)
167	Orthocaine	168(100)	136(10)		
90	Oxalic acid	91(100)			
157	Oxanamide	158(100)	131(10)	158(68)	140(100); 141(48); 113(24); 97(17)
286	Oxazepam	287(100)	269(99); 271(40); 289(35); 288(20); 270(20)	287(40)	269(100); 271(35); 257(5)
467	Oxethazine			468(100)	145(14); 277(12); 466(10)
315	Oxycodone	316(100)		316(100)	298(6)
260	Oxymetazocine	261(100)			
324	Oxyphenbutazone	325(100)	324(70); 199(35)	325(99)	120(7); 309(4); 190(4)
460	Oxytetracycline	461(100)	443(30); 198(30); 460(16)		
301	Oxymorphone	302(100)			
159	Pagyline	160(100)	159(20)		
256	Palmitic acid	257(100)	256(90); 239(85)	257(100)	255(30); 239(22); 237(22)
339	Papaverine	340(100)		340(100)	151(3)
				340(100)	341(13); 339(7)
157	Paramethadione			158(100)	
266	Pentachlorophenol	267(100)	269(67); 265(67); 271(20)		
102	Pentanediamine	103(100)	86(39)		
285	Pentazocine	286(100)	284(50); 285(40)	286(78)	217(100); 230(96); 284(41); 200(20)
301	Pentazocine mtb.			302(35)	284(100); 217(30); 230(28); 300(24)
301	Pentazocine mtb.		302(100)	217(55); 284(43); 300(33); 230(22)	
226	Pentobarbital	227(100)		227(80)	157(100); 185(1); 156(8)
254	Pentobarbital dimethyl deriv.			255(100)	185(45); 169(7); 253(4)
138	Pentylenetrazole	139(100)		139(100)	96(8)
403	Perphenazine	404(100)	406(30); 234(19)		
298	Phenacaine	299(100)	262(18)		
179	Phenacetin	180(100)	179(12)	180(100)	152(11)
178	Phenacemide	179(100)	136(40)	179(100)	136(74); 118(9); 91(7)
137	Phenacetin mtb.			138(100)	121(4)

CI DATA—ALPHABETICAL INDEX (continued)

		CI			
		Isobutane		Methane	
Mol wt	Substance	MH⁺ (relative abundance)	Major peaks	MH⁺ (relative abundance)	Major peaks
214	Phenaglycodol	215(-)	197(100); 155(45); 199(35); 157(18); 198(15); 156(15)	215(2)	197(100); 199(3); 103(16); 121(4)
321	Phenazocine	322(100)	230(80)	322(18)	230(100); 320(5); 306(3)
213	Phenazopyridine	214(100)			
243	Phencyclidine	244(100)	242(63); 159(32)244(43)	159(100); 243(9); 242(45); 200(23)	
158	Phencyclidine contaminant			159(100)	157(23); 119(16)
191	Phendimetrazine	192(100)		192(100)	105(31); 190(22); 114(13)
136	Phenelzine	137(100)	135(34); 122(22); 105(12)		
121	Phenethylamine	122(100)	121(35); 105(15)		
261	Phenindamine	262(100)		262(100)	261(58); 260(45); 219(23); 247(7)
150	Pheniprazine			151(100)	91(98); 119(93); 145(26); 149(11)
240	Pheniramine	241(100)		241(29)	196(100); 239(2)
177	Phenmetrazine	178(100)		178(100)	100(18); 176(10); 91(8); 119(5)
232	Phenobarbital	233(100)		233(100)	204(6)
260	Phenobarbital dimethyl deriv.			261(100)	232(16)
318	Phenolphthalein	319(100)		319(100)	255(30)
199	Phenothiazine	200(45)	199(100)	200(100)	199(57)
189	Phensuximide			190(100)	104(5)
149	Phentermine	150(100)			
308	Phenylbutazone	309(100)	190(50)	309(100)	120(15); 183(8)
324	Phenylbutazone-alcohol mtb.			94(100)	324(35); 185(30); 185(30); 234(20); 122(20); 115(18)
380	Phenylbutazone trimethylsilyl ether	381(100)	309(85); 35(5)		
229	1-(1-Phenylcyclopentyl)-Piperadine	230(100)	229(75); 145(43); 228(40); 200(12); 187(11); 201(9)		
229	1-(1-Phenylcyclohexyl)-Pyrrolidine	230(100)	229(65); 228(40); 159(22); 186(13)		
108	p-Phenylenediamine			109(83)	108(108); 107(12)
167	Phenylephrine	168(100)	150(33); 123(10)	168(15)	150(100); 166(5)
214	Phenylsalicylate	215(100)	215(67)	121(100); 149(20)	
255	Phenyltoloxamine	256(100)		256(100)	254(60)
180	o-Phentroline	181(100)			
214	Phenyramidol	215(65)	107(100); 95(23); 108(13); 197(10)	215(59)	197(100); 107(30); 108(22); 137(5)
126	Phloroglucinol	127(100)	126(25)		
166	Phthalic acid	167(25)	167(25); 123(12); 149(100)		
275	Physostigmine			276(100)	275(75); 219(55); 274(25); 162(5)
275	Physostigmine salicylate	276(100)	139(38); 219(18)		
228	Picric acid	200(100)	230(70)		
264	Picrolonic acid	233(100)	249(65); 234(45); 232(15); 265(10)		
602	Picrotoxin	293(100)	585(80); 603(70)		
208	Pilocarpine	209(100)	208(15)	209(100)	95(4)
366	Piminodine	367(100)	234(12)		
85	Piperidine	86(100)	85(20); 84(18)		
192	1-Piperidine-cyclohexane-carbonitrile (PCC)	193(-)	166(100);192(10); 149(10);124(5)		
169	Piperidione			170(100)	152(86); 141(10); 168(7)

CI DATA—ALPHABETICAL INDEX (continued)

		CI			
		Isobutane		Methane	
Mol wt	Substance	MH^+ (relative abundance)	Major peaks	MH^+ (relative abundance)	Major peaks
323	Piperidolate	324(37)	112(100); 114(40)		
261	Piperocaine			262(100)	140(40); 112(2); 269(21); 246(1)
267	Pipradrol			268(21)	250(100); 190(1); 266(7)
189	a-Phenylglutaramide			190(100)	102(15); 104(5)
151	Phenylpropanolamine	152(100)	134(50); 107(25)	152(5)	134(100); 117(4); 150(2)
149	Phenylpropylmethylamine	150(100)		150(90)	119(100); 91(63); 148(55); 134(8)
248	Piridocaine			249(100)	112(88); 120(45); 110(18)
206	Pivalylbenzhydrazine			207(100)	91(50); 106(10); 205(4)
439	Polythiazide	440(90)	300(100); 302(45); 442(40); 266(40); 406(30); 361(30)		
360	Prednisolone			361(16)	61(100); 343(54); 325(49)
358	Prednisone			359(100)	341(57); 299(18); 329(15)
220	Prilocaine	221(100)		221(100)	164(7); 219(4); 136(2)
206	Primidone mtb.			207(14)	162(100); 163(20); 190(5)
216	Primidone mtb.			217(100)	146(22); 161(16); 189(13)
198	Probarbital	199(100)			
285	Probenecid	286(100)		286(100)	256(23); 268(4); 185(3); 284(3)
236	Procaine	237(100)	100(16); 99(12)	237(12)	100(100); 120(4); 164(9); 125(5)
				237(100)	100(80); 120(18); 86(12); 221(8); 164(5)
235	Procaine Amide	236(100)	99(18); 136(12)		
373	Prochlorperazine	374(100)	376(30); 373(27); 234(20)	374(7)	99(100); 373(5); 113(4)
287	Procyclidine			288(100)	286(50); 204(36); 210(8); 270(3)
284	Promazine	285(100)	284(15)	285(100)	284(96); 212(12); 199(7); 240(6)
				285(100)	86(35); 200(15); 226(10); 254(2)
300	Promazine mtb.			301(-%)	300(100)
284	Promethazine	285(100)	284(27)	285(100)	198(93); 283(58); 284(54); 240(34)
286	Promethazine mtb.			287(55)	213(100); 230(25); 200(15)
300	Promethazine mtb.			301(100)	
294	Proparacaine			294(47)	100(100); 99(76); 178(19)
340	Propoxycaine			295(4)	100(100); 99(30); 178(24)
339	Propoxyphene			340(16)	131(100); 266(28); 338(19(
208	Propoxyphene decomp. prod.			209(4)	131(100); 159(2); 105(6); 208(3)
251	Propoxyphene mtb.			252(64)	143(100); 250(6); 221(24)
265	Propoxyphene mtb.			266(100)	264(98); 143(52); 221(16)
307	Propoxyphene mtb.			308(100)	143(20); 220(18); 306(4)
259	Propranolol	260(100)	72(60)		
285	Prothipendyl			286(100)	241(66); 285(65); 284(7); 269(3)
263	Protriptyline	264(100)			
180	Propylaraben			181(90)	139(100); 121(60); 95(18); 167(10)
155	Propylhexedrine			156(100)	154(39); 140(9); 125(4)
8	Putrescine	89(100)	72(60); 73(30)		
109	3-Pyridine methanol	110(100)	92(14)	110(100)	92(45)
169	Pyridoxine	170(100)	152(15)		
285	Pyrilamine	286(100)	241(10)	286(24) 121(100); 117(30); 241(22)	

CI DATA—ALPHABETICAL INDEX (continued)

		CI			
		Isobutane		Methane	
Mol wt	Substance	MH⁺ (relative abundance)	Major peaks	MH⁺ (relative abundance)	Major peaks
126	Pyrogallol	127(100)	126(40)		
173	Quinaldic acid	174(100)			
272	Quinalizarin	273(100)	257(90); 241(80); 272(40); 256(30); 240(30)		
324	Quinidine	325(100)	136(18)	325(100)	307(12); 136( ); 295(4); 323(4)
324	Quinine	325(100)	136(20)	325(100)	136(37); 307( ); 323(7)
145	Quinolinol			146(100)	145(7)
608	Reserpine			609(92)	213(100); 607(24); 195(24)
				609(100)	213(70); 195(25); 129(20)
110	Resorcinol	111(100)	110(100)	111(100)	110(9)
478	Rhodamine	479(-)	443(100); 398(20)		
376	Riboflavin			377(1)	243(100); 244(15); 362(13)
220	Rompun			221(100)	90(20); 205(10)
183	Saccharin	184(100)		184(14)	
137	Salicylaldoxine	138(100)	137(22); 120(15); 91(10)		
137	Salicylamide	138(100)	137(11)	138(100)	121(20); 95(4)
138	Salicylic acid	139(100)		139(100)	121(90); 95(8)
282	Salicylic acid bis-TMS deriv.			283(17)	193(100); 267(89)
89	Sarcosine	90(100)			
303	Scopolamine	304(16)	138(100)	304(4)	138(100); 156(6)
238	Secobarbital	239(100)	199(36)	239(51)	169(100); 197(51); 168(11)
266	Secobarbital dimethyl deriv.			267(73)	197(100); 225(16); 265(5)
255	Solfathiazole			256(100)	156(71); 101(3)
182	Sorbitol	183(100)	165(20); 147(10); 129(10)		
234	Sparteine			235(42)	233(100); 98(21); 137(8)
284	Stearic acid	285(100)	284(70); 267(45); 283(26)	285(100)	283(19); 265(13); 267(8)
334	Strychnine	335(100)	334(25)	335(100)	334(30); 333(22)
350	Strychnine-N-oxide	351(20)	335(100); 333(25)		
181	Styramate			182(1)	121(100); 164(43); 107(4)
342	Sucrose	343(−%)	163(100); 145(85); 127(25)		
284	Sulfachlorpyritazine	285(100)	220(60); 287(40); 251(20)		
250	Sulfadiazine	251(100)	186(60)	251(100)	156(16); 96(5)
310	Sulfadimethoxine			311(100)	156(90); 184(20); 140(20); 246(10)
241	Sulfaguanidine			242(1)	215(100); 173(58); 156(42)
264	Sulfamerazine			265(100)	156(21); 110(4)
280	Sulfameter	281(100)	216(10)		
278	Sulfamethazine			279(100)	156(8); 214(7); 124(2)
270	Sulfamethizole	271(100)			
253	Sulfamethoxazole	254(100)			
280	Sulfamethoxypyridazine			281(100)	156(3); 216(3); 126(2); 188(2)
172	Sulfanilamide			173(100)	156(67); 93(3)
249	Sulfapyridine	250(100)	185(20)		
300	Sulfaquinoxaline			301(10)	146(100); 237(78); 210(22)
404	Sulfinpyrazone	405(35)	279(100); 278(55); 280(20)		
278	Sulfisomidine	279(100)	214(40)	279(100)	124(6); 215(5); 156(3)
267	Sulfisoxazole	268(100)			
218	Sulfosalicylic acid	219(100)	218(65); 201(25); 200(20)		
224	Talbutal	225(100)	185(22)	225(100)	169(40)
150	Tartaric acid	151(100)	105(33); 123(15)		
190	Terpin			191(<1)	137(100); 96(3)
288	Testosterone			289(100)	271(22); 272(5)
264	Tetracaine	265(100)	194(14)	265(32)	176(100); 263(14); 220(5)
444	Tetracycline	445(40)	427(100); 428(30); 426(28); 257(25)		
200	Tetrahydrazoline	201(100)			
314	Tetrahydrocannabinol	315(100)		315(100)	313(55); 299(15); 135(7); 193(5)
330	Tetrahydrocannabinol mtb.			331(48)	313(100); 330(47); 329(25); 190(9)
408	γ-Tetrahydropyranyl ether of phenylbutazone alcohol mtb.			325(100)	85(48); 101(25); 120(10); 409(2)

CI DATA—ALPHABETICAL INDEX (continued)

		CI			
		Isobutane		Methane	
Mol wt	Substance	MH^+ (relative abundance)	Major peaks	MH^+ (relative abundance)	Major peaks
254	Tetra-methyldiaminodiphenylmethane	255(100)	134(40)		
332	Tetraphenylethylene	333(100)	389(100)		
261	Thenyldiamine			262(18)	217(100); 260(12); 191(6); 245(6)
180	Theobromine	181(100)	180(10)	181(100)	138(2)
180	Theophylline	181(100)	180(12)	181(100)	124(5); 95(2)
		181(100)	180(15)		
264	Thiamine			265(3)	144(100); 126(99); 158(38)
254	Thiamylal	255(100)		255(84)	185(100); 213(33)
399	Thiethylperazine	400(100)	401(95); 402(25)		
75	Thioacetamide	76(100)	75(80)		
200	Thiobarbituric acid	201(−%)	145(100)		
242	Thiopental	243(100)	227(25)	243(100)	173(38); 227(20); 157(10); 201(10)
370	Thioridazine	371(100)	370(77)	371(100)	370(84); 98(81); 126(38)
76	Thiourea	77(100)			
286	Thonzylamine			287(29)	121(100); 242(14); 285(14); 216(8)
204	Thozalinone			205(100)	136(53)
150	Thymol			151(62)	109(100); 149(46); 135(37); 137(35)
160	Tolazoline	161(100)		161(100)	159(12)
270	Tolbutamide	271(32)	172(100); 198(25)		
133	Tranylcypromine	134(100)	132(25); 133(20)	134(25)	117(100); 91(88); 133(52); 119(26)
394	Triamcinalone			395(16) 254(100)	347(100); 327(74); 375(54)
398	Tributoxyethyl phosphate			399(100)	299(68); 199(47)
379	Trichlormethazide	380(95)	382(100); 384(33)		
407	Trifluoperazine	408(100)	407(30); 278(13); 302(10); 279(10)	408(100)	407(67); 388(52); 406(39)
352	Triflupromazine	353(100)		353(72)	352(100); 280(5) 282(4)
301	Trihexyphenidyl	302(100)	300(12)		
298	Trimeprazine	299(100)	200(55); 298(50); 240(42); 239(20); 199(20)		
143	Trimethadione	144(100)		144(100)	100(10); 128(2)
388	Trimethobenzamide	389(100)			
255	Tripelenamine			256(51)	211(100); 254(41); 185(27); 239(13)
278	Triprolidine			279(48)	208(100); 277(29); 236(7)
121	Tromethamine	122(100)	104(15)		
141	Tropine	142(60)	124(100); 140(50); 141(20)		
694	Tubocurarine chloride			695(−%)	595(100); 596(43); 609(25)
274	Tybamate	275(100)	232(40); 158(30); 176(30); 214(20)		
137	Tyramine	138(100)	121(50); 108(30); 107(10)		
166	Vanillal			167(100)	139(25); 111(6)
223	Vanillic acid diethylamide			224(100)	100(22); 151(9)
152	Vanillin	153(100)	152(15)	153(100)	125(12)
224	Vinbarbital	225(100)		225(90)	157(100)
157	Violuric acid	158(100)	142(33)		
308	Warfarin	309(100)	251(15); 291(10)		
198	Xanthydrol	181(100)	197(25)		
195	*p*-Xenylcarbimide	196(40)	170(100); 195(30); 169(30)		
244	Xylometazoline	245(100)		245(100)	244(23); 243(23); 229(14)
354	Yohimbine	355(100)	354(75); 353(40); 337(10); 352(10)	355(100)	354(43); 337(24); 323(10)
168	Zoxazolamine			169(100)	133(12)

EI DATA—ALPHABETICAL INDEX

Substance	Mol wt	Base peak
Acetaminophen	151	109
Acetaminophen	151	109
Acetaminophen glucuronide methyl ester TMS ether	557	317
Acetaminophen methyl ether	165	108
Acetaminophen TMS ether	223	181
Acetophenazine	411	254
Acetophenetidin	179	108
3-Acetoxychlorpromazine	376	58
8-Acetoxychlorpromazine	376	58
4-Acetyl-aminoantipyrine	245	56
Acetylcarbromal	27	41
Acetylmethadol	353	72
Acetylpromazine	326	58
Acetylsalicylic acid	180	120
Acetylsalicylic acid methyl ester	194	120
Acetylsalicylic acid TMS ester	252	195
Allobarbital	208	167
Allobarbital-1,2 (or 1,4)-dimethyl deriv.	236	195
Allobarbital-1,3-dimethyl deriv.	236	195
Allopurinol	136	136
Allylprodine	287	172
Alpha chloralose	308	71
Alphaprodine	287	172
Alphenal	244	215
Alphenal-1,3-dimethyl deriv.	272	118
4-Aminoantipyrine	203	56
p-Aminobenzoic acid methyl ester	151	120
Aminoglutethimide	232	203
Aminopromazine	327	198
Aminopyrine	231	56
Amitriptyline	277	58
Amobarbital	226	156
Amobarbital-1,2 (or 4)-dimethyl deriv.	254	169
Amobarbital-1,3-dimethyl deriv.	254	169
Amobarbital mtb. 1	242	59
Amphetamine (also D-amphetamine)	135	44
Ampyrone	203	56
Anileridine	352	246
Anisotropine	267	124
Antazoline	265	84
Antipyrine	188	188
Apomorphine	267	266
Aprobarbital	210	167
Aprobarbital-1,3 dimethyl deriv.	238	195
Ascorbic acid	176	116
Atropine (see hyoscyamine)	289	124
Atropine TMS ether	361	124
Azacyclonol	267	85
Azathioprine	277	42

EI DATA—ALPHABETICAL INDEX (continued)

Substance	Mol wt	Base peak
Banthine® (methantheline bromide)	420	181
Barbital	184	156
Barbital-1,2 (or 1,4)-dimethyl deriv.	212	184
Barbital-1,3-dimethyl deriv.	212	169
Benzocaine	165	120
Benztropine	307	140
Benzyl alcohol	108	79
Benzylmorphine	375	284
5-(2-Bromoallyl) barbituric acid	246	168
5-(2-Bromoallyl)-5-(1-methylbutyl) barbituric acid	316	43
Bromodiphenhydramine	333	58
D-Bromopheniramine (D-para bromolylamine)	319	58
Buclizine	432	147
Bufotenine	204	58
Butalbital	224	168
Butabarbital-1,3-dimethyl deriv.	240	169
Butalbital-1,2 (or 4)-dimethyl deriv.	252	196
Butalbital-1,3-dimethyl deriv.	252	196
Butalbital-2,4 (or 4,6)-dimethyl deriv.	252	95
Butethal	212	141
Butethal-1,3-dimethyl deriv.	240	184
Butyl butoxyethyl phthalate	322	149
Butyl carbobutoxymethyl phthalate	336	149
Caffeine	194	194
Camphor	152	126
Cannabidiol		231
Cannabinol	310	295
Captodiamine	359	58
Caraminephen	289	86
Carbamazepine	236	193
Carbinoxamine	290	58
Carbromal	237	44
Carisoprodol	260	58
Carphenazine	425	268
Chelidonine	353	332
Chloral betaine	282	82
Chloral hydrate	165	82
Chloral hydrate mtb. 1	148	31
α-Chloralose		71
Chlorcyclizine	300	99
Chlordiazepoxide	299	282
Chlordiazepoxide (GC)	299	282
Chlordiazepoxide mtb. 1	285	268
Chlormezanone	273	152
4-Chlorobenzoic acid methyl ester	170	139
4-Chlorobenzoic acid TMS ether	228	213
1-(4-Chlorobenzoyl)-2-methyl-5-trimethylsilyloxyindole-3-acetic acid methyl ester	429	139

EI DATA—ALPHABETICAL INDEX (continued)

Substance	Mol wt	Base peak
1-(4-Chlorobenzoyl)-2-methyl-5-trimethylsilyloxyindole-3-acetic acid trimethylsilyoxy ester	487	139
7-Chloro-1,3-dihydro-5-phenyl-2H-1,4-dibenzodiazepin-2-one TMS deriv.	342	73
7-Chloro-1,3-dihydro-3-hydroxy-1-methyl-5-phenyl-2H-1,4-benzodiazepin-2-one TMS ester	372	73
7-Chloro-1,3-dihydro-3-hydroxy-1-methyl-5-phenyl-2H-1,4-benzodiazepin-2-one TMS ether	372	334
2-(4-Chlorophenoxy)-2-methyl propionic TMS ester	286	128
2-(4-Chlorophenoxy)-2-methyl propionic acid methyl ester	228	128
7-Chloro-1,3-dihydro-2H-1,4-benzodiazepin-2-one	270	242
Chlorophenothane (DDT)	354	235
Chloroquine	319	86
Chlorothen	295	58
Chlorpheniramine	274	203
7-Chloro-3-hydroxy-5-phenyl-1,3-dihydro-2H-1,4-benzodiazepin- 2-one-di TMS deriv.	430	73
Chlorpromazine	318	58
Chlorpromazine mtb. 1	233	233
Chlorpromazine mtb. 2	304	233
Chlorpromazine mtb. 4	334	58
Chlorprothixene	315	58
Chlorothiazide	295	295
Chlorzoxazone	169	169
Cholesta-3,5-diene	368	43
3,5-Cholestadiene-7-one	382	174
Cholesterol	386	368, 386
Clemizole	325	255
Clidinium bromide	432	105
Clofibrate	242	128
Clofibrate glucuronide methyl ester TMS ether	620	73
Clonitazene	386	86
Cocaine	303	82/182
Codeine	299	299
Codeine *N*-oxide	315	299
Colchicine	399	312
Cotinine (from nicotine)	176	98
p-Cresol	108	107
Cyclizine	266	99
Cyclobarbital	236	207
Cyclopentolate	291	58
Cycrimine	287	98
Cyproheptadine	287	287
Deserpidine	578	365
Desipramine	266	195
Diampromide	324	162
Diazepam	284	256
p-Dibromobenzene	235	236
Dibutyl adipate	258	185

EI DATA—ALPHABETICAL INDEX (continued)

Substance	Mol wt	Base peak
Dibutyl phthalate	278	149
Di-*n*-butyl phthalate	278	
2,4-Dichlorophenol	162	162
Dicyclomine	309	86
Diethazine	298	86
3,7-Diethyl-1-methyl-xanthine	222	222
1,7-Diethyl-3-methyl-xanthine	222	222
1,3-Diethyl-7-methyl-xanthine	222	222
1,3-Diethyl-9-methyl-xanthine	222	123
Diethylpropion	205	100
Dihydrocodeine	301	301
Dihydrocodeinone	299	299
Dihydromorphine	287	287
Dihydronoroxymorphone-2-TMS		73
Dihydronoroxymorphone-3-TMS		73
3,4-Dihydroxybenzoic acid methyl ester trimethylsilyl ether		338
5-(3,4-Dihydroxycyclohexa-1,5-dienyl)-3-methyl-5-phenylhydantoin TMS deriv.	516	73
Dihydroxy secobarbital-di-TMS ether-1,3-dimethyl deriv.	444	73
Dihydroxy secobarbital-di-HFB derivative-1,3-dimethyl deriv.	692	169
Dimenoxadol	327	57
2,5-Dimethoxyamphetamine	196	44
2,5-Dimethoxy-4-methyl-amphetamine (STP)	209	44
N,N′-Dimethoxymethylphenobarbital		146
1,2-Dimethyl-5-methoxyindole-3-acetic acid methyl ester	247	174
N,N-Dimethylphenobarbital	260	232
Dimethylsulphone	94	79
Dimethylthiambutene	263	248
DMT (N,N-dimethyltryptamine)	188	58
Dioctyl adipate	370	129
Dioctyl phthalate	390	149
Diphenhydramine	255	58
Diphenoxylate	452	246
Diphenylhydantoin	252	180
Diphenylhydantoin (D5-Ph)	257	185
Diphenylhydantoin mtb.	266	180
5,5-Diphenylhydantoin 1,3-diethyl deriv.	308	180
5,5-Diphenylhydantoin 2,3-diethyl deriv.	308	279
Diphenylhydantoin 3-methyl deriv.	266	180
Diphenylpyraline	281	99
Dipipanone	349	112
Dixyrazine	270	58
Doxylamine	270	58
Doxylamine mtb. 1	199	182
Doxylamine mtb. 2	181	180
Doxylamine (GC)		58
Dramamine (dimenhydrinate)	469	58
Ectyurea	156	41
Emylcamate (1-ethyl-1-methyl propylcarbal)	145	73

EI DATA—ALPHABETICAL INDEX (continued)

Substance	Mol wt	Base peak
Ephedrine TMS ether	237	58
Erythromycin A	733	158
Ethchlorvynol	144	115
Ethinazone	264	249
Ethoheptazine	261	107
Ethopropazine	312	100
Ethosuximide	141	113
Ethosuximide-n-methyl deriv.	155	127
Ethosuximide-N-ethyl deriv.	169	141
Ethotoin	204	204
Ethyl-methyl thiambutene	277	262
Ethylmorphine	313	313
Ethylphenylmalondiamide-di-TMS deriv.	350	235
Ethinamate (ethynylcyclohexyl carbamate)	167	91
Etorphine	411	411
Fentanyl	336	245
Fluphenazine	437	42
Flurazepam	387	86
Furethidine	361	246
Fursemide	330	81
Glutethimide (doriden)	217	117/189
Heptabarbital (mendomin)	250	221
Heroin	369	327
Hexethal	226	156
Hexobarbital (cyclonal,hexanal,hexobarbitone)	236	221
Hexobarbital-2- (or 4)-methyl deriv.	250	235
Hexobarbital-3-methyl deriv.	250	235
Hippuric acid	179	105
Histamine	111	82
Hydrochlorothiazide	297	269
Hydromorphine	285	285
Hydroxyamobarbital-1,3-dimethyl deriv. TMS ether	342	131
Hydroxyamobarbital glucuronide methyl-TMS deriv. (peak 2)	676	73
Hydroxyamobarbital glucuronide methyl-TMS deriv. (peak 1)	676	73
2-Hydroxybenzoyl glycine-methyl ester methyl ether	223	135
2-Hydroxybenzoyl glycine-TMS ester TMS ether	339	324
3-Hydroxy 2-chlorpromazine	334	58
8-Hydroxy-2-chlorpromazine	334	58
alpha-Hydroxy *naltrexone*	343	343
3-Hydroxyphenylhydantoin-methyl ether-3-methyl deriv.	296	296
p-Hydroxyphenylhydantoin-methyl ether-3-methyl	296	296
Hydroxy hexobarbital-3-methyl deriv. TMS ether	338	169
Hydroxypentobarbital-1,3-dimethyl deriv. TMS ether	342	117
Hydroxypentobarbital-glucuronide-1,3-dimethyl deriv. methyl ester TMS ether	676	73
Hydroxypentobarbital-1,3-dimethyl deriv. HFB deriv.	466	184
Hydroxypentobarbital	242	156

EI DATA—ALPHABETICAL INDEX (continued)

Substance	Mol wt	Base peak
α-Hydroxynaltrexone	343	343
α-Hydroxynaltrexone-3-TMS		73
Hydroxypethidine	263	71
Hydroxyphenobarbital-1,3-diethyl deriv. TMS ether	376	376
Hydroxyphenobarbital-1,3-dimethyl deriv. methyl ether	290	261
Hydrophenobarbital-1,3-diethyl deriv. ethyl ether	332	332
Hydroxyphenobarbital-1,3-dimethyl deriv. HFB deriv.	472	444
Hydroxyphenobarbital-1,3-dimethyl deriv. TMS ether	348	73
p-Hydroxyphenylbutazone	324	93
5-(4-Hydroxyphenyl)-3-methyl-5-phenylhydantoin glucuronide methyl ester TMS ether	688	73
5-(4-Hydroxyphenyl)-3-methyl-5-phenylhydantoin TMS ether	354	354
Hydroxysecobarbital HFB deriv.-1,3-dimethyl deriv.	478	195
Hydroxysecobarbital glucuronide-1,3-dimethyl deriv. methyl ester TMS ether	688	73
Hydroxyzine	374	201
Hyoscyamine	289	124
Ibogaine		136
2-Imino-5-phenyl-4-oxazoldinone (see pemoline)	176	176
Imipramine	280	235
Impurity from b-d vacutainer (tri-2-butoxyethyl phosphate)	398	57
Indole	117	117
Indomethacin	357	44
Indomethacin methyl ester	371	139
Indomethacin TMS ester	429	139
Ionol	220	205
Isocarboxazid	231	91
Isomethadone	309	58
Isoniazid	137	78
Iso-octyl phthalate	390	149
Ketobemidone	247	70
Levallorphan	283	283
Levulinic acid	116	43
Lidocaine	234	86
LSD (lysergide)	323	323
Mebutamate	232	97
Mebutamate (GC)	232	57
Meclizne	390	189
Meconin	194	165
Medazepam	270	242
Mepazine	310	310
Meponzolate bromide	325	97
Meperidine	247	71
Meperidine acid methyl ester	233	71
Meperidinic acid TMS ester	291	71
N-Demethyl meperidine	233	57
Mephenisin	182	108
Mephenoxalone (trepidone)	223	124

EI DATA—ALPHABETICAL INDEX (continued)

Substance	Mol wt	Base peak
Mephenytoin	218	189
Mephenytoin mtb. 1	204	104
Mephobarbital (mebaral, prominal)	246	218
Mephobarbital 3-ethyl deriv.	274	246
Mephobarbital 4-ethyl deriv.	274	217
Meprobamate	218	83
Mescaline	211	182
Mestranol	310	227
Metazocine	231	231
Methadone	309	72
Methadone mtb. 1	277	276
Methamphetamine	149	58
Methapyrone	333	56
Methantholine bromide	420	181
Methapyrilene	261	58
Methapyrilene mtb. 1	247	97
Methapyrilene mtb. 2	165	58
Methaqualone	250	235
Methaqualone mtb. 2	266	251
Methaqualone mtb. 3	266	251
Methaqualone mtb. 5	266	251
Methaqualone mtb. 5 (GC)	266	251
Methaqualone mtb. 1 TMS		73
Methaqualone mtb. 3 TMS		323
Methaqualone mtb. 4 TMS		73
Methaqualone mtb. 5 TMS		323
Metharbital (germonil)	198	170
Methdilazine	296	97
Methenamine	140	42
Methorphan (dormethan, racemeth)	271	59
Methotrimeprazine	328	58
p-Methoxy phenylacetic acid methyl ester		121
Methoxypromazine (methopromazine, teutone)	314	58
Methsuximide	203	
Methsuximide mtb. 1	189	118
Methyclothiazide (endurone)	360	310
2-Methyl-5-methoxyindole-3-acetic acid methyl ester	291	174
2-Methyl-5-methoxyindole-3-acetic acid TMS ester	291	174
8-Methylxanthine 1,3,7-triethyl deriv.	250	250
N-Methyl, N-Methoxy methyl phenobarbital	290	146
Methyl palmitate	270	74
Methyl phenidate	233	84
Methyl phenidate mtb. 1	150	91
2 Methyl-2-propyl-1,3-propanediol carbamate	218	83
Methyl salicylate	152	120
Methyl stearate	298	74
Methyprylon	183	155
Methyprylon mtb. 1	199	83
Methoxyflurone		81

EI DATA—ALPHABETICAL INDEX (continued)

Substance	Mol wt	Base peak
Metopon	299	299
Moramide	392	100
Morpheridine	346	246
Morphine	285	285
Morphine N-oxide	301	285
Nalorphine	311	271
Naltrexone	341	341
Naltrexone-1 TMS		55
Naltrexone-2 TMS		73
Naltrexone mtb. 1-3 TMS		73
Narceine	445	58
Narcotine (noscapine, gnoscopine)	413	220
Nialamide	298	91
Nicotine	162	84
2-Nitro-imidazole	113	113
Norethindrone	298	91
Norlevorphanol	243	243
Nor-meperidinic acid ethyl ester N-TMS deriv.	305	73
Nor-meperidinic acid methyl ester	219	57
Nor-meperidinic acid TMS ester N-TMS deriv.	349	73
Noroxymorphone-1 TMS		73
Noroxymorphone-2 TMS		73
Noroxymorphone-3 TMS		73
Nortriptyline	263	44
Noscapine	413	220
Oleic acid	282	41
Oxacillin mtb. 1	161	146
Oxacillin mtb. 2	202	144
Oxacillin mtb. 3	103	103
Oxanamide	157	36
Oxazepam	286	77
Oxycodeine	315	315
Oxycodone		315
Oxymorphine	301	69
Oxymorphone	233	57
Oxyphenbutazone (p-Hydroxyphenylbutazone) (see hydroxyphenylbutatone)	324	93
Palmitic acid	256	73
Papaverine	339	324
Parabromdylamine	318	249
Paraldehyde	132	45
Peganone	204	204
Pemoline	176	176
Pentachlorethane	202	167
Pentazocine	285	217
Pentobarbital	226	156
Pentobarbital 1,3 dimethyl deriv.	254	169
Pentobarbital mtb. 1	242	156
Perazil	300	183

EI DATA—ALPHABETICAL INDEX (continued)

Substance	Mol wt	Base peak
Pericyazine	365	114
Perphenazine	403	42
Phenaglycodol	214	59
Phenazocine	321	230
Phencyclidine (D5-PH)	248	205
Phencyclidine	243	200
Phenelzine	136	105
Phenetsal (salophen)	271	121
Phenindamine	261	260
Phenmetrazine	177	71
Phenobarbital	232	204
Phenobarbital 1,2 (or 4)-dimethyl deriv.	260	203
Phenobarbital 1,3-dimethyl deriv.	288	146
Phenobarbital 1,3-dimethyl deriv.	260	232
Phenobarbital mtb. 1	145	117
N-03-methyl phenobarbital (D5-ethyl)	254	222
N,N-03-methyl phenobarbital (D5-ethyl)	271	239
N-methyl,N-03-methyl phenobarbital	263	235
Phenol	94	94
Phenomorphan	347	256
Phenothiazine	199	199
Phensuximide	189	104
Phenylbutazone	308	183
1-Phenylcyclohexene	158	129
5-(4-Hydroxyphenyl)-3-methyl-5-phenylhydantoin glucuronide methyl ester	176	176
Phenyltoloxamine	255	58
Phenyramidol	214	107
Pholcodine	398	114
Piminodine (alvodine, pimadin, cimadon)	366	246
Pinene (from turpentine)	136	93
Piperamazine	401	141
Piperidone	99	
Primidone	218	117
Primidone mtb. 1	206	163
Primidone mtb. 2	163	91
Primidone	218	146
Primidone 1,3-di TMS deriv.	362	146
Probarbital (ethylisopropylbarbituric acid)	198	156
Probenecid	285	256
Procaine	236	86
Prochlorperazine	373	113
Proheptazine	275	57
Proketazine	425	
Promazine	284	58
Promethazine	284	72
Pronethal TMS ether	301	72
Propiomazine	340	72

EI DATA—ALPHABETICAL INDEX (continued)

Substance	Mol wt	Base peak
Propoxyphene HCl	339	58
Propoxyphene mtb. 1	325	44
Propoxyphene decomp (peak 1)	208	105
Propoxyphene decomp (peak 2)	208	105
Propoxyphene mtb. 3	325	44
Propylparaben methyl ether	194	135
Propylparaben TMS ether	252	73
Proquamazine	327	
Prostaglandin A1 methyloxime TMS ester TMS ether	509	73
Prostaglandin A1 TMS ester TMS ether	480	73
Prostaglandin B1 methyloxine TMS ester TMS ether	509	73
Prostaglandin E2 methyloxime TMS ester di-TMS ether	597	73
Prostaglandin F1A TMS ester tri-TMS ether	644	73
Prostaglandin F1B TMS ester tri-TMS ether	644	73
Prostaglandin F2A TMS ester tri-TMS ether	642	73
Prostaglandin F2B TMS ester tri-TMS ether	642	73
Prostaglandin 8-ISO-E1 methoxime TMS ester TMS ether	599	73
Pseudoephedrin	265	58
Psilocin (4-hydroxy-N-dimethyltryptamine)	204	58
Psilocin-1 TMS		58
Psilocin-2 TMS		58
Psilocybin	284	58
Pyrathiazine	296	84
Pyrilamine (antihistamine, mepyramine)	285	121
Pyrimethamine	248	247
Pyrobutamine	311	205
Quinacrine	400	86
Quinidine	324	136
Rotoxamine	290	58
Salicylic acid	138	92
Salicylic acid methyl ester TMS ether	224	209
Salicyluric acid methyl ester methyl ether	223	135
Scopolamine	303	94
Secobarbital	238	41/168
Secobarbital 1,3-diethyl deriv.	294	224
Secobarbital 1	254	168
Stearic acid	284	57/44
STP (see 2,5-dimethoxy-4-methylamphetamine)	209	44
Strychnine	334	334
Sucrose	342	73
Sulfamethoxazole	253	92
Sulphapyridine	249	184
Talbutal	224	41/167
Tetrahydrocannabinol-Δ-8	314	231
Tetrahydrocannibinol-Δ-9	314	
Tetrahydrocannibinol mtb. 1	330	299
1,3,7 (or 9), 8 Tetramethyl xanthine	208	208
Thebacon	341	341
Thebaine	311	311

EI DATA—ALPHABETICAL INDEX (continued)

Substance	Mol wt	Base peak
Thenyldiamine	216	58
Theobromine	180	180
Theobromine 1-ethyl deriv.	208	208
Theophylline	180	180
Theophylline 7-ethyl deriv.	208	208
Thiambutene	291	276
Thiamylal	254	43
Thiopental	242	172
Thiopropazate	446	246
Thioproperazine	446	70
Thioridazine	370	98
Thioridazine mtb. 1	255	245
Thioridazine mtb. 2	277	198
Thonzylamine	286	121
Torecane	399	399
Tranylcypromine	133	132
Tranylcypromine sulfate	133	132
Triamterene	253	253
Trifluoperazine	407	113
Trifluopromazine	352	58
Trimeprazine	298	58
Trimethobenzamide	388	58
Trimethobenzamide mtb.	343	195
Trimethoprim	290	290
Trimethoprim mtb. 1	306	289
Trimethoprim mtb. 2	306	290
Trimethoprim mtb. 3	276	276
Trimethoprim mtb. 4	276	276
Tripelennamine	255	58
Triprolidine	278	209
Tropine TMS ether	213	83
Tubocurarine chloride	681	298
Tybamate	274	55
Tybamate (GC)	274	41
Tyramine (tocosine, systogene)	137	30
Urea di-TMS deriv.	204	147
Warfarin mtb. 1	292	292
Zactane (see ethoheptazine)	261	107
Zoxazolamine	168	168

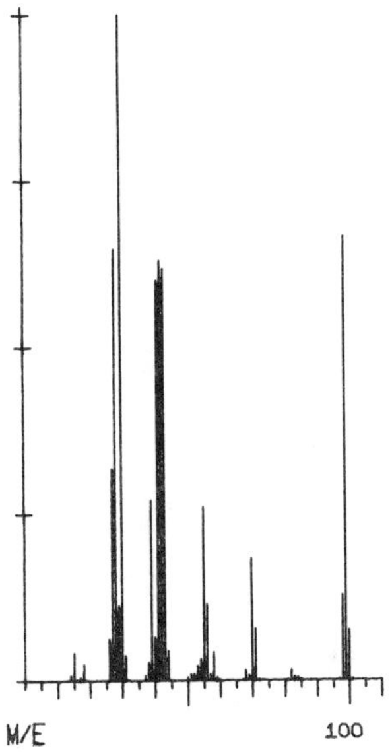
M/E
100

PIPERIDONE

C_5H_9NO MW = 99

Base Peak = 30 12 Peaks

Mass	Int.	Mass	Int.	Mass	Int.	Mass	Int.	Mass	Int.	Mass	Int.	Mass	Int.	Mass	Int.	Mass	Int.
28	647	30	999	42	629	43	617	55	259	56	114	70	182	71	77	82	15
83	4	98	128	99	665												

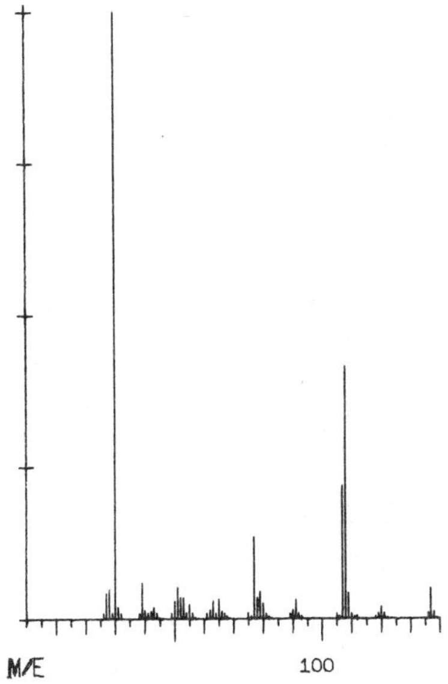
M/E
100

TYRAMINE

$C_8H_{11}NO$ MW = 137

Base Peak = 30 18 Peaks

Mass	Int.	Mass	Int.	Mass	Int.	Mass	Int.	Mass	Int.	Mass	Int.	Mass	Int.	Mass	Int.	Mass	Int.
28	50	30	999	39	60	43	20	51	52	53	36	63	31	65	32	77	136
79	44	90	14	91	32	107	220	108	416	119	10	120	20	137	50	138	12

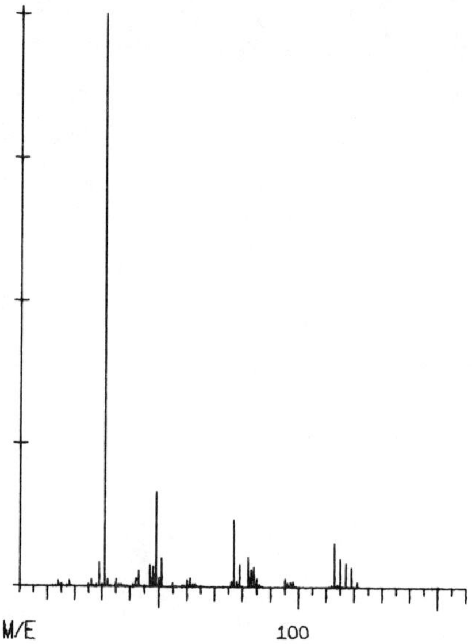
M/E
100

CHLORAL HYDRATE MTB. 1

$C_2H_3Cl_3O$ MW = 148

Base Peak = 31 19 Peaks

Mass	Int.	Mass	Int.	Mass	Int.	Mass	Int.	Mass	Int.	Mass	Int.	Mass	Int.	Mass	Int.	Mass	Int.
29	43	31	999	43	28	47	37	49	165	51	50	62	4	63	6	77	118
82	52	95	15	97	10	113	77	115	50	119	34	121	11	133	1	148	3
150	3																

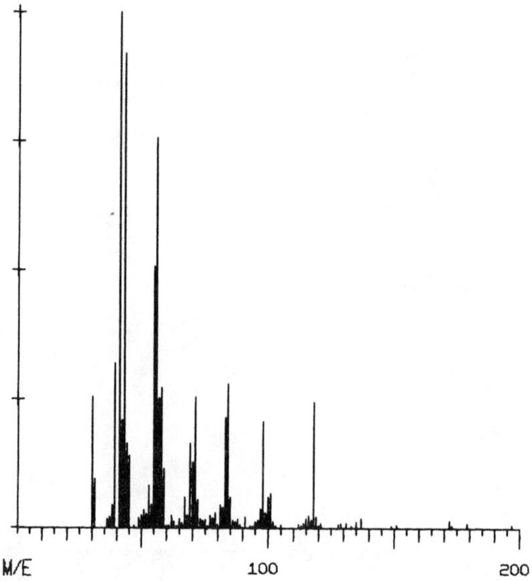
M/E
100
200

TYBAMATE (GC)

$C_{13}H_{26}N_2O_4$ MW = 274

Base Peak = 41 24 Peaks

Mass	Int.	Mass	Int.	Mass	Int.	Mass	Int.	Mass	Int.	Mass	Int.	Mass	Int.	Mass	Int.	Mass	Int.
30	254	31	95	41	999	43	921	55	507	56	758	69	165	71	254	83	215
84	280	98	207	101	67	115	17	116	24	118	244	119	22	135	12	137	21
149	6	151	8	172	14	173	6	179	9	197	7						

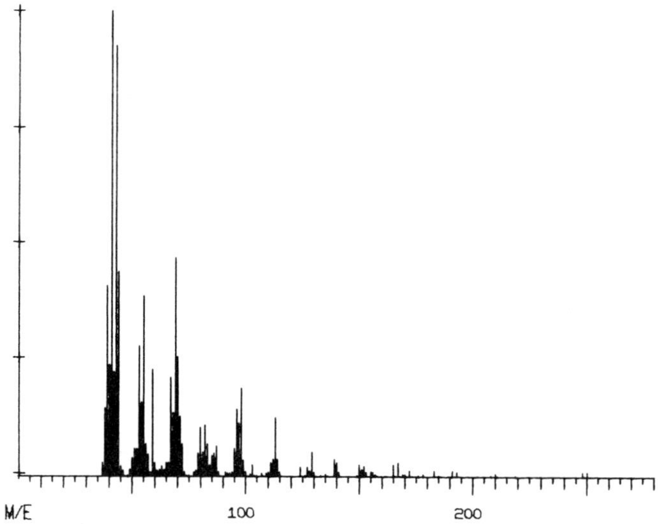
M/E
100
200

ACETYLCARBROMAL

$C_9H_{15}BrN_2O_3$ MW = 278

Base Peak = 41 30 Peaks

Mass	Int.	Mass	Int.	Mass	Int.	Mass	Int.	Mass	Int.	Mass	Int.	Mass	Int.	Mass	Int.	Mass	Int.
41	999	43	924	53	281	55	387	69	469	70	257	80	104	82	109	96	145
98	191	112	38	113	127	127	21	129	52	139	37	140	31	150	24	152	22
165	25	167	29	178	4	183	12	191	13	193	11	208	3	210	7	218	1
219	3	248	9	250	9												

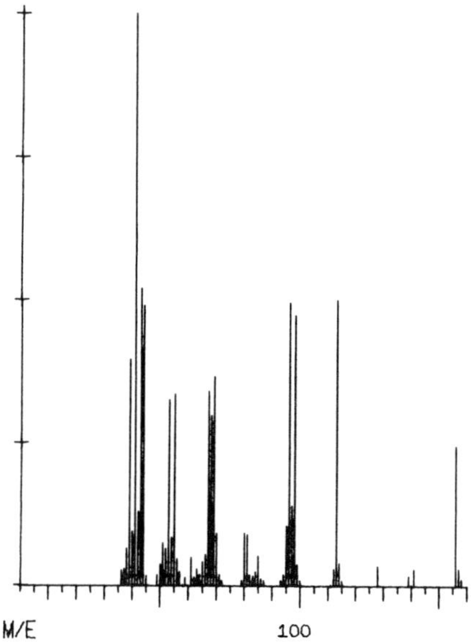
M/E
100

ECTYLUREA

$C_7H_{12}N_2O_2$ MW = 156

Base Peak = 41 17 Peaks

Mass	Int.	Mass	Int.	Mass	Int.	Mass	Int.	Mass	Int.	Mass	Int.	Mass	Int.	Mass	Int.	Mass	Int.
41	999	43	520	53	325	55	334	67	339	69	366	80	92	81	89	96	494
98	472	113	499	114	41	128	36	139	17	141	29	156	246	157	29		

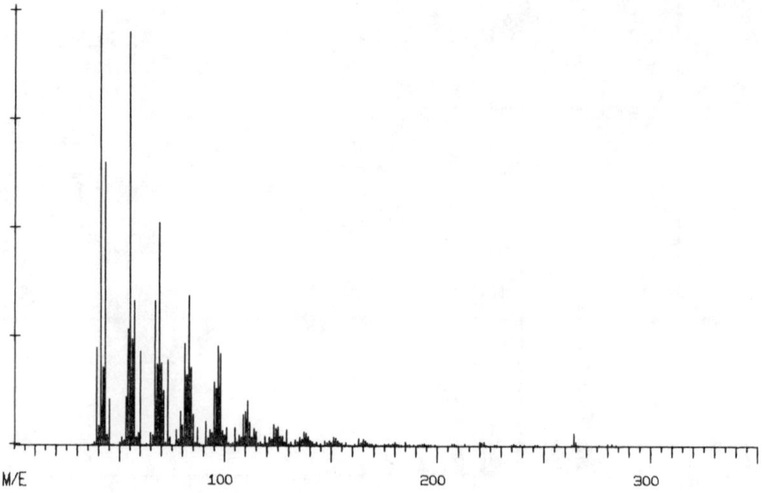
M/E
100
200
300

OLEIC ACID

$C_{18}H_{34}O_2$ MW = 282

Base Peak = 41 39 Peaks

Mass	Int.	Mass	Int.	Mass	Int.	Mass	Int.	Mass	Int.	Mass	Int.	Mass	Int.	Mass	Int.	Mass	Int.
41	999	43	649	55	949	57	333	67	333	69	513	81	236	83	344	97	231
98	213	110	79	111	104	123	50	225	46	137	32	138	29	151	21	152	18
163	17	165	14	180	10	185	10	193	6	194	6	207	5	208	4	220	10
222	11	235	3	236	5	246	2	256	4	264	30	265	10	280	5	282	4
293	1	333	1	347	1												

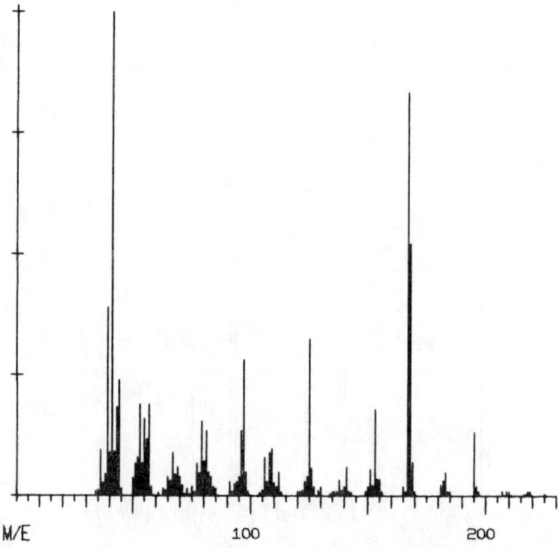
M/E
100
200

TALBUTAL

$C_{11}H_{16}N_2O_3$ MW = 224

Base Peak = 41 28 Peaks

Mass	Int.	Mass	Int.	Mass	Int.	Mass	Int.	Mass	Int.	Mass	Int.	Mass	Int.	Mass	Int.	Mass	Int.
39	389	41	999	53	191	57	189	67	89	69	59	79	154	81	134	96	136
97	283	108	89	109	98	125	326	126	58	138	33	141	59	151	56	153	180
167	834	168	523	182	33	183	51	195	133	196	21	207	11	209	9	218	9
219	10																

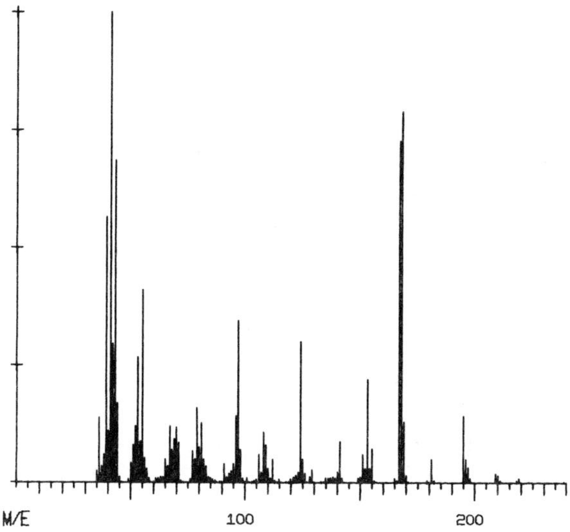
M/E
100
200

SECOBARBITAL

$C_{12}H_{18}N_2O_3$ MW = 238

Base Peak = 41 30 Peaks

Mass	Int.	Mass	Int.	Mass	Int.	Mass	Int.	Mass	Int.	Mass	Int.	Mass	Int.	Mass	Int.	Mass	Int.
41	999	43	684	53	268	55	410	67	120	70	118	79	160	81	128	96	143
97	345	108	108	109	80	124	299	125	49	140	23	141	87	153	220	155	73
167	727	168	791	179	6	181	51	195	142	196	51	209	20	210	15	218	4
219	9	238	1	239	2												

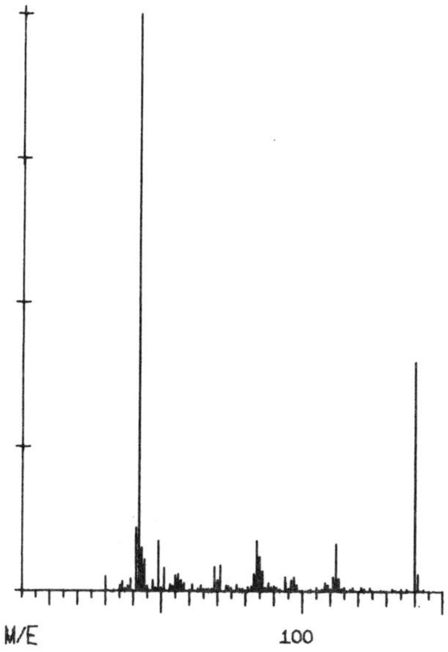
M/E
100

METHENAMINE

$C_6H_{12}N_4$ MW = 140

Base Peak = 42 18 Peaks

Mass	Int.	Mass	Int.	Mass	Int.	Mass	Int.	Mass	Int.	Mass	Int.	Mass	Int.	Mass	Int.	Mass	Int.
30	25	33	3	41	109	42	999	49	87	51	39	69	42	71	44	84	87
85	59	94	26	97	25	111	24	112	83	118	7	124	8	140	397	141	30

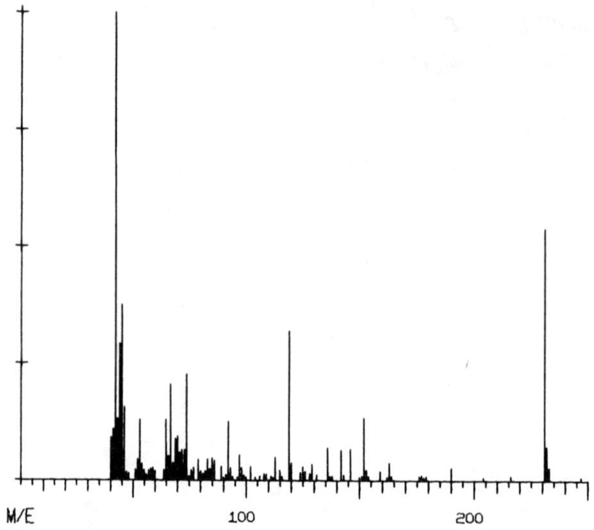
M/E
100
200

AZATHIOPRINE

$C_9H_7N_7O_2S$ MW = 277

Base Peak = 42 28 Peaks

Mass	Int.	Mass	Int.	Mass	Int.	Mass	Int.	Mass	Int.	Mass	Int.	Mass	Int.	Mass	Int.	Mass	Int.
42	999	45	375	52	44	53	129	67	206	74	228	79	44	85	47	92	128
97	54	113	50	115	23	119	321	120	38	136	69	142	64	146	67	152	134
163	38	164	9	176	10	177	13	190	27	204	8	216	11	231	540	232	72
247	8																

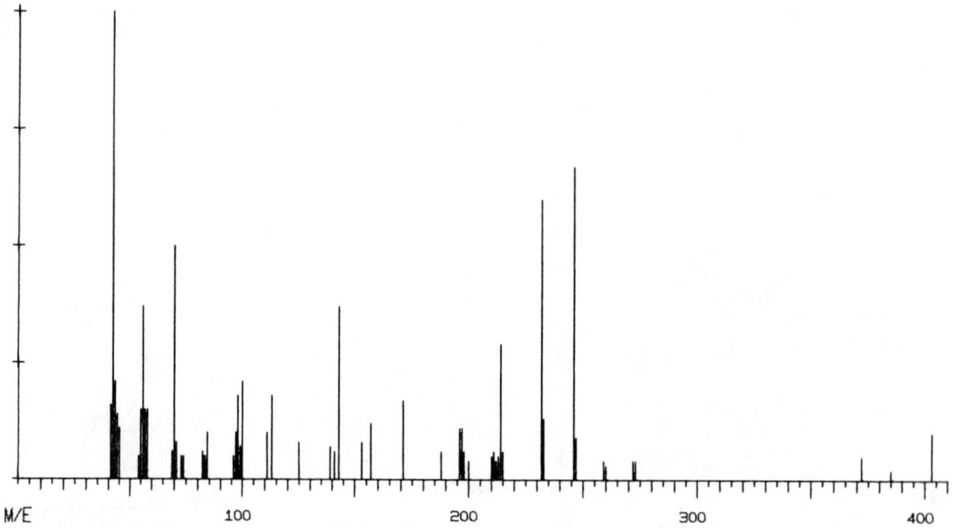
M/E
100
200
300
400

PERPHENAZINE

$C_{21}H_{26}ClN_3OS$ MW = 403

Base Peak = 42 33 Peaks

Mass	Int.	Mass	Int.	Mass	Int.	Mass	Int.	Mass	Int.	Mass	Int.	Mass	Int.	Mass	Int.	Mass	Int.
42	999	43	209	55	149	56	369	70	499	71	79	82	59	84	99	98	179
100	209	111	99	113	179	125	79	139	69	143	369	153	79	157	119	171	169
196	109	197	109	211	59	214	289	232	599	233	129	246	669	247	89	259	39
260	29	272	39	273	39	372	49	385	19	403	99						

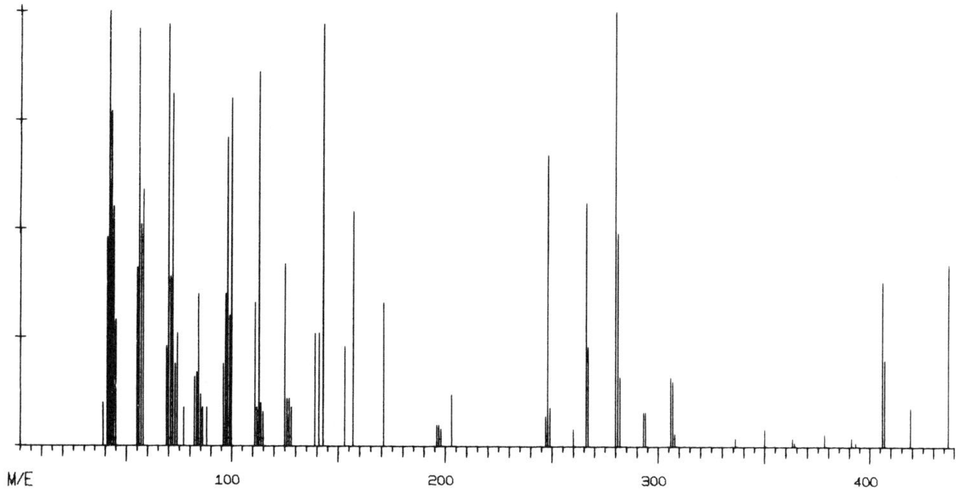
M/E
100
200
300
400

FLUPHENAZINE

$C_{22}H_{26}F_3N_3OS$ MW = 437

Base Peak = 42 43 Peaks

Mass	Int.	Mass	Int.	Mass	Int.	Mass	Int.	Mass	Int.	Mass	Int.	Mass	Int.	Mass	Int.	Mass	Int.
42	999	43	769	56	959	58	589	70	969	72	809	83	169	84	349	98	709
100	799	111	329	113	859	125	419	126	109	139	259	143	969	153	229	157	539
171	329	196	49	197	49	203	119	248	669	249	89	266	559	267	229	280	999
281	489	293	79	294	79	306	159	307	149	336	19	350	39	363	19	364	9
378	29	391	19	393	9	406	379	407	199	419	89	437	419				

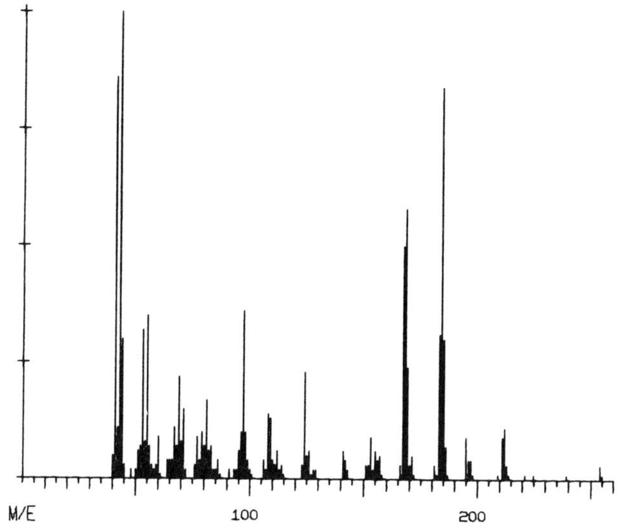
M/E
100
200

THIAMYLAL

$C_{12}H_{18}N_2O_2S$ MW = 254

Base Peak = 43 31 Peaks

Mass	Int.	Mass	Int.	Mass	Int.	Mass	Int.	Mass	Int.	Mass	Int.	Mass	Int.	Mass	Int.	Mass	Int.
41	860	43	999	53	320	55	350	69	220	71	150	79	100	81	170	96	100
97	360	108	140	109	130	124	230	126	60	141	60	142	40	153	90	155	60
167	500	168	580	183	310	184	840	195	90	196	40	211	90	212	110	221	10
225	10	239	10	254	30	255	10										

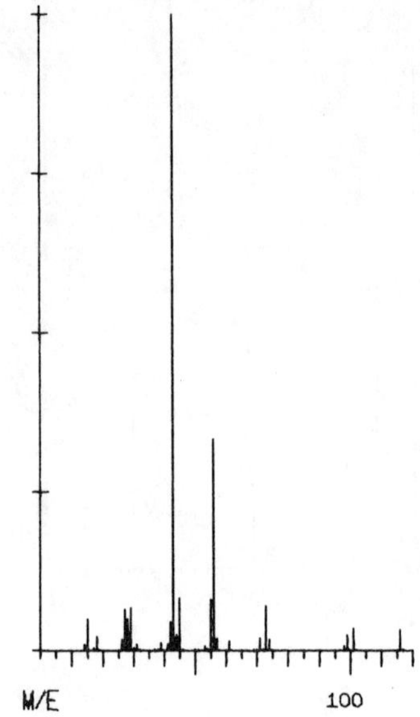
M/E
100

LEVULINIC ACID

$C_5H_8O_3$ MW = 116

Base Peak = 43 12 Peaks

Mass	Int.	Mass	Int.	Mass	Int.	Mass	Int.	Mass	Int.	Mass	Int.	Mass	Int.	Mass	Int.	Mass	Int.
27	66	29	68	43	999	45	83	55	79	56	332	71	20	73	69	99	24
101	35	116	32	117	2												

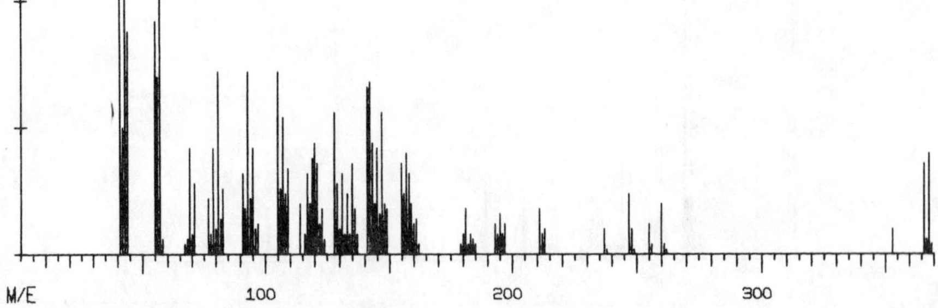
M/E
100
200
300

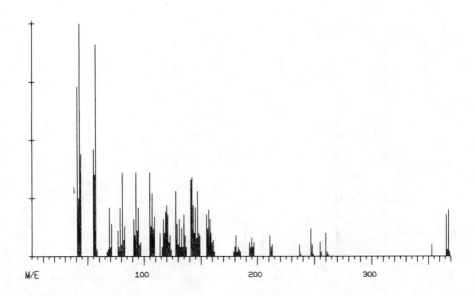

CHOLESTA-3,5-DIENE

$C_{27}H_{44}$ MW = 368

Base Peak = 43 35 Peaks

Mass	Int.	Mass	Int.	Mass	Int.	Mass	Int.	Mass	Int.	Mass	Int.	Mass	Int.	Mass	Int.	Mass	Int.
41	730	43	999	55	460	57	910	69	210	71	140	79	210	81	360	93	360
95	210	105	360	107	270	120	220	128	280	141	330	142	340	147	280	157	200
160	60	161	70	180	40	181	90	193	60	195	80	211	90	213	50	237	50
238	10	247	120	255	60	260	100	261	20	353	50	366	180	368	200		

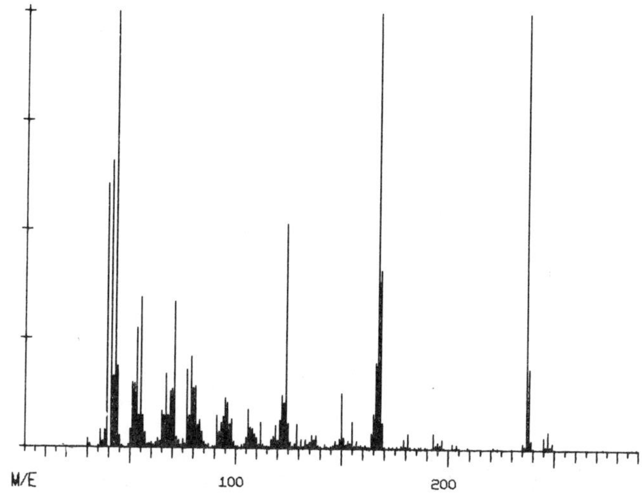
M/E
100
200

5-(-2-BROMOALLYL)-5-(1-METHYL BUTYL) BARBITURIC ACID

$C_{12}H_{17}BrN_2O_3$ MW = 316

Base Peak = 43 37 Peaks

Mass	Int.	Mass	Int.	Mass	Int.	Mass	Int.	Mass	Int.	Mass	Int.	Mass	Int.	Mass	Int.	Mass	Int.
30	21	31	11	41	657	43	999	53	274	55	346	67	171	71	335	77	181
79	210	95	116	96	105	106	91	112	61	122	122	124	516	136	30	138	29
150	127	155	63	167	999	168	410	179	22	181	36	193	34	197	23	202	12
204	10	221	4	223	3	137	999	238	186	245	27	247	39	273	2	287	3
289	3																

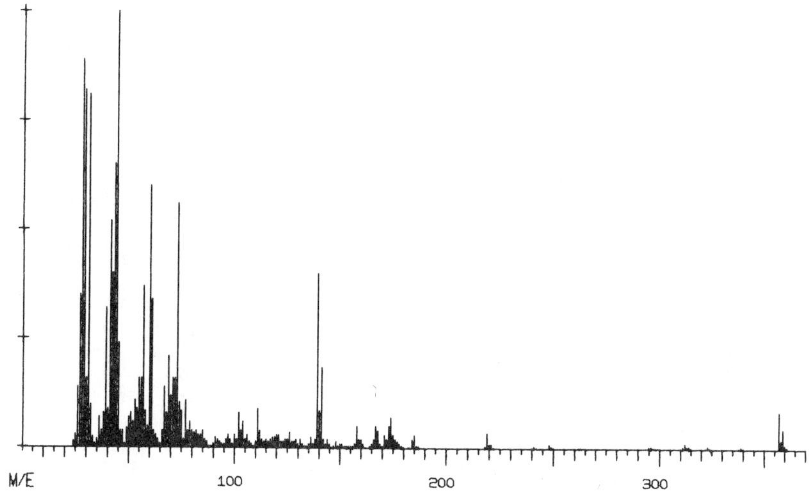
M/E
100
200
300

INDOMETHACIN

$C_{19} \cdot H_{16} \cdot Cl \cdot N \cdot O_4$ MW = 357

Base Peak = 44 42 Peaks

Mass	Int.	Mass	Int.	Mass	Int.	Mass	Int.	Mass	Int.	Mass	Int.	Mass	Int.	Mass	Int.	Mass	Int.
28	890	29	820	43	650	44	999	57	370	60	600	69	210	73	560	77	110
79	60	102	80	103	40	104	60	111	90	121	31	126	35	139	401	141	185
158	50	159	20	167	50	173	50	174	70	175	30	219	35	220	10	240	2
241	4	248	10	249	4	262	3	263	2	295	4	296	6	312	12	313	6
314	4	323	7	338	3	339	5	357	85	359	45						

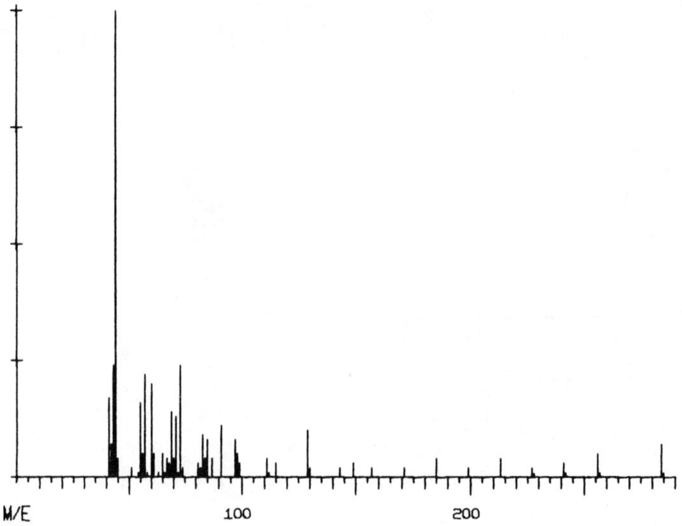
M/E
100
200

STEARIC ACID

$C_{18}H_{36}O_2$ MW = 284

Base Peak = 44 29 Peaks

Mass	Int.	Mass	Int.	Mass	Int.	Mass	Int.	Mass	Int.	Mass	Int.	Mass	Int.	Mass	Int.	Mass	Int.
43	240	44	999	57	220	60	200	69	140	73	240	83	90	85	80	91	110
97	80	111	40	115	30	129	100	130	20	143	20	149	30	157	20	171	20
185	40	199	20	213	40	227	20	228	10	241	30	242	10	256	50	257	10
284	70	285	10														

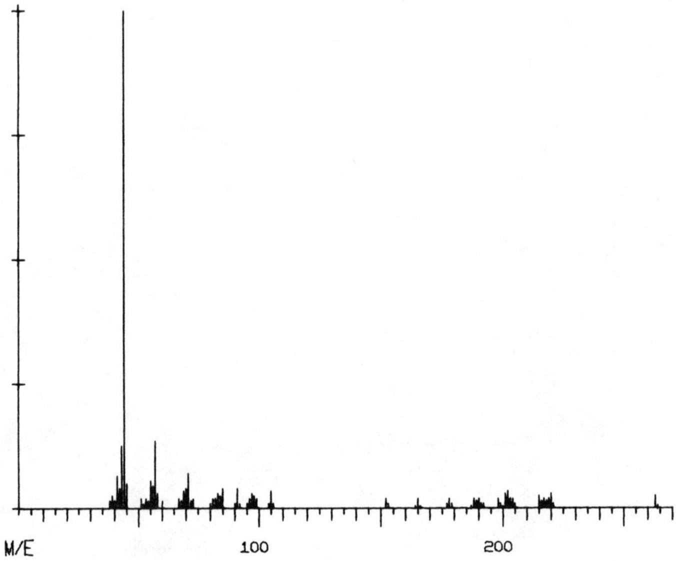
M/E
100
200

NORTRIPTYLINE

$C_{19}H_{21}N$ MW = 263

Base Peak = 44 26 Peaks

Mass	Int.	Mass	Int.	Mass	Int.	Mass	Int.	Mass	Int.	Mass	Int.	Mass	Int.	Mass	Int.	Mass	Int.
43	124	44	999	55	54	57	134	70	39	71	69	83	29	85	39	91	39
97	29	104	9	105	34	152	19	153	9	164	4	165	19	177	9	178	21
188	19	201	29	202	34	215	24	219	21	220	29	263	24	264	4		

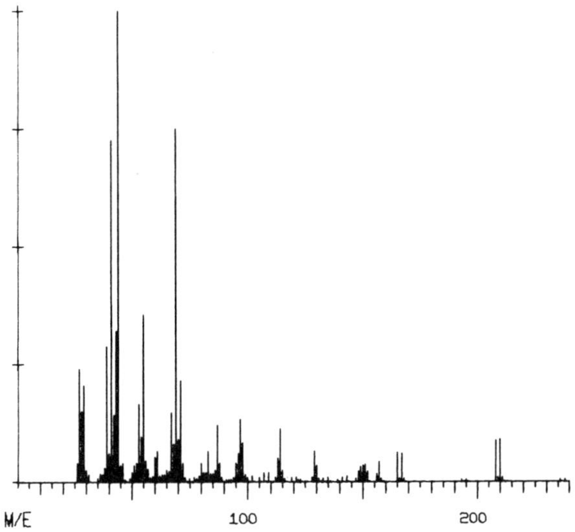
M/E
100
200

CARBROMAL

$C_7H_{13}BrN_2O_2$ MW = 236

Base Peak = 44 28 Peaks

Mass	Int.	Mass	Int.	Mass	Int.	Mass	Int.	Mass	Int.	Mass	Int.	Mass	Int.	Mass	Int.	Mass	Int.
27	241	29	204	41	725	44	999	53	166	55	356	69	749	71	214	83	64
87	120	97	132	98	82	113	51	114	112	129	64	130	36	141	11	143	12
151	37	157	43	165	62	167	60	193	4	195	4	208	88	210	89	236	4
238	4																

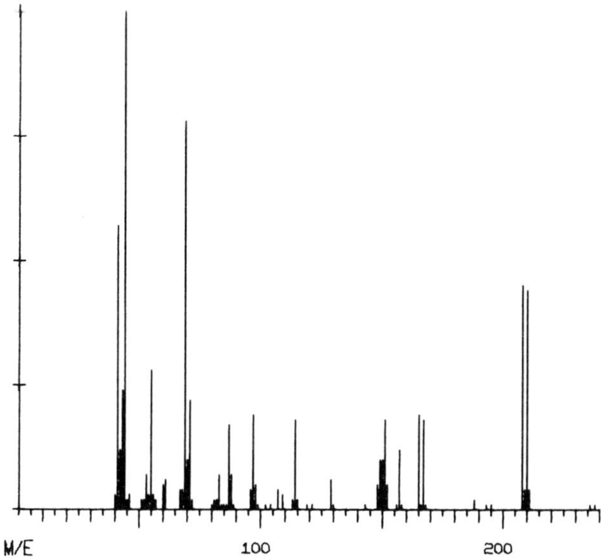
M/E
100
200

CARBROMAL

$C_7H_{13}BrN_2O_2$ MW = 236

Base Peak = 44 25 Peaks

Mass	Int.	Mass	Int.	Mass	Int.	Mass	Int.	Mass	Int.	Mass	Int.	Mass	Int.	Mass	Int.	Mass	Int.
41	570	44	999	53	70	55	280	69	780	71	220	83	70	87	170	97	190
98	50	107	40	114	180	119	10	129	60	143	10	151	180	157	120	165	190
167	180	188	20	193	10	208	450	210	440	236	10	238	10				

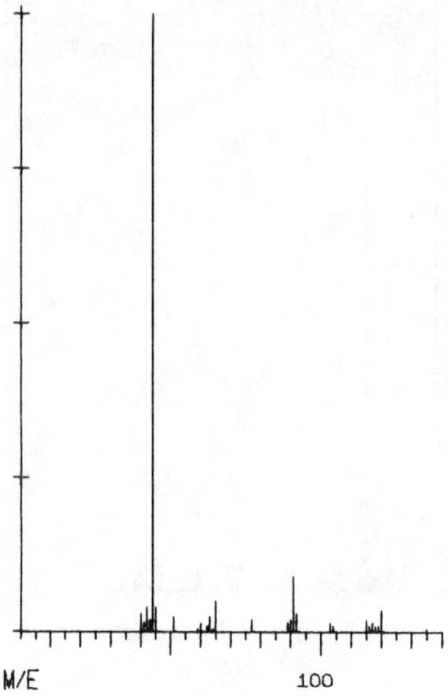
M/E
100

AMPHETAMINE

$C_9H_{13}N$ MW = 135

Base Peak = 44 15 Peaks

Mass	Int.	Mass	Int.	Mass	Int.	Mass	Int.	Mass	Int.	Mass	Int.	Mass	Int.	Mass	Int.	Mass	Int.
42	39	44	999	51	24	60	14	63	24	65	49	77	19	89	14	91	89
92	29	115	19	117	14	118	9	120	34	135	4						

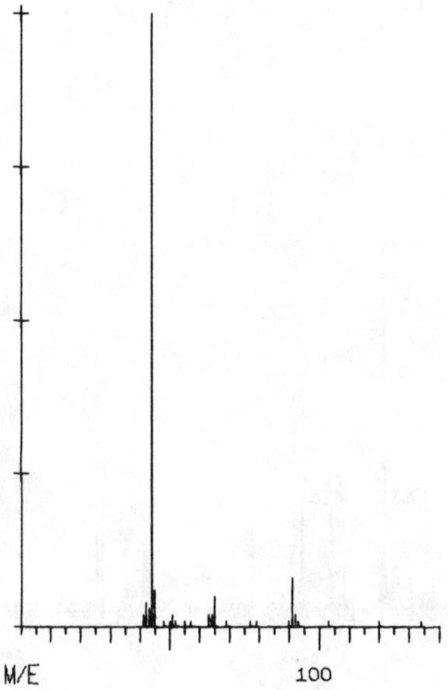
M/E
100

AMPHETAMINE

$C_9H_{13}N$ MW = 135

Base Peak = 44 12 Peaks

Mass	Int.	Mass	Int.	Mass	Int.	Mass	Int.	Mass	Int.	Mass	Int.	Mass	Int.	Mass	Int.	Mass	Int.
44	999	45	60	48	10	51	20	63	20	65	50	77	10	79	10	91	80
92	20	120	10	134	10												

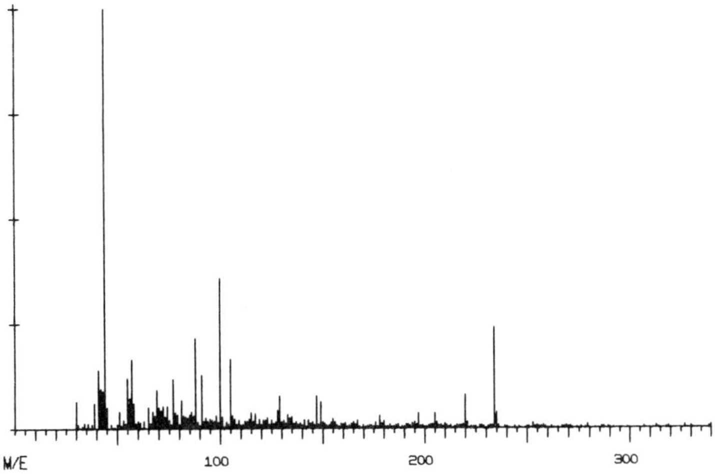
M/E
100
200
300

PROPOXYPHENE MTB. 3

$C_{21}H_{27}NO_2$ MW = 325

Base Peak = 44 46 Peaks

Mass	Int.	Mass	Int.	Mass	Int.	Mass	Int.	Mass	Int.	Mass	Int.	Mass	Int.	Mass	Int.	Mass	Int.
30	64	31	10	41	141	44	999	55	120	57	164	69	92	72	53	77	118
88	216	91	128	100	357	105	165	115	37	128	43	129	78	133	32	135	27
147	78	149	62	165	15	167	19	178	30	180	17	195	14	197	35	205	34
206	14	220	81	221	14	234	241	235	38	253	13	255	8	258	7	269	7
276	4	279	9	286	5	287	4	307	5	313	7	319	4	327	8	330	4
340	7																

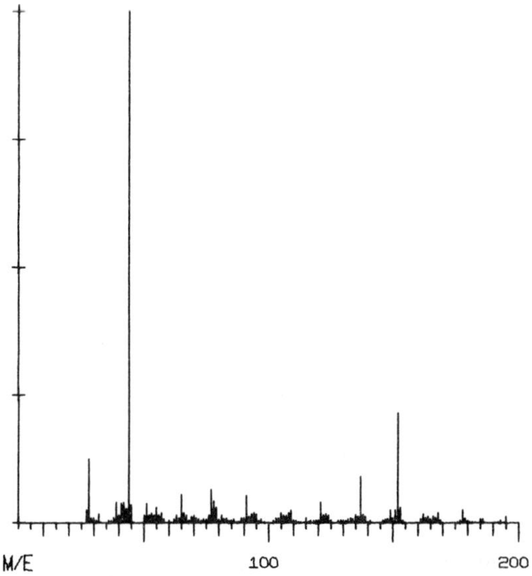
M/E
100
200

2,5-DIMETHOXYAMPHETAMINE

$C_{11}H_{18}NO_2$ MW = 196

Base Peak = 44 26 Peaks

Mass	Int.	Mass	Int.	Mass	Int.	Mass	Int.	Mass	Int.	Mass	Int.	Mass	Int.	Mass	Int.	Mass	Int.
27	25	28	124	39	40	44	999	51	37	55	30	65	54	66	20	77	64
78	42	91	52	94	19	108	21	109	24	121	41	123	17	135	14	137	89
152	214	153	29	162	18	168	21	178	24	179	10	193	4	195	12		

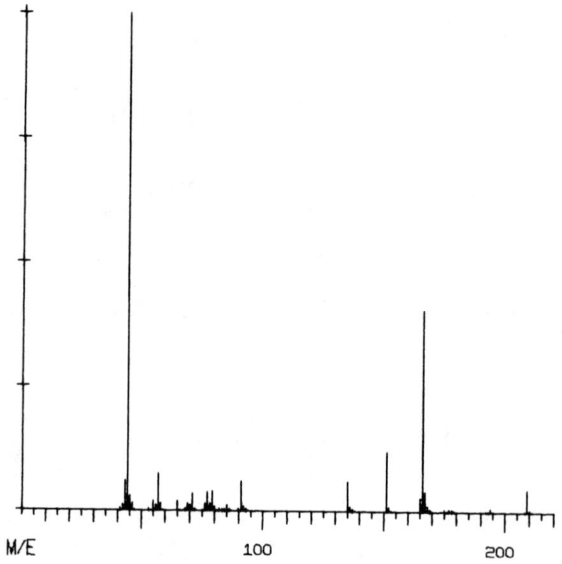
M/E
100
200

2,5-DIMETHOXY-4-METHYLAMPHETAMINE

$C_{12}H_{19}NO_2$ MW = 209

Base Peak = 44 22 Peaks

Mass	Int.	Mass	Int.	Mass	Int.	Mass	Int.	Mass	Int.	Mass	Int.	Mass	Int.	Mass	Int.	Mass	Int.
43	61	44	999	55	20	57	76	65	19	71	36	77	38	79	39	91	59
92	11	135	59	136	11	151	121	152	11	166	404	167	39	175	5	177	5
192	3	194	7	209	44	210	4										

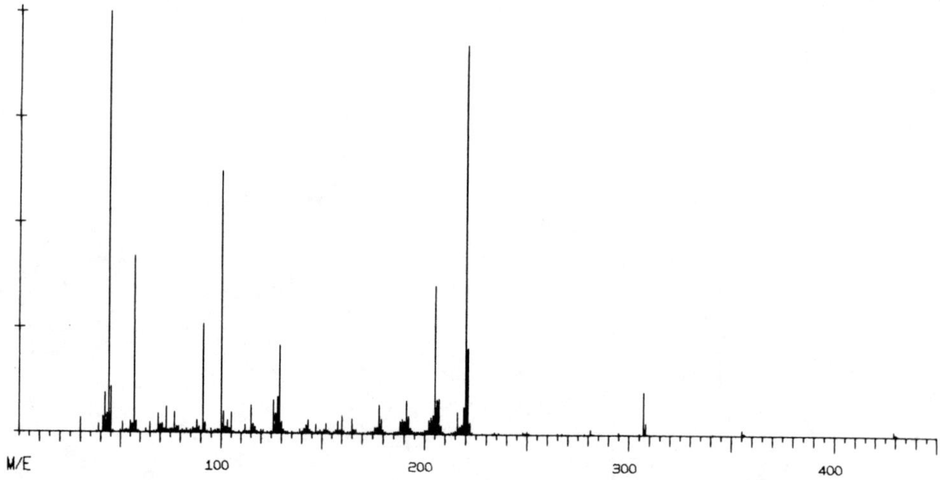
M/E
100
200
300
400

PROPOXYPHENE MTB. 1

$C_{21}H_{27}N_1O_2$ MW = 325

Base Peak = 44 51 Peaks

Mass	Int.	Mass	Int.	Mass	Int.	Mass	Int.	Mass	Int.	Mass	Int.	Mass	Int.	Mass	Int.	Mass	Int.
30	34	31	1	44	999	45	110	55	28	57	421	69	45	73	63	77	51
88	29	91	261	100	622	105	50	115	67	128	87	129	209	142	21	143	33
152	24	158	29	160	42	165	35	178	69	179	36	191	80	192	42	205	352
207	84	220	925	221	204	233	2	234	5	248	6	250	7	265	2	267	2
281	13	282	3	293	1	295	4	307	102	308	28	319	1	325	1	341	2
342	1	355	13	356	6	370	1	429	13	430	6						

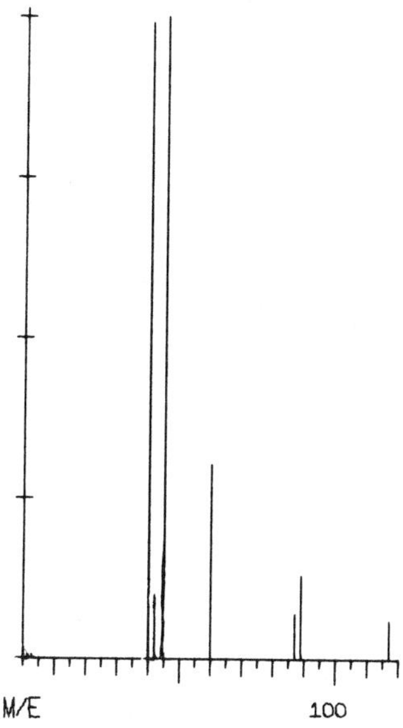
M/E
100

PARALDEHYDE

$C_6H_{12}O_3$ MW = 132

Base Peak = 45 6 Peaks

Mass	Int.	Mass	Int.	Mass	Int.	Mass	Int.	Mass	Int.	Mass	Int.	Mass	Int.	Mass	Int.	Mass	Int.
40	989	45	999	60	303	87	70	89	131	117	60						

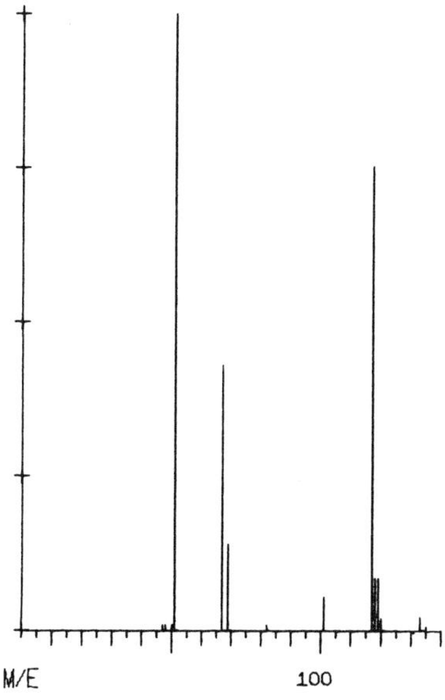
M/E
100

2-CHLORO-1,1,2-TRIFLUORO-ETHYL-DIFLUOROMETHYL ETHER

$C_3 \cdot H_2 \cdot Cl \cdot F_5 \cdot O$ MW = 184

Base Peak = 51 12 Peaks

Mass	Int.	Mass	Int.	Mass	Int.	Mass	Int.	Mass	Int.	Mass	Int.	Mass	Int.	Mass	Int.	Mass	Int.
47	11	48	11	51	999	67	430	69	140	82	11	101	54	117	753	118	86
119	86	133	22	135	7												

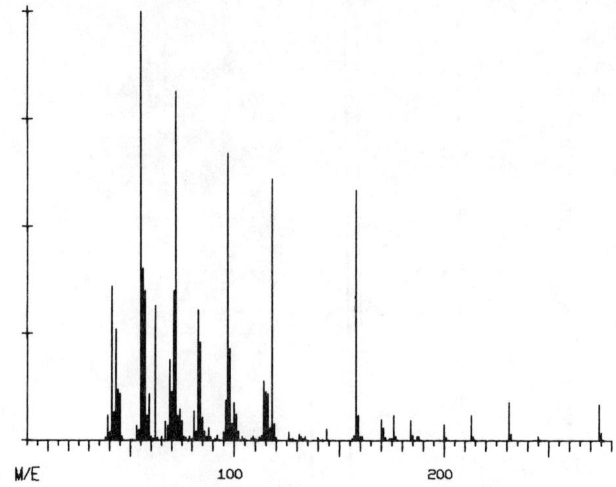
M/E
100
200

TYBAMATE

$C_{13}H_{26}N_2O_4$ MW = 274

Base Peak = 55   32 Peaks

Mass	Int.	Mass	Int.	Mass	Int.	Mass	Int.	Mass	Int.	Mass	Int.	Mass	Int.	Mass	Int.	Mass	Int.
41	360	43	261	55	999	56	403	71	349	72	814	83	306	84	229	97	669
98	214	114	139	115	116	118	611	119	39	132	10	144	27	158	584	159	61
170	49	171	30	176	60	184	47	188	10	200	38	213	59	214	11	231	89
232	14	245	11	246	2	274	86	275	18								

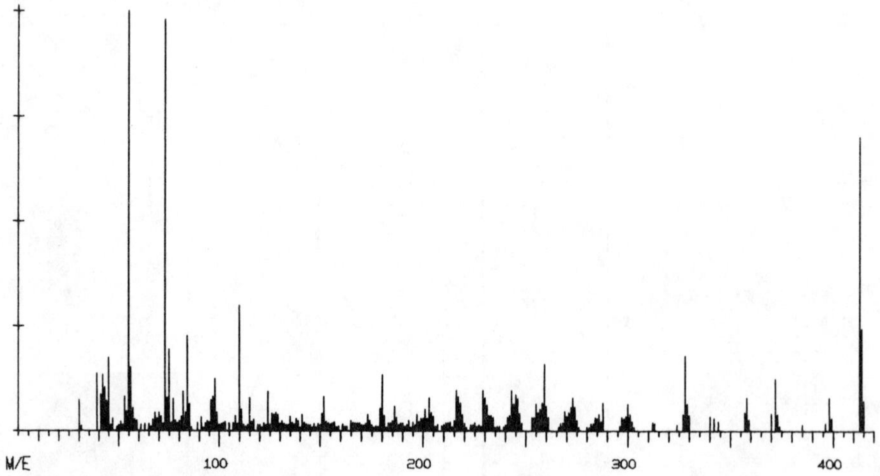
M/E
100
200
300
400

NALTREXONE-1 TMS

$C_{23} \cdot H_{31} \cdot N \cdot O_4 \cdot SI$ MW = 413

Base Peak = 55   58 Peaks

Mass	Int.	Mass	Int.	Mass	Int.	Mass	Int.	Mass	Int.	Mass	Int.	Mass	Int.	Mass	Int.	Mass	Int.
30	74	31	12	39	137	45	174	55	999	56	153	73	981	75	196	82	96
84	228	97	83	98	125	110	299	115	81	124	94	128	45	135	36	141	39
151	43	152	83	165	26	173	41	180	134	186	61	197	41	201	52	203	79
204	44	216	98	229	97	230	81	243	97	245	87	246	78	258	68	259	161
272	57	273	79	286	41	288	67	300	66	301	40	324	15	327	40	328	181
329	65	342	29	344	22	357	45	358	81	370	43	372	125	385	14	396	17
398	80	399	31	413	702	414	245										

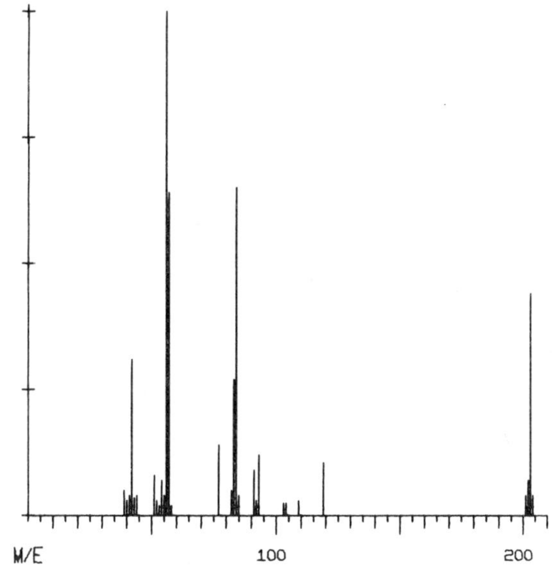
M/E
100
200

AMPYRONE

$C_{11}H_{13}N_3O_1$ MW = 203

Base Peak = 56 12 Peaks

Mass	Int.	Mass	Int.	Mass	Int.	Mass	Int.	Mass	Int.	Mass	Int.	Mass	Int.	Mass	Int.	Mass	Int.
39	49	42	309	56	999	57	639	83	269	84	649	91	89	93	119	104	24
109	29	119	104	201	39	202	69	203	439								

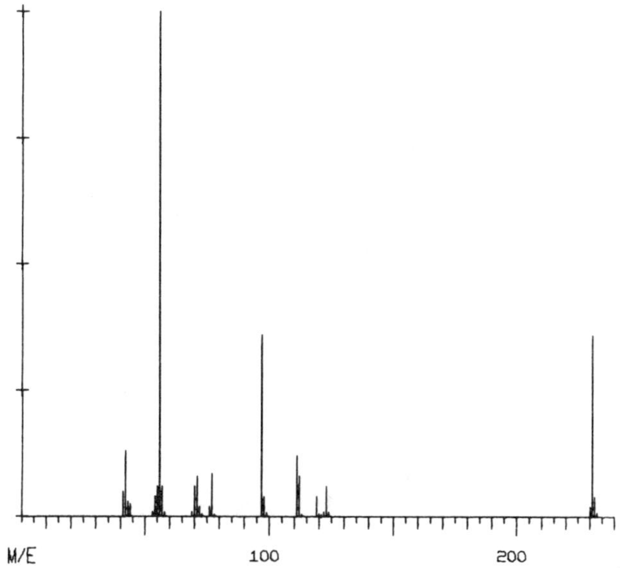
M/E
100
200

AMINOPYRINE

$C_{13} \cdot H_{17} \cdot N_3 \cdot O$ MW = 231

Base Peak = 56 16 Peaks

Mass	Int.	Mass	Int.	Mass	Int.	Mass	Int.	Mass	Int.	Mass	Int.	Mass	Int.	Mass	Int.	Mass	Int.
41	49	42	129	55	59	56	999	70	59	71	79	76	19	77	84	97	359
98	39	111	119	112	79	119	39	123	59	231	359	232	39				

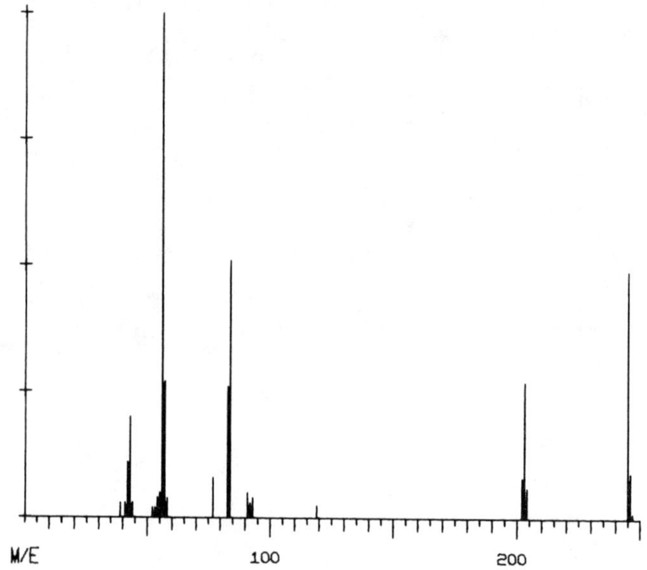
M/E
100
200

4-ACETYL-AMINOANTIPYRINE

$C_{13}H_{15}N_3O_2$ MW = 245

Base Peak = 56 13 Peaks

Mass	Int.	Mass	Int.	Mass	Int.	Mass	Int.	Mass	Int.	Mass	Int.	Mass	Int.	Mass	Int.	Mass	Int.
42	109	43	199	56	999	57	269	83	259	84	509	91	49	93	39	119	24
202	79	203	269	245	489	246	89										

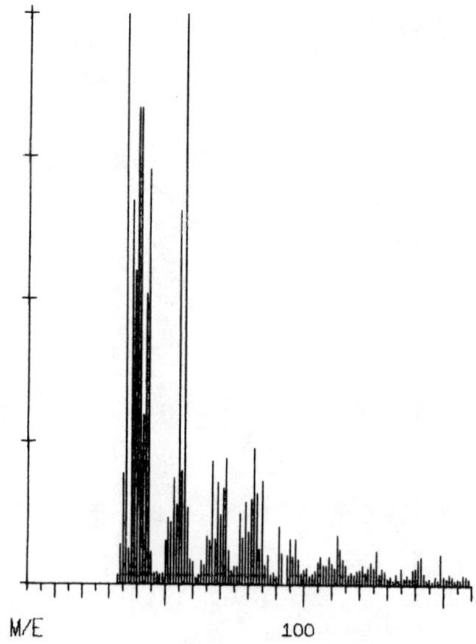
M/E
100

OXANAMIDE

$C_8H_{15}N_1O_2$ MW = 157

Base Peak = 57 20 Peaks

Mass	Int.	Mass	Int.	Mass	Int.	Mass	Int.	Mass	Int.	Mass	Int.	Mass	Int.	Mass	Int.	Mass	Int.
33	17	36	999	41	835	55	655	57	999	67	217	72	223	82	241	85	182
91	103	97	81	112	87	113	62	124	41	126	61	141	44	142	50	149	54
152	19	160	2														

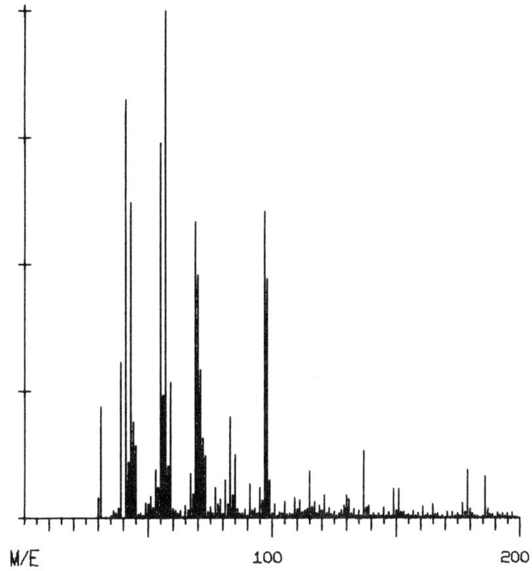
M/E
100
200

MEBUTAMATE (GC)

$C_{10}H_2ON_2O_4$ MW = 232

Base Peak = 57 26 Peaks

Mass	Int.	Mass	Int.	Mass	Int.	Mass	Int.	Mass	Int.	Mass	Int.	Mass	Int.	Mass	Int.	Mass	Int.
30	39	31	219	41	825	43	622	55	740	57	999	69	584	70	480	83	200
85	125	97	606	98	473	109	39	115	92	121	44	130	44	137	133	139	24
140	57	151	58	161	22	165	28	179	96	186	82	191	10	193	10		

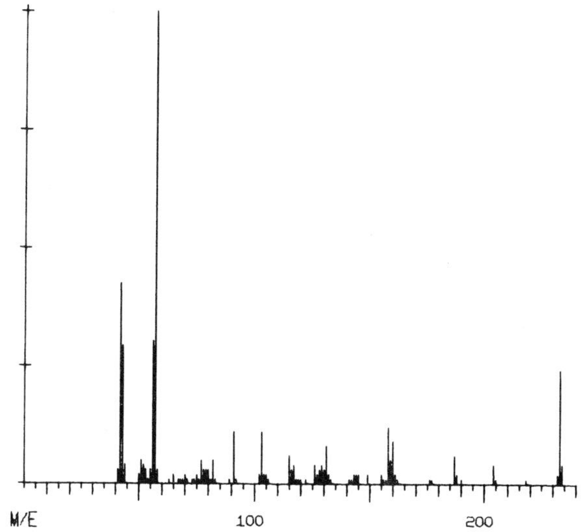
M/E
100
200

MEPERIDINE MTB. 1

$C_{14} \cdot H_{19} \cdot N \cdot O_2$ MW = 233

Base Peak = 57 29 Peaks

Mass	Int.	Mass	Int.	Mass	Int.	Mass	Int.	Mass	Int.	Mass	Int.	Mass	Int.	Mass	Int.	Mass	Int.
42	424	43	292	56	303	57	999	65	20	70	20	77	50	82	50	91	111
103	111	115	60	117	40	126	40	131	80	132	20	143	20	158	121	159	50
160	90	161	20	176	10	187	60	188	20	190	10	204	40	205	10	218	10
233	242	234	40														

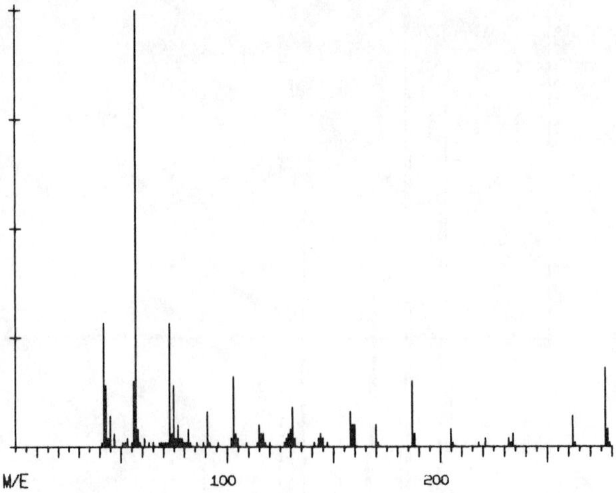
M/E
100
200

NOR-MEPERIDINIC ACID TMS ESTER

$C_{15} \cdot H_{23} \cdot N \cdot O_2 \cdot SI$ MW = 277

Base Peak = 57 32 Peaks

Mass	Int.	Mass	Int.	Mass	Int.	Mass	Int.	Mass	Int.	Mass	Int.	Mass	Int.	Mass	Int.	Mass	Int.
42	282	43	141	56	151	57	999	73	282	75	141	77	50	82	40	91	80
103	161	104	30	115	50	130	40	131	90	132	20	144	30	158	80	159	50
160	50	170	50	187	151	188	30	205	40	206	10	218	10	221	20	232	20
234	30	262	70	263	10	277	181	278	40								

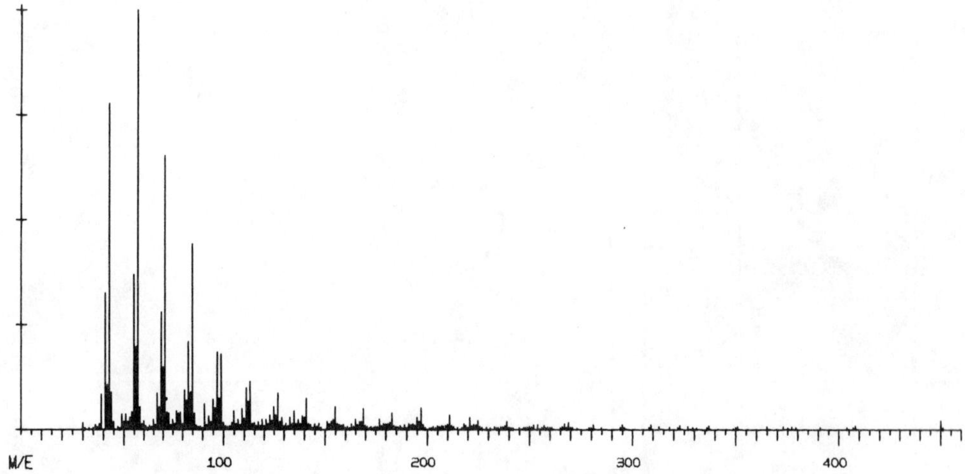
M/E
100
200
300
400

C32HGG STANDARD

$C_{32}H_{66}$ MW = 450

Base Peak = 57 56 Peaks

Mass	Int.	Mass	Int.	Mass	Int.	Mass	Int.	Mass	Int.	Mass	Int.	Mass	Int.	Mass	Int.	Mass	Int.
30	18	31	6	41	325	43	777	55	369	57	999	69	281	71	653	83	209
85	442	97	184	99	180	111	101	113	116	125	56	127	88	135	46	141	75
154	24	155	56	165	26	169	49	177	24	183	41	195	29	197	53	210	13
211	35	221	30	225	23	238	9	239	23	252	13	254	12	267	15	269	17
280	6	281	13	294	6	295	12	309	12	313	8	323	9	327	7	331	6
337	9	350	5	351	8	365	7	373	8	377	8	391	5	404	5	408	10
450	23	451	6														

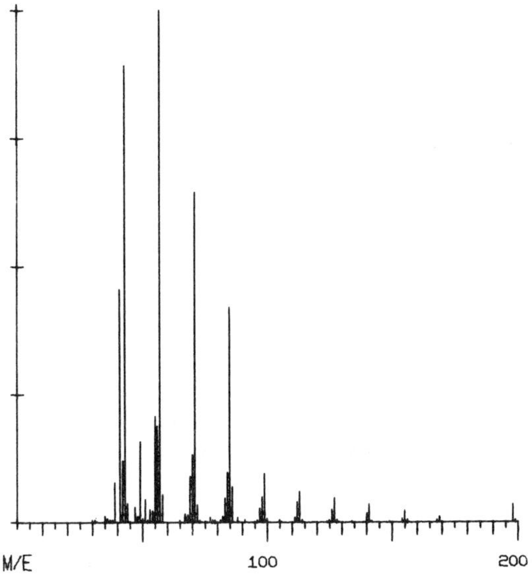
M/E
100
200

C14H30 STANDARD

$C_{14}H_{30}$ MW = 198

Base Peak = 57 24 Peaks

Mass	Int.	Mass	Int.	Mass	Int.	Mass	Int.	Mass	Int.	Mass	Int.	Mass	Int.	Mass	Int.
30	6	31	4	41	454	43	892	55	207	57	999	70	132	71	645	84	97
85	420	98	50	99	96	112	39	113	59	126	24	127	48	140	17	141	36
154	8	155	23	168	5	169	13	198	36	199	5						

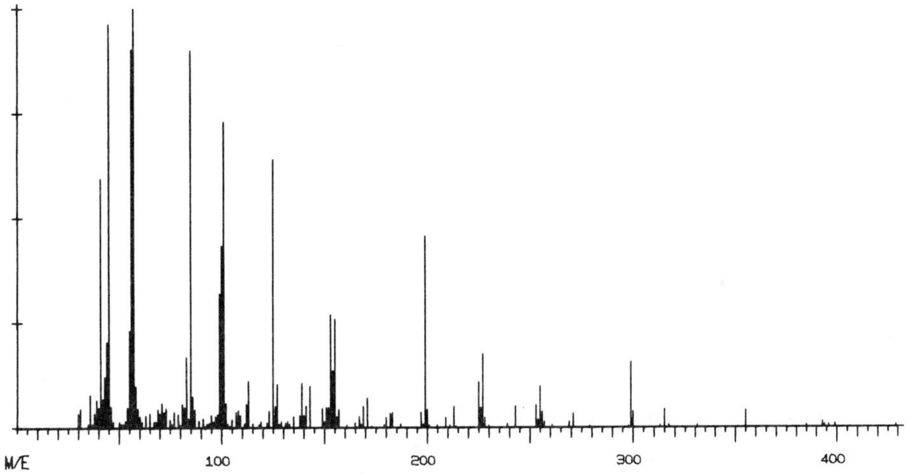
M/E
100
200
300
400

B-D VACUTAINER IMPURITY

Tri-2-Butoxyethyl Phosphate

$C_{18}H_{39}O_7P$ MW = 398

Base Peak = 57 49 Peaks

Mass	Int.	Mass	Int.	Mass	Int.	Mass	Int.	Mass	Int.	Mass	Int.	Mass	Int.	Mass	Int.	Mass	Int.
30	34	31	47	41	595	45	962	56	902	57	999	71	61	73	47	83	170
85	900	100	434	101	729	112	57	113	113	125	639	127	106	130	108	143	100
153	270	155	259	169	53	171	72	182	34	183	37	199	457	200	45	209	25
213	53	225	109	227	177	239	11	243	52	253	54	255	100	269	16	271	35
279	6	298	6	299	157	300	40	313	8	315	44	317	8	331	7	355	43
385	7	393	14	399	9	429	7										

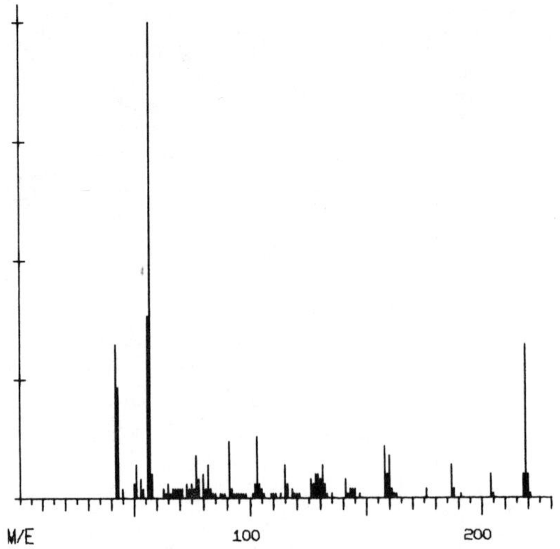
M/E
100
200

NOR-MEPERIDINIC ACID METHYL ESTER

$C_{13}H_{17}NO_2$ MW = 219

Base Peak = 57 28 Peaks

Mass	Int.	Mass	Int.	Mass	Int.	Mass	Int.	Mass	Int.	Mass	Int.	Mass	Int.	Mass	Int.	Mass	Int.
42	323	43	232	56	383	57	999	65	30	73	30	77	90	82	70	91	121
103	131	104	30	115	70	128	50	131	70	132	30	141	40	158	111	159	50
160	90	161	20	176	20	187	70	188	20	191	10	204	50	205	10	218	50
219	323																

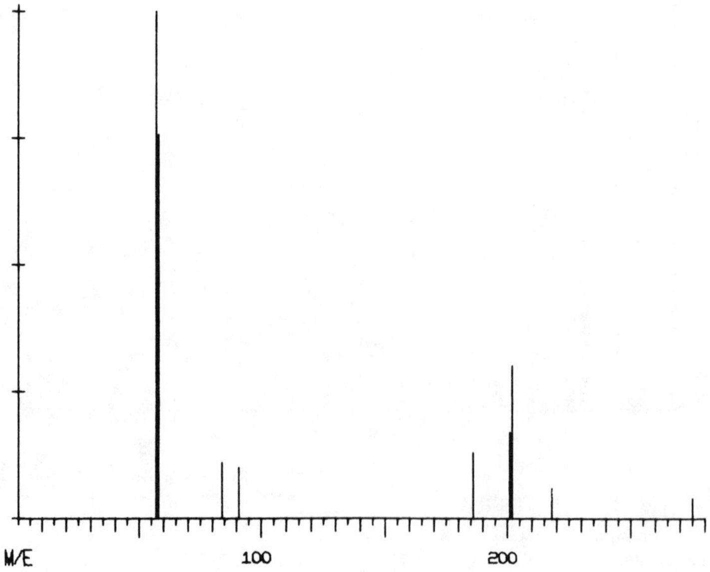
M/E
100
200

PROHEPTAZINE

$C_{17}H_{25}NO_2$ MW = 275

Base Peak = 57 9 Peaks

Mass	Int.	Mass	Int.	Mass	Int.	Mass	Int.	Mass	Int.	Mass	Int.	Mass	Int.	Mass	Int.	Mass	Int.
57	999	58	757	84	111	91	101	186	131	201	171	202	303	218	60	275	40

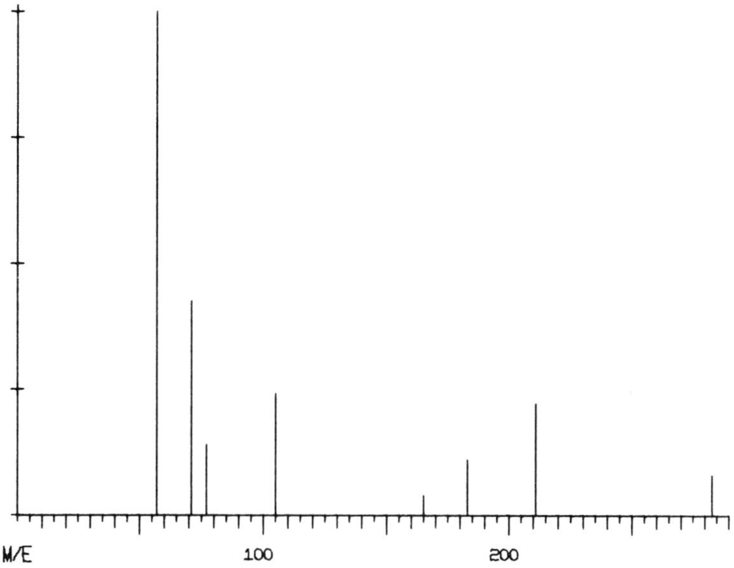

M/E
100
200

DIMENOXADOL

$C_{20}H_{25}NO_3$ MW = 327

Base Peak = 57 8 Peaks

Mass	Int.	Mass	Int.	Mass	Int.	Mass	Int.	Mass	Int.	Mass	Int.	Mass	Int.	Mass	Int.	Mass	Int.
57	999	71	424	77	141	105	242	165	40	183	111	211	222	283	80		

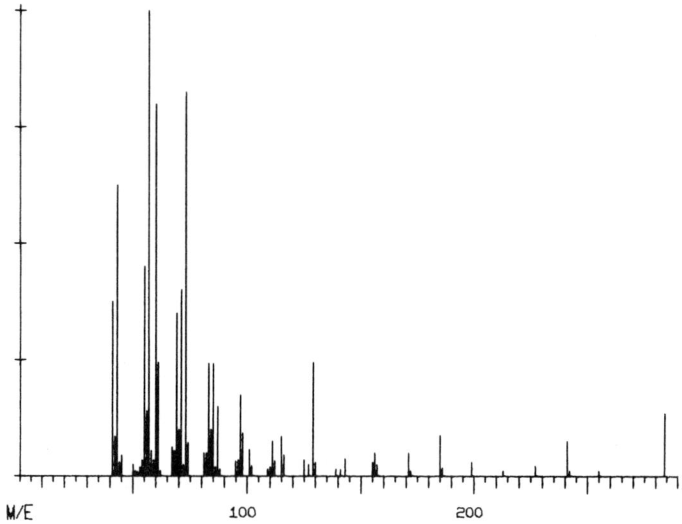

M/E
100
200

STEARIC ACID

$C_{18}H_{36}O_2$ MW = 284

Base Peak = 57 29 Peaks

Mass	Int.	Mass	Int.	Mass	Int.	Mass	Int.	Mass	Int.	Mass	Int.	Mass	Int.	Mass	Int.	Mass	Int.
41	375	43	625	57	999	60	800	71	400	73	825	83	242	85	242	97	175
98	93	111	75	115	85	125	34	129	245	139	14	143	37	155	29	156	50
171	50	172	12	185	87	186	17	199	30	213	12	227	23	241	75	242	12
255	12	284	135														

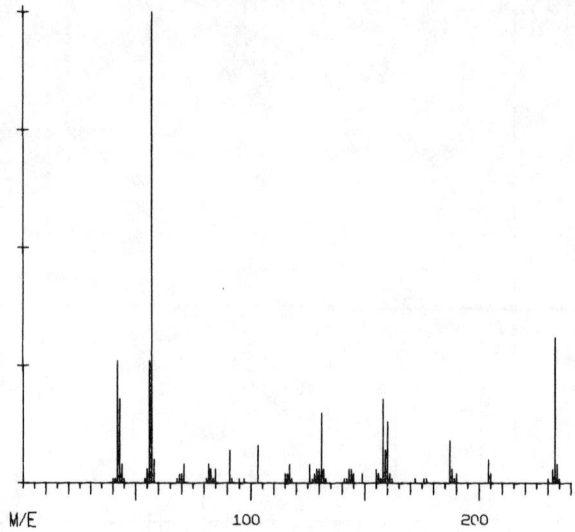
M/E
100
200

MEPERIDINE MTB. 1

$C_{14} \cdot H_{19} \cdot N \cdot O_2$ MW = 233

Base Peak = 57 28 Peaks

Mass	Int.	Mass	Int.	Mass	Int.	Mass	Int.	Mass	Int.	Mass	Int.	Mass	Int.	Mass	Int.	Mass	Int.
42	260	43	180	56	260	57	999	69	20	71	40	82	40	83	30	91	70
103	80	115	20	117	40	126	40	131	150	132	30	143	30	158	180	159	70
160	130	161	20	176	10	187	90	188	30	190	20	204	50	205	20	233	310
234	40																

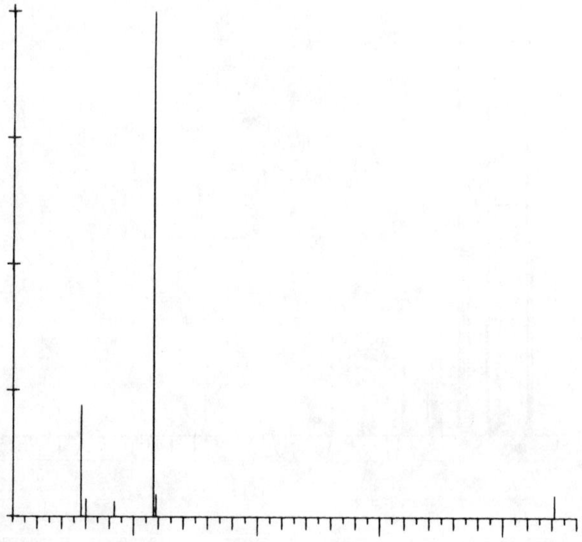
M/E
100
200

CHLORPROTHIXENE

$C_{18} \cdot H_{18} \cdot Cl \cdot N \cdot S$ MW = 315

Base Peak = 58 6 Peaks

Mass	Int.	Mass	Int.	Mass	Int.	Mass	Int.	Mass	Int.	Mass	Int.
28	220	30	35	42	31	58	999	59	46	221	44

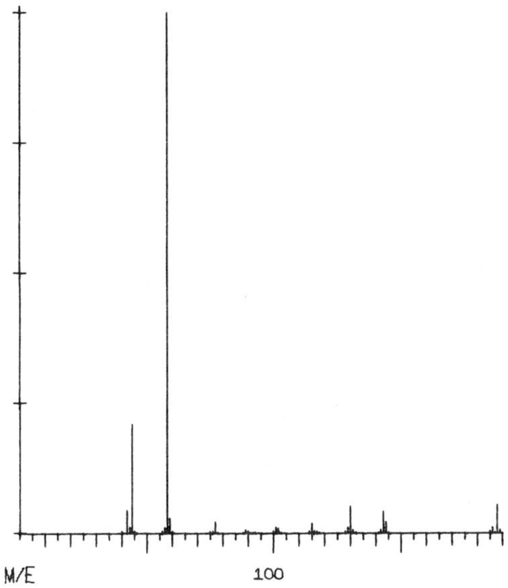
M/E
100

N,N-DIMETHYLTRYPTAMINE

$C_{12}H_{16}N_2$ MW = 188

Base Peak = 58 19 Peaks

Mass	Int.	Mass	Int.	Mass	Int.	Mass	Int.	Mass	Int.	Mass	Int.	Mass	Int.	Mass	Int.	Mass	Int.
42	44	44	211	58	999	59	29	75	4	77	22	89	7	101	12	102	10
115	21	117	6	129	12	130	52	143	42	144	22	185	4	186	13	188	56
189	7																

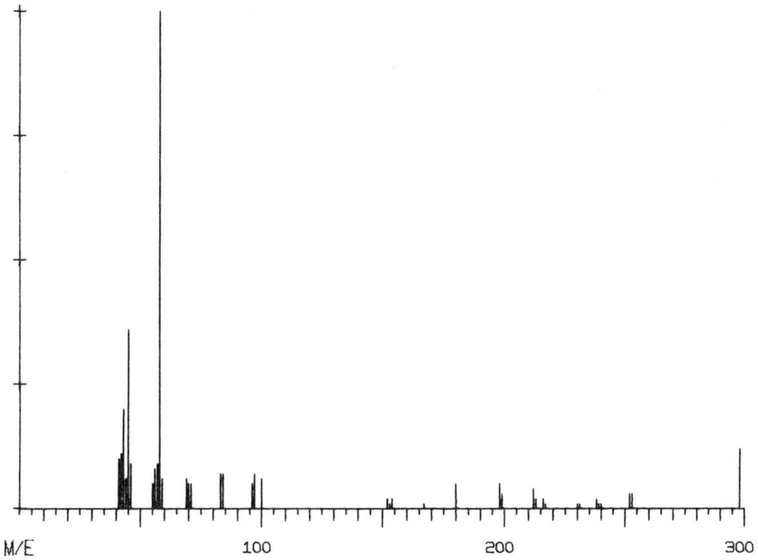
M/E
100
200
300

TRIMEPRAZINE

$C_{18}H_{22}N_2S$ MW = 298

Base Peak = 58 25 Peaks

Mass	Int.	Mass	Int.	Mass	Int.	Mass	Int.	Mass	Int.	Mass	Int.	Mass	Int.	Mass	Int.	Mass	Int.
43	199	45	359	57	89	58	999	69	59	70	49	83	69	84	69	97	69
100	59	152	19	154	19	167	9	180	49	198	49	199	29	212	39	213	19
216	19	217	9	230	9	238	19	252	29	253	29	298	119				

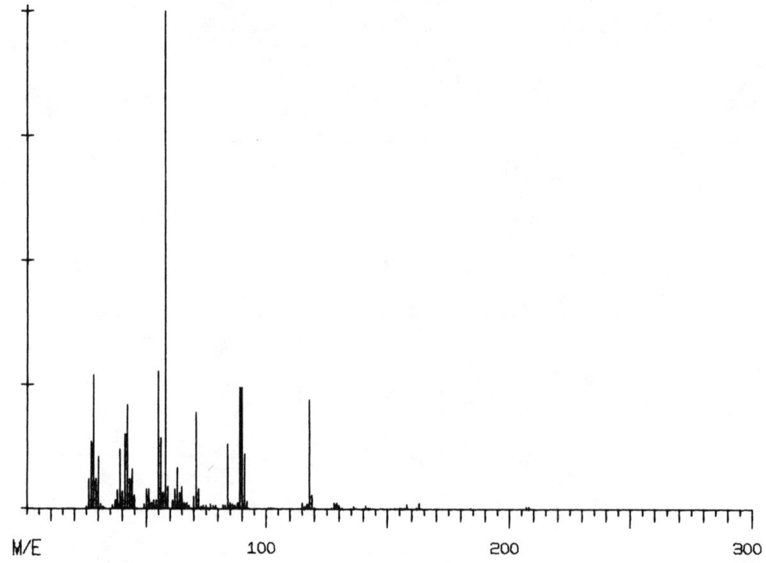
M/E
100
200
300

CYCLOPENTOLATE

$C_{17}H_{25}N_1O_3$ MW = 291

Base Peak = 58 25 Peaks

Mass	Int.	Mass	Int.	Mass	Int.	Mass	Int.	Mass	Int.	Mass	Int.	Mass	Int.	Mass	Int.	Mass	Int.
27	135	28	270	41	149	42	209	55	277	58	999	63	83	71	196	84	130
89	244	90	244	91	111	115	13	117	9	118	219	119	27	136	4	141	8
155	3	158	11	162	3	163	13	184	3	207	6	208	4				

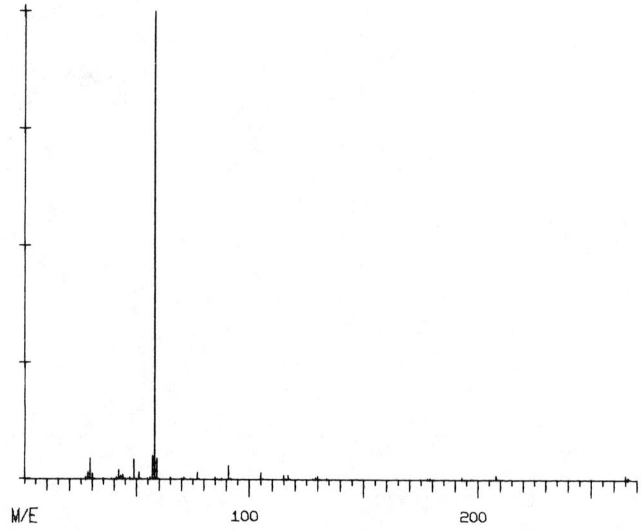
M/E
100
200

PROPOXYPHENE

$C_{22} \cdot H_{29} \cdot N \cdot O_2$ MW = 339

Base Peak = 58 28 Peaks

Mass	Int.	Mass	Int.	Mass	Int.	Mass	Int.	Mass	Int.	Mass	Int.	Mass	Int.	Mass	Int.	Mass	Int.
28	15	29	44	42	19	44	9	57	49	58	999	65	5	71	5	77	14
85	4	91	31	92	2	105	16	115	11	129	4	130	7	132	1	134	2
165	1	178	3	179	3	193	5	197	2	208	10	209	1	250	3	265	9
266	4																

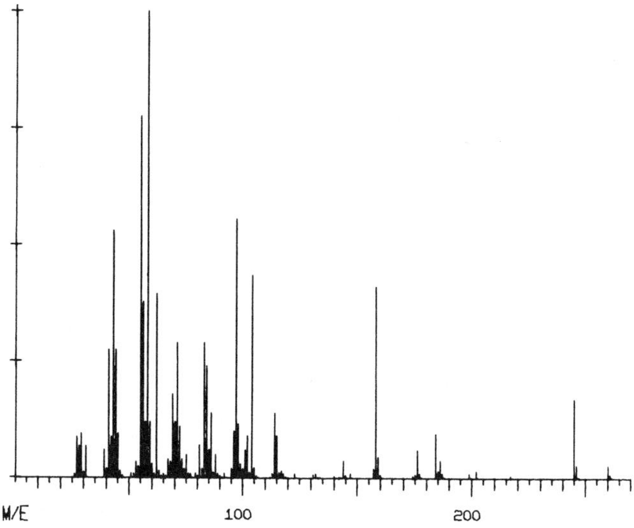
M/E
100
200

CARISOPRODOL

$C_{12} \cdot H_{24} \cdot N_2 \cdot O_4$ MW = 260

Base Peak = 58 31 Peaks

Mass	Int.	Mass	Int.	Mass	Int.	Mass	Int.	Mass	Int.	Mass	Int.	Mass	Int.	Mass	Int.	Mass	Int.
27	88	29	94	41	274	43	530	55	774	58	999	62	394	71	289	83	289
84	239	97	554	98	114	104	435	114	139	118	9	123	9	132	9	144	37
158	409	159	44	160	7	176	59	184	94	199	11	200	1	202	16	217	5
245	171	246	27	260	28	261	8										

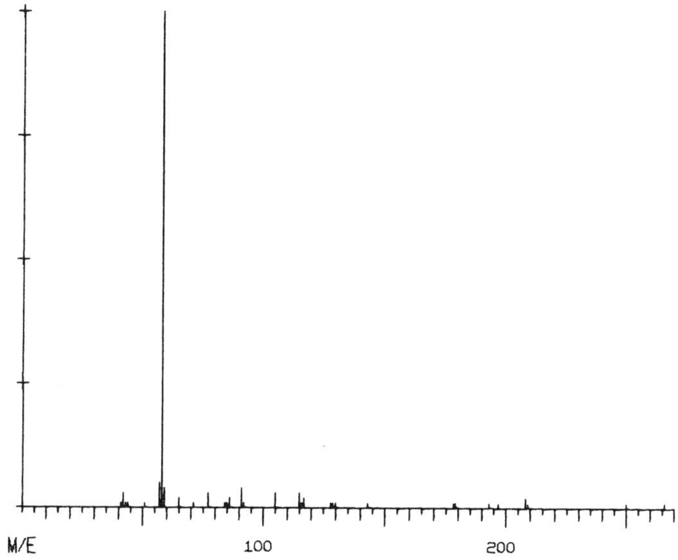
M/E
100
200

PROPOXYPHENE

$C_{22} \cdot H_{29} \cdot N \cdot O_2$ MW = 339

Base Peak = 58 23 Peaks

Mass	Int.	Mass	Int.	Mass	Int.	Mass	Int.	Mass	Int.	Mass	Int.	Mass	Int.	Mass	Int.	Mass	Int.
41	10	42	30	57	50	58	999	65	20	71	10	77	30	86	20	91	40
92	10	105	30	115	30	128	10	129	10	143	10	178	10	179	10	193	10
197	10	208	20	209	10	250	10	266	10								

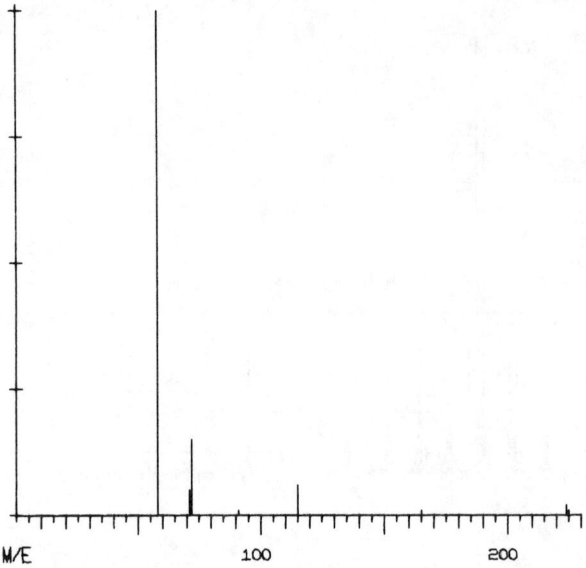
M/E
100
200

NORMETHADONE

$C_{20} \cdot H_{25} \cdot N \cdot O$ MW = 295

Base Peak = 58 8 Peaks

Mass	Int.	Mass	Int.	Mass	Int.	Mass	Int.	Mass	Int.	Mass	Int.	Mass	Int.	Mass	Int.	Mass	Int.
58	999	71	50	72	151	91	10	115	60	165	10	224	20	225	10		

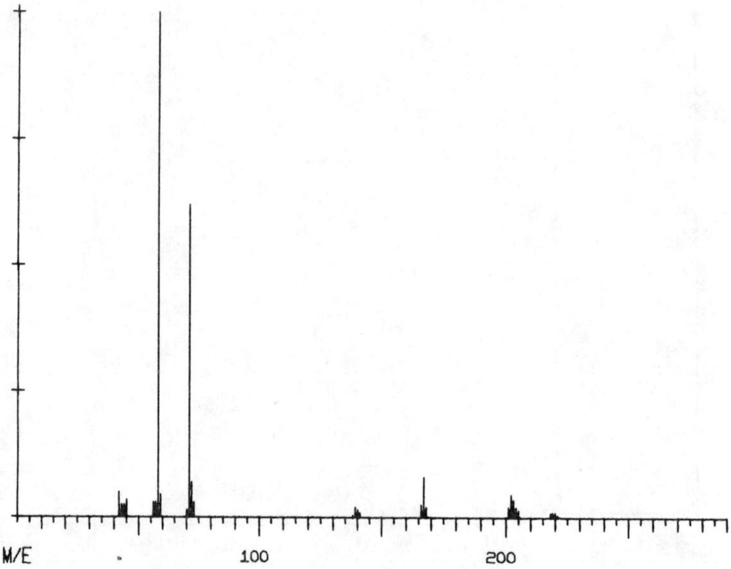
M/E
100
200

CARBINOXAMINE

$C_{16}H_{19}Cl \cdot N_2 \cdot O$ MW = 290

Base Peak = 58 15 Peaks

Mass	Int.	Mass	Int.	Mass	Int.	Mass	Int.	Mass	Int.	Mass	Int.	Mass	Int.	Mass	Int.	Mass	Int.
42	49	45	34	58	999	59	44	71	619	72	69	139	19	140	14	166	24
167	79	201	19	202	44	203	34	218	9	219	9						

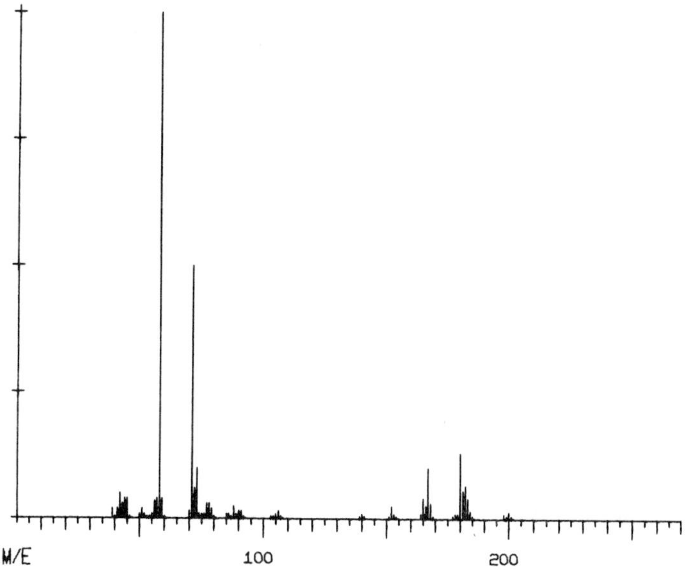
M/E
100
200

DOXYLAMINE

$C_{17} \cdot H_{22} \cdot N_2 \cdot O$ MW = 270

Base Peak = 58 22 Peaks

Mass	Int.	Mass	Int.	Mass	Int.	Mass	Int.	Mass	Int.	Mass	Int.	Mass	Int.	Mass	Int.	Mass	Int.
42	49	44	39	57	39	58	999	71	499	73	99	77	29	78	29	90	14
91	14	105	9	106	14	139	4	140	9	152	24	153	9	165	39	167	99
180	129	182	64	198	9	200	14										

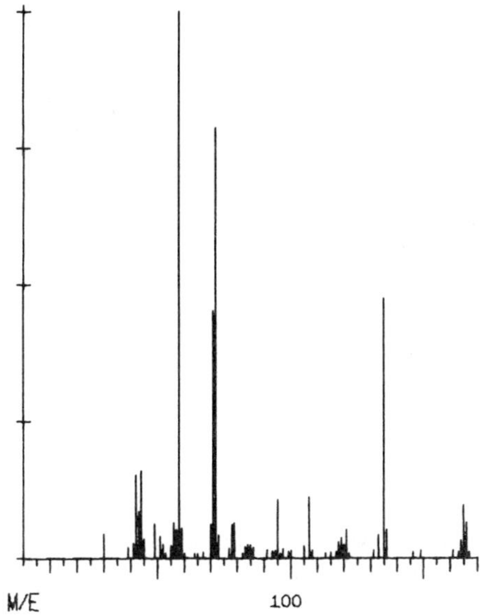
M/E
100

METHAPYRILENE MTB. 1

$C_9H_{15}N_3$ MW = 165

Base Peak = 58 21 Peaks

Mass	Int.	Mass	Int.	Mass	Int.	Mass	Int.	Mass	Int.	Mass	Int.	Mass	Int.	Mass	Int.	Mass	Int.
30	46	42	152	44	161	56	66	58	999	71	453	72	788	78	63	79	66
95	107	97	17	105	22	107	112	119	38	121	53	135	474	136	53	146	12
149	15	165	97	166	65												

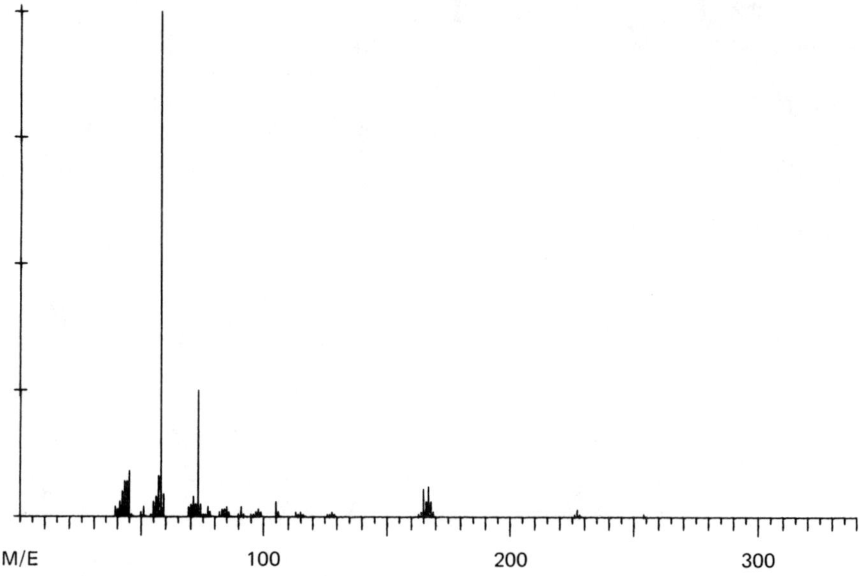
M/E
100
200
300

BROMODIPHENHYDRAMINE

$C_{17}H_{20}BrNO$ MW = 333

Base Peak = 58 19 Peaks

Mass	Int.	Mass	Int.	Mass	Int.	Mass	Int.	Mass	Int.	Mass	Int.	Mass	Int.	Mass	Int.	Mass	Int.
43	69	45	89	57	81	58	999	71	39	73	249	77	21	85	19	91	19
98	14	105	29	115	11	126	4	128	9	165	54	167	59	226	4	227	14
254	4																

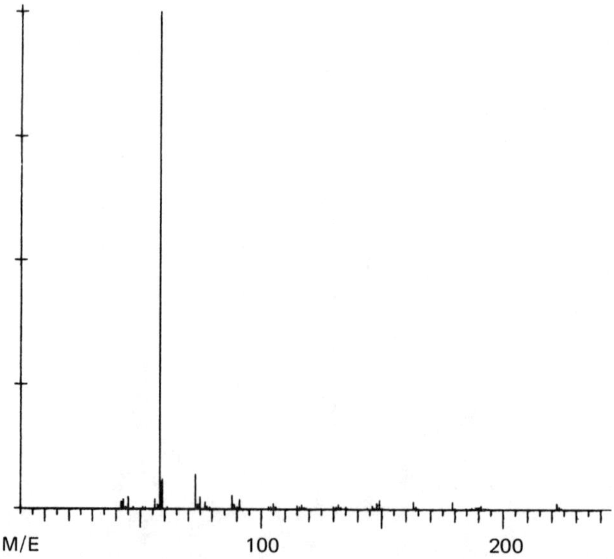
M/E
100
200

EPHEDRINE TMS ETHER

$C_{13}H_{23}NOSi$ MW = 237

Base Peak = 58 27 Peaks

Mass	Int.	Mass	Int.	Mass	Int.	Mass	Int.	Mass	Int.	Mass	Int.	Mass	Int.	Mass	Int.	Mass	Int.
43	20	45	25	58	999	58	60	73	70	75	25	77	15	88	28	91	20
103	5	105	12	117	11	130	5	131	5	132	10	133	5	148	12	149	18
163	15	164	5	179	15	180	3	191	8	193	3	222	12	223	5	237	1

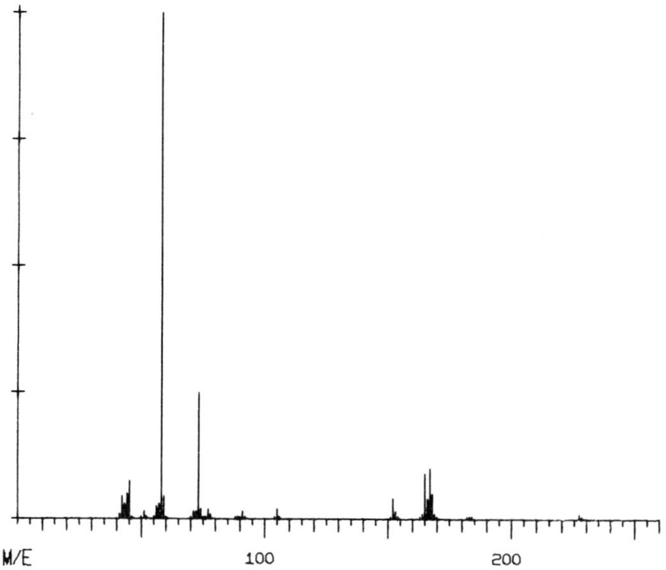
M/E
100
200

DIPHENHYDRAMINE

$C_{17}H_{21}NO$ MW = 255

Base Peak = 58 21 Peaks

Mass	Int.	Mass	Int.	Mass	Int.	Mass	Int.	Mass	Int.	Mass	Int.	Mass	Int.	Mass	Int.	Mass	Int.
44	49	45	74	58	999	59	44	73	249	74	19	77	19	78	9	90	4
91	14	104	4	105	19	152	39	153	14	165	89	167	99	182	4	183	6
227	9	228	4	255	1												

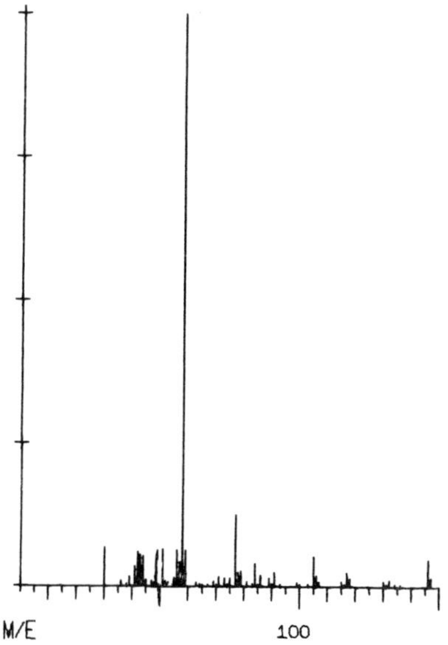
M/E
100

PSEUDOEPHEDRINE

$C_{10}H_{15}NO$ MW = 165

Base Peak = 58 19 Peaks

Mass	Int.	Mass	Int.	Mass	Int.	Mass	Int.	Mass	Int.	Mass	Int.	Mass	Int.	Mass	Int.	Mass	Int.
30	68	42	59	43	56	51	65	58	999	71	17	73	16	77	124	84	41
91	24	99	8	105	52	117	26	118	16	130	9	132	12	134	5	146	48
147	15																

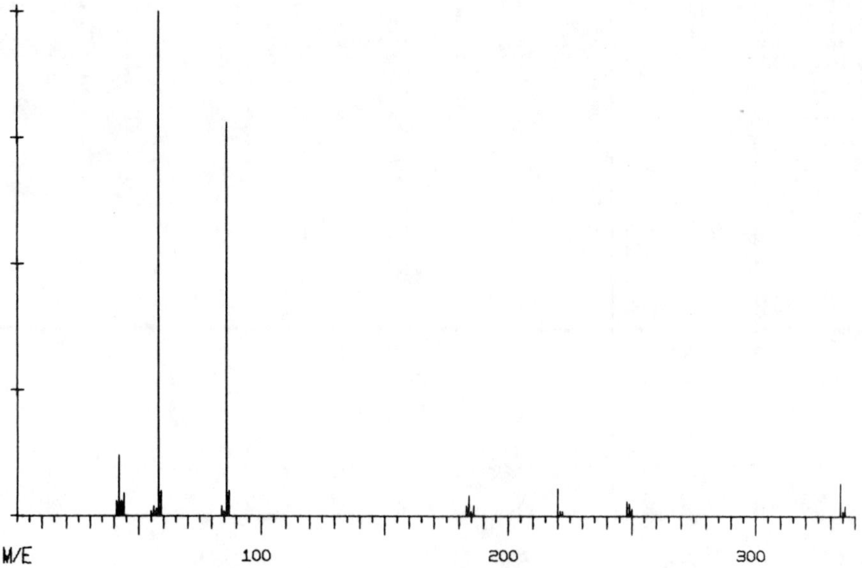
M/E
100
200
300

3-HYDROXY-CHLORPROMAZINE

$C_{17}H_{19}ClN_2OS$ MW = 334

Base Peak = 58 14 Peaks

Mass	Int.	Mass	Int.	Mass	Int.	Mass	Int.	Mass	Int.	Mass	Int.	Mass	Int.	Mass	Int.	Mass	Int.
42	119	44	44	58	999	59	49	86	779	87	49	183	19	184	39	220	54
221	9	248	29	249	24	334	64	336	19								

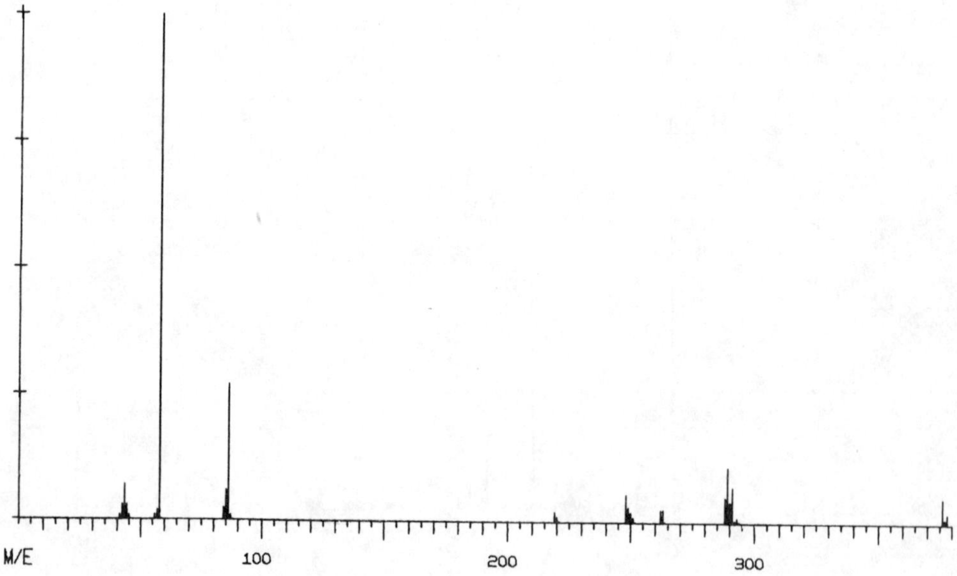
M/E
100
200
300

8-ACETOXYCHLORPROMAZINE

$C_{19}H_{21}Cl_1N_1N_2O_2S_1$ MW = 376

Base Peak = 58 16 Peaks

Mass	Int.	Mass	Int.	Mass	Int.	Mass	Int.	Mass	Int.	Mass	Int.	Mass	Int.	Mass	Int.	Mass	Int.
42	29	43	69	57	19	58	999	85	59	86	269	219	19	220	9	248	54
249	29	262	24	263	24	289	109	291	69	376	49	378	19				

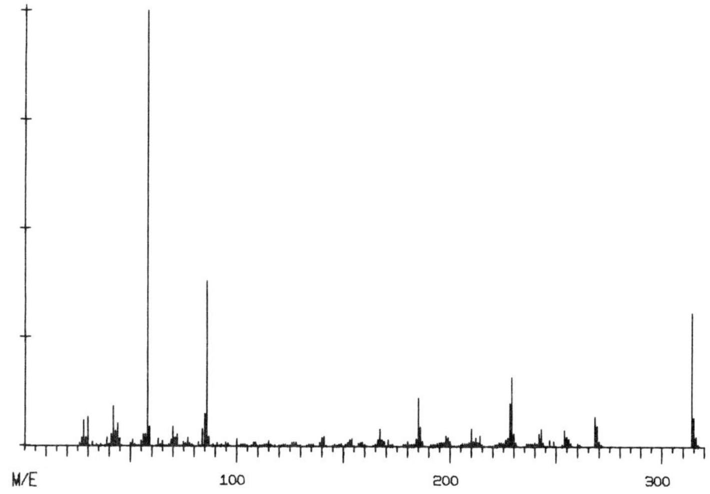
M/E
100
200
300

METHOXYPROMAZINE

$C_{18}H_{22}N_2OS$ MW = 314

Base Peak = 58 38 Peaks

Mass	Int.	Mass	Int.	Mass	Int.	Mass	Int.	Mass	Int.	Mass	Int.	Mass	Int.	Mass	Int.	Mass	Int.
28	61	30	67	42	93	44	52	58	999	59	45	70	45	72	28	85	74
86	380	95	9	100	18	109	11	115	13	127	11	128	10	140	19	141	23
153	15	154	17	167	39	168	16	185	112	186	45	198	24	199	21	210	39
214	25	228	101	229	161	242	31	243	39	254	37	255	22	268	71	269	48
314	307	315	67														

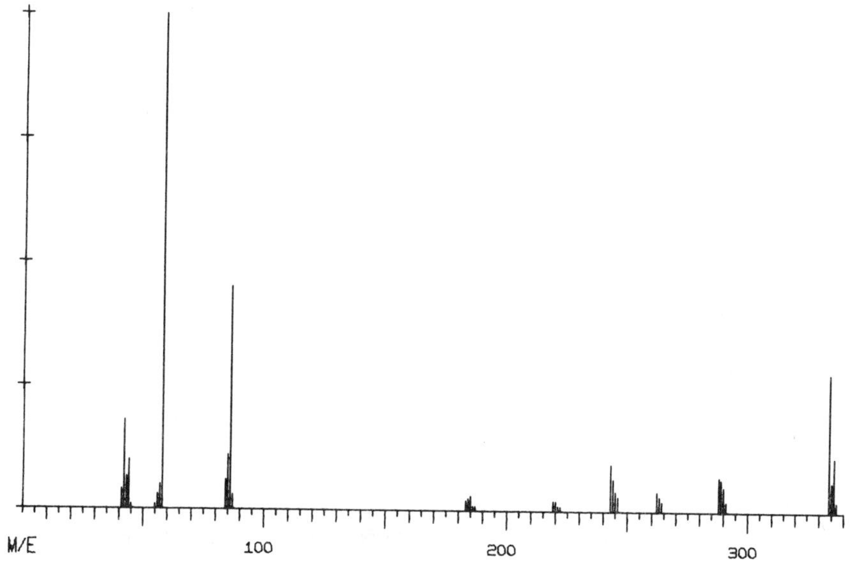
M/E
100
200
300

β-HYDROXY-CHLORPROMAZINE

$C_{17}H_{19}ClN_2OS$ MW = 334

Base Peak = 58 19 Peaks

Mass	Int.	Mass	Int.	Mass	Int.	Mass	Int.	Mass	Int.	Mass	Int.	Mass	Int.	Mass	Int.	Mass	Int.
42	179	44		57	49	58	999	85	109	86	449	184	24	185	29	219	19
220	19	243	94	244	64	245	39	262	39	263	29	288	69	289	64	334	279
336	109																

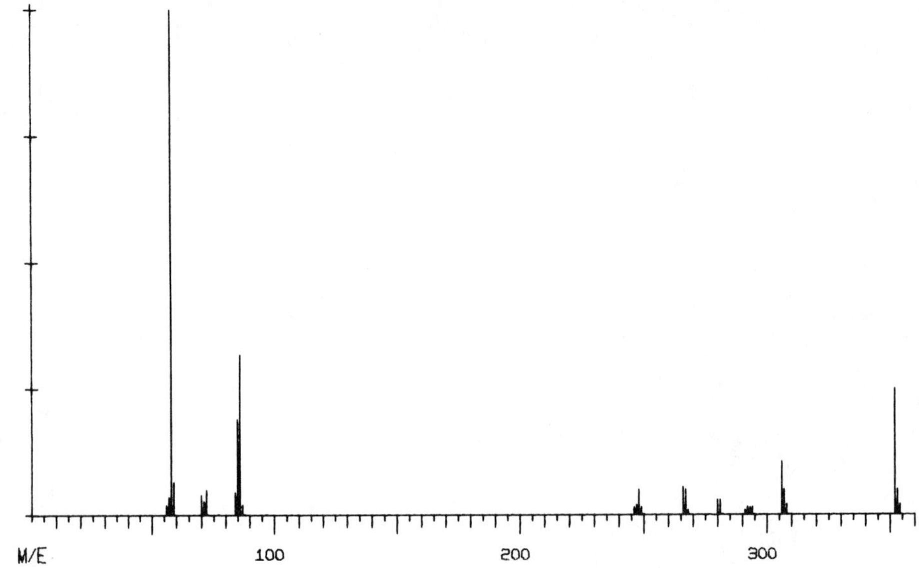
M/E
100
200
300

TRIFLUPROMAZINE

$C_{18}H_{19}F_3N_2S$ MW = 352

Base Peak = 58 18 Peaks

Mass	Int.	Mass	Int.	Mass	Int.	Mass	Int.	Mass	Int.	Mass	Int.	Mass	Int.	Mass	Int.	Mass	Int.
58	999	59	64	70	39	72	50	85	189	86	318	247	19	248	51	266	56
267	50	280	30	281	30	293	15	294	16	306	104	307	49	352	249	353	51

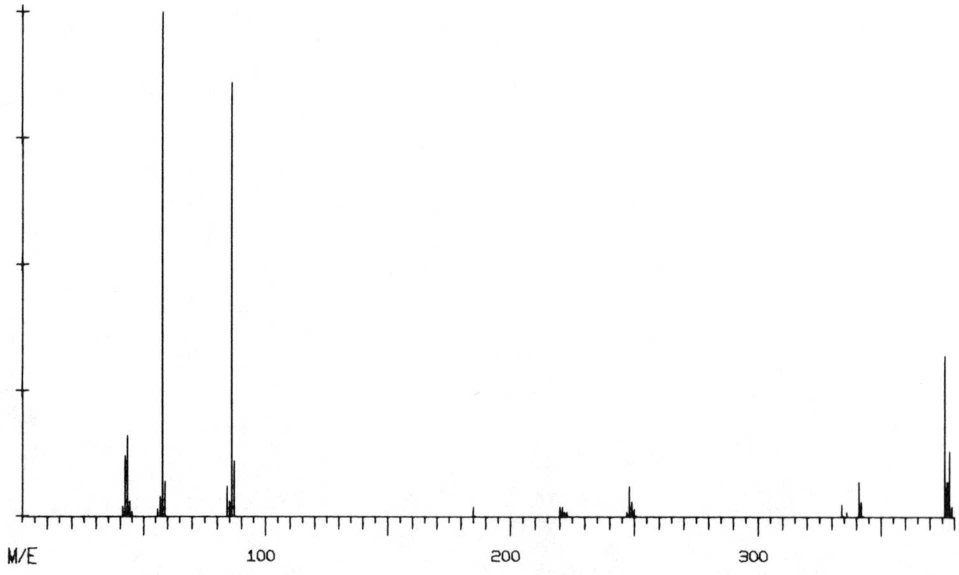
M/E
100
200
300

3-ACETOXYCHLORPROMAZINE

$C_{19}H_{21}Cl_1N_2O_2S_1$ MW = 376

Base Peak = 58 16 Peaks

Mass	Int.	Mass	Int.	Mass	Int.	Mass	Int.	Mass	Int.	Mass	Int.	Mass	Int.	Mass	Int.	Mass	Int.
42	119	43	159	58	999	59	69	86	859	87	109	185	19	220	19	221	19
248	59	249	29	334	24	341	69	342	29	376	319	378	129				

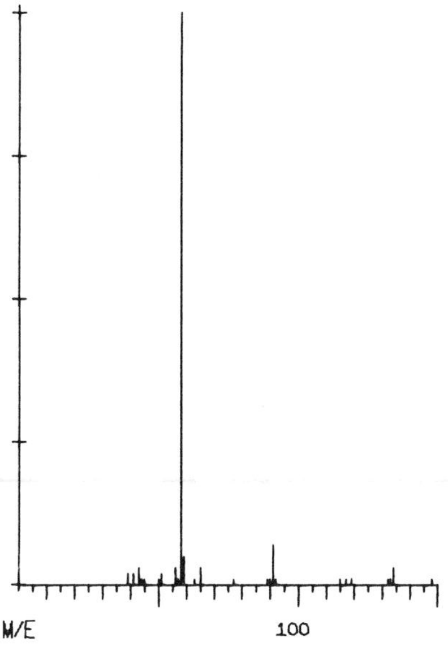
M/E
100

METHAMPHETAMINE

$C_{10}H_{15}N$ MW = 149

Base Peak = 58 16 Peaks

Mass	Int.	Mass	Int.	Mass	Int.	Mass	Int.	Mass	Int.	Mass	Int.	Mass	Int.	Mass	Int.	Mass	Int.
39	20	43	30	58	999	59	50	63	10	65	30	77	10	89	10	90	10
91	70	115	10	117	10	119	10	132	10	134	30	148	10				

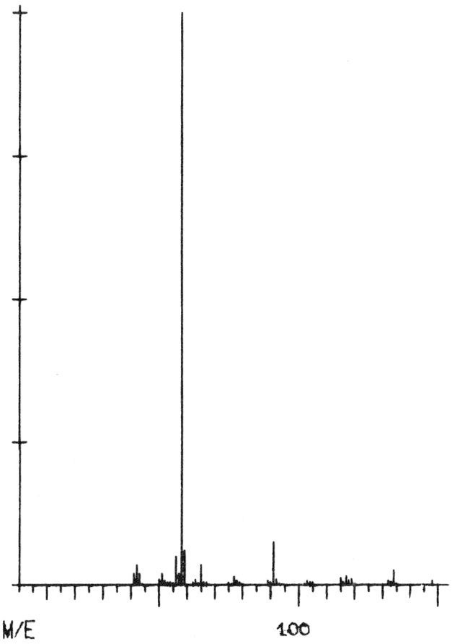
M/E
100

METHAMPHETAMINE

$C_{10}H_{15}N$ MW = 149

Base Peak = 58 17 Peaks

Mass	Int.	Mass	Int.	Mass	Int.	Mass	Int.	Mass	Int.	Mass	Int.	Mass	Int.	Mass	Int.	Mass	Int.
41	20	42	35	58	999	59	60	63	10	65	35	77	15	89	8	91	75
92	11	115	12	117	15	118	8	119	11	132	7	134	25	148	7		

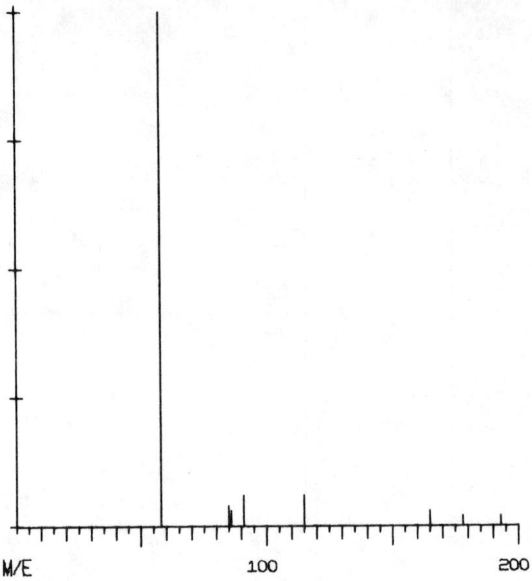
M/E
100
200

ISOMETHADONE

$C_{21}H_{27}NO$ MW = 309

Base Peak = 58 8 Peaks

Mass	Int.	Mass	Int.	Mass	Int.	Mass	Int.	Mass	Int.	Mass	Int.	Mass	Int.	Mass	Int.
58	999	85	40	86	30	91	60	115	60	165	30	178	20	193	20

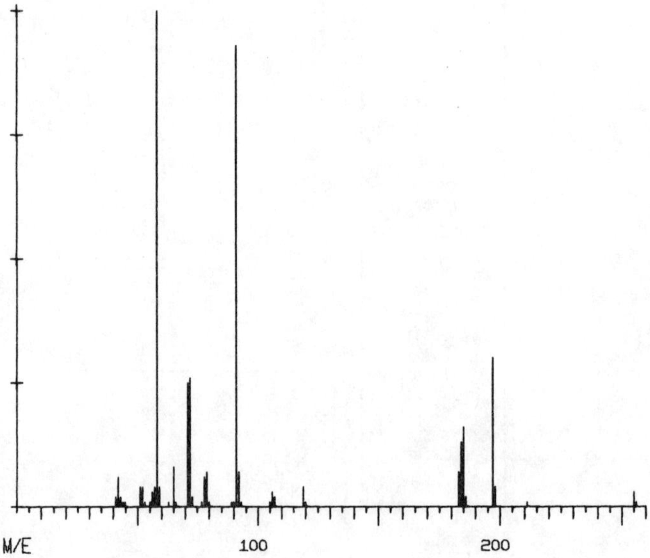
M/E
100
200

TRIPELENNAMINE

$C_{16}H_{21}N_3$ MW = 255

Base Peak = 58 21 Peaks

Mass	Int.	Mass	Int.	Mass	Int.	Mass	Int.	Mass	Int.	Mass	Int.	Mass	Int.	Mass	Int.	Mass	Int.
41	20	42	60	51	40	58	999	71	250	72	260	78	60	79	70	91	930
92	70	106	30	107	20	119	40	120	10	184	130	185	160	197	300	198	40
211	10	255	30	256	10												

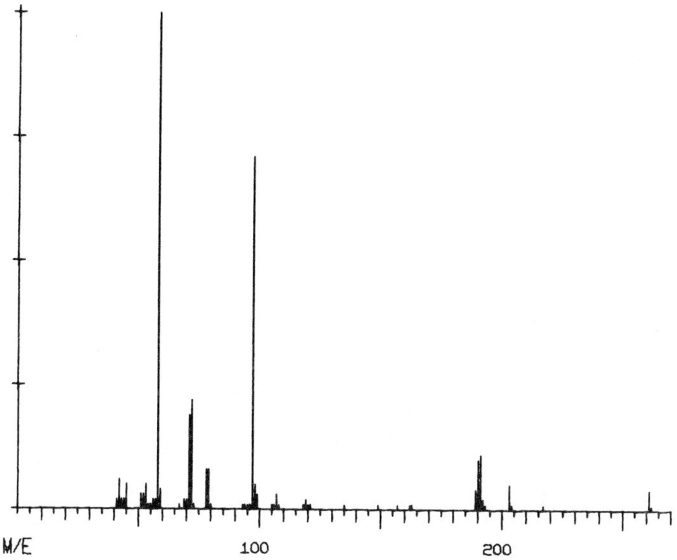
M/E
100
200

METHAPYRILENE

$C_{14}H_{19}N_3S$ MW = 261

Base Peak = 58 26 Peaks

Mass	Int.	Mass	Int.	Mass	Int.	Mass	Int.	Mass	Int.	Mass	Int.	Mass	Int.	Mass	Int.
42	60	45	50	53	50	58	999	71	190	72	220	78	80	79	80	97	710
98	50	105	10	107	30	118	10	119	20	135	10	149	10	157	10	162	10
163	10	190	100	191	110	203	50	204	10	217	10	261	40	262	10		

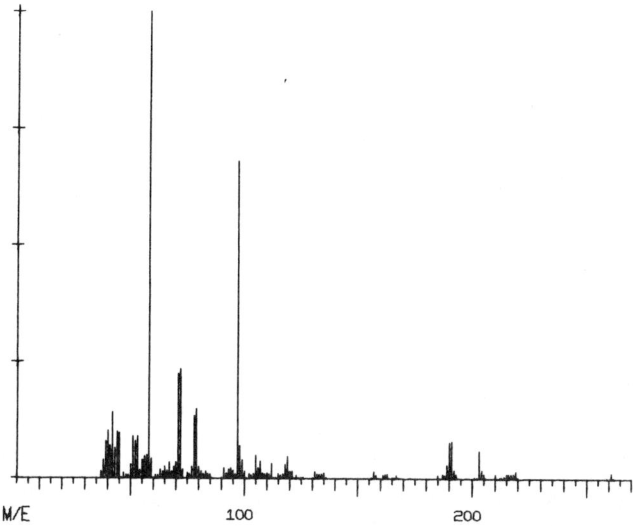
M/E
100
200

THENYLDIAMINE

$C_{14}H_{19}N_3S$ MW = 261

Base Peak = 58 30 Peaks

Mass	Int.	Mass	Int.	Mass	Int.	Mass	Int.	Mass	Int.	Mass	Int.	Mass	Int.	Mass	Int.	Mass	Int.
40	103	42	142	53	91	58	999	71	224	72	236	78	135	79	149	97	681
98	70	105	51	107	38	118	29	119	48	133	10	135	12	157	16	158	7
162	11	163	9	185	7	187	10	190	78	191	79	203	61	204	17	217	11
219	15	259	3	261	13												

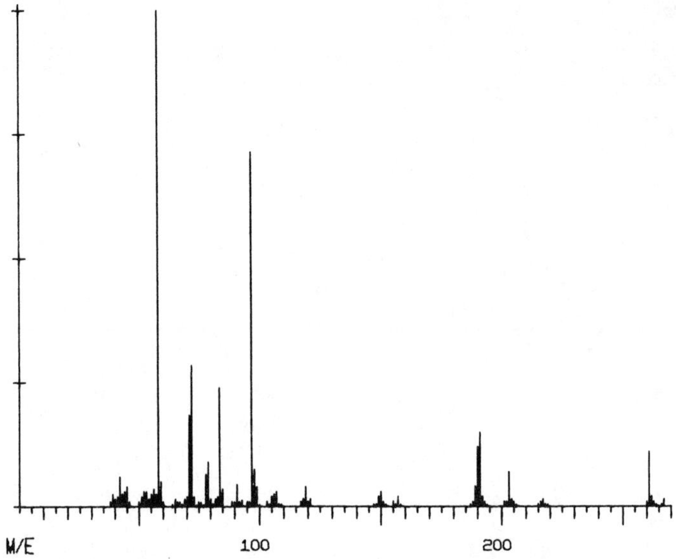
M/E
100
200

METHAPYRILENE

$C_{14}H_{19}N_3S$ MW = 261

Base Peak = 58 25 Peaks

Mass	Int.	Mass	Int.	Mass	Int.	Mass	Int.	Mass	Int.	Mass	Int.	Mass	Int.	Mass	Int.	Mass	Int.
42	59	45	39	58	999	59	49	71	184	72	284	79	89	84	239	97	714
98	74	106	24	107	29	118	14	119	39	149	19	150	29	187	4	190	119
191	149	203	69	204	14	216	9	217	14	261	109	262	19				

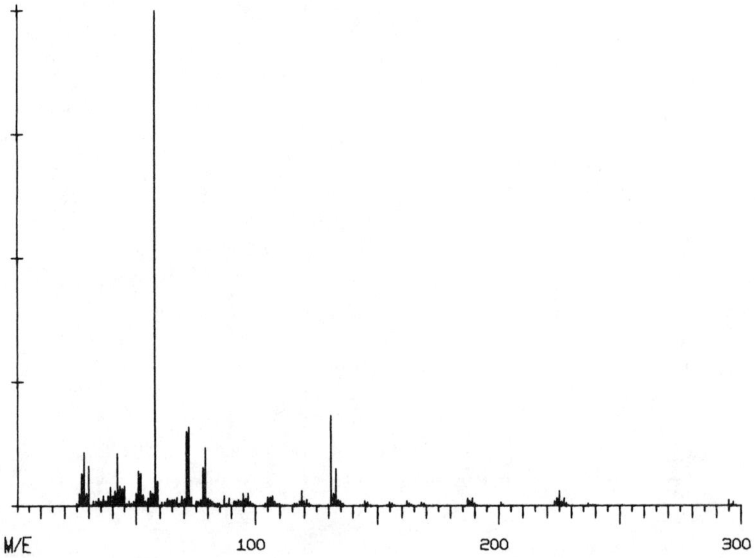
M/E
100
200
300

CHLOROTHEN

$C_{14}H_{18}ClN_3S$ MW = 295

Base Peak = 58 35 Peaks

Mass	Int.	Mass	Int.	Mass	Int.	Mass	Int.	Mass	Int.	Mass	Int.	Mass	Int.	Mass	Int.	Mass	Int.
28	107	30	80	42	106	43	40	51	69	58	999	71	151	72	160	78	77
79	118	95	24	97	26	105	17	107	21	119	31	131	182	132	26	133	74
146	8	155	8	162	9	168	7	187	14	188	11	189	15	202	2	203	1
224	15	225	30	237	6	239	3	251	3	252	1	295	13	297	8		

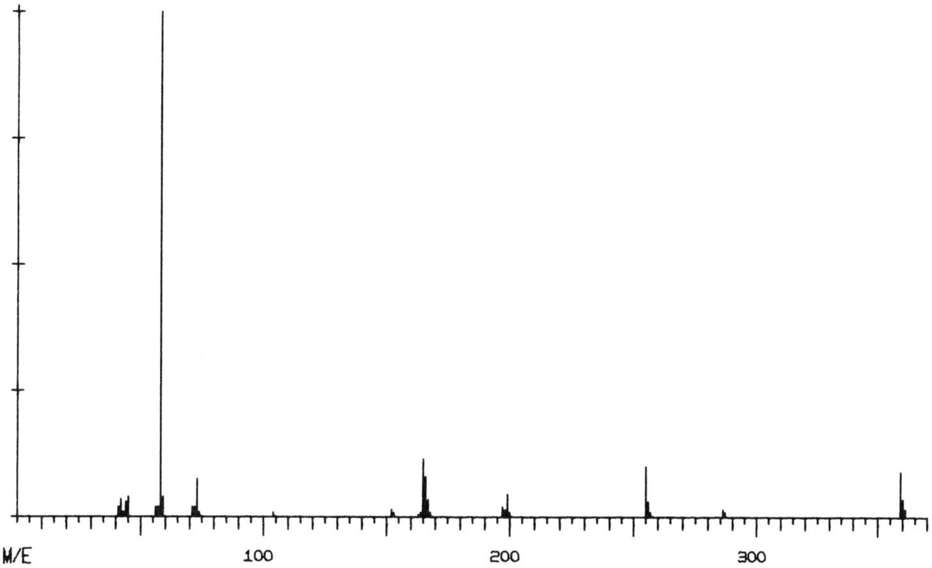
M/E
100
200
300

CAPTODIAMINE

$C_{21}H_{29}NS_2$ MW = 359

Base Peak = 58 10 Peaks

Mass	Int.	Mass	Int.	Mass	Int.	Mass	Int.	Mass	Int.	Mass	Int.	Mass	Int.	Mass	Int.	Mass	Int.
42	34	45	39	58	999	59	39	71	19	73	74	104	9	152	14	153	9
165	114	166	79	107	19	199	44	255	99	256	29	286	14	287	9	359	89
360	34																

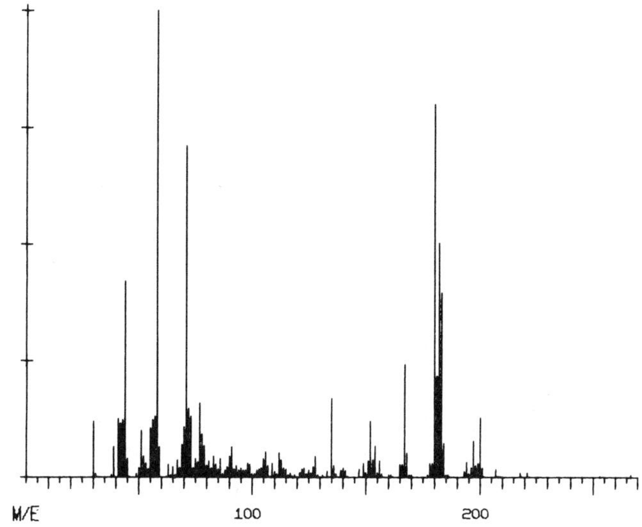
M/E
100
200

DOXYLAMINE (GC)

$C_{17}H_{22}N_2O$ MW = 270

Base Peak = 58 30 Peaks

Mass	Int.	Mass	Int.	Mass	Int.	Mass	Int.	Mass	Int.	Mass	Int.	Mass	Int.	Mass	Int.	Mass	Int.
30	119	31	7	41	125	44	419	57	129	58	999	71	709	72	147	77	161
78	93	90	46	91	66	106	55	112	52	127	23	128	46	135	170	136	26
152	119	154	68	167	243	168	53	180	799	182	503	197	78	200	127	207	17
208	3	218	11	221	11	253	2	267	4								

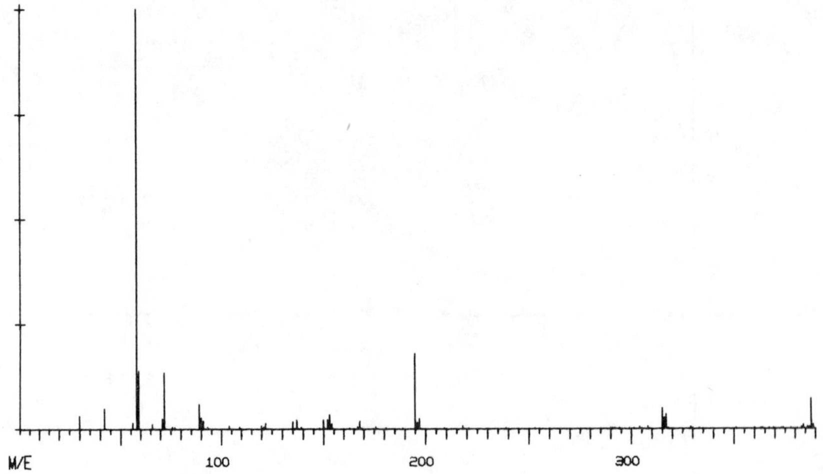
M/E
100
200
300

TRIMETHOBENZAMIDE

$C_{21}H_{28}N_2O_5$ MW = 388

Base Peak = 58 48 Peaks

Mass	Int.	Mass	Int.	Mass	Int.	Mass	Int.	Mass	Int.	Mass	Int.	Mass	Int.	Mass	Int.	Mass	Int.
30	33	38	1	42	50	58	999	59	140	71	24	72	136	77	6	89	61
90	28	91	19	104	8	109	5	120	7	122	16	135	17	137	22	150	23
153	34	167	4	168	20	176	6	181	2	195	181	197	25	209	2	210	1
218	8	221	2	237	1	239	3	253	3	290	3	291	3	304	4	308	4
315	51	317	36	329	5	330	2	349	1	351	3	360	3	363	3	373	3
383	4	388	72	389	11												

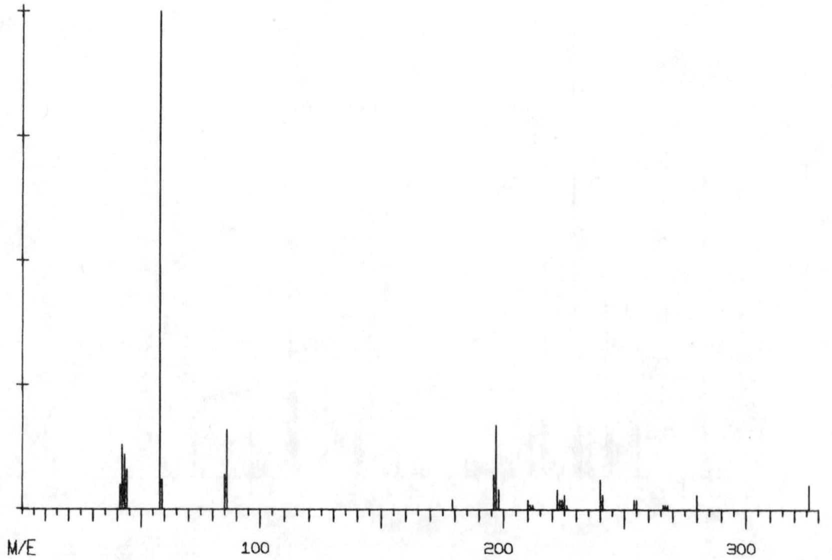
M/E
100
200
300

ACETYLPROMAZINE

$C_{19}H_{22}N_2OS$ MW = 326

Base Peak = 58 21 Peaks

Mass	Int.	Mass	Int.	Mass	Int.	Mass	Int.	Mass	Int.	Mass	Int.	Mass	Int.	Mass	Int.	Mass	Int.
42	129	43	109	58	999	59	59	85	69	86	159	179	19	196	69	197	169
210	19	211	9	222	39	225	29	240	59	241	29	254	19	255	19	266	9
267	9	280	29	326	49												

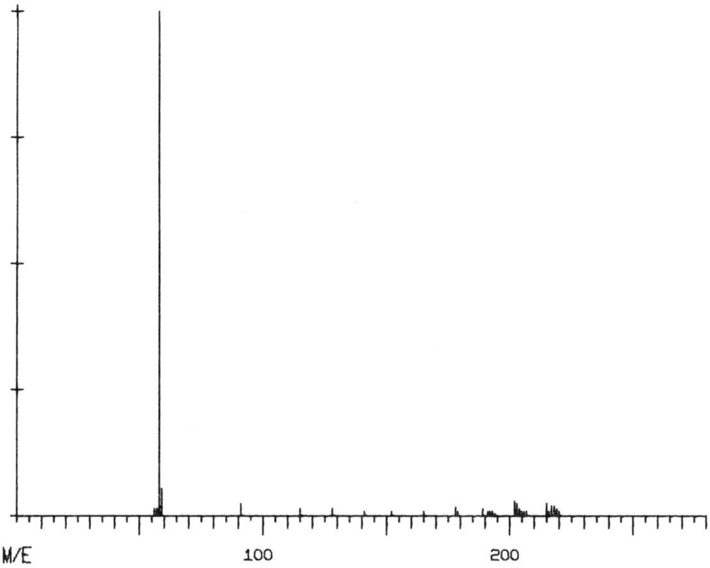
M/E
100
200

AMITRIPTYLINE

$C_{20}H_{23}N$ MW = 277

Base Peak = 58 16 Peaks

Mass	Int.	Mass	Int.	Mass	Int.	Mass	Int.	Mass	Int.	Mass	Int.	Mass	Int.	Mass	Int.	Mass	Int.
58	999	59	54	91	24	115	14	128	14	141	9	152	9	165	11	178	17
179	9	189	14	191	11	202	31	203	24	217	19	218	21				

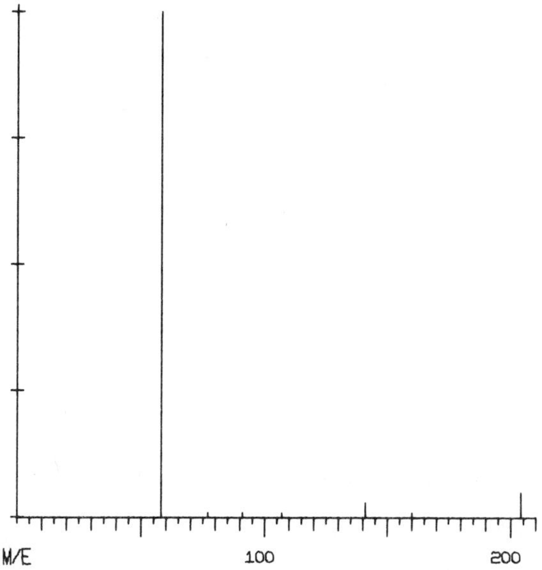
M/E
100
200

BUFOTENINE

$C_{12}H_{16}N_2O$ MW = 204

Base Peak = 58 7 Peaks

Mass	Int.	Mass	Int.	Mass	Int.	Mass	Int.	Mass	Int.	Mass	Int.	Mass	Int.
58	999	77	10	91	10	107	10	141	30	160	10	204	50

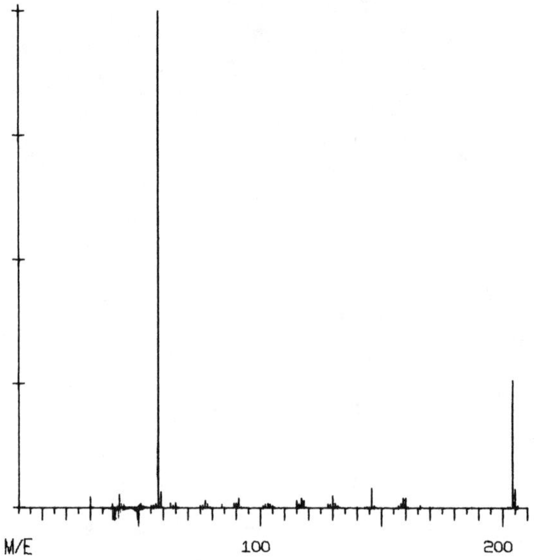
M/E
100
200

PSILOCIN

$C_{12}H_{16}N_2O$ MW = 204
Base Peak = 58 24 Peaks

Mass	Int.	Mass	Int.	Mass	Int.	Mass	Int.	Mass	Int.	Mass	Int.	Mass	Int.	Mass	Int.	Mass	Int.
30	23	42	28	44	12	58	999	59	33	63	10	65	11	77	15	89	11
91	20	103	10	115	15	117	20	118	16	130	24	132	4	140	3	146	40
159	19	160	20	166	6	187	3	204	258	205	38						

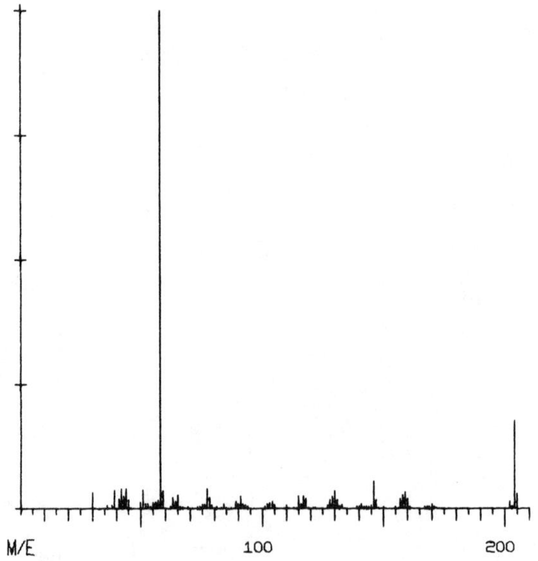
M/E
100
200

PSILOCYBIN

$C_{12}H_{17}N_2O_4P$ MW = 284
Base Peak = 58 23 Peaks

Mass	Int.	Mass	Int.	Mass	Int.	Mass	Int.	Mass	Int.	Mass	Int.	Mass	Int.	Mass	Int.	Mass	Int.
30	32	42	41	44	40	51	37	58	999	63	22	65	28	77	39	78	22
91	24	103	13	115	25	117	25	129	26	130	34	133	7	141	11	146	54
159	32	160	20	170	10	204	177	205	30								

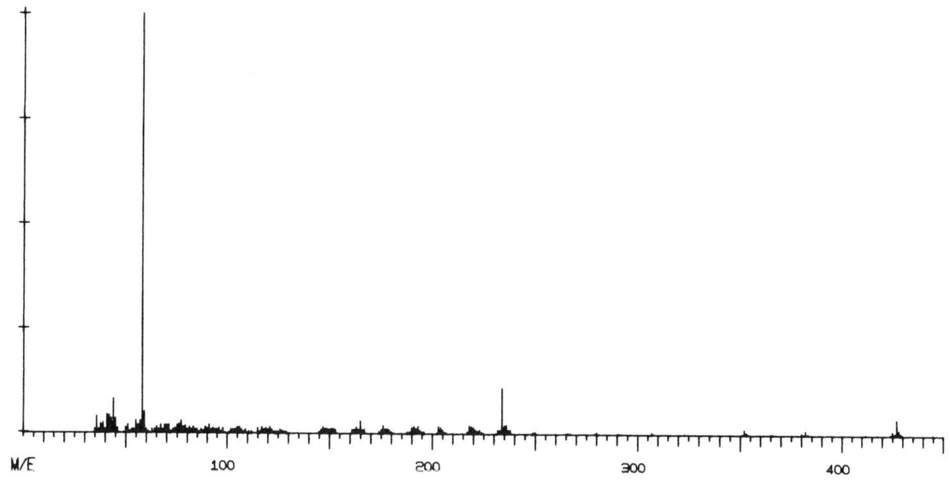
M/E
100
200
300
400

NARCEINE

$C_{23}H_{27}NO_8$ MW = 445

Base Peak = 58 52 Peaks

Mass	Int.	Mass	Int.	Mass	Int.	Mass	Int.	Mass	Int.	Mass	Int.	Mass	Int.	Mass	Int.	Mass	Int.
41	44	44	82	58	999	59	52	67	20	69	21	76	22	77	31	90	12
91	21	105	16	117	16	119	12	121	14	145	4	147	14	148	12	163	16
165	29	176	21	178	13	191	17	193	18	203	17	204	15	218	21	219	18
234	109	236	22	249	5	250	4	265	3	266	3	279	2	280	4	290	2
291	2	306	1	307	4	323	2	324	1	352	12	353	4	365	2	366	2
380	4	382	9	412	2	425	10	427	38	428	12	445	2				

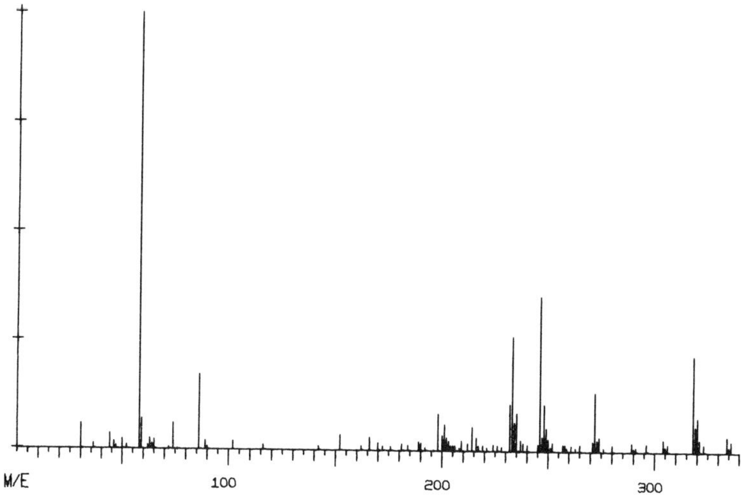
M/E
100
200
300

CHLORPROMAZINE MTB 4

$C_{17}H_{19}CLN_2OS$ MW = 334

Base Peak = 58 40 Peaks

Mass	Int.	Mass	Int.	Mass	Int.	Mass	Int.	Mass	Int.	Mass	Int.	Mass	Int.	Mass	Int.	Mass	Int.
30	57	44	34	46	17	58	999	59	70	63	25	74	60	86	172	89	21
90	8	102	21	116	13	142	11	152	36	166	29	170	17	181	15	184	12
198	85	201	60	202	29	214	56	216	31	224	14	232	107	233	262	246	355
248	108	258	16	271	22	272	135	274	32	289	19	296	18	304	28	306	17
318	220	320	78	334	35	336	25										

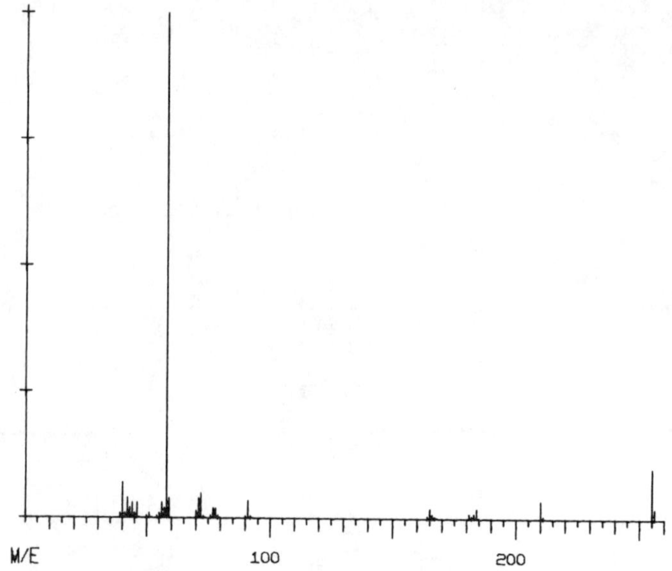
M/E
100
200

PHENYLTOLOXAMINE

$C_{17}H_{21}N_1O_1$ MW = 255

Base Peak = 58 18 Peaks

Mass	Int.	Mass	Int.	Mass	Int.	Mass	Int.	Mass	Int.	Mass	Int.	Mass	Int.	Mass	Int.	Mass	Int.
40	69	42	39	58	999	59	39	71	39	72	49	77	19	78	19	90	4
91	34	165	19	166	9	181	9	184	19	210	34	211	4	255	99	256	19

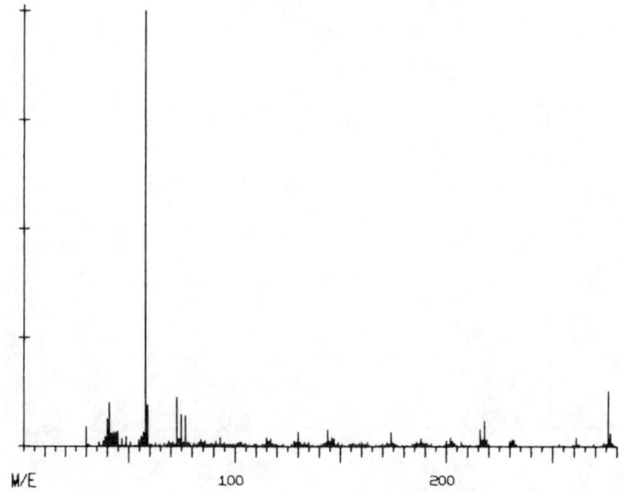
M/E
100
200

PSILOCIN-1 TMS

$C_{15}H_{24}N_2OSI$ MW = 276

Base Peak = 58 28 Peaks

Mass	Int.	Mass	Int.	Mass	Int.	Mass	Int.	Mass	Int.	Mass	Int.	Mass	Int.	Mass	Int.	Mass	Int.
30	45	31	4	40	62	41	100	58	999	59	96	73	112	75	72	77	69
84	15	91	13	93	20	115	21	117	17	128	12	130	32	144	38	145	12
146	21	147	17	160	11	172	11	174	33	186	11	188	18	200	13	202	20
203	12	216	38	218	58	231	16	232	16	253	4	255	2	261	21	262	5
276	125	277	30														

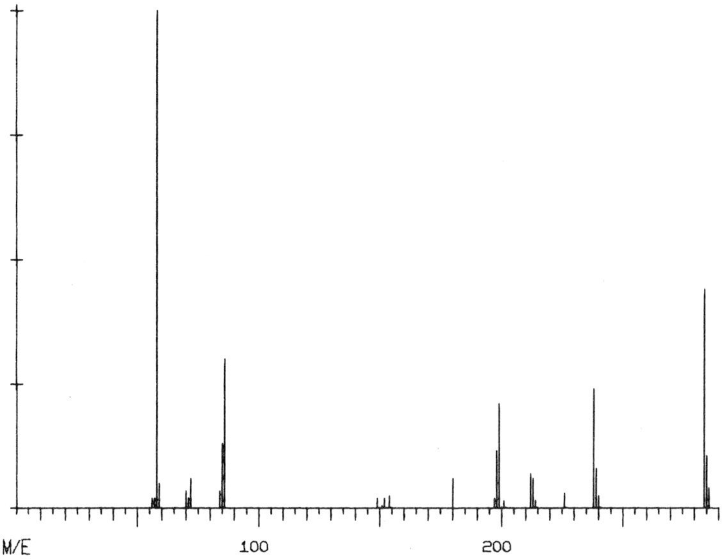
M/E
100
200

PROMAZINE

$C_{17}H_{20}N_2S$ MW = 284

Base Peak = 58 19 Peaks

Mass	Int.	Mass	Int.	Mass	Int.	Mass	Int.	Mass	Int.	Mass	Int.	Mass	Int.	Mass	Int.	Mass	Int.
58	999	59	49	70	34	72	59	85	129	86	299	149	19	154	24	180	59
198	114	199	209	212	69	213	59	226	29	238	239	239	79	284	439	285	104
286	39																

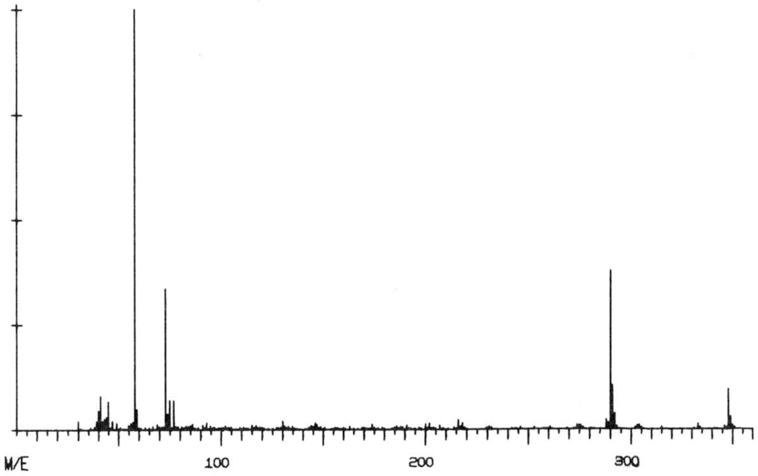
M/E
100
200
300

PSILOCIN-2 TMS

$C_{18}H_{32}N_2OSI_2$ MW = 348

Base Peak = 58 47 Peaks

Mass	Int.	Mass	Int.	Mass	Int.	Mass	Int.	Mass	Int.	Mass	Int.	Mass	Int.	Mass	Int.	Mass	Int.
30	21	31	3	41	79	45	67	58	999	59	47	73	336	75	70	77	70
86	12	91	10	93	14	115	9	117	11	130	21	131	7	133	8	144	10
146	15	147	13	160	5	163	7	174	12	186	9	191	11	200	12	202	14
207	10	216	22	218	15	231	7	232	6	246	5	253	5	258	3	261	4
274	10	276	11	290	377	291	104	303	9	304	10	315	4	333	13	334	4
348	95	349	31														

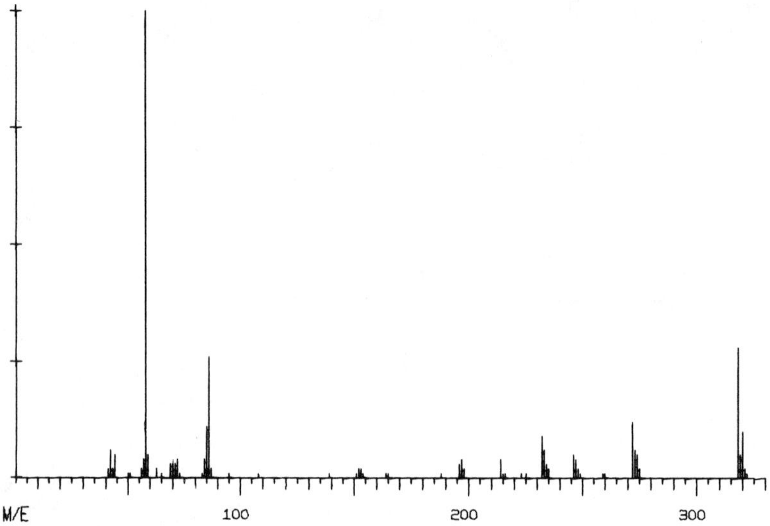
M/E
100
200
300

CHLORPROMAZINE

$C_{17}H_{19}CLN_2S$ MW = 318

Base Peak = 58 31 Peaks

Mass	Int.	Mass	Int.	Mass	Int.	Mass	Int.	Mass	Int.	Mass	Int.	Mass	Int.	Mass	Int.	Mass	Int.
42	60	44	50	58	999	59	50	70	40	72	40	85	110	86	260	95	10
108	10	139	10	152	20	153	20	164	10	165	10	196	30	197	40	214	40
215	10	216	10	223	10	232	90	233	60	246	50	247	40	259	10	260	10
272	120	273	60	318	280	320	100										

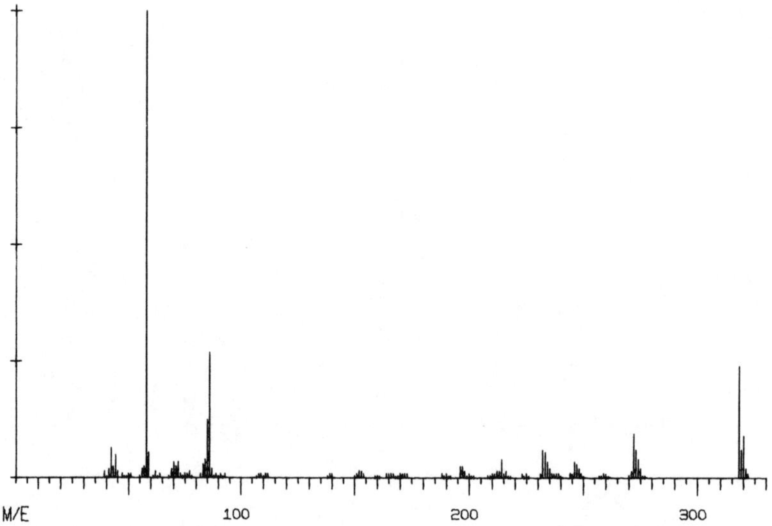
M/E
100
200
300

CHLORPROMAZINE

$C_{17}H_{19}CLN_2S$ MW = 318

Base Peak = 58 34 Peaks

Mass	Int.	Mass	Int.	Mass	Int.	Mass	Int.	Mass	Int.	Mass	Int.	Mass	Int.	Mass	Int.	Mass	Int.
42	64	44	49	58	999	59	54	70	34	72	34	85	124	86	269	91	9
93	9	108	9	109	9	139	9	140	9	152	14	153	14	164	9	165	9
196	24	197	24	212	14	214	39	216	14	223	9	232	59	233	54	246	34
247	29	259	9	271	14	272	94	273	59	318	239	320	89				

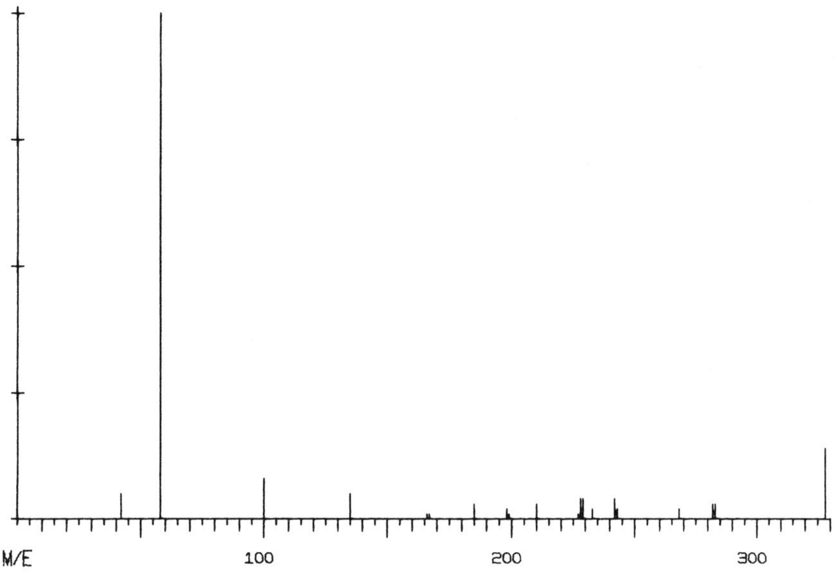
M/E
100
200
300

METHOTRIMEPRAZINE

$C_{19}H_{24}N_2OS$ MW = 328

Base Peak = 58 18 Peaks

Mass	Int.	Mass	Int.	Mass	Int.	Mass	Int.	Mass	Int.	Mass	Int.	Mass	Int.	Mass	Int.	Mass	Int.
42	49	58	999	100	79	135	49	166	9	167	9	185	29	198	19	199	9
210	29	228	39	229	39	233	19	242	39	268	19	282	29	283	29	328	139

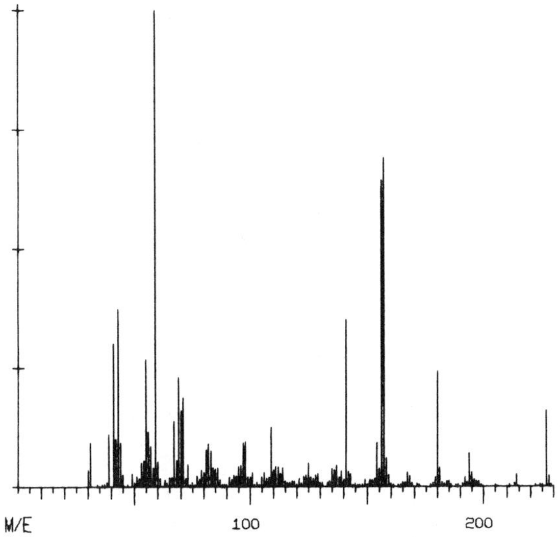
M/E
100
200

AMOBARBITAL MTB. 1

$C_{11}H_{18}N_2O_4$ MW = 242

Base Peak = 59 30 Peaks

Mass	Int.	Mass	Int.	Mass	Int.	Mass	Int.	Mass	Int.	Mass	Int.	Mass	Int.	Mass	Int.	Mass	Int.
30	35	31	92	41	300	43	372	55	267	59	999	69	231	71	188	81	77
82	90	97	92	98	95	109	126	111	42	125	49	129	27	137	44	141	349
156	643	157	698	167	31	168	23	180	243	181	39	194	70	195	31	213	8
214	26	227	161	228	22												

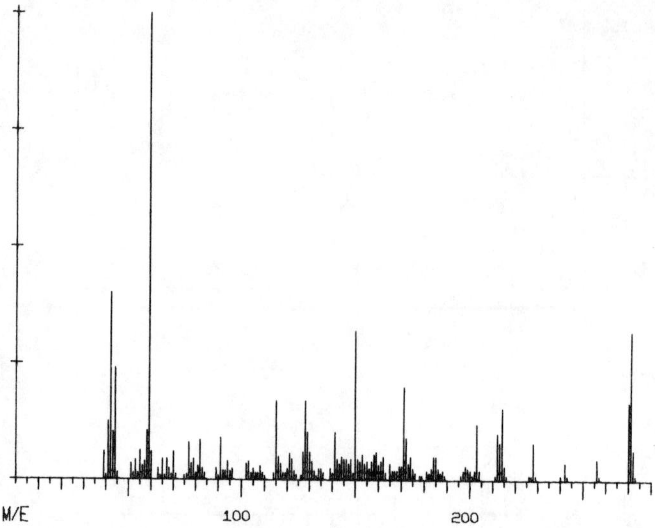
M/E
100
200

METHORPHAN

$C_{18}H_{25}N_{1}O_{1}$ MW = 271

Base Peak = 59 36 Peaks

Mass	Int.	Mass	Int.	Mass	Int.	Mass	Int.	Mass	Int.	Mass	Int.	Mass	Int.	Mass	Int.	Mass	Int.
42	399	44	239	58	105	59	999	65	44	70	59	77	81	82	84	91	90
94	39	115	169	116	49	128	169	129	103	141	102	144	49	150	319	159	61
171	199	172	89	174	49	184	49	198	29	199	24	203	119	214	155	228	79
229	10	242	37	243	10	256	44	257	9	270	167	271	319	272	64	273	9

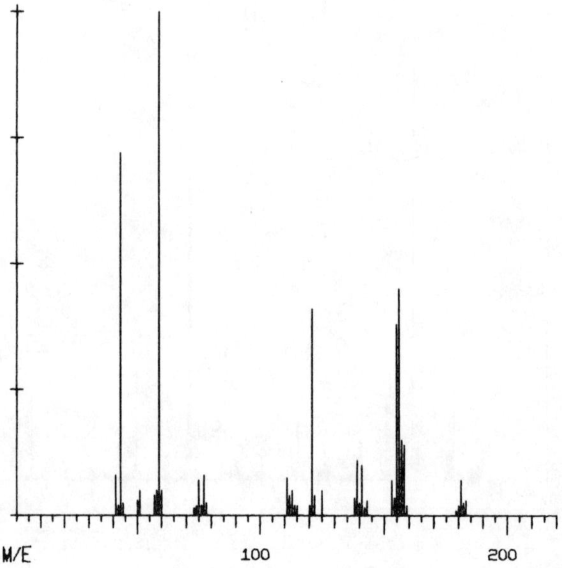
M/E
100
200

PHENAGLYCODOL

$C_{11}H_{15}CLO_{2}$ MW = 214

Base Peak = 59 18 Peaks

Mass	Int.	Mass	Int.	Mass	Int.	Mass	Int.	Mass	Int.	Mass	Int.	Mass	Int.	Mass	Int.	Mass	Int.
41	79	43	719	51	49	59	999	74	19	75	69	77	79	78	24	111	74
113	49	121	409	125	49	139	109	141	99	155	379	156	449	181	69	183	29

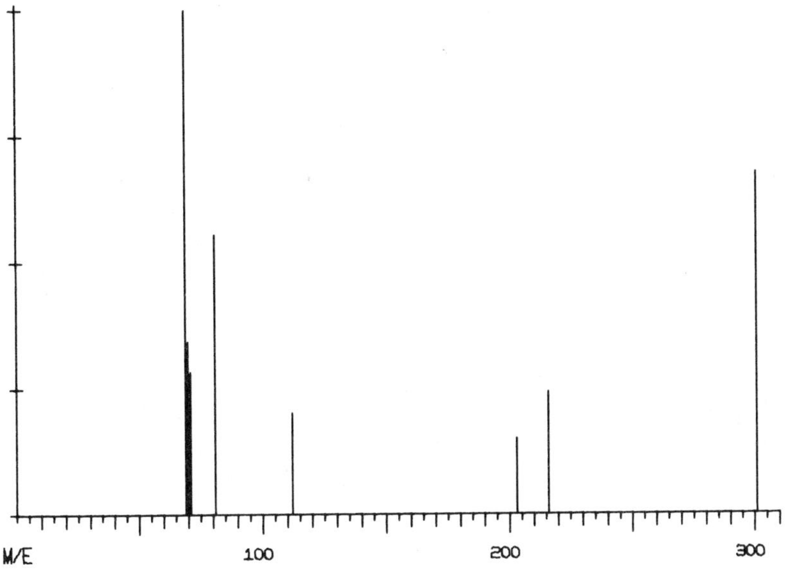
M/E
100
200
300

OXYMORPHONE

$C_{17}H_{19}NO_4$ MW = 301

Base Peak = 69 7 Peaks

Mass	Int.	Mass	Int.	Mass	Int.	Mass	Int.	Mass	Int.	Mass	Int.	Mass	Int.	Mass	Int.	Mass	Int.
69	999	70	343	81	555	112	202	203	151	216	242	301	676				

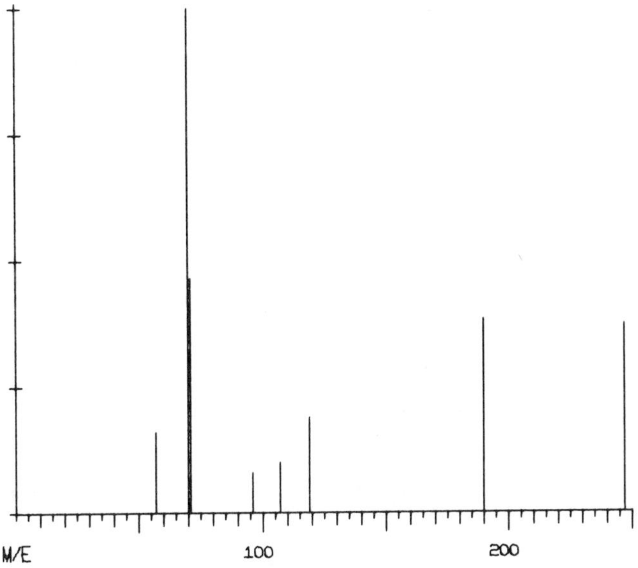
M/E
100
200

KETOBEMIDONE

$C_{15}H_{21}NO_2$ MW = 247

Base Peak = 70 8 Peaks

Mass	Int.	Mass	Int.	Mass	Int.	Mass	Int.	Mass	Int.	Mass	Int.	Mass	Int.	Mass	Int.	Mass	Int.
57	161	70	999	71	464	96	80	107	101	119	191	190	383	247	373		

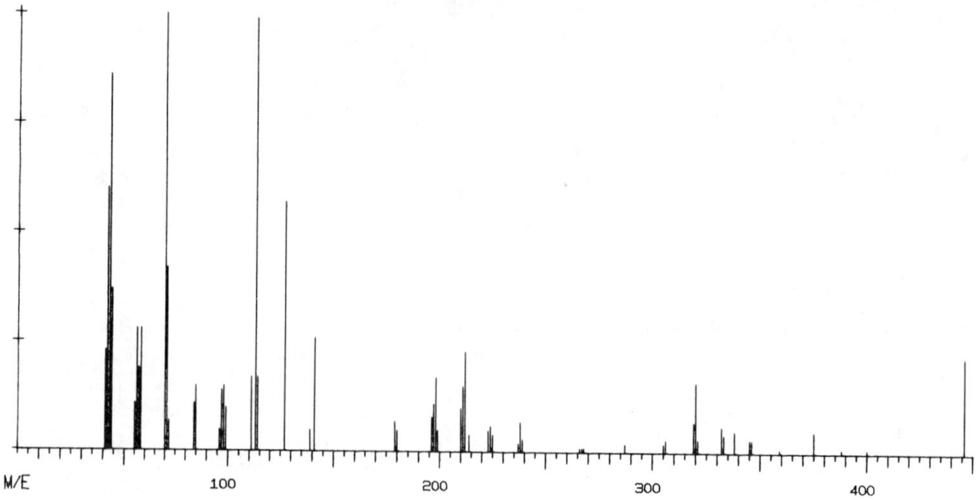

M/E
100
200
300
400

THIOPROPERAZINE

$C_{22}H_{30}N_4O_2S_2$ MW = 446
Base Peak = 70 41 Peaks

Mass	Int.	Mass	Int.	Mass	Int.	Mass	Int.	Mass	Int.	Mass	Int.	Mass	Int.	Mass	Int.	Mass	Int.
42	599	43	859	56	279	58	279	70	999	71	419	84	109	85	149	97	139
98	149	111	169	113	989	127	569	139	49	141	259	179	69	180	49	197	109
198	169	211	149	212	229	223	49	224	59	238	69	239	29	266	9	267	9
287	19	305	19	306	29	319	69	320	159	332	59	338	49	345	29	346	29
359	9	375	49	388	9	400	9	446	219								

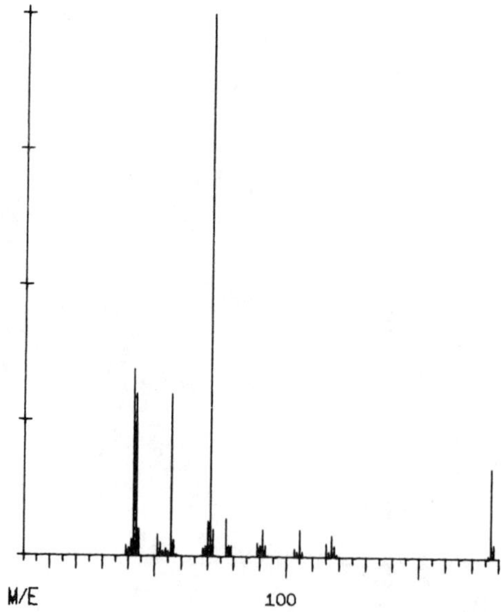

M/E
100

PHENMETRAZINE

$C_{11}H_{15}NO$ MW = 177
Base Peak = 71 16 Peaks

Mass	Int.	Mass	Int.	Mass	Int.	Mass	Int.	Mass	Int.	Mass	Int.	Mass	Int.	Mass	Int.	Mass	Int.
42	344	43	299	51	39	56	299	70	64	71	999	77	69	89	24	90	19
91	49	105	49	117	39	118	19	119	4	177	164	178	24				

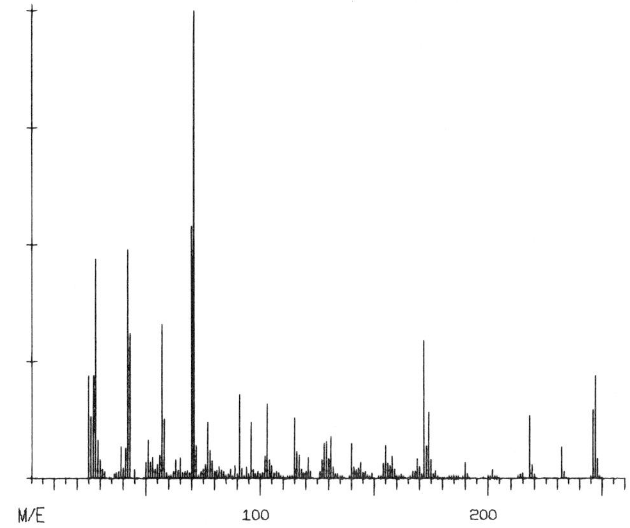
M/E
100
200

MEPERIDINE

$C_{15}H_{21}NO_2$ MW = 247

Base Peak = 71 34 Peaks

Mass	Int.	Mass	Int.	Mass	Int.	Mass	Int.	Mass	Int.	Mass	Int.	Mass	Int.	Mass	Int.	Mass	Int.
25	220	28	469	42	490	43	310	57	331	58	128	70	540	71	999	77	120
78	60	91	180	103	160	115	129	116	58	129	81	131	91	140	74	144	36
155	71	158	47	172	295	173	69	174	142	175	41	190	34	191	10	202	21
215	12	218	134	219	31	232	68	233	16	246	147	247	220				

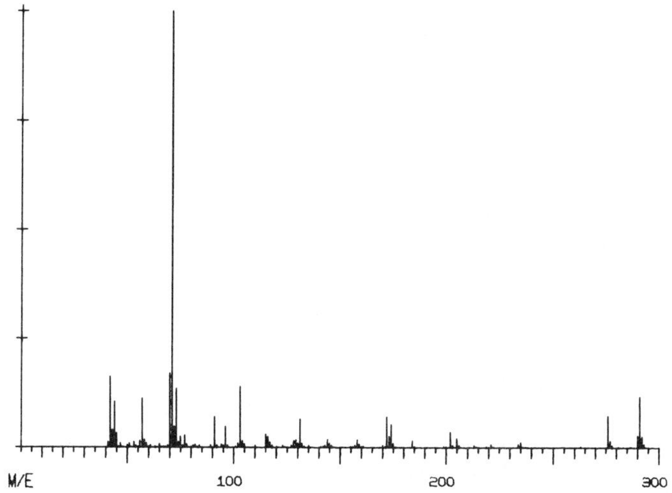
M/E
100
200
300

MEPERIDINIC ACID TMS ESTER

$C_{16}H_{25}NO_2Si$ MW = 291

Base Peak = 71 38 Peaks

Mass	Int.	Mass	Int.	Mass	Int.	Mass	Int.	Mass	Int.	Mass	Int.	Mass	Int.	Mass	Int.	Mass	Int.
42	162	44	104	57	112	58	17	70	170	71	999	77	27	78	7	91	69
103	140	115	31	116	25	129	18	131	64	132	9	144	18	158	18	159	7
172	69	173	26	174	53	184	14	199	2	200	2	202	34	205	20	219	3
221	7	234	8	235	12	246	1	247	1	262	1	263	2	276	72	277	16
290	28	291	118														

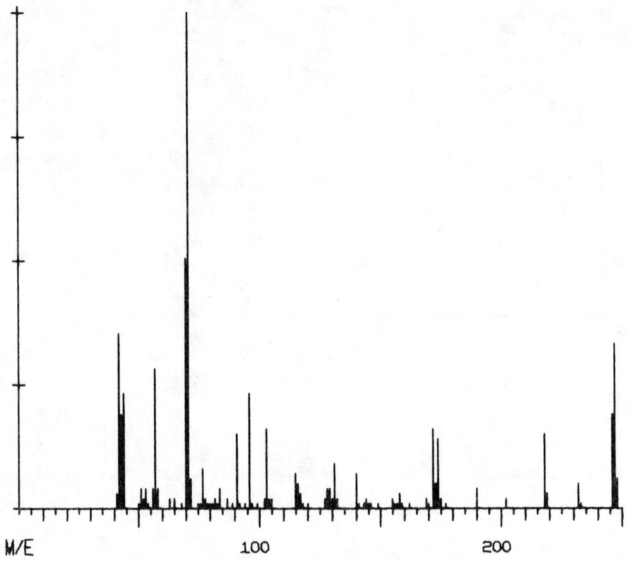

M/E
100
200

MEPERIDINE

$C_{15}H_{21}NO_2$ MW 247

Base Peak = 71 27 Peaks

Mass	Int.	Mass	Int.	Mass	Int.	Mass	Int.	Mass	Int.	Mass	Int.	Mass	Int.	Mass	Int.	Mass	Int.
42	353	44	232	51	40	57	282	70	505	71	999	77	80	84	40	96	232
103	161	115	70	116	50	128	40	131	90	132	20	140	70	155	20	158	30
172	161	173	50	174	141	175	20	190	40	202	20	218	151	219	30	232	50
233	10	246	191	247	333												

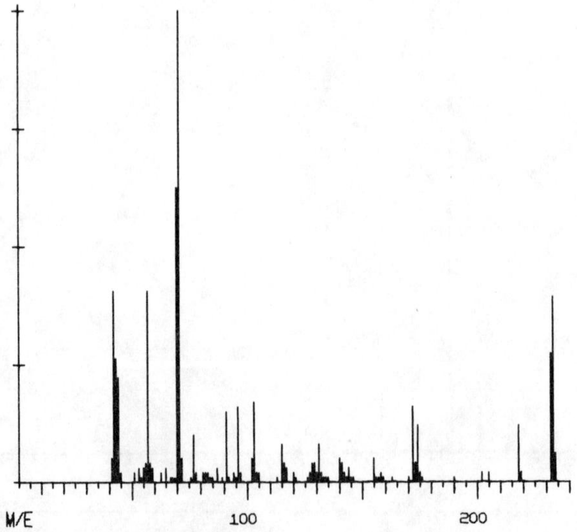

M/E
100
200

MEPERIDINE ACID METHYL ESTER

$C_{14}H_{19}NO_2$ MW 233

Base Peak = 71 29 Peaks

Mass	Int.	Mass	Int.	Mass	Int.	Mass	Int.	Mass	Int.	Mass	Int.	Mass	Int.	Mass	Int.	Mass	Int.
42	404	43	262	56	40	57	404	70	626	71	999	77	101	87	30	96	161
103	171	115	80	116	40	128	40	131	80	140	50	141	40	155	50	158	20
172	161	173	40	174	121	175	20	190	10	202	20	205	20	218	121	219	20
232	272	233	393														

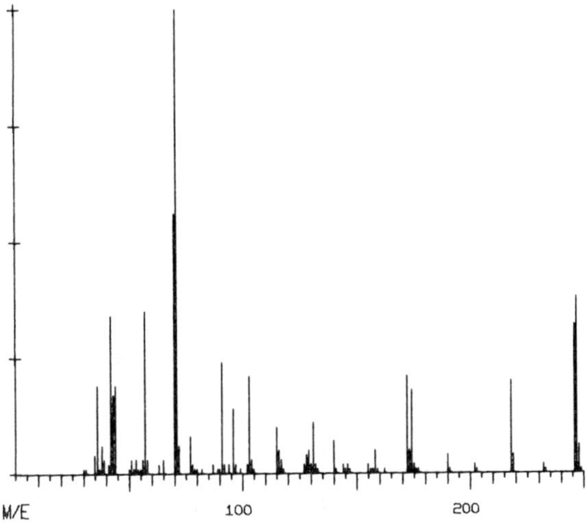
M/E
100
200

MEPERIDINE

$C_{15}H_{21}NO_2$ MW = 247

Base Peak = 71 34 Peaks

Mass	Int.	Mass	Int.	Mass	Int.	Mass	Int.	Mass	Int.	Mass	Int.	Mass	Int.	Mass	Int.	Mass	Int.
30	10	31	10	36	190	42	340	51	30	57	350	70	560	71	999	77	80
78	20	91	240	103	210	115	100	116	50	129	50	131	110	132	20	140	70
146	20	158	50	172	210	173	50	174	180	175	20	190	40	191	10	202	20
203	10	218	200	219	40	232	20	233	10	246	320	247	380				

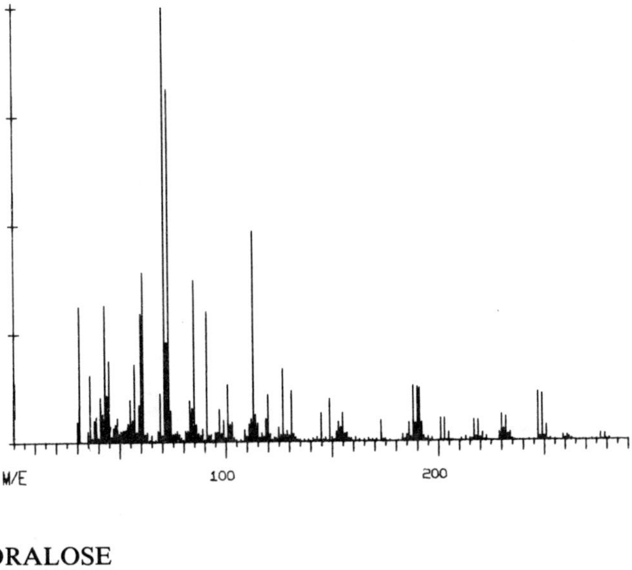
M/E
100
200

ALPHA CHLORALOSE

$C_8H_{11}Cl_3O_6$ MW = 308

Base Peak = 71 38 Peaks

Mass	Int.	Mass	Int.	Mass	Int.	Mass	Int.	Mass	Int.	Mass	Int.	Mass	Int.	Mass	Int.	Mass	Int.
30	47	31	312	43	314	45	187	60	296	61	390	71	999	73	812	83	95
85	372	91	301	101	132	113	485	114	62	127	167	131	117	132	18	145	64
149	97	155	64	161	9	173	47	183	15	186	43	188	127	190	126	203	52
205	20	217	48	219	47	230	60	232	56	247	113	249	107	259	13	261	13
277	15	279	15														

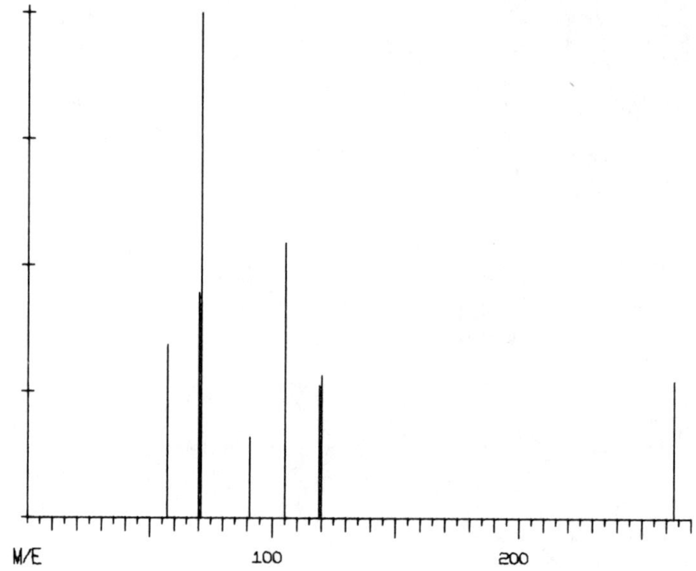
M/E
100
200

HYDROXYPETHIDINE

$C_{15}H_{21}NO_3$ MW = 263

Base Peak = 71 8 Peaks

Mass	Int.	Mass	Int.	Mass	Int.	Mass	Int.	Mass	Int.	Mass	Int.	Mass	Int.	Mass	Int.
57	343	70	444	71	999	91	161	105	545	119	262	120	282	263	272

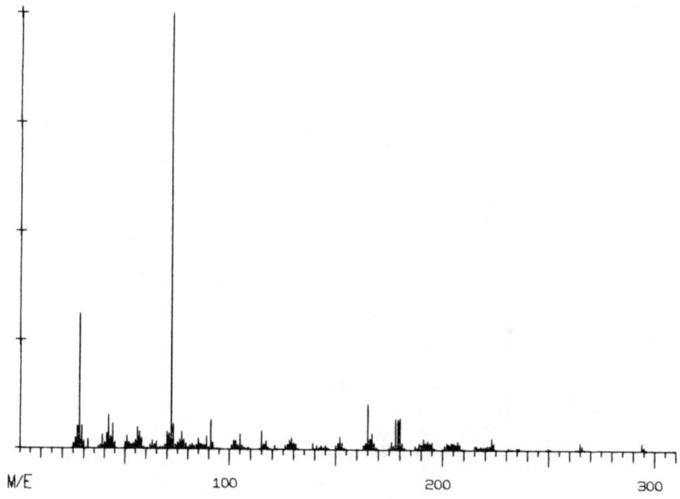
M/E
100
200
300

METHADONE

$C_{21}H_{27}NO$ MW = 309

Base Peak = 72 40 Peaks

Mass	Int.	Mass	Int.	Mass	Int.	Mass	Int.	Mass	Int.	Mass	Int.	Mass	Int.	Mass	Int.	Mass	Int.
28	309	29	53	42	78	44	57	56	50	57	40	72	999	73	58	77	40
89	29	91	67	102	19	105	34	115	42	128	22	129	27	139	14	143	11
151	14	152	29	165	104	167	37	178	70	180	72	191	24	193	20	205	16
207	19	223	27	224	14	231	6	235	4	249	3	250	4	265	18	266	9
294	17	295	8	309	2	310	1										

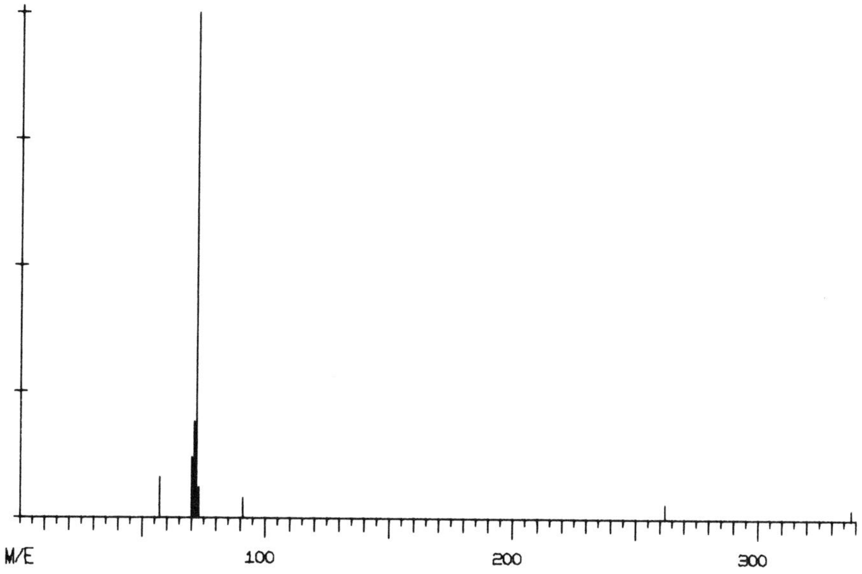
M/E
100
200
300

ACETYLMETHADOL

$C_{23}H_{31}NO_2$ MW = 353

Base Peak = 72 6 Peaks

Mass	Int.	Mass	Int.	Mass	Int.	Mass	Int.	Mass	Int.	Mass	Int.
57	80	71	191	72	999	91	40	262	30	338	20

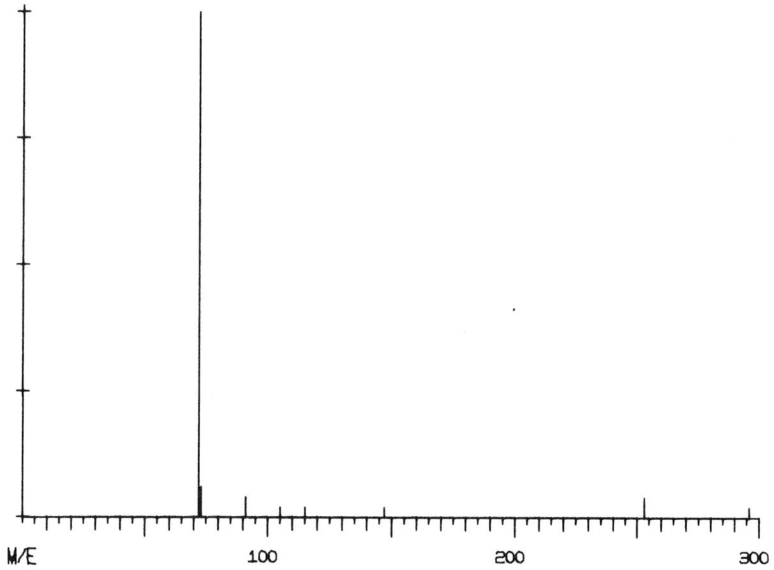
M/E
100
200
300

METHADOL

$C_{21}H_{29}NO$ MW = 311

Base Peak = 72 8 Peaks

Mass	Int.	Mass	Int.	Mass	Int.	Mass	Int.	Mass	Int.	Mass	Int.	Mass	Int.	Mass	Int.
72	999	73	60	91	40	105	20	115	20	147	20	253	40	296	20

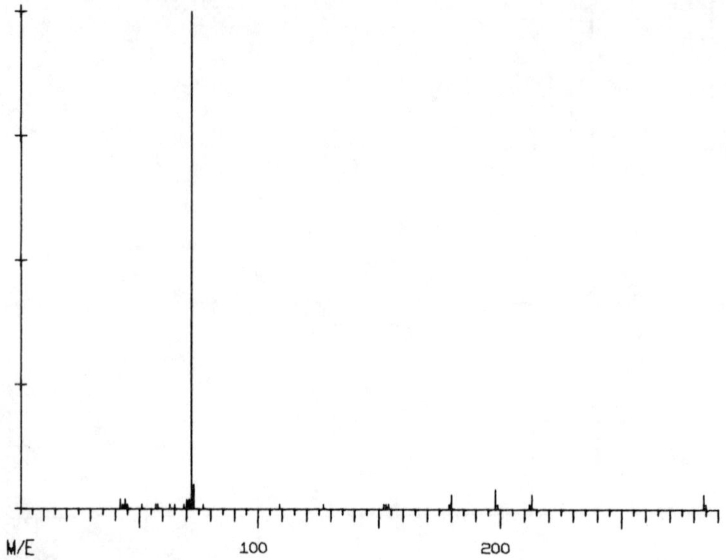
M/E
100
200

PROMETHAZINE

$C_{17}H_{20}N_2S$ MW = 284

Base Peak = 72 19 Peaks

Mass	Int.	Mass	Int.	Mass	Int.	Mass	Int.	Mass	Int.	Mass	Int.	Mass	Int.	Mass	Int.	Mass	Int.
42	20	44	20	51	10	57	10	72	999	73	50	77	10	109	10	127	10
152	10	153	10	179	10	180	30	198	40	199	10	212	10	213	30	284	30
285	10																

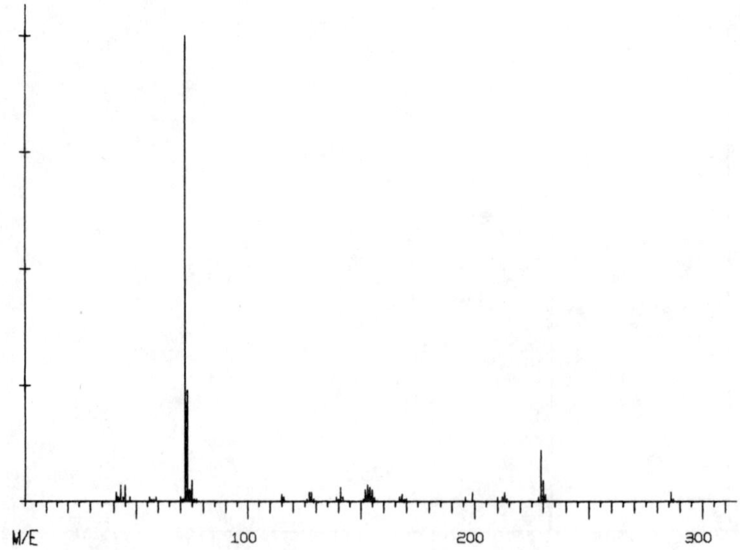
M/E
100
200
300

PRONETHALOL TMS ETHER

$C_{18}H_{27}NOSi$ MW = 301

Base Peak = 72 29 Peaks

Mass	Int.	Mass	Int.	Mass	Int.	Mass	Int.	Mass	Int.	Mass	Int.	Mass	Int.	Mass	Int.	Mass	Int.
43	35	45	35	56	10	59	10	72	999	73	240	76	5	77	5	115	15
116	10	127	20	128	20	139	10	141	30	153	35	154	30	167	10	168	15
196	10	199	20	210	10	213	20	228	10	229	110	230	45	231	15	286	20
287	5	301	1														

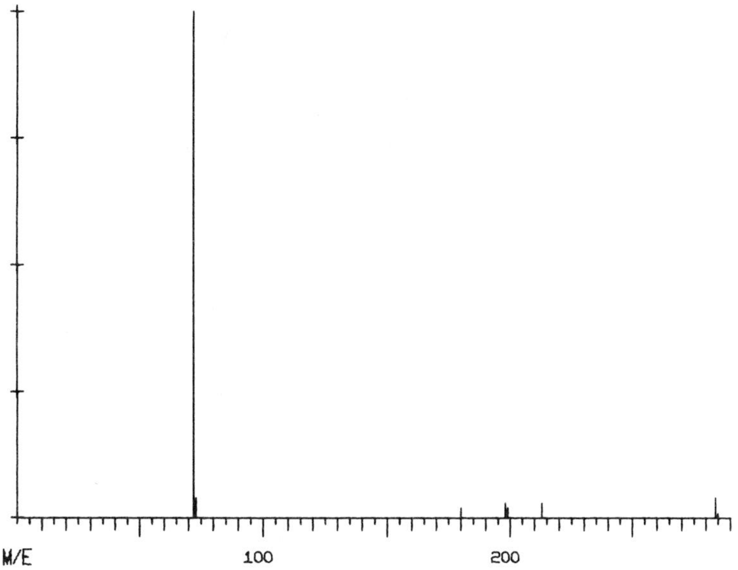
M/E
100
200

PROMETHAZINE

$C_{17}H_{20}N_2S$ MW = 284

Base Peak = 72 8 Peaks

Mass	Int.	Mass	Int.	Mass	Int.	Mass	Int.	Mass	Int.	Mass	Int.	Mass	Int.	Mass	Int.
72	999	73	39	180	19	198	29	199	19	213	29	284	39	285	9

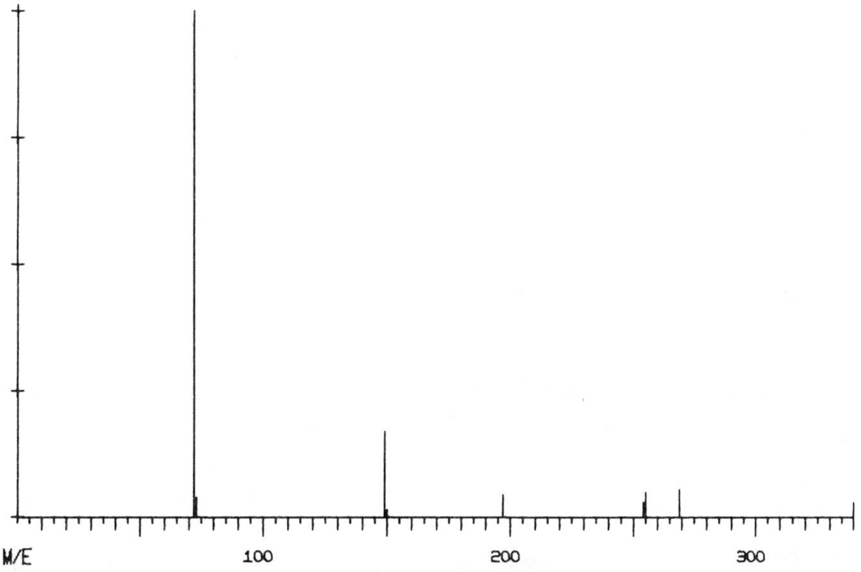
M/E
100
200
300

PROPIOMAZINE

$C_{20}H_{24}N_2OS$ MW = 340

Base Peak = 72 9 Peaks

Mass	Int.	Mass	Int.	Mass	Int.	Mass	Int.	Mass	Int.	Mass	Int.	Mass	Int.	Mass	Int.	Mass	Int.
72	999	73	39	149	169	150	14	197	44	254	29	255	49	269	54	340	29

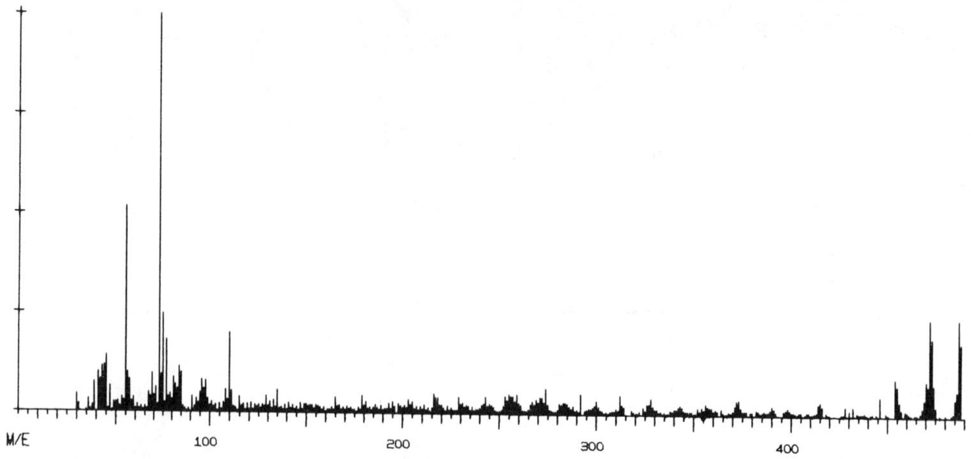
M/E
100
200
300
400

α-HYDROXY NALTREXONE-2 TMS

$C_{26}H_{41}NO_4SI_2$ MW = 487

Base Peak = 73 68 Peaks

Mass	Int.	Mass	Int.	Mass	Int.	Mass	Int.	Mass	Int.	Mass	Int.	Mass	Int.	Mass	Int.	Mass	Int.
30	44	31	21	44	117	45	143	55	518	56	101	73	999	75	247	77	182
84	115	96	83	98	79	108	57	110	199	129	42	131	27	133	33	135	57
147	26	149	23	165	40	167	19	179	45	181	30	193	26	195	29	203	36
205	29	216	49	229	43	231	28	243	42	253	45	255	49	259	49	271	42
272	42	274	66	288	23	292	52	300	34	312	50	326	27	327	27	328	43
329	22	342	20	343	24	356	30	357	23	372	38	373	39	390	25	391	17
398	19	399	14	414	30	415	34	428	26	432	26	446	49	449	11	454	94
455	78	472	246	473	197	487	246	488	186								

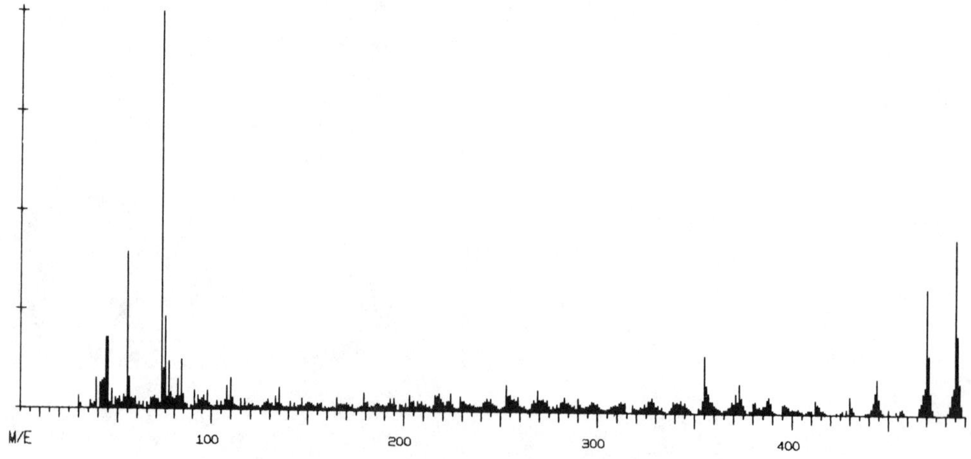
M/E
100
200
300
400

NALTREXONE-2 TMS

$C_{26}H_{39}NO_4SI_2$ MW = 485

Base Peak = 73 68 Peaks

Mass	Int.	Mass	Int.	Mass	Int.	Mass	Int.	Mass	Int.	Mass	Int.	Mass	Int.	Mass	Int.	Mass	Int.
30	32	31	13	44	181	45	180	55	396	56	80	73	999	75	232	77	121
84	125	91	47	98	47	108	61	110	81	128	19	129	27	133	34	135	57
147	30	151	20	165	32	170	21	179	46	181	23	193	31	195	33	203	40
205	26	218	45	224	44	230	25	243	32	253	67	255	42	259	38	269	56
272	32	283	37	290	35	297	28	311	24	312	28	326	29	327	30	328	38
339	30	343	32	355	146	356	71	357	51	371	49	373	75	387	35	388	42
398	18	400	13	412	35	413	26	430	44	431	21	443	56	444	91	466	20
467	29	470	318	471	149	485	443	486	201								

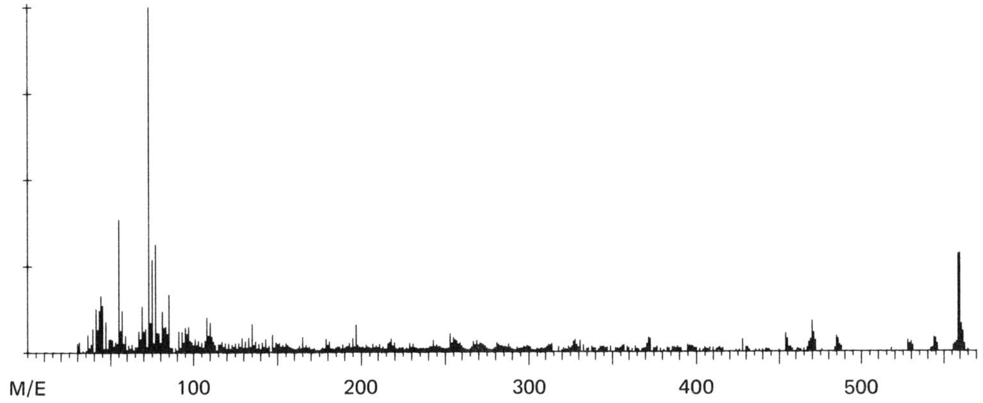
M/E
100
200
300
400
500

α-HYDROXY NALTREXONE-3 TMS

$C_{29}H_{49}NO_4SI_3$ MW = 559

Base Peak = 73 75 Peaks

Mass	Int.	Mass	Int.	Mass	Int.	Mass	Int.	Mass	Int.	Mass	Int.	Mass	Int.	Mass	Int.	Mass	Int.
30	25	31	31	44	163	45	134	55	386	57	121	73	999	75	268	77	313
85	167	95	70	97	72	108	99	110	84	129	41	131	29	133	41	135	81
147	50	149	28	165	42	167	21	179	37	181	29	195	39	197	77	207	22
211	22	217	27	218	37	231	23	243	32	253	52	255	40	267	30	269	33
272	23	281	25	288	22	299	17	312	19	313	21	314	23	327	29	328	36
331	32	344	15	349	15	356	17	357	20	372	39	373	37	395	19	396	17
398	15	408	13	413	10	414	13	428	36	430	13	442	7	443	6	454	53
455	37	470	87	471	55	485	46	486	38	518	8	528	32	530	28	544	39
545	38	559	282	560	286												

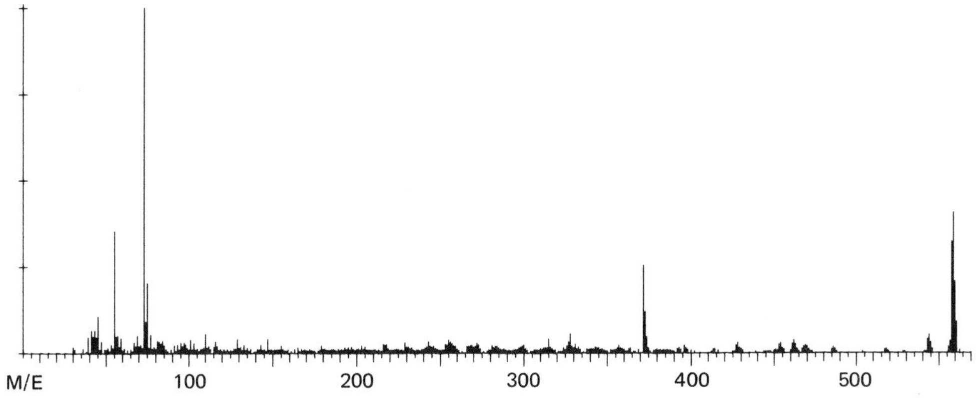
M/E
100
200
300
400
500

NALTREXONE MTB. 1-3 TMS

$C_{29}H_{49}NO_4SI_3$ MW = 559

Base Peak = 73 87 Peaks

Mass	Int.	Mass	Int.	Mass	Int.	Mass	Int.	Mass	Int.	Mass	Int.	Mass	Int.	Mass	Int.	Mass	Int.
30	18	31	9	41	66	45	105	55	352	57	49	73	999	75	203	77	53
84	35	95	30	101	38	110	55	116	33	129	39	131	17	133	23	143	23
147	40	155	20	165	14	169	11	179	20	181	13	193	15	197	17	203	19
205	17	218	25	229	30	242	18	243	32	255	37	256	33	268	22	269	22
272	27	273	22	298	18	299	20	300	22	312	14	315	40	327	32	328	55
331	25	343	18	344	15	356	16	357	22	372	254	373	120	393	14	396	23
398	11	400	8	413	11	414	13	427	23	428	29	451	10	453	24	454	31
462	38	468	20	469	22	486	18	487	12	504	5	517	10	518	13	528	5
529	6	543	40	544	52	558	322	559	407								

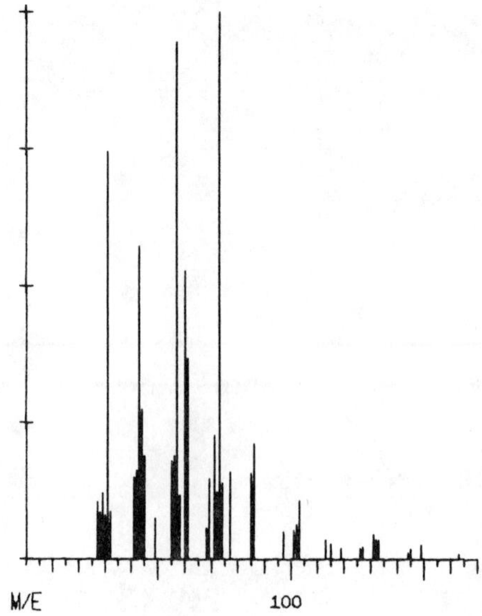
M/E
100

SUCROSE

$C_{12}H_{22}O_{11}$ MW = 342

Base Peak = 73 20 Peaks

Mass	Int.	Mass	Int.	Mass	Int.	Mass	Int.	Mass	Int.	Mass	Int.	Mass	Int.	Mass	Int.	Mass	Int.
29	121	31	746	43	573	44	276	57	946	60	528	71	227	73	999	77	159
85	212	102	62	103	107	113	35	115	28	127	22	131	44	132	35	133	36
149	24	163	10														

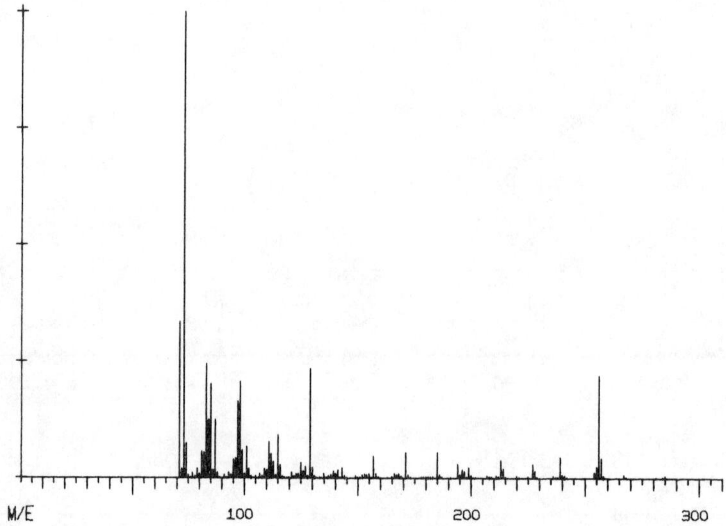
M/E
100
200
300

PALMITIC ACID

$C_{16}H_{32}O_2$ MW = 256

Base Peak = 73 36 Peaks

Mass	Int.	Mass	Int.	Mass	Int.	Mass	Int.	Mass	Int.	Mass	Int.	Mass	Int.	Mass	Int.	Mass	Int.
71	335	73	999	83	246	85	202	97	166	98	208	111	77	115	93	125	33
129	234	141	18	143	23	155	10	157	48	166	9	171	54	185	55	186	8
194	30	199	22	213	37	214	21	227	30	228	15	239	44	240	8	256	220
257	42	258	5	267	7	284	3	285	4	286	1	287	1	300	1	302	1

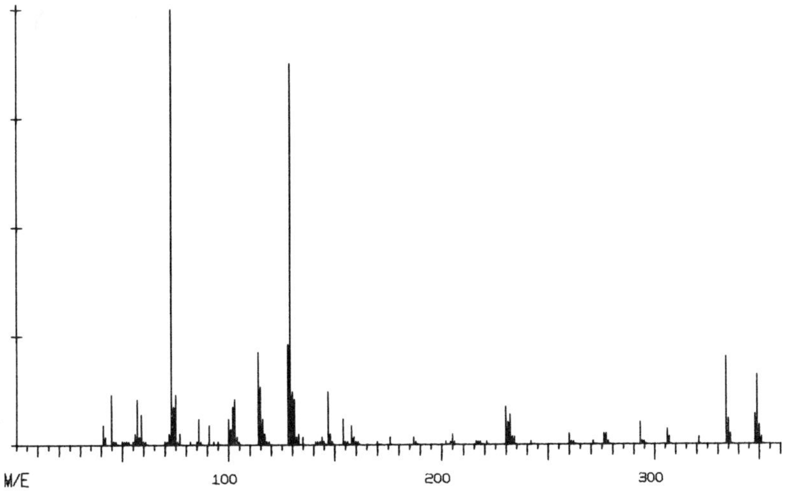
M/E
100
200
300

NOR-MEPERIDINIC ACID TMS ESTER N-TMS DERIVATIVE

$C_{18}H_{31}NO_2Si_2$ MW = 349

Base Peak = 73 42 Peaks

Mass	Int.	Mass	Int.	Mass	Int.	Mass	Int.	Mass	Int.	Mass	Int.	Mass	Int.	Mass	Int.	Mass	Int.
41	44	45	115	57	106	59	70	73	999	75	115	77	26	86	61	102	88
103	106	114	212	115	132	128	230	129	876	132	17	133	26	147	123	154	61
160	8	161	8	176	17	187	17	188	8	202	8	205	26	216	8	217	8
230	88	232	70	260	26	261	8	276	26	277	26	293	53	294	8	306	35
307	17	321	17	334	203	335	61	348	70	349	159						

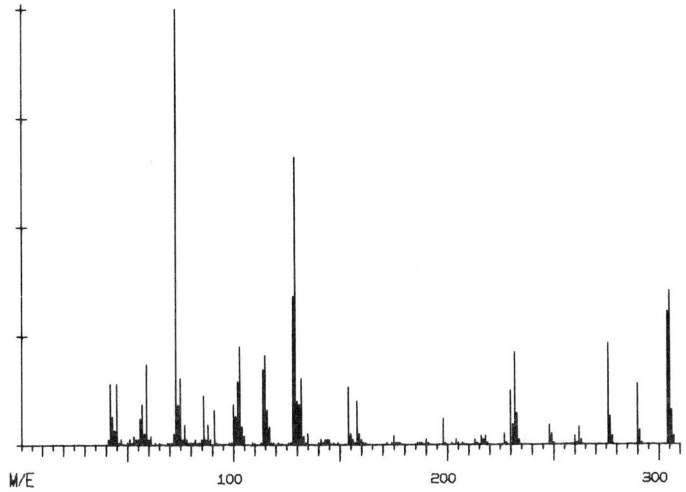
M/E
100
200
300

NOR-MEPERIDINIC ACID ETHYL ESTER N-TMS DERIVATIVE

$C_{17}H_{27}NO_2Si$ MW = 305

Base Peak = 73 40 Peaks

Mass	Int.	Mass	Int.	Mass	Int.	Mass	Int.	Mass	Int.	Mass	Int.	Mass	Int.	Mass	Int.	Mass	Int.
42	139	45	139	57	93	59	186	73	999	75	153	77	46	86	113	102	146
103	226	114	173	115	206	128	339	129	659	132	153	135	26	154	133	158	99
160	13	161	6	174	6	175	19	190	13	198	59	204	13	213	13	216	19
227	26	230	126	232	213	248	46	249	26	260	19	262	39	276	233	277	66
290	139	291	33	304	306	305	353										

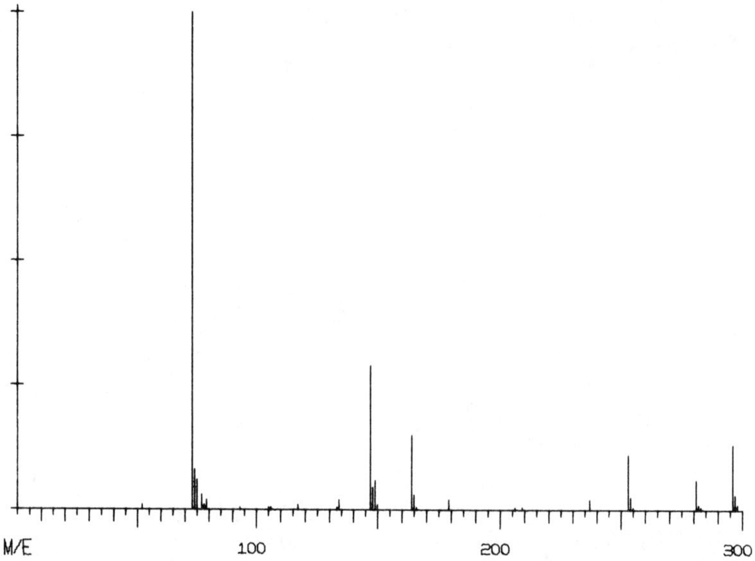
M/E
100
200
300

O-HYDROXYPHENYLACETIC ACID TRIMETHYLSILYL ETHER

$C_{14}H_{24}O_3SI_2$ MW = 296

Base Peak = 73 24 Peaks

Mass	Int.	Mass	Int.	Mass	Int.	Mass	Int.	Mass	Int.	Mass	Int.	Mass	Int.	Mass	Int.	Mass	Int.
52	9	73	999	74	79	77	29	79	19	93	4	105	4	117	9	133	4
134	19	147	289	149	59	164	149	165	29	179	19	206	4	209	4	237	19
253	109	254	24	281	59	282	9	296	129	297	29						

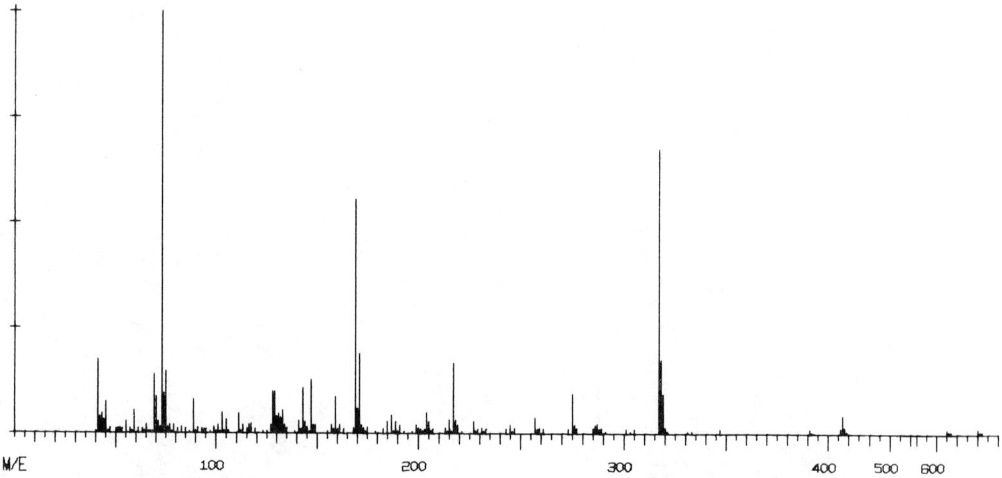
M/E
100
200
300
400
500
600

CLOFIBRATE GLUCURONIDE METHYL ESTER TMS ETHER

$C_{26}H_{45}O_9Si_3Cl$ MW = 620

Base Peak = 73 54 Peaks

Mass	Int.	Mass	Int.	Mass	Int.	Mass	Int.	Mass	Int.	Mass	Int.	Mass	Int.	Mass	Int.	Mass	Int.
41	175	45	74	55	27	59	54	73	999	75	148	77	20	89	81	101	20
103	50	105	33	111	47	128	101	129	101	133	54	143	108	147	128	159	87
169	554	171	189	185	27	187	43	189	27	191	20	204	50	215	33	217	168
218	33	231	13	233	10	245	20	257	37	258	10	259	13	275	94	276	20
286	20	287	23	301	10	305	10	317	674	318	175	331	6	333	6	347	10
391	10	392	3	407	43	408	16	605	10	606	6	620	13	621	6	622	6

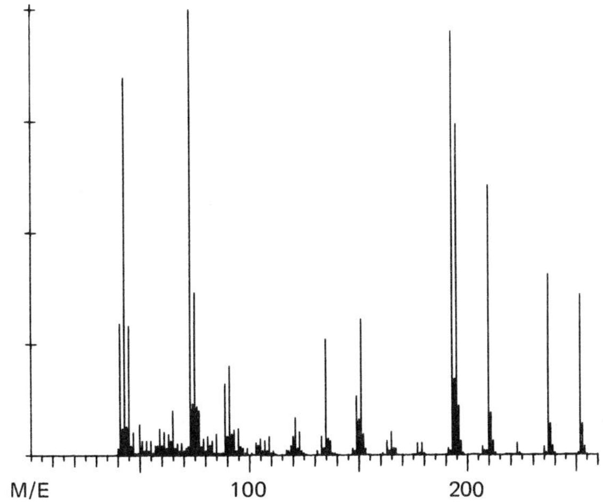
M/E
100
200

PROPYLPARABEN TMS ETHER

$C_{13}H_{20}O_3SI$ MW = 252

Base Peak = 73 42 Peaks

Mass	Int.	Mass	Int.	Mass	Int.	Mass	Int.	Mass	Int.	Mass	Int.	Mass	Int.	Mass	Int.	Mass	Int.
41	295	43	847	50	71	59	61	73	999	75	366	76	109	89	161	91	199
95	61	105	38	109	42	121	85	123	52	133	42	135	261	149	133	151	304
163	33	165	52	177	28	179	28	193	951	195	742	210	604	211	95	223	28
224	4	237	404	238	71	252	361	253	71								

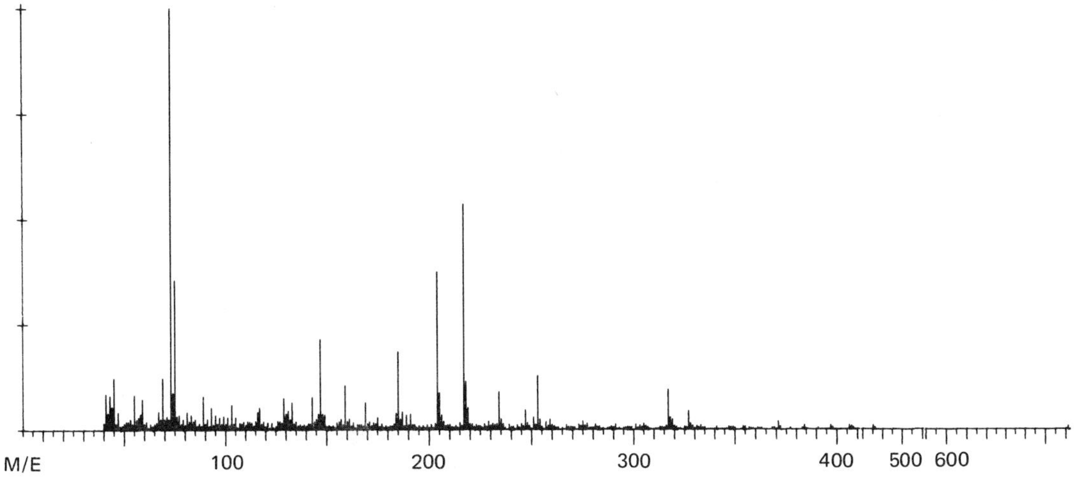
M/E
100
200
300
400
500
600

HYDROXYAMOBARBITAL GLUCURONIDE ME TMS DERIVATIVE (PEAK 2)

$C_{29}H_{56}N_2O_{10}Si_3$ MW = 676

Base Peak = 73 57 Peaks

Mass	Int.	Mass	Int.	Mass	Int.	Mass	Int.	Mass	Int.	Mass	Int.	Mass	Int.	Mass	Int.	Mass	Int.
41	85	45	122	55	82	59	72	73	999	75	355	81	42	89	80	93	53
103	61	116	42	117	53	129	74	131	45	133	66	143	77	147	216	159	106
161	26	169	66	185	184	187	42	189	34	191	37	204	374	205	88	217	534
218	114	234	90	235	26	247	48	253	128	259	24	260	13	275	21	277	16
289	8	291	13	301	13	305	16	317	96	327	45	328	16	329	10	347	8
348	8	357	5	361	5	371	21	372	8	384	10	397	10	406	10	407	10
385	10	486	5	661	8												

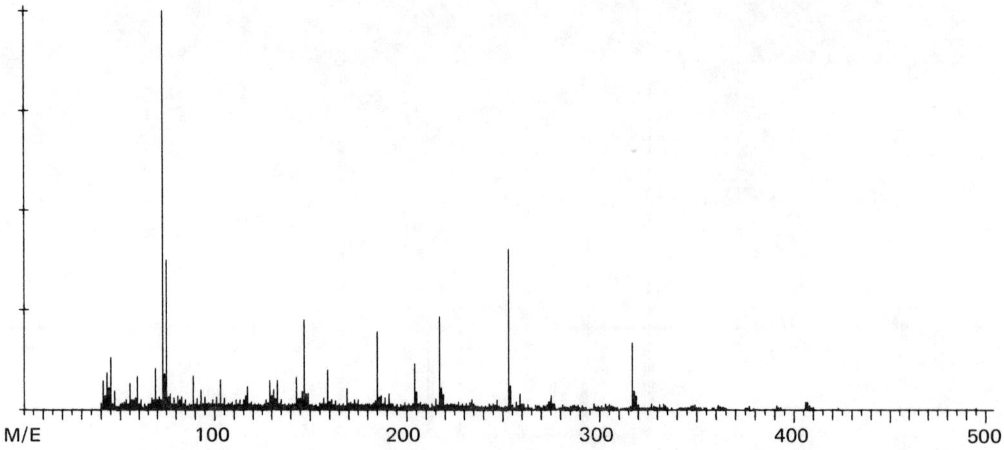
M/E
100
200
300
400
500

HYDROXYAMOBARBITAL GLUCORONIDE ME TMS DERIVATIVE (PEAK 1)

$C_{29}H_{56}N_2O_{10}Si_3$ MW = 676
Base Peak = 73 58 Peaks

Mass	Int.	Mass	Int.	Mass	Int.	Mass	Int.	Mass	Int.	Mass	Int.	Mass	Int.	Mass	Int.	Mass	Int.
43	92	45	131	55	64	59	82	73	999	75	374	77	40	89	86	93	50
103	74	116	34	117	58	129	72	131	50	133	72	143	80	147	225	159	100
161	24	169	52	185	195	187	36	189	30	191	40	204	116	205	46	217	233
218	56	233	18	234	24	253	402	254	60	259	40	260	14	274	20	275	34
287	12	297	16	303	16	305	14	317	167	318	48	331	12	333	16	348	10
349	12	361	12	362	8	375	4	377	8	391	10	392	6	406	20	407	20
420	2	423	4	495	6	661	6										

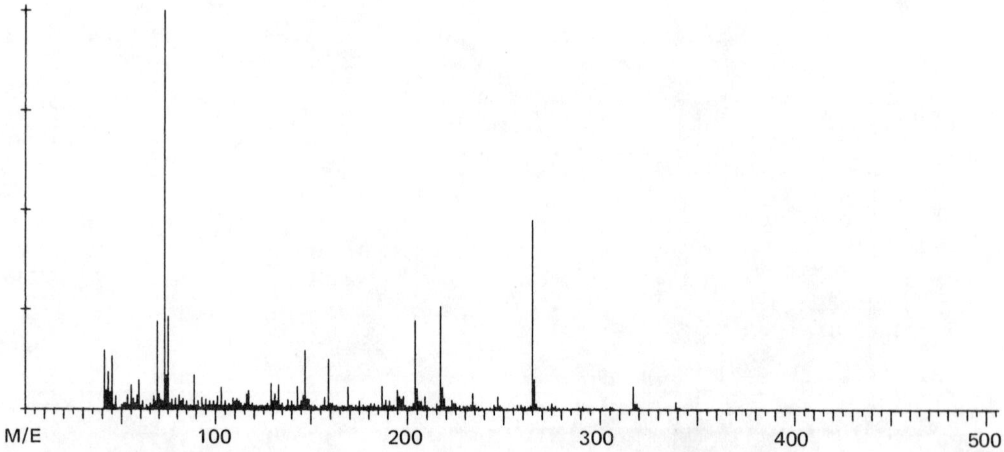
M/E
100
200
300
400
500

HYDROXYSECOBARBITAL GLUCURONIDE 1,3-DI-ME DERIV ME ESTER TMS ETHER

$C_{30}H_{56}N_2O_{10}Si_3$ MW = 688
Base Peak = 73 57 Peaks

Mass	Int.	Mass	Int.	Mass	Int.	Mass	Int.	Mass	Int.	Mass	Int.	Mass	Int.	Mass	Int.	Mass	Int.
41	147	45	133	55	61	59	73	73	999	75	233	81	35	89	85	101	33
103	54	116	38	117	47	129	64	131	38	133	61	143	57	147	147	159	126
160	16	169	54	175	21	187	57	195	47	196	33	204	223	205	52	217	257
218	54	234	40	235	11	245	9	247	33	265	475	266	76	273	7	275	14
289	4	291	7	301	4	305	7	317	57	318	16	339	19	340	4	342	2
347	2	359	2	361	2	375	2	383	2	391	2	393	2	406	4	407	4
498	2	673	2	674	2												

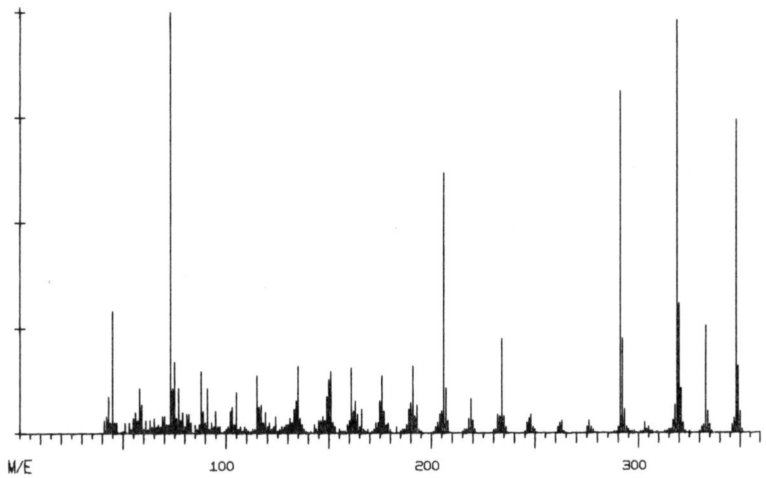
M/E
100
200
300

HYDROXYPHENOBARBITAL 1,3-DIMETHYL DERIVATIVE TMS ETHER

$C_{17}H_{24}N_2O_4Si$ MW = 348

Base Peak = 73 46 Peaks

Mass	Int.	Mass	Int.	Mass	Int.	Mass	Int.	Mass	Int.	Mass	Int.	Mass	Int.	Mass	Int.	Mass	Int.
43	88	45	289	58	107	59	68	73	999	75	171	77	107	88	147	91	107
103	63	105	98	115	137	119	49	124	39	134	78	135	161	150	127	151	147
161	156	163	78	175	78	176	137	190	73	191	161	206	617	207	107	218	34
219	83	232	44	234	225	247	34	248	44	262	24	263	29	275	14	276	29
291	813	292	225	303	24	305	14	319	979	320	308	333	254	334	53	348	745
349	161																

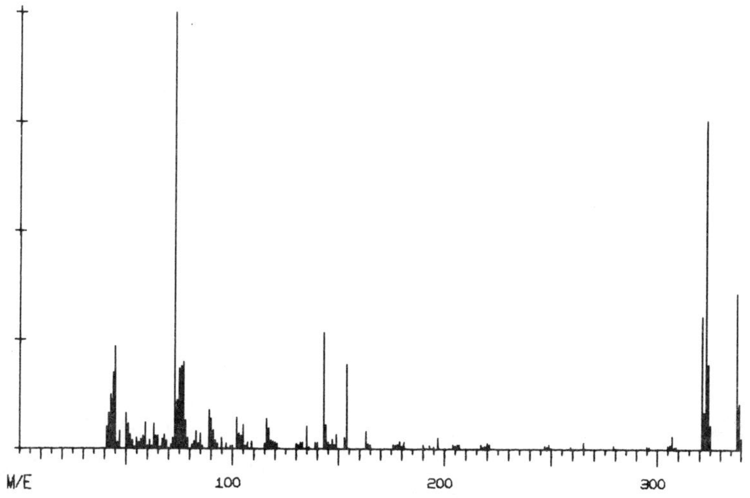
M/E
100
200
300

METHAQUALONE MTB 4 TMS

$C_{19}H_{22}N_2O_2SI$ MW = 338

Base Peak = 73 41 Peaks

Mass	Int.	Mass	Int.	Mass	Int.	Mass	Int.	Mass	Int.	Mass	Int.	Mass	Int.	Mass	Int.	Mass	Int.
44	176	45	235	50	83	59	61	73	999	75	185	76	190	77	199	90	69
102	72	105	55	116	70	118	21	119	18	143	268	144	54	149	33	154	194
163	40	164	12	179	18	181	14	190	9	197	25	204	9	206	11	220	16
221	11	247	7	249	10	259	6	265	16	279	8	295	6	296	5	306	9
307	30	321	304	323	754	338	358	339	106								

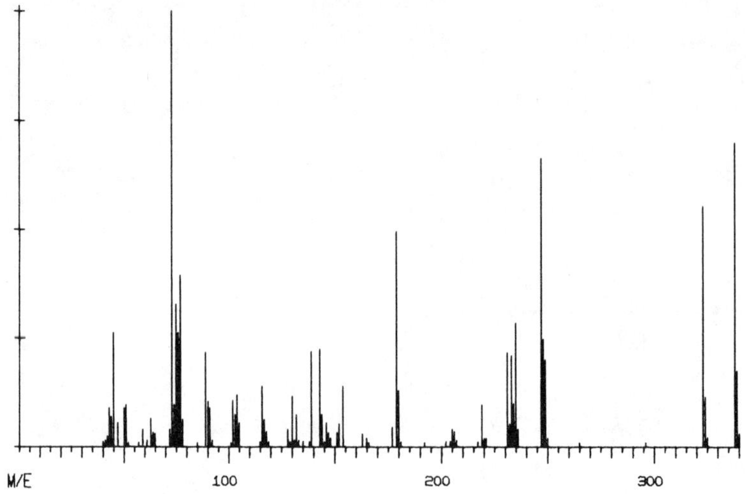
M/E
100
200
300

METHAQUALONE MTB 1 TMS

$C_{19}H_{22}N_2O_2SI$ MW = 338

Base Peak = 73 37 Peaks

Mass	Int.	Mass	Int.	Mass	Int.	Mass	Int.	Mass	Int.	Mass	Int.	Mass	Int.	Mass	Int.	Mass	Int.
43	91	45	263	50	91	51	97	73	999	75	328	76	263	77	396	90	104
102	107	104	120	116	140	128	39	130	117	139	221	143	224	146	55	154	140
163	29	165	19	179	494	180	130	192	10	205	39	206	36	219	97	220	19
231	218	235	286	247	662	248	247	265	10	296	10	323	552	324	114	338	698
339	175																

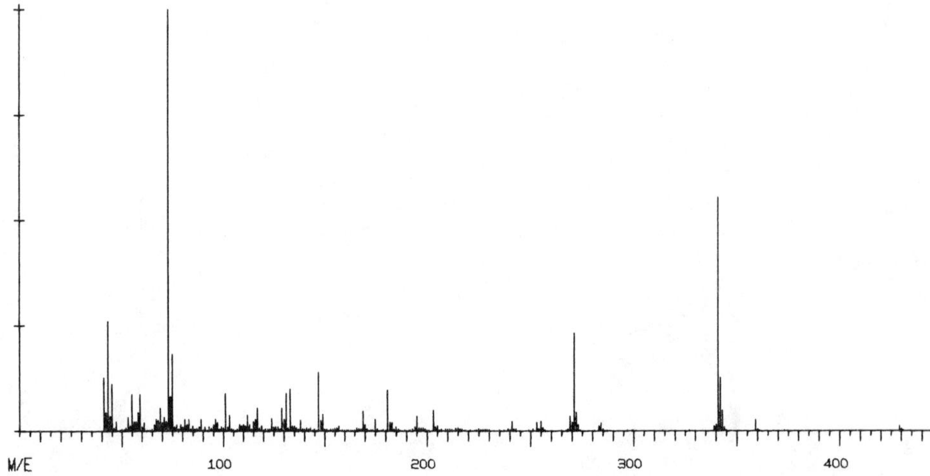
M/E
100
200
300
400

DIHYDROXYSECOBARBITAL DI-TMS ETHER 1,3-DIMETHYL DERIVATIVE

$C_{20}H_{40}N_2O_5Si_2$ MW = 444

Base Peak = 73 51 Peaks

Mass	Int.	Mass	Int.	Mass	Int.	Mass	Int.	Mass	Int.	Mass	Int.	Mass	Int.	Mass	Int.	Mass	Int.
41	127	43	261	55	88	59	88	73	999	75	183	81	27	83	27	101	91
103	38	112	38	117	55	129	55	131	91	133	99	138	24	147	141	149	41
169	47	170	16	175	27	181	97	194	11	195	36	203	49	205	13	216	5
217	5	239	5	241	22	253	19	255	22	269	36	271	233	272	44	284	19
286	2	287	2	300	2	314	2	327	2	340	16	341	554	342	127	343	49
359	27	360	5	373	2	429	13	430	5								

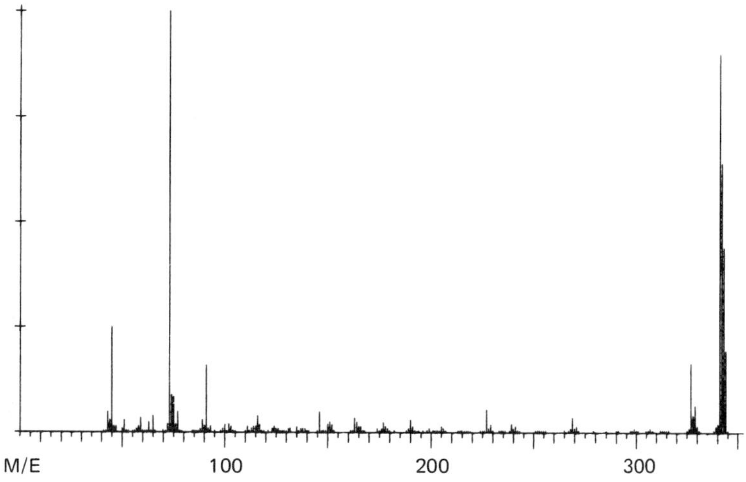
M/E
100
200
300

7-CHLORO-1,3-DIHYDRO-5-PHENYL-2H-1,4-DIBENZODIAZEPIN-2-ONE TMS DERIVATIVE

$C_{18}H_{19}N_2OSiCl$ MW = 342

Base Peak = 73 46 Peaks

Mass	Int.	Mass	Int.	Mass	Int.	Mass	Int.	Mass	Int.	Mass	Int.	Mass	Int.	Mass	Int.	Mass	Int.
43	50	45	250	51	30	59	35	73	999	74	90	77	50	89	30	91	160
100	20	116	40	117	20	123	10	124	15	132	10	135	15	146	50	151	25
163	35	164	25	177	25	178	15	190	30	191	15	205	15	206	10	227	55
229	20	239	20	241	15	251	5	252	5	269	35	271	15	272	5	279	5
290	5	299	10	300	5	307	10	326	10	327	165	329	65	341	900	342	640
343	440																

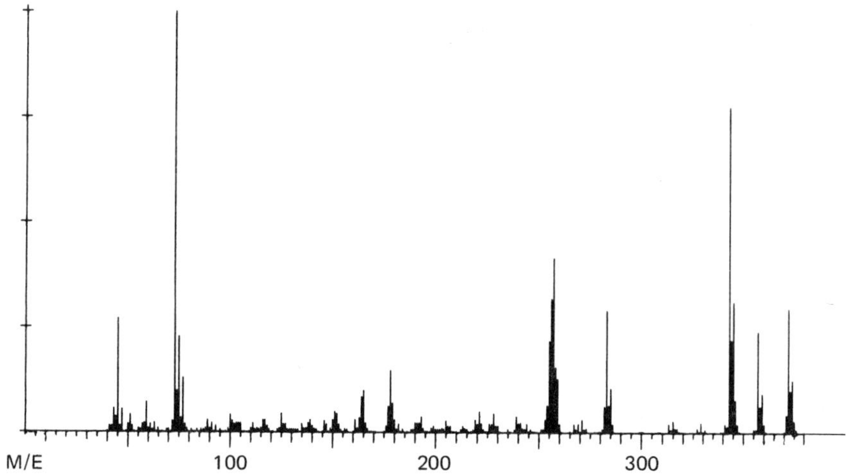
M/E
100
200
300

7-CHLORO-1,3-DIHYDRO-3-HYDROXY-1-ME-5-PH-2H-1,4-BENZODIAZEPIN-2-ONE TMS ETHER

$C_{19}H_{21}N_2O_2SiCl$ MW = 322

Base Peak = 73 50 Peaks

Mass	Int.	Mass	Int.	Mass	Int.	Mass	Int.	Mass	Int.	Mass	Int.	Mass	Int.	Mass	Int.	Mass	Int.
43	58	45	271	51	42	59	73	73	999	75	228	76	34	77	131	100	42
101	27	116	31	117	31	125	46	126	19	138	23	139	31	151	50	152	46
164	85	165	100	178	147	179	69	190	23	193	38	205	27	206	15	221	50
228	46	239	38	240	23	256	317	257	414	258	155	259	127	283	290	285	104
286	19	299	3	300	3	313	19	315	27	316	11	329	23	341	19	343	774
345	310	357	240	359	93	372	294	374	124								

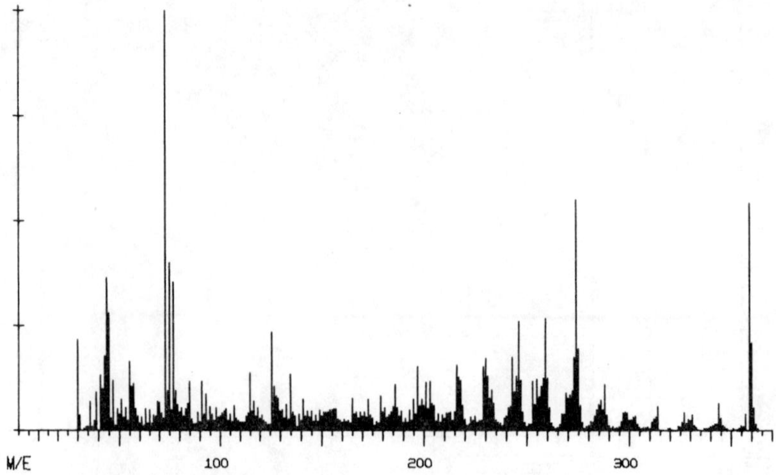
M/E
100
200
300

NOROXYMORPHONE-1 TMS

$C_{19}H_{25}NO_4SI$ MW = 359

Base Peak = 73 50 Peaks

Mass	Int.	Mass	Int.	Mass	Int.	Mass	Int.	Mass	Int.	Mass	Int.	Mass	Int.	Mass	Int.	Mass	Int.
30	218	31	37	44	365	45	280	55	164	57	112	73	999	75	400	77	354
85	118	91	117	93	88	115	137	117	71	126	235	127	105	135	134	141	74
156	53	157	52	165	78	173	76	179	83	186	109	197	152	201	114	203	118
204	66	216	156	229	153	230	172	243	175	245	130	246	261	258	125	259	267
274	550	275	196	286	72	288	110	303	36	313	33	314	58	327	42	329	28
331	38	344	64	345	29	359	542	360	211								

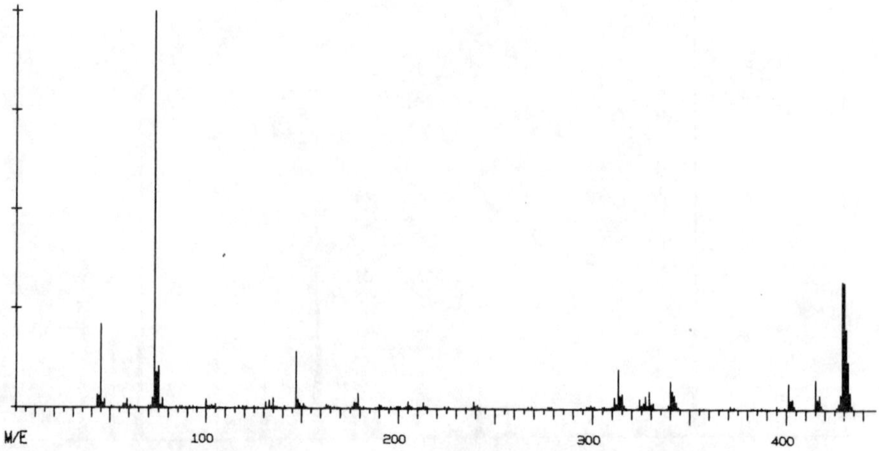
M/E
100
200
300
400

7-Cl-3-OH-5-Ph-1,3-DIHYDRO-2H-1,4-BENZODIAZEPIN-2-ONE-DI TMS DERIVATIVE

$C_{21}H_{27}ClN_2O_2Si_2$ MW = 430

Base Peak = 73 58 Peaks

Mass	Int.	Mass	Int.	Mass	Int.	Mass	Int.	Mass	Int.	Mass	Int.	Mass	Int.	Mass	Int.	Mass	Int.
43	32	45	210	58	11	59	22	73	999	75	106	76	8	77	25	100	22
103	9	104	4	105	8	119	4	131	14	133	21	135	24	147	142	148	22
163	11	164	8	177	16	179	38	190	9	191	8	205	17	213	16	224	4
226	4	239	17	241	9	244	1	250	1	267	4	269	4	277	4	279	6
297	4	299	11	311	29	313	100	314	32	315	37	329	45	340	71	342	35
343	17	357	3	367	3	371	8	373	4	387	6	395	8	401	64	403	24
415	74	417	35	429	323	430	320										

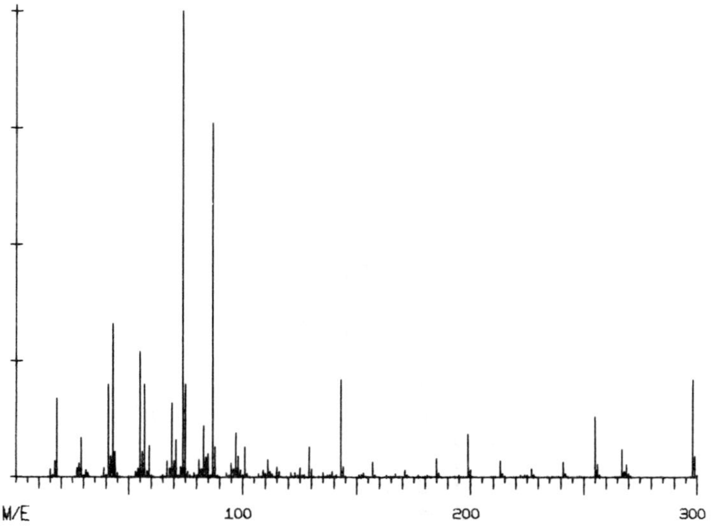
M/E
100
200
300

METHYL STEARATE

$C_{19}H_{38}O_2$ MW = 298

Base Peak = 74 40 Peaks

Mass	Int.	Mass	Int.	Mass	Int.	Mass	Int.	Mass	Int.	Mass	Int.	Mass	Int.	Mass	Int.	Mass	Int.
28	29	29	86	41	200	43	330	55	270	57	200	74	999	75	200	83	110
87	760	97	94	101	64	111	37	115	22	125	20	129	64	143	210	144	23
153	10	157	32	167	7	171	16	185	40	186	11	199	93	200	14	213	36
214	8	224	6	227	17	241	32	242	8	255	130	256	27	267	61	269	28
284	3	298	210	299	45	300	6										

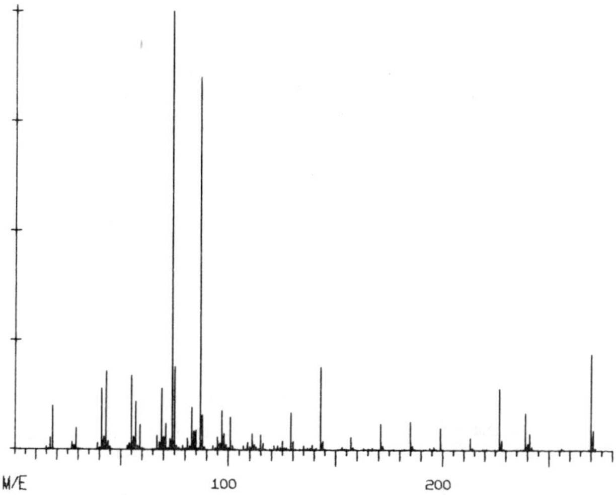
M/E
100
200

METHYL PALMITATE

$C_{17}H_{34}O_2$ MW = 270

Base Peak = 74 37 Peaks

Mass	Int.	Mass	Int.	Mass	Int.	Mass	Int.	Mass	Int.	Mass	Int.	Mass	Int.	Mass	Int.	Mass	Int.
27	17	29	50	41	140	43	180	55	170	57	110	74	999	75	190	83	97
87	850	97	89	101	76	111	37	115	36	125	20	129	85	143	190	144	20
153	7	157	30	171	60	172	11	185	64	186	10	196	7	199	50	213	27
214	6	227	140	228	23	239	84	241	38	256	5	257	2	270	220	271	45
272	6																

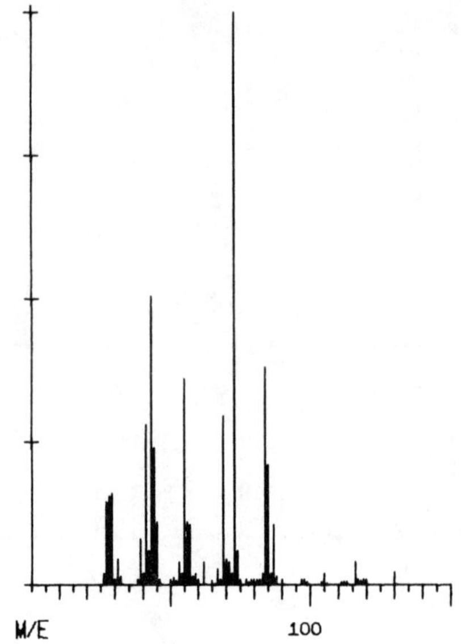
M/E
100

EMYLCAMATE

$C_7H_{15}NO_2$ MW = 145

Base Peak = 73 16 Peaks

Mass	Int.	Mass	Int.	Mass	Int.	Mass	Int.	Mass	Int.	Mass	Int.	Mass	Int.	Mass	Int.	Mass	Int.
28	154	29	159	41	281	43	504	55	359	56	109	69	294	73	999	84	379
85	209	90	9	97	9	105	19	116	39	119	9	130	23				

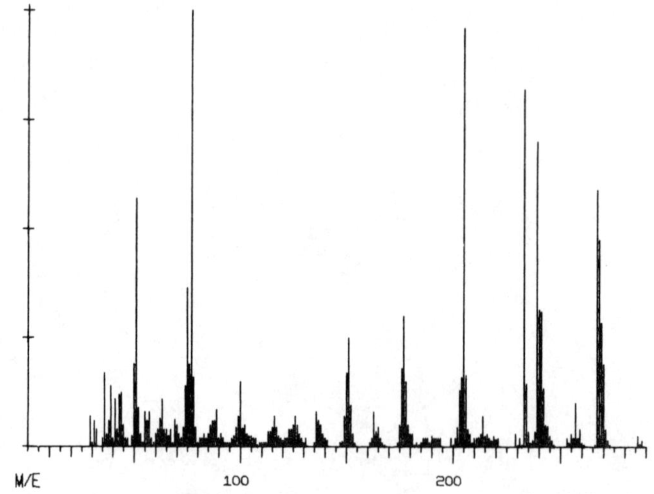
M/E
100
200

OXAZEPAM

$C_{15}H_{11}ClN_2O_2$ MW = 286

Base Peak = 77 40 Peaks

Mass	Int.	Mass	Int.	Mass	Int.	Mass	Int.	Mass	Int.	Mass	Int.	Mass	Int.	Mass	Int.	Mass	Int.
29	70	31	60	36	170	39	140	50	190	51	570	74	140	75	365	76	190
77	999	99	70	100	150	115	45	116	70	125	50	126	70	136	80	137	60
150	170	151	250	163	80	165	45	176	181	177	300	188	20	190	26	205	960
206	165	216	29	229	30	233	820	239	700	244	48	257	100	267	590	268	475
272	15	273	5	286	25	288	15										

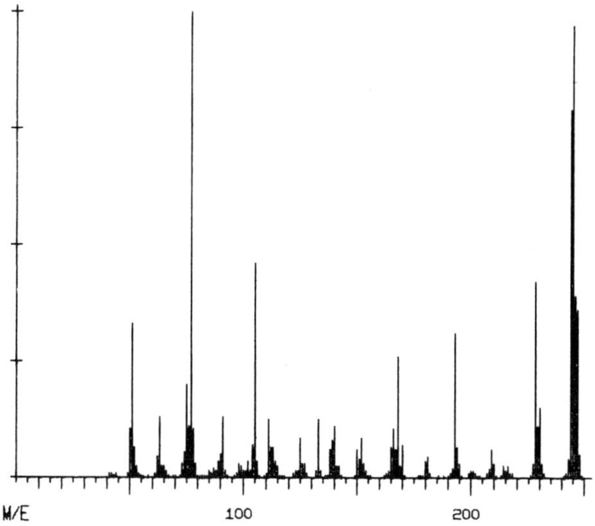
M/E
100
200

2-METHYLAMINO-5-CHLOROBENZOPHENONE

$C_{14}H_{12}ClNO$ MW = 245

Base Peak = 77 32 Peaks

Mass	Int.	Mass	Int.	Mass	Int.	Mass	Int.	Mass	Int.	Mass	Int.	Mass	Int.	Mass	Int.	Mass	Int.
41	10	42	10	50	105	51	330	63	130	75	200	76	110	77	999	90	50
91	130	105	460	111	125	125	85	126	30	133	125	140	110	150	60	152	85
166	105	168	260	180	35	181	45	193	310	194	65	209	60	210	30	228	420
229	110	230	150	243	40	244	790	245	970								

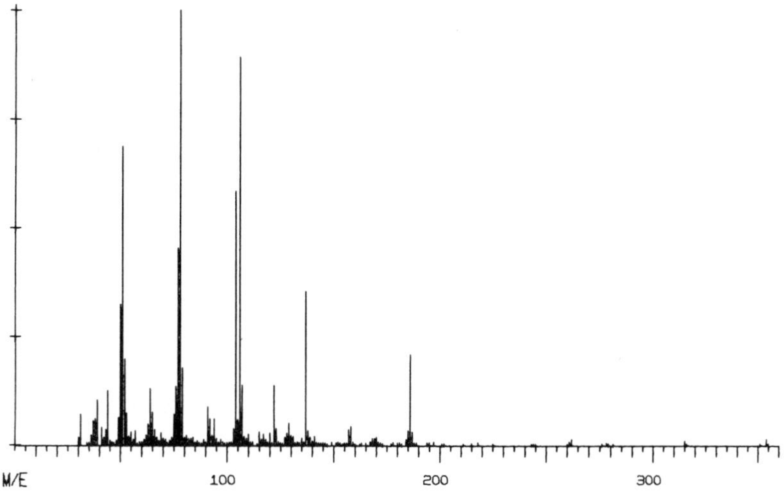
M/E
100
200
300

ISONIAZID

$C_6H_7N_3O$

Base Peak = 78 41 Peaks

Mass	Int.	Mass	Int.	Mass	Int.	Mass	Int.	Mass	Int.	Mass	Int.	Mass	Int.	Mass	Int.	Mass	Int.
30	20	31	73	39	104	44	128	50	324	51	688	64	132	65	78	77	456
78	999	91	90	92	63	104	585	106	893	122	140	129	52	137	355	138	34
157	38	158	45	168	17	170	21	185	34	186	211	188	8	197	9	202	6
211	6	218	7	225	6	243	5	244	5	245	6	261	11	262	14	276	6
278	8	315	13	316	6	351	6	354	14								

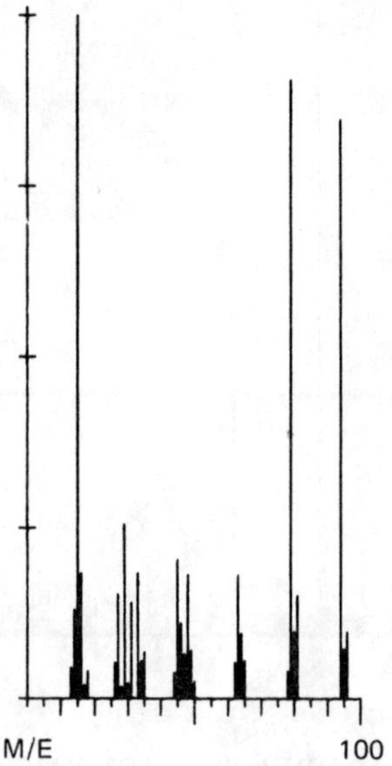
M/E
100

DIMETHYL SULFONE

$C_2H_6O_2S$ MW = 94

Base Peak = 79 12 Peaks

Mass	Int.	Mass	Int.	Mass	Int.	Mass	Int.	Mass	Int.	Mass	Int.	Mass	Int.	Mass	Int.	Mass	Int.
29	281	33	201	45	222	46	120	48	197	49	76	63	197	64	105	79	999
81	164	94	935	96	108												

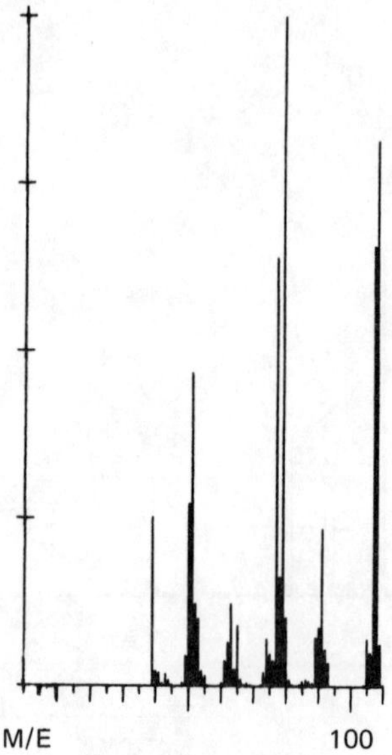
M/E
100

BENZYL ALCOHOL

C_7H_8O MW = 108

Base Peak = 79 7 Peaks

Mass	Int.	Mass	Int.	Mass	Int.	Mass	Int.	Mass	Int.	Mass	Int.	Mass	Int.	Mass	Int.	Mass	Int.
39	253	40	22	50	272	51	467	63	123	65	90	77	641	79	999	90	88
91	236	107	657	108	816												

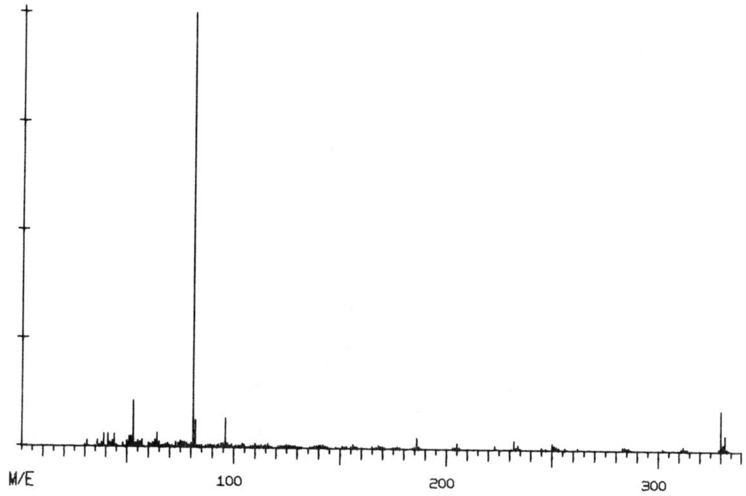
M/E
100
200
300

FUROSEMIDE

$C_{12}H_{11}CLN_2O_5S$ MW = 330

Base Peak = 81 42 Peaks

Mass	Int.	Mass	Int.	Mass	Int.	Mass	Int.	Mass	Int.	Mass	Int.	Mass	Int.	Mass	Int.	Mass	Int.
30	4	31	16	39	30	44	31	51	24	53	108	63	17	64	32	81	999
82	63	96	68	97	11	104	11	110	11	125	7	126	8	140	7	141	8
156	10	157	6	168	8	170	6	177	5	186	26	188	2	205	14	206	6
223	10	232	19	234	10	250	14	251	10	262	4	283	8	284	8	286	5
287	2	302	5	312	10	314	5	330	93	332	36						

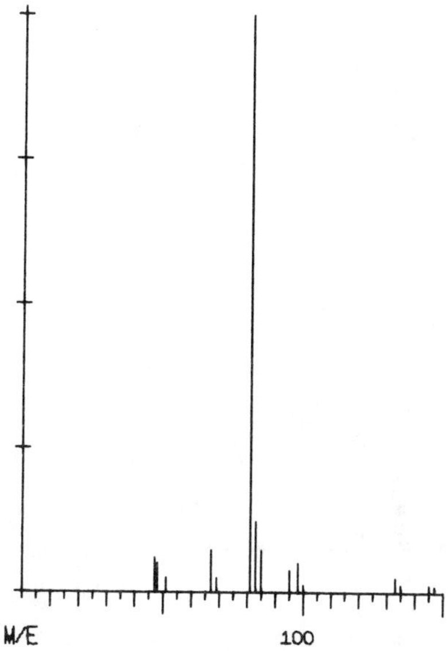
M/E
100

2,2-DICHLORO-1,1-DIFLUOROETHYL METHYL ETHER

$C_3H_4CL_2F_2O$ MW = 164

Base Peak = 81 12 Peaks

Mass	Int.	Mass	Int.	Mass	Int.	Mass	Int.	Mass	Int.	Mass	Int.	Mass	Int.	Mass	Int.	Mass	Int.
47	61	48	49	51	24	67	73	69	24	81	999	83	122	95	37	98	49
133	24	135	12	147	9												

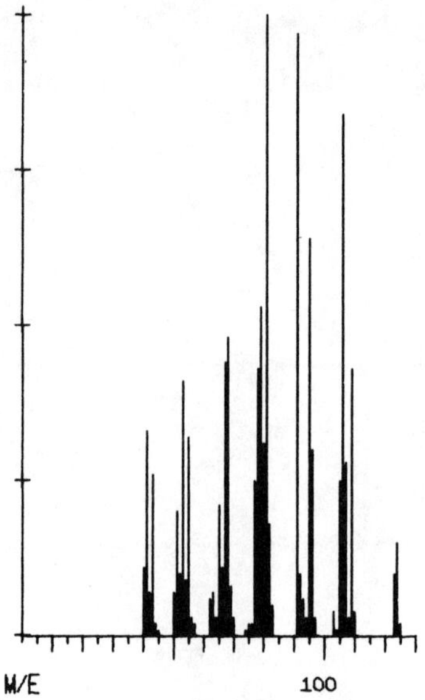
M/E
100

ETHINAMATE

$C_9H_{13}NO_2$ MW = 167

Base Peak = 81 14 Peaks

Mass	Int.	Mass	Int.	Mass	Int.	Mass	Int.	Mass	Int.	Mass	Int.	Mass	Int.	Mass	Int.	Mass	Int.
41	330	43	260	53	410	55	320	67	440	68	480	79	530	81	999	91	970
95	640	106	840	109	430	123	100	124	150								

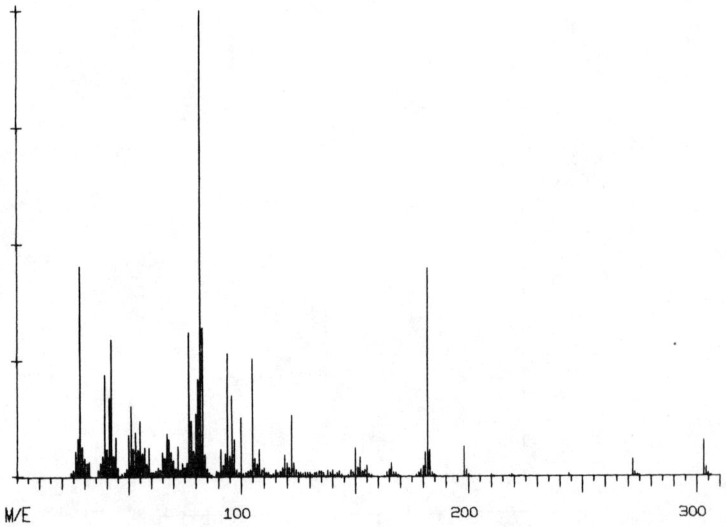
M/E
100
200
300

COCAINE

$C_{17}H_{21}NO_4$ MW = 303

Base Peak = 82 36 Peaks

Mass	Int.	Mass	Int.	Mass	Int.	Mass	Int.	Mass	Int.	Mass	Int.	Mass	Int.	Mass	Int.	Mass	Int.
27	82	28	452	39	220	42	295	51	152	55	120	67	93	68	81	82	999
83	321	94	264	96	174	105	254	108	59	119	47	122	132	138	16	140	16
150	62	152	42	165	14	166	31	182	447	183	58	198	65	199	14	210	4
211	2	218	3	219	4	244	8	245	1	272	38	273	10	303	78	304	20

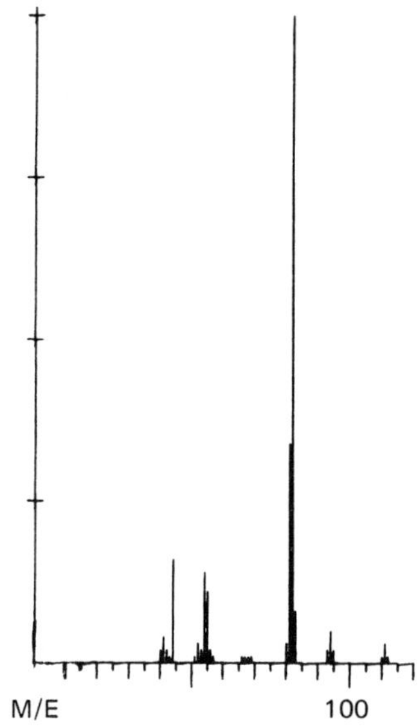
M/E
100

HISTAMINE

$C_5H_9N_3$ MW = 111

Base Peak = 82 12 Peaks

Mass	Int.	Mass	Int.	Mass	Int.	Mass	Int.	Mass	Int.	Mass	Int.	Mass	Int.	Mass	Int.	Mass	Int.
41	40	44	160	54	140	55	110	66	10	67	10	81	340	82	999	93	2(
94	50	110	10	111	30												

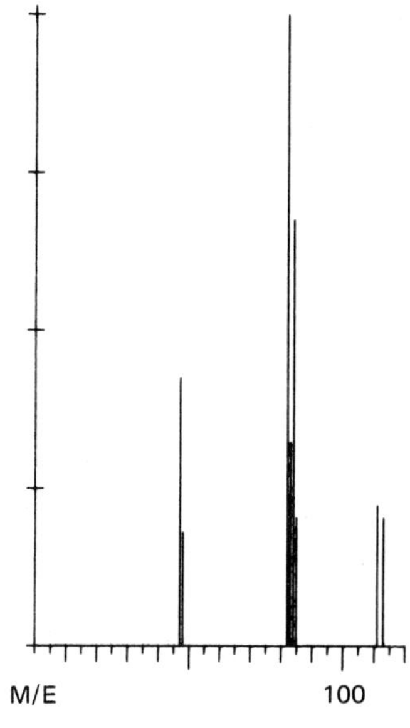
M/E
100

CHLORAL HYDRATE

$C_2H_3CL_3O_2$ MW = 164

Base Peak = 82 6 Peaks

Mass	Int.	Mass	Int.	Mass	Int.	Mass	Int.	Mass	Int.	Mass	Int.
47	424	48	181	82	999	84	676	111	222	113	202

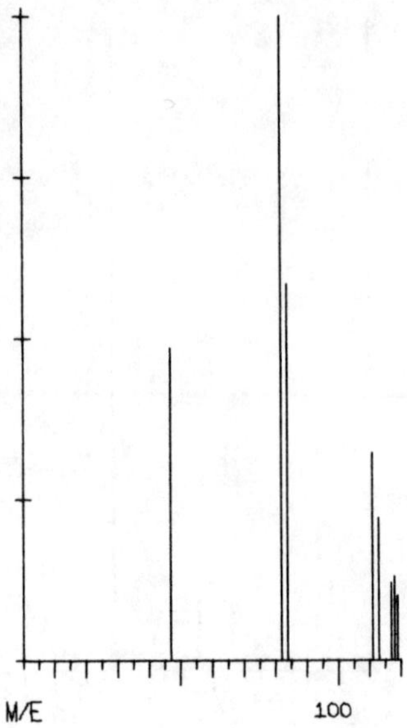
M/E
100

CHLORAL BETAINE

$C_7H_{14}Cl_3NO_4$ MW = 281

Base Peak = 82 7 Peaks

Mass	Int.	Mass	Int.	Mass	Int.	Mass	Int.	Mass	Int.	Mass	Int.	Mass	Int.
47	484	82	999	84	585	111	323	113	222	118	131	119	101

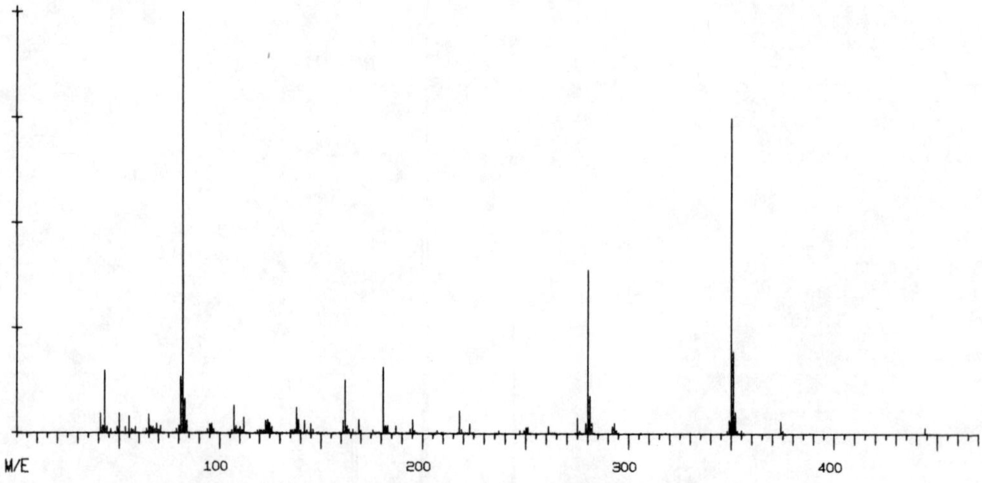
M/E
100
200
300
400

DIHYDROXYSECOBARBITAL DI-D9-TMS ETHER 1,3-DIMETHYL DERIV.

$C_{20}H_{22}D_{18}O_5N_2Si_2$ MW = 462

Base Peak = 82 56 Peaks

Mass	Int.	Mass	Int.	Mass	Int.	Mass	Int.	Mass	Int.	Mass	Int.	Mass	Int.	Mass	Int.	Mass	Int.
40	0	41	48	42	18	43	150	44	15	45	3	46	11	48	3	49	15
50	48	51	3	52	3	53	15	54	3	55	41	56	11	57	7	58	15
59	3	62	3	63	3	64	11	65	45	66	15	67	15	68	11	69	22
70	7	71	18	72	3	73	3	7	3	78	3	79	11	80	18	81	135
82	999	83	82	84	30	85	3	88	3	89	3	90	3	91	3	92	3
93	3	94	11	95	22	96	22	97	11	98	3	99	3	104	3	105	3
106	7	107	67														

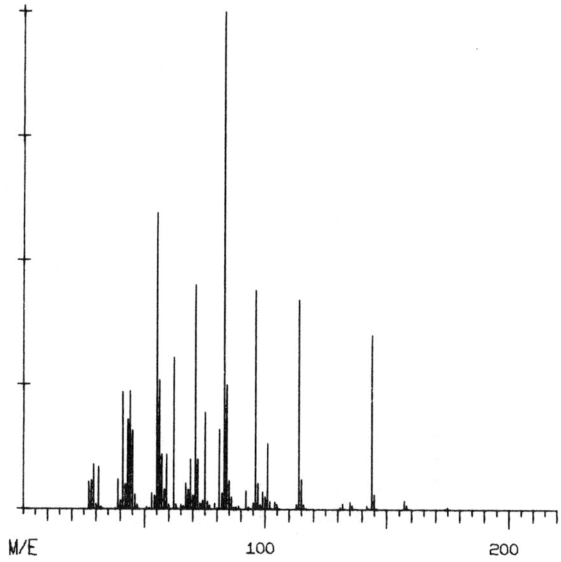
M/E
100
200

MEPROBAMATE

$C_9H_{18}N_2O_4$ MW = 218

Base Peak = 83 21 Peaks

Mass	Int.	Mass	Int.	Mass	Int.	Mass	Int.	Mass	Int.	Mass	Int.	Mass	Int.	Mass	Int.	Mass	Int.
29	89	31	84	41	235	44	237	55	594	56	259	62	304	71	449	83	999
84	249	96	439	101	132	114	421	115	59	131	5	144	349	145	31	157	18
158	9	174	3	175	6												

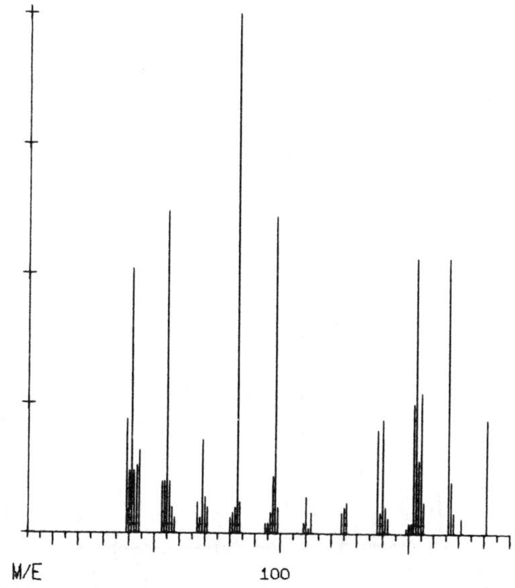
M/E
100

METHYPRYLON METABOLITE 1

$C_{10}H_{17}N_1O_3$ MW = 199

Base Peak = 83 21 Peaks

Mass	Int.	Mass	Int.	Mass	Int.	Mass	Int.	Mass	Int.	Mass	Int.	Mass	Int.	Mass	Int.	Mass	Int.
39	219	41	509	53	99	55	619	69	179	70	69	83	999	84	59	97	109
98	609	110	69	112	39	125	49	126	59	138	199	140	219	153	529	155	269
166	529	167	99	181	219												

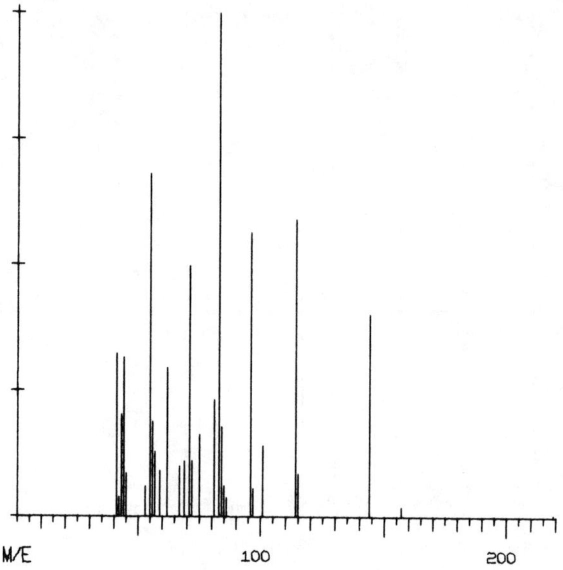
M/E
100
200

2-METHYL-2-PROPYL-1,3-PROPANE DIOL CARBAMATE

$C_9H_{18}N_2O_4$ MW = 218

Base Peak = 83 15 Peaks

Mass	Int.	Mass	Int.	Mass	Int.	Mass	Int.	Mass	Int.	Mass	Int.	Mass	Int.	Mass	Int.	Mass	Int.
41	323	44	314	55	680	56	187	62	296	71	497	81	233	83	999	96	565
101	140	114	589	115	84	144	403	157	20	219	4						

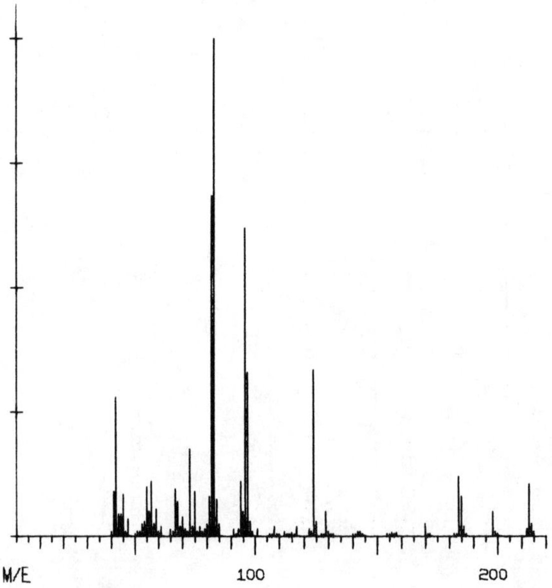
M/E
100
200

TROPINE TMS ETHER

$C_{11}H_{23}NOSI$ MW = 213

Base Peak = 83 26 Peaks

Mass	Int.	Mass	Int.	Mass	Int.	Mass	Int.	Mass	Int.	Mass	Int.	Mass	Int.	Mass	Int.	Mass	Int.
41	90	42	280	55	100	57	110	67	95	73	175	82	685	83	999	96	620
97	330	108	20	117	20	124	335	129	50	142	10	143	10	156	10	158	10
170	25	171	5	184	120	185	80	198	50	199	10	213	105	214	25		

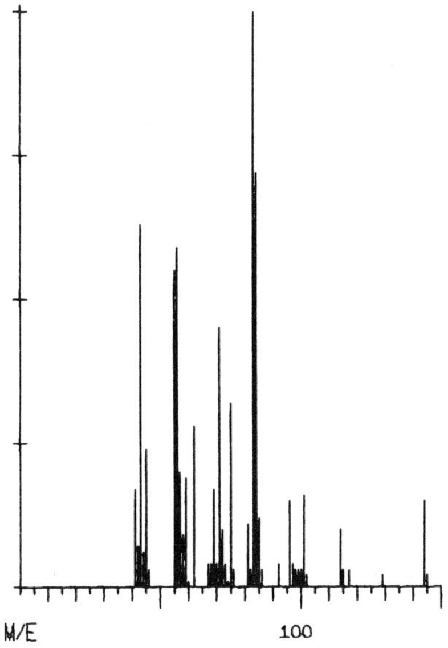
M/E
100

MEPROBAMATE

$C_9H_{18}N_2O_4$ MW = 218

Base Peak = 83 15 Peaks

Mass	Int.	Mass	Int.	Mass	Int.	Mass	Int.	Mass	Int.	Mass	Int.	Mass	Int.	Mass	Int.	Mass	Int.
43	630	45	240	55	550	56	590	71	450	75	320	83	999	84	720	96	150
101	160	114	100	115	30	129	20	144	150	145	20						

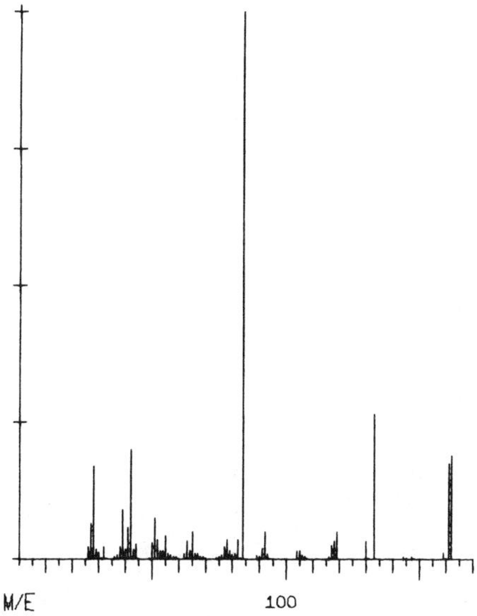
M/E
100

NICOTINE

$C_{10}H_{14}N_2$ MW = 162

Base Peak = 84 22 Peaks

Mass	Int.	Mass	Int.	Mass	Int.	Mass	Int.	Mass	Int.	Mass	Int.	Mass	Int.	Mass	Int.	Mass	Int.
27	64	28	170	39	90	42	201	51	74	55	43	63	33	65	49	78	35
84	999	91	19	92	50	104	15	117	26	118	33	119	51	133	266	144	4
147	4	159	12	161	176	162	191										

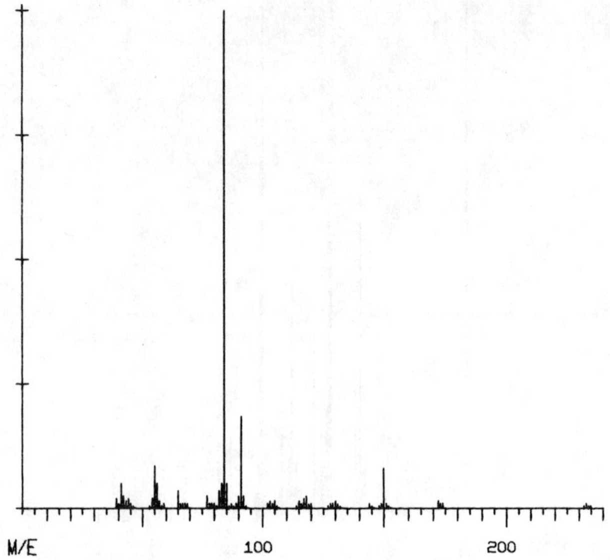
M/E
100
200

METHYLPHENIDATE

$C_{14}H_{19}NO_2$ MW = 233

Base Peak = 84 23 Peaks

Mass	Int.	Mass	Int.	Mass	Int.	Mass	Int.	Mass	Int.	Mass	Int.	Mass	Int.	Mass	Int.	Mass	Int.
41	49	42	24	55	84	56	49	65	34	66	9	83	49	84	999	90	24
91	184	115	16	117	19	118	24	130	14	132	4	144	9	149	9	150	79
172	14	173	9	174	9	232	4	233	9								

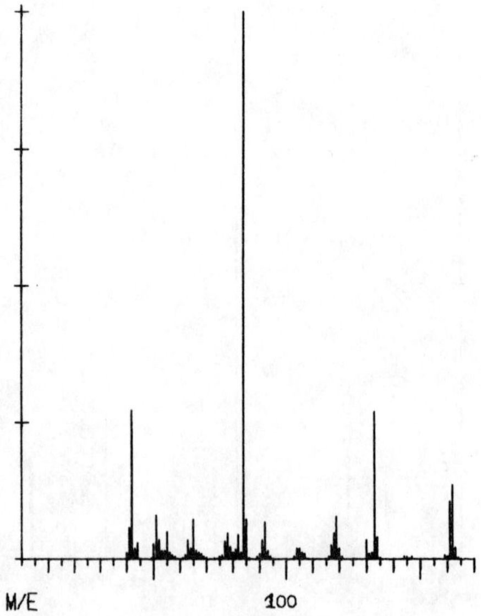
M/E
100

NICOTINE

$C_{10}H_{14}N_2$ MW = 162

Base Peak = 84 20 Peaks

Mass	Int.	Mass	Int.	Mass	Int.	Mass	Int.	Mass	Int.	Mass	Int.	Mass	Int.	Mass	Int.	Mass	Int.
41	58	42	273	51	80	55	50	63	34	65	73	84	999	85	73	91	34
92	68	105	21	117	24	118	47	119	78	133	269	134	41	147	4	159	8
161	106	162	135														

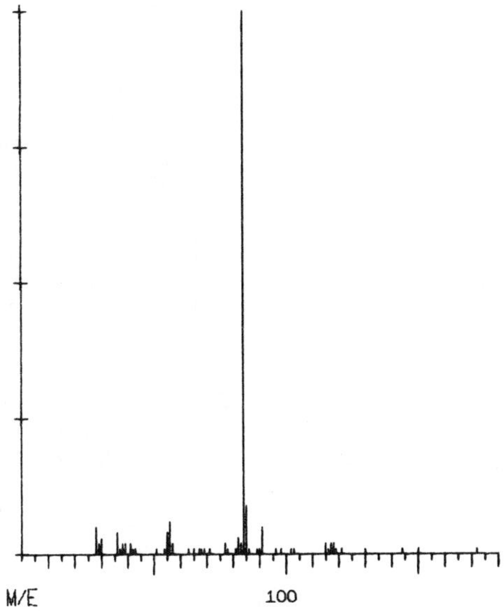
M/E
100

METHYLPHENIDATE

$C_{14}H_{19}NO_2$ MW = 233

Base Peak = 84   19 Peaks

Mass	Int.	Mass	Int.	Mass	Int.	Mass	Int.	Mass	Int.	Mass	Int.	Mass	Int.	Mass	Int.	Mass	Int.
28	50	30	30	36	40	38	20	55	40	56	60	63	10	65	10	84	999
85	90	90	10	91	50	115	20	117	20	118	20	119	10	144	10	150	10
172	10																

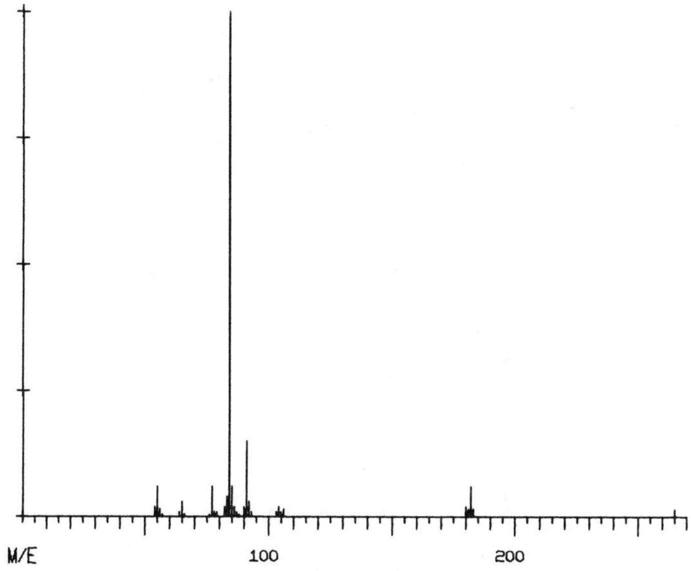
M/E
100
200

ANTAZOLINE

$C_{17}H_{19}N_3$ MW = 265

Base Peak = 84   13 Peaks

Mass	Int.	Mass	Int.	Mass	Int.	Mass	Int.	Mass	Int.	Mass	Int.	Mass	Int.	Mass	Int.	Mass	Int.
54	19	55	59	64	9	65	29	77	59	84	999	91	149	92	29	104	19
106	14	180	19	182	59	265	14										

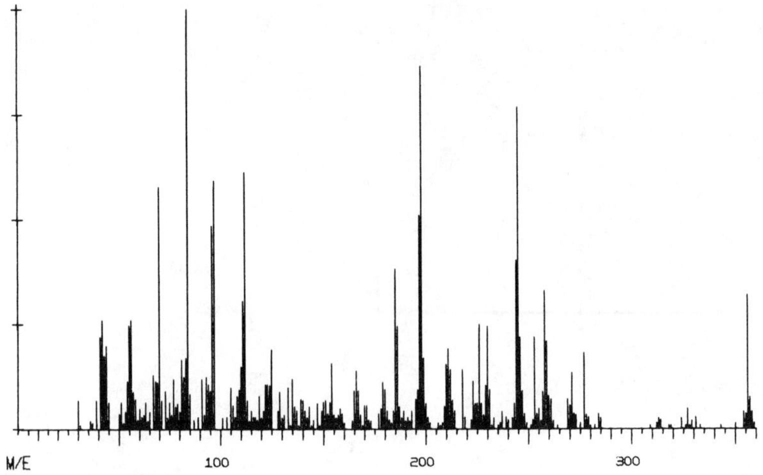

M/E
100
200
300

THIORIDAZINE MTB. 3

$C_{20}H_{24}N_2S_2$ MW = 356

Base Peak = 84 48 Peaks

Mass	Int.	Mass	Int.	Mass	Int.	Mass	Int.	Mass	Int.	Mass	Int.	Mass	Int.	Mass	Int.	Mass	Int.
30	69	31	11	41	220	42	259	55	248	56	259	67	131	70	577	83	170
84	999	96	484	97	593	111	306	112	612	122	108	125	189	133	100	135	120
151	69	154	158	165	93	166	139	185	383	186	244	197	511	198	864	210	155
211	193	218	143	226	251	230	244	231	96	244	403	245	767	258	329	259	209
272	38	277	182	312	15	313	27	324	27	327	50	329	19	331	31	354	42
355	38	356	321	357	77												

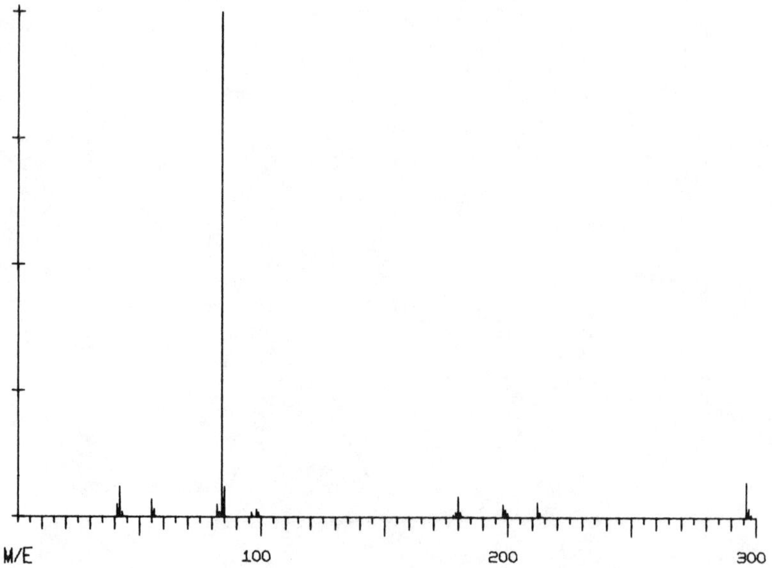

M/E
100
200
300

PYRATHIAZINE

$C_{18}H_{20}N_2S$ MW = 296

Base Peak = 84 16 Peaks

Mass	Int.	Mass	Int.	Mass	Int.	Mass	Int.	Mass	Int.	Mass	Int.	Mass	Int.	Mass	Int.	Mass	Int.
41	24	42	61	55	34	56	14	84	999	85	59	98	14	99	11	180	39
181	10	198	24	199	14	212	31	213	9	296	71	297	17				

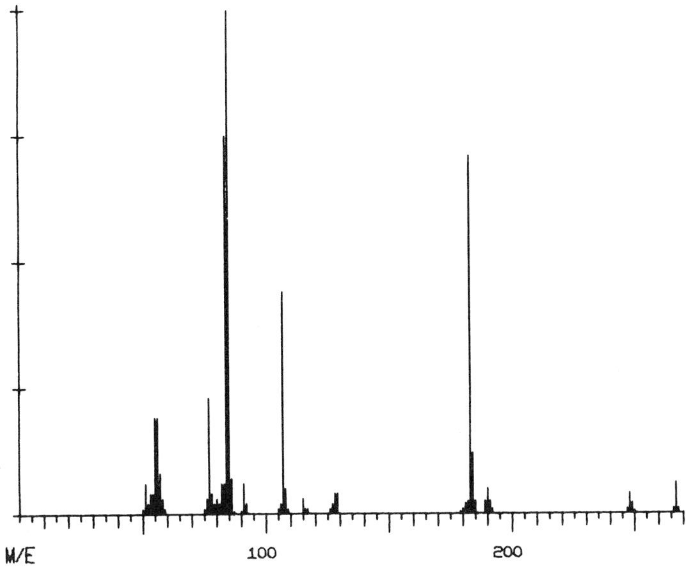
M/E
100
200

AZACYCLONOL

$C_{18}H_{21}NO$ MW = 267

Base Peak = 85   19 Peaks

Mass	Int.	Mass	Int.	Mass	Int.	Mass	Int.	Mass	Int.	Mass	Int.	Mass	Int.	Mass	Int.	Mass	Int.
55	189	56	189	75	9	84	749	85	999	91	59	92	19	107	439	108	49
128	39	129	39	183	709	184	119	189	24	190	49	248	39	249	19	266	9
267	59																

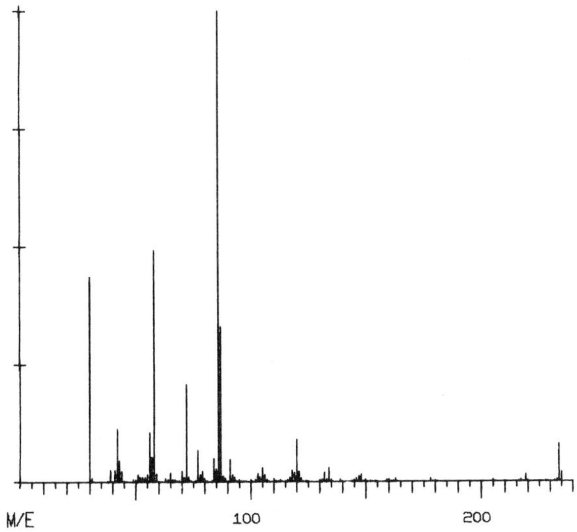
M/E
100
200

LIDOCAINE

$C_{14}H_{22}N_2O$ MW = 234

Base Peak = 86   29 Peaks

Mass	Int.	Mass	Int.	Mass	Int.	Mass	Int.	Mass	Int.	Mass	Int.	Mass	Int.	Mass	Int.	Mass	Int.
30	436	31	8	42	113	43	44	56	106	58	493	70	23	72	207	86	999
87	330	91	48	103	17	105	31	106	14	118	25	120	91	132	20	134	29
147	13	148	16	160	4	163	7	178	7	180	3	205	4	217	5	219	14
234	79	235	21														

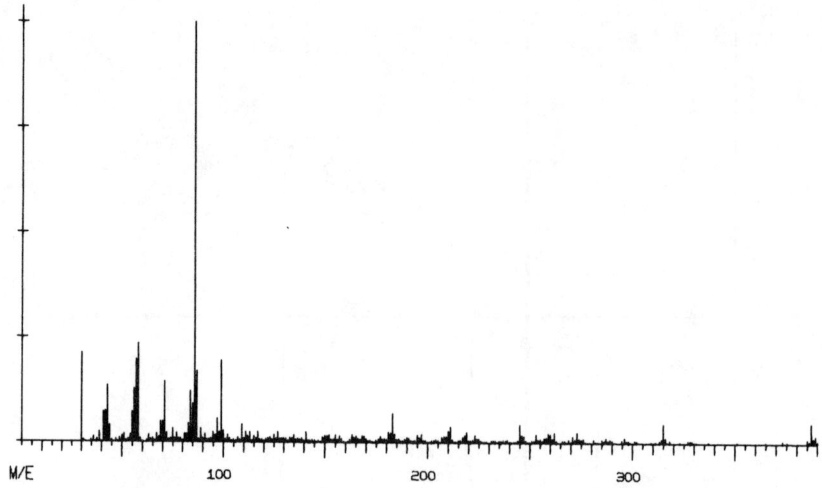
M/E
100
200
300

FLURAZEPAM

$C_{21}H_{23}CLFN_3O$ MW = 387

Base Peak = 86 52 Peaks

Mass	Int.	Mass	Int.	Mass	Int.	Mass	Int.	Mass	Int.	Mass	Int.	Mass	Int.	Mass	Int.	Mass	Int.
30	212	31	4	42	76	43	136	57	197	58	236	69	51	71	144	86	999
87	169	97	58	99	194	109	42	113	26	125	18	127	26	135	18	141	26
152	18	155	17	163	18	169	16	181	26	183	69	195	18	197	19	210	27
211	37	218	17	219	24	237	7	239	7	245	42	253	21	259	23	262	24
273	24	274	11	287	12	296	13	301	4	302	5	315	45	316	12	328	4
329	4	351	2	365	2	373	4	375	2	387	47	389	18				

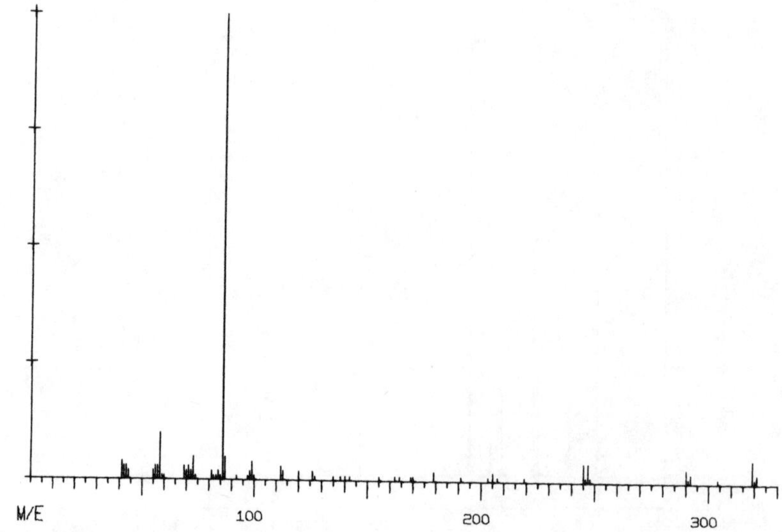
M/E
100
200
300

CHLOROQUINE

$C_{18}H_{26}CLN_3$ MW = 319

Base Peak = 86 31 Peaks

Mass	Int.	Mass	Int.	Mass	Int.	Mass	Int.	Mass	Int.	Mass	Int.	Mass	Int.	Mass	Int.	Mass	Int.
41	40	42	30	56	30	58	100	69	30	73	50	86	999	87	50	98	20
99	40	112	30	113	20	120	20	126	20	135	10	138	10	155	10	162	10
164	10	179	20	191	10	203	10	205	20	219	10	245	40	247	40	290	30
292	20	304	10	319	50	321	20										

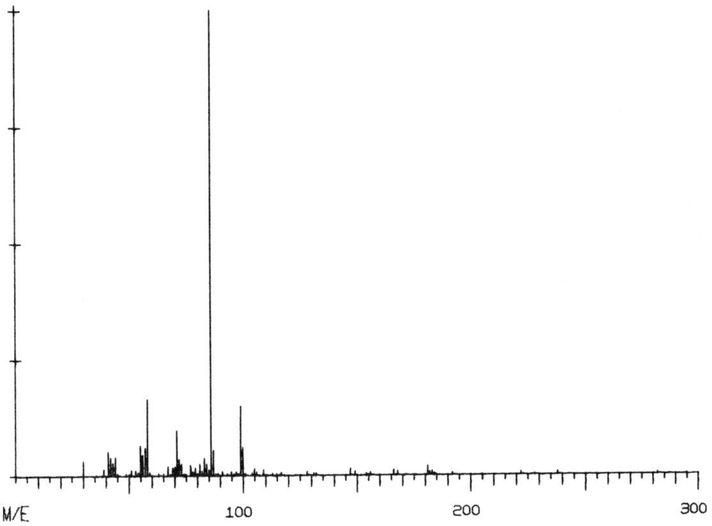
M/E
100
200
300

DICYCLOMINE

$C_{19}H_{35}NO_2$ MW = 309

Base Peak = 86 30 Peaks

Mass	Int.	Mass	Int.	Mass	Int.	Mass	Int.	Mass	Int.	Mass	Int.	Mass	Int.	Mass	Int.	Mass	Int.
30	32	31	3	41	53	44	41	55	64	58	165	71	98	72	34	86	999
87	56	99	150	100	60	105	15	109	12	128	7	131	4	132	6	147	14
149	9	166	13	168	10	181	20	183	9	192	6	205	3	222	7	228	3
238	8	282	5	295	3												

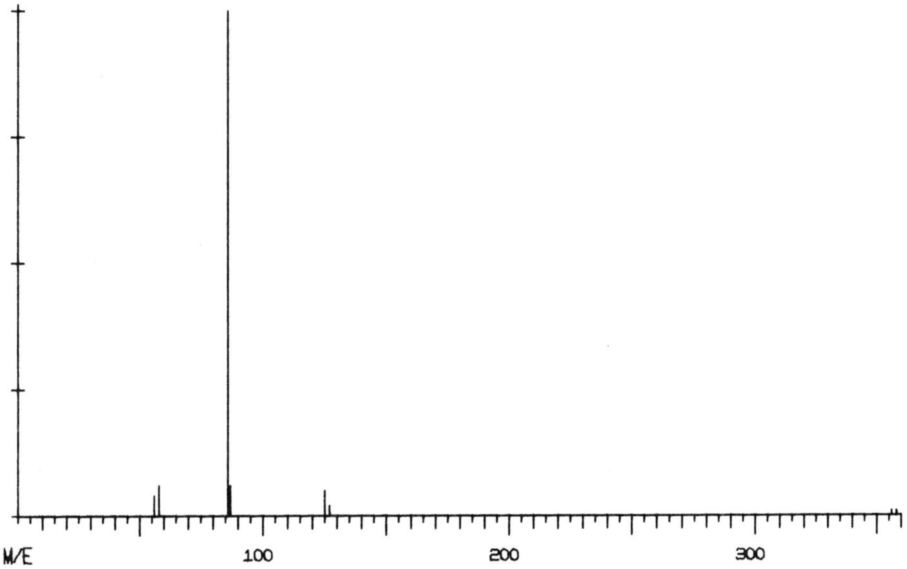
M/E
100
200
300

CLONITAZENE

$C_{20}H_{23}CLN_4O_2$ MW = 386

Base Peak = 86 8 Peaks

Mass	Int.	Mass	Int.	Mass	Int.	Mass	Int.	Mass	Int.	Mass	Int.	Mass	Int.	Mass	Int.
56	40	58	60	86	999	87	60	125	50	127	20	356	10	358	10

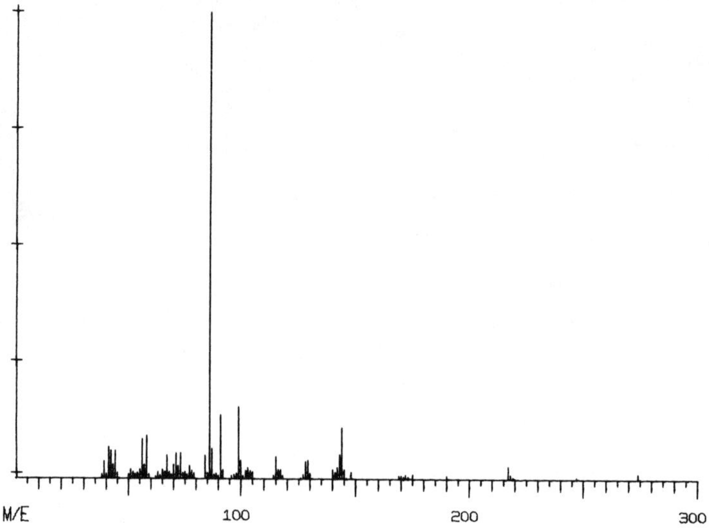
M/E
100
200
300

CARAMIPHEN

$C_{18}H_{27}NO_2$ MW = 289

Base Peak = 86 29 Peaks

Mass	Int.	Mass	Int.	Mass	Int.	Mass	Int.	Mass	Int.	Mass	Int.	Mass	Int.	Mass	Int.	Mass	Int.
41	68	42	61	56	85	58	93	71	56	73	54	86	999	87	64	91	138
99	156	115	47	117	21	128	38	129	41	143	52	144	111	146	2	148	15
169	8	172	11	175	10	190	8	191	1	217	28	218	11	246	1	247	5
274	12	275	1														

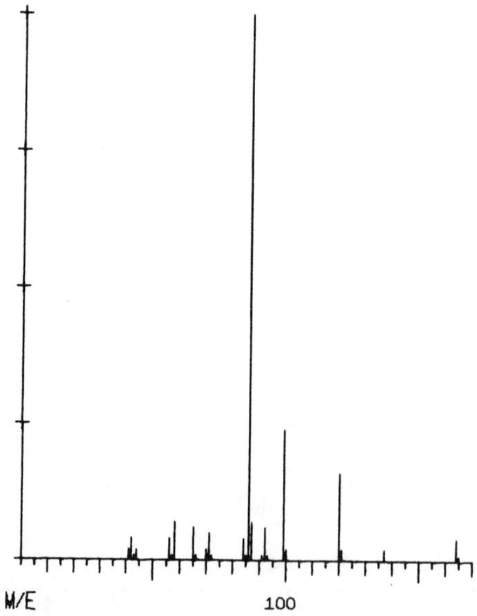
M/E
100

PROCAINE

$C_{13}H_{20}N_2O_2$ MW = 236

Base Peak = 86 15 Peaks

Mass	Int.	Mass	Int.	Mass	Int.	Mass	Int.	Mass	Int.	Mass	Int.	Mass	Int.	Mass	Int.	Mass	Int.
41	20	42	40	56	40	58	70	65	60	71	50	86	999	87	70	92	60
99	240	120	160	121	20	137	20	164	40	165	10						

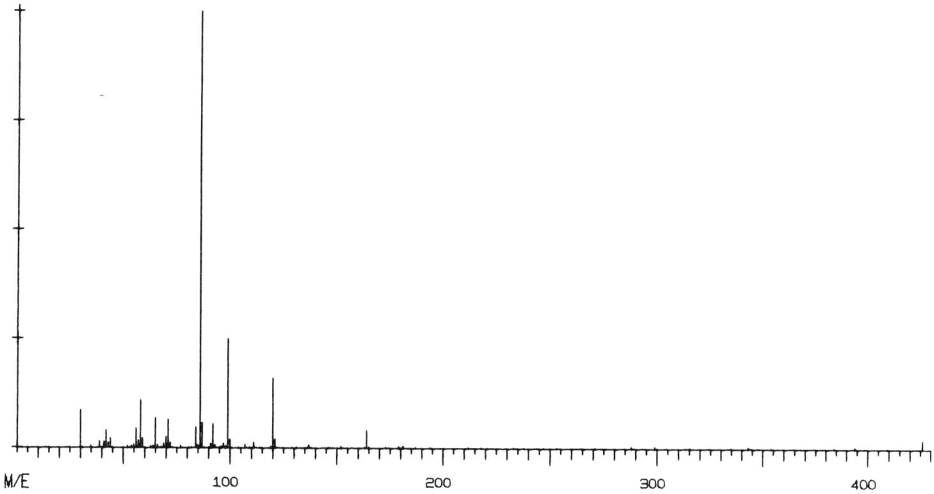
M/E
100
200
300
400

PROCAINE

$C_{13}H_{20}N_2O_2$ MW = 236

Base Peak = 86   44 Peaks

Mass	Int.	Mass	Int.	Mass	Int.	Mass	Int.	Mass	Int.	Mass	Int.	Mass	Int.	Mass	Int.	Mass	Int.
30	88	31	3	42	39	44	22	56	46	58	109	65	70	71	66	86	999
87	57	92	55	99	251	107	7	111	12	120	161	121	19	136	2	137	8
146	2	152	5	164	41	165	3	179	4	181	4	194	3	212	2	218	2
221	2	233	3	238	2	247	3	253	2	271	3	288	4	299	4	313	3
315	3	335	3	343	4	354	3	384	2	394	6	409	3	426	19		

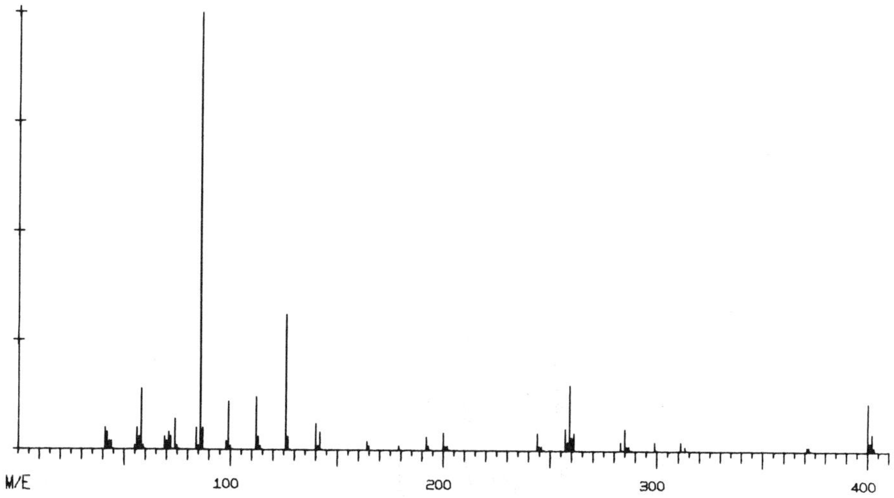
M/E
100
200
300
400

QUINACRINE

$C_{23}H_{30}CLN_3O$ MW = 399

Base Peak = 86   36 Peaks

Mass	Int.	Mass	Int.	Mass	Int.	Mass	Int.	Mass	Int.	Mass	Int.	Mass	Int.	Mass	Int.	Mass	Int.
41	50	42	40	56	50	58	140	71	40	74	70	84	50	86	999	98	20
99	110	112	120	113	30	126	310	127	30	140	60	142	40	164	20	165	10
179	10	192	30	200	40	202	10	244	40	257	50	259	150	261	40	283	20
285	50	286	10	299	20	311	20	313	10	371	10	372	10	400	110	402	40

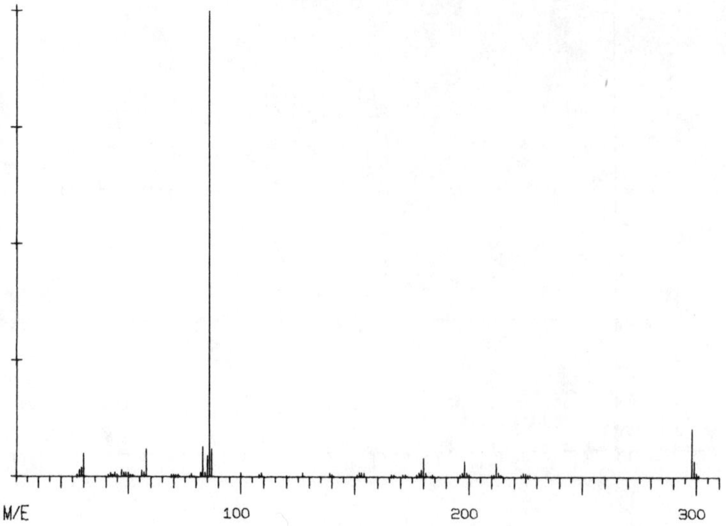
M/E
100
200
300

DIETHAZINE

$C_{18}H_{22}N_2S$ MW = 298

Base Peak = 86 34 Peaks

Mass	Int.	Mass	Int.	Mass	Int.	Mass	Int.	Mass	Int.	Mass	Int.	Mass	Int.	Mass	Int.	Mass	Int.
29	19	30	49	42	11	47	14	56	14	58	59	69	4	70	4	83	64
86	999	99	1	100	9	108	4	109	9	120	2	127	9	139	9	140	6
152	9	154	11	166	6	167	6	179	14	180	39	198	34	199	11	212	29
213	9	224	7	225	8	298	104	299	34	300	9	301	4				

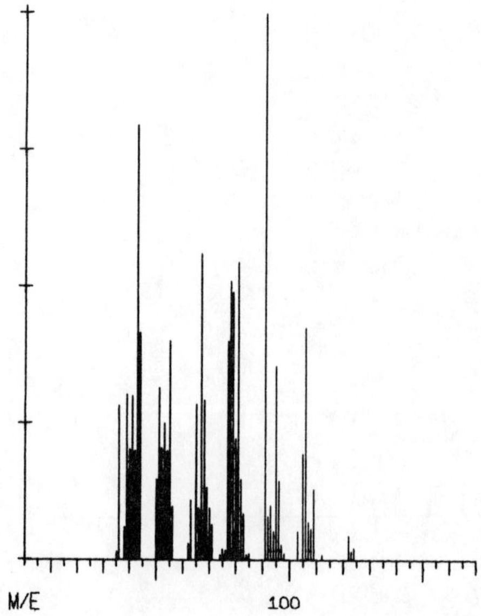
M/E
100

ETHINAMATE

$C_9H_{13}NO_2$ MW = 167

Base Peak = 91 14 Peaks

Mass	Int.	Mass	Int.	Mass	Int.	Mass	Int.	Mass	Int.	Mass	Int.	Mass	Int.	Mass	Int.	Mass	Int.
43	796	44	415	51	314	55	401	67	559	68	293	78	509	81	545	91	999
95	355	105	196	106	425	122	46	124	23								

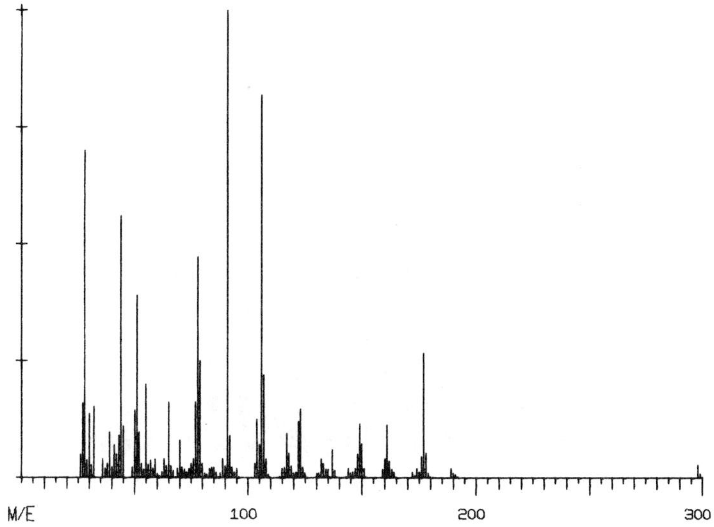
M/E
100
200
300

NIALAMIDE

$C_{16}H_{18}N_4O_2$ MW = 298

Base Peak = 91 29 Peaks

Mass	Int.	Mass	Int.	Mass	Int.	Mass	Int.	Mass	Int.	Mass	Int.	Mass	Int.	Mass	Int.	Mass	Int.
27	159	28	699	44	559	45	111	51	391	55	199	65	163	70	79	78	473
79	250	91	999	92	89	106	821	107	220	122	119	123	148	132	39	137	60
149	116	150	73	160	39	161	113	177	267	178	53	189	19	190	9	298	27
299	11	300	2														

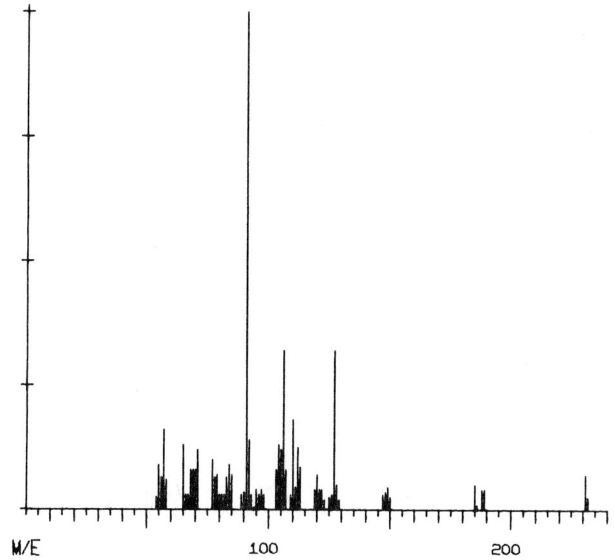
M/E
100
200

ISOCARBOXAZID

$C_{12}H_{13}N_3O_2$ MW = 231

Base Peak = 91 20 Peaks

Mass	Int.	Mass	Int.	Mass	Int.	Mass	Int.	Mass	Int.	Mass	Int.	Mass	Int.	Mass	Int.	Mass	Int.
55	89	57	159	65	129	71	119	77	99	84	91	91	999	92	139	106	319
110	179	120	69	127	319	148	34	149	44	185	49	186	9	188	39	189	39
231	69	232	24														

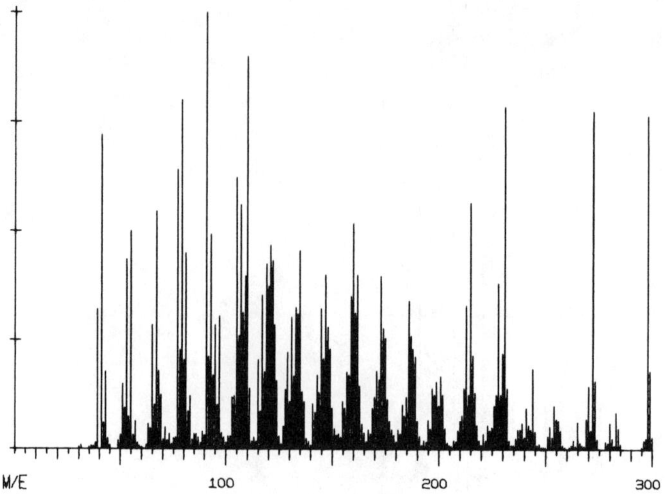
M/E
100
200
300

NORETHINDRONE

$C_{20}H_{26}O_2$ MW = 298

Base Peak = 91 41 Peaks

Mass	Int.	Mass	Int.	Mass	Int.	Mass	Int.	Mass	Int.	Mass	Int.	Mass	Int.	Mass	Int.	Mass	Int.
30	6	31	10	39	319	41	720	53	434	55	501	65	286	67	546	77	640
79	800	91	999	93	492	105	623	110	901	121	468	122	433	133	324	135	455
147	400	159	349	160	518	162	401	174	278	186	341	188	229	189	212	213	331
215	566	216	214	228	381	230	221	231	786	244	184	254	100	269	71	270	145
272	776	273	158	298	765	299	181	300	27								

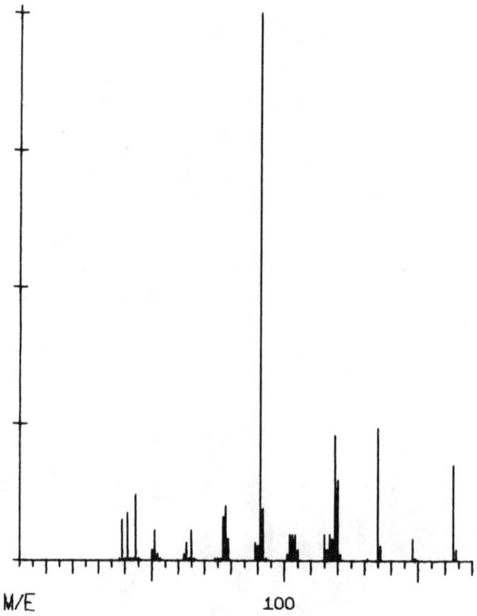
M/E
100

PRIMACLONE MTB. 2

$C_{10}H_{13}NO$ MW = 163

Base Peak = 91 20 Peaks

Mass	Int.	Mass	Int.	Mass	Int.	Mass	Int.	Mass	Int.	Mass	Int.	Mass	Int.	Mass	Int.	Mass	Int.
41	87	44	121	50	20	51	54	63	33	65	54	77	81	78	101	91	999
92	94	104	47	115	47	119	229	120	148	135	243	136	27	148	40	149	6
163	175	164	20														

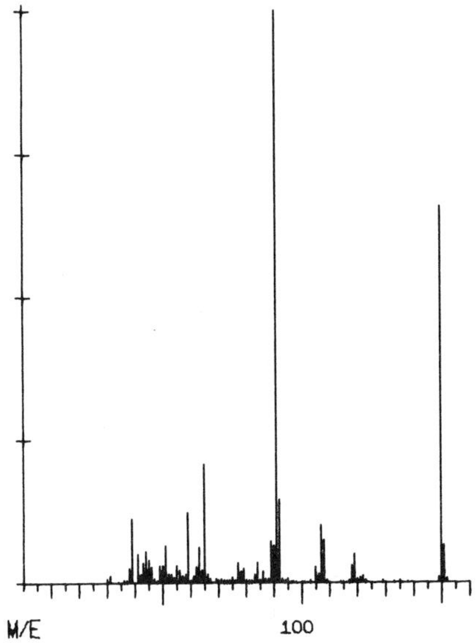
M/E
100

METHYLPHENIDATE MTB. 1

$C_9H_{10}O_2$ MW = 150

Base Peak = 91 20 Peaks

Mass	Int.	Mass	Int.	Mass	Int.	Mass	Int.	Mass	Int.	Mass	Int.	Mass	Int.	Mass	Int.
30	8	31	13	39	113	44	55	51	66	59	123	63	63	65	207
77	34	89	73	91	999	92	144	107	99	108	74	118	30	119	51
133	3	135	6	150	658	151	66								

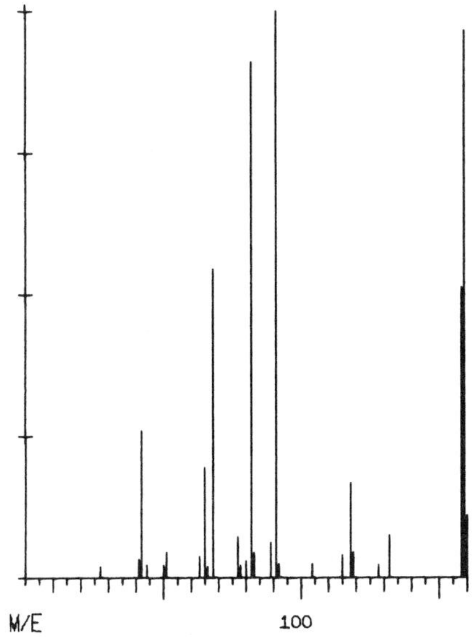
M/E
100

N-METHYL-N-2-PROPYNYL-BENZYLAMINE

$C_{11}H_{13}N$ MW = 159

Base Peak = 91 19 Peaks

Mass	Int.	Mass	Int.	Mass	Int.	Mass	Int.	Mass	Int.	Mass	Int.	Mass	Int.	Mass	Int.
27	21	41	33	42	260	50	22	51	44	65	196	68	546	77	73
82	909	91	999	92	24	104	24	115	41	118	168	119	44	132	74
158	513	159	966	160	110										

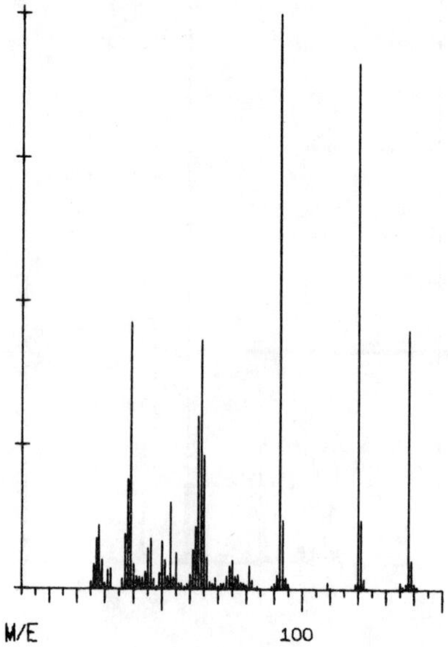
M/E
100

SALICYLIC ACID

$C_7H_6O_3$ MW = 138

Base Peak = 92 18 Peaks

Mass	Int.	Mass	Int.	Mass	Int.	Mass	Int.	Mass	Int.	Mass	Int.	Mass	Int.	Mass	Int.	Mass	Int.
27	88	28	110	38	191	39	462	50	82	53	151	63	301	64	432	77	26
81	41	92	999	93	120	109	12	110	2	120	915	121	121	138	451	139	51

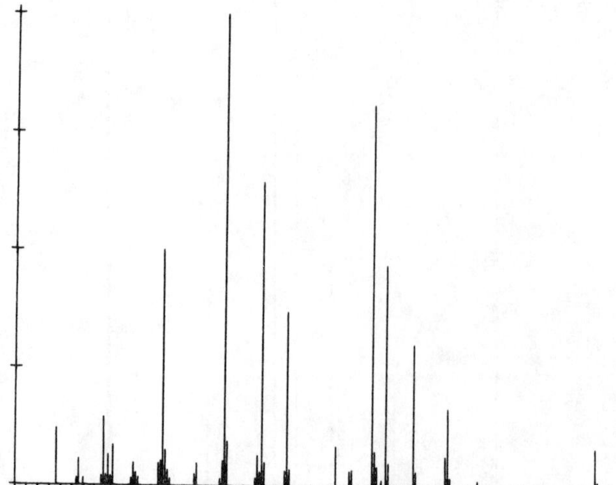
M/E
100
200

SULFAMETHOXAZOLE

$C_{10}H_{11}N_3O_3S$ MW = 253

Base Peak = 92 28 Peaks

Mass	Int.	Mass	Int.	Mass	Int.	Mass	Int.	Mass	Int.	Mass	Int.	Mass	Int.	Mass	Int.	Mass	Int.
27	15	28	54	39	146	43	84	52	48	53	27	65	500	66	76	79	26
80	47	92	999	93	96	106	64	108	646	119	370	120	35	140	86	156	811
157	76	162	469	163	51	174	303	175	33	188	65	189	167	202	15	253	84
254	16																

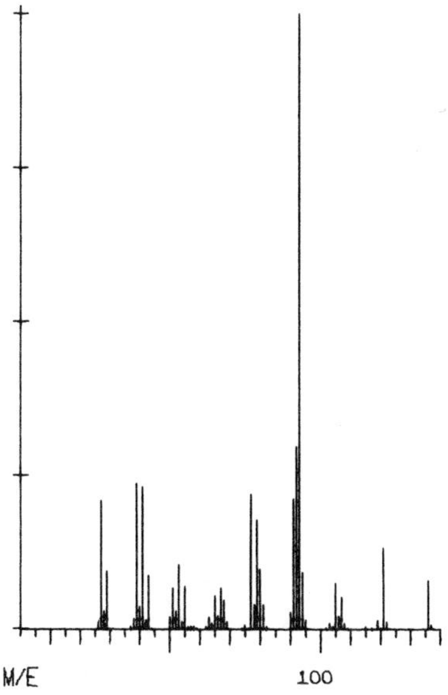
M/E
100

PINENE

$C_{10}H_{16}$ MW = 136

Base Peak = 93 18 Peaks

Mass	Int.	Mass	Int.	Mass	Int.	Mass	Int.	Mass	Int.	Mass	Int.	Mass	Int.	Mass	Int.	Mass	Int.
27	211	29	94	39	237	41	232	53	105	55	71	65	54	67	68	77	221
79	177	92	297	93	999	105	75	107	53	119	15	121	132	136	80	137	8

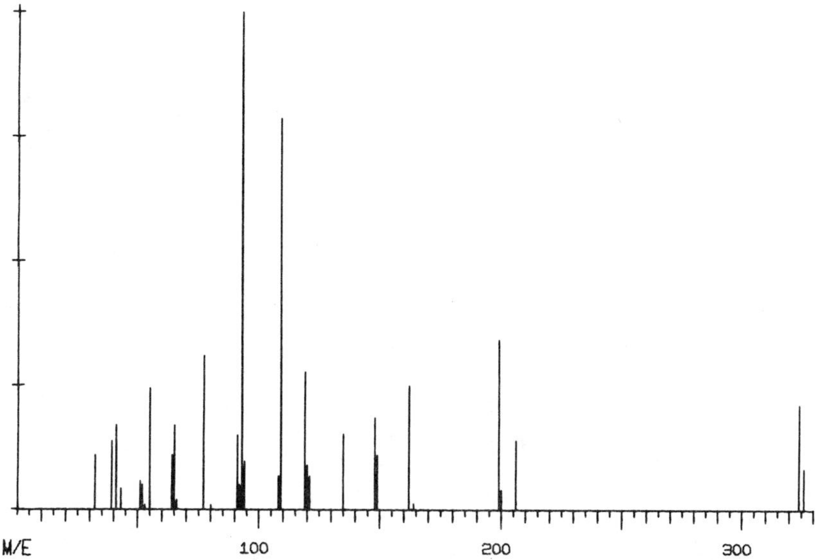
M/E
100
200
300

P-HYDROXYPHENYLBUTAZONE

$C_{19}H_{20}N_2O_3$ MW = 324

Base Peak = 93 25 Peaks

Mass	Int.	Mass	Int.	Mass	Int.	Mass	Int.	Mass	Int.	Mass	Int.	Mass	Int.	Mass	Int.	Mass	Int.
32	111	39	137	41	169	51	57	55	244	64	111	65	170	77	311	80	9
91	151	93	999	108	67	109	788	119	277	120	91	135	153	148	184	149	109
162	250	164	13	199	343	200	41	206	141	324	213	326	82				

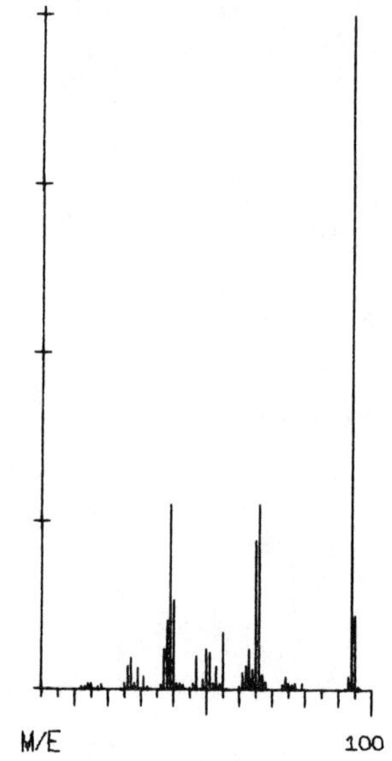

M/E
100

PHENOL

C_6H_6O MW = 94

Base Peak = 94 12 Peaks

Mass	Int.	Mass	Int.	Mass	Int.	Mass	Int.	Mass	Int.	Mass	Int.	Mass	Int.	Mass	Int.	Mass	Int.
26	34	27	47	39	275	40	132	50	60	55	84	65	223	66	276	77	9
79	9	94	999	95	109												

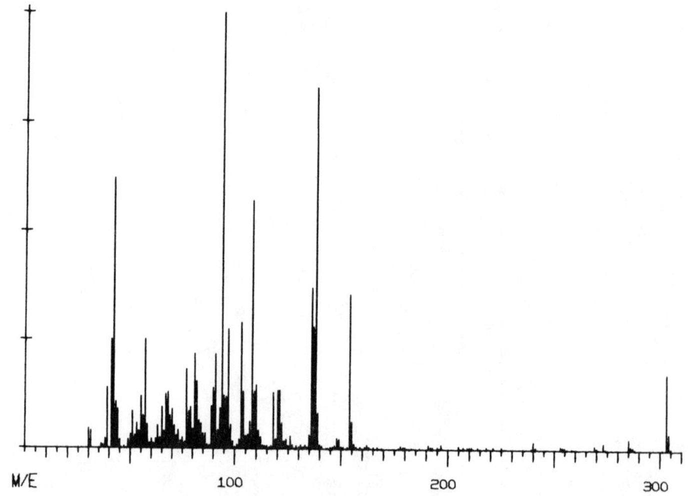

M/E
100
200
300

SCOPOLAMINE

$C_{17}H_{21}NO_4$ MW 303

Base Peak = 94 42 Peaks

Mass	Int.	Mass	Int.	Mass	Int.	Mass	Int.	Mass	Int.	Mass	Int.	Mass	Int.	Mass	Int.	Mass	Int.
30	46	31	40	41	250	42	621	55	121	57	251	67	124	68	129	77	183
81	217	94	999	103	289	108	571	110	147	120	135	121	135	136	371	138	829
154	356	155	63	162	9	165	6	178	7	179	6	191	9	197	9	205	7
211	6	218	5	225	5	240	18	241	6	253	7	255	5	269	9	270	4
273	14	285	26	286	10	287	8	303	175	304	37						

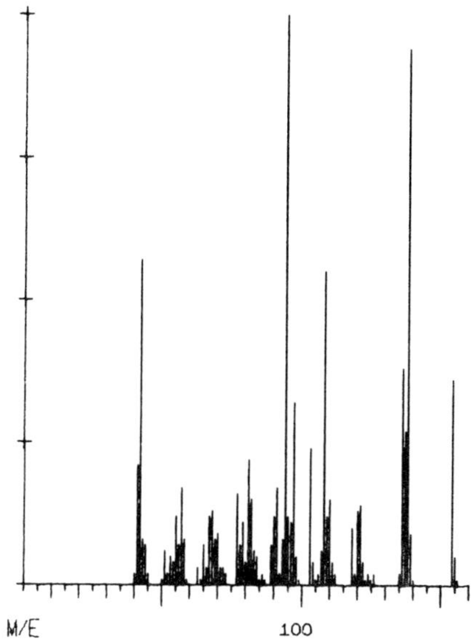
M/E
100

SCOPOLAMINE

$C_{17}H_{21}NO_4$ MW = 303

Base Peak = 94 18 Peaks

Mass	Int.	Mass	Int.	Mass	Int.	Mass	Int.	Mass	Int.	Mass	Int.	Mass	Int.	Mass	Int.	Mass	Int.
41	210	42	570	55	120	57	170	67	120	68	130	77	160	81	220	94	999
97	320	108	550	110	150	120	130	121	140	136	380	138	940	154	360	155	50

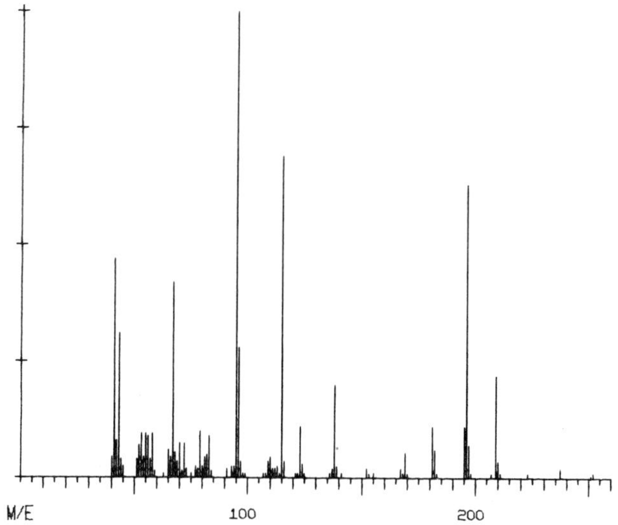
M/E
100
200

BUTALBITAL 2,4(OR 4,6)-DIMETHYL DERIVATIVE

$C_{13}H_{20}N_2O_3$ MW = 252

Base Peak = 95 30 Peaks

Mass	Int.	Mass	Int.	Mass	Int.	Mass	Int.	Mass	Int.	Mass	Int.	Mass	Int.	Mass	Int.	Mass	Int.
41	470	43	310	53	95	55	95	67	420	70	75	79	100	83	90	95	999
96	280	110	45	115	690	123	110	124	30	138	200	139	25	152	20	153	10
167	20	169	55	181	110	182	60	195	110	196	630	209	220	210	35	223	10
237	20	251	5	252	10	181	110	182	60	195	110	196	630	209	220	210	35
223	10	237	20	251	5												

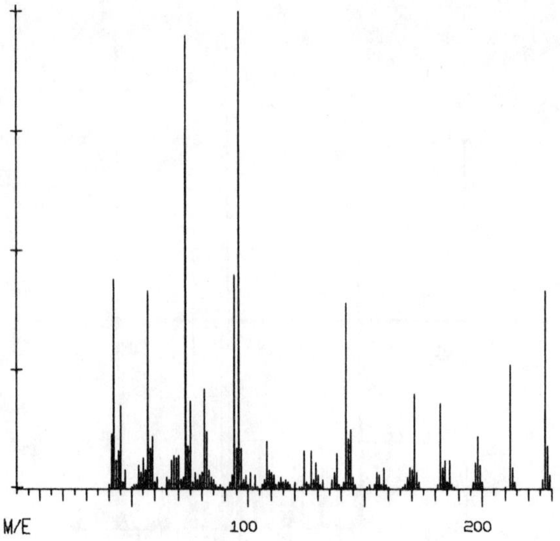

M/E
100
200

SCOPOLINE TMS ETHER

$C_{11}H_{21}NO_2SI$ MW = 227
Base Peak = 96 28 Peaks

Mass	Int.	Mass	Int.	Mass	Int.	Mass	Int.	Mass	Int.	Mass	Int.	Mass	Int.	Mass	Int.	Mass	Int.
42	440	45	175	57	415	59	110	73	950	75	185	81	210	82	120	94	450
96	999	108	100	109	40	124	80	127	80	142	390	144	125	155	35	158	45
169	45	171	200	182	180	184	60	197	55	198	110	212	260	213	45	227	415
228	90																

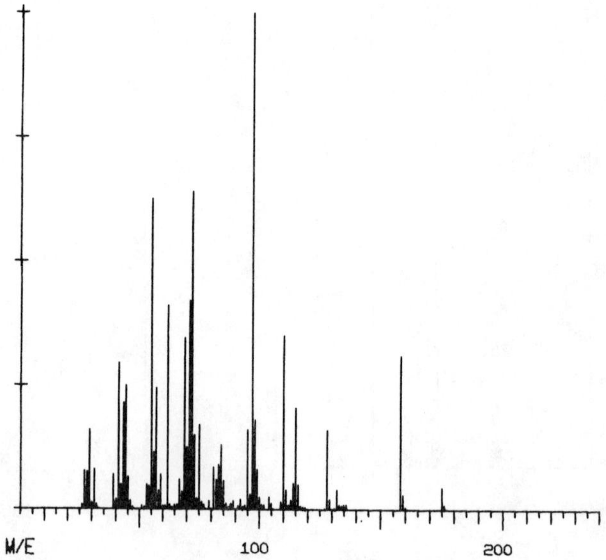

M/E
100
200

MEBUTAMATE

$C_{10}H_{20}N_2O_4$ MW = 232
Base Peak = 97 23 Peaks

Mass	Int.	Mass	Int.	Mass	Int.	Mass	Int.	Mass	Int.	Mass	Int.	Mass	Int.	Mass	Int.	Mass	Int.
29	159	31	79	41	294	44	249	55	624	57	245	71	421	72	640	83	89
84	129	97	999	98	179	110	349	115	204	128	159	129	19	132	39	135	11
158	309	159	29	160	4	175	44	176	9								

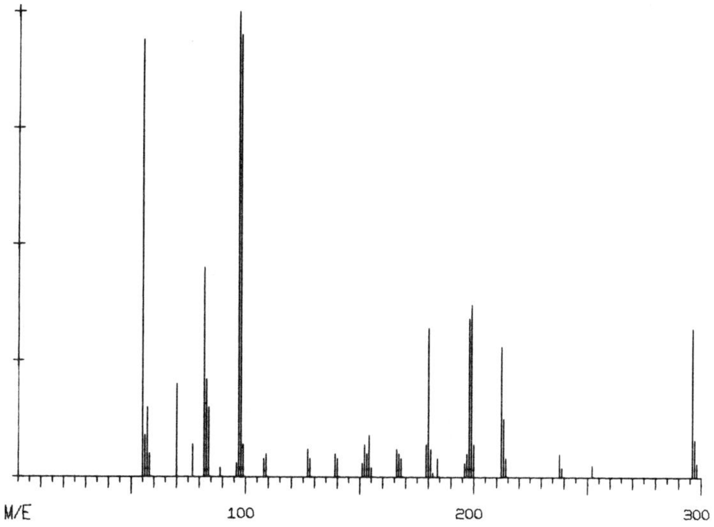
M/E
100
200
300

METHDILAZINE

$C_{18}H_{20}N_2S_1$ MW = 296

Base Peak = 97   28 Peaks

Mass	Int.	Mass	Int.	Mass	Int.	Mass	Int.	Mass	Int.	Mass	Int.	Mass	Int.	Mass	Int.	Mass	Int.
55	939	57	149	70	199	82	449	83	209	97	999	98	949	108	39	109	49
127	59	128	39	139	49	140	39	152	69	154	89	166	59	167	49	179	69
180	319	198	339	199	369	212	279	213	124	238	49	239	19	252	24	296	319
297	79																

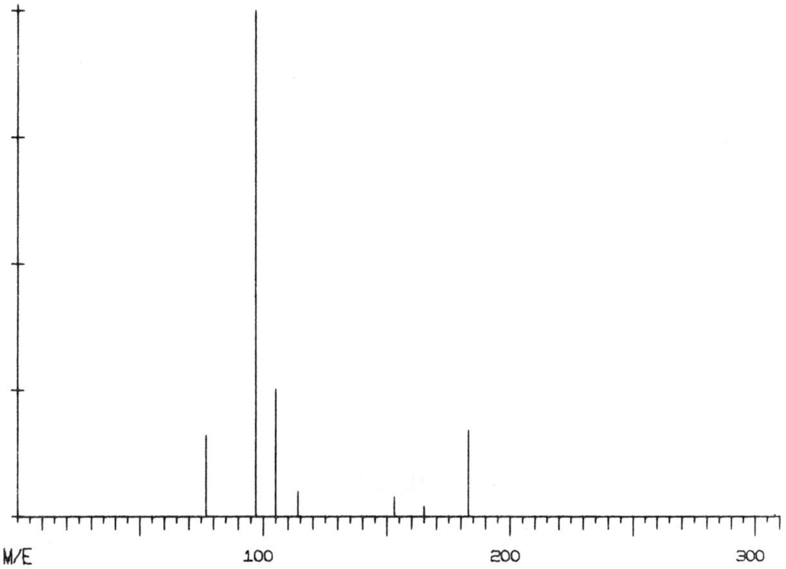
M/E
100
200
300

MEPENZOLATE BROMIDE

$C_{20}H_{23}NO_3$ MW = 325

Base Peak = 97   8 Peaks

Mass	Int.	Mass	Int.	Mass	Int.	Mass	Int.	Mass	Int.	Mass	Int.	Mass	Int.	Mass	Int.
77	161	97	999	105	252	114	50	153	40	165	20	183	171	308	10

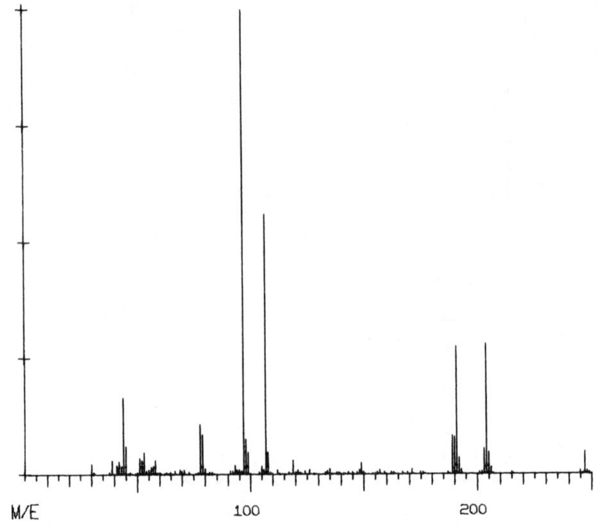

M/E
100
200

METHAPYRILENE MTB. 1

$C_{13}H_{17}N_3S$ MW = 247

Base Peak = 97 30 Peaks

Mass	Int.	Mass	Int.	Mass	Int.	Mass	Int.	Mass	Int.	Mass	Int.	Mass	Int.	Mass	Int.	Mass	Int.
30	23	31	4	44	166	45	59	51	36	53	47	69	11	71	11	78	107
79	86	97	999	98	75	107	560	108	48	119	29	121	11	134	7	135	12
148	11	149	26	162	5	171	10	175	5	187	5	189	83	191	274	203	54
204	279	247	47	248	8												

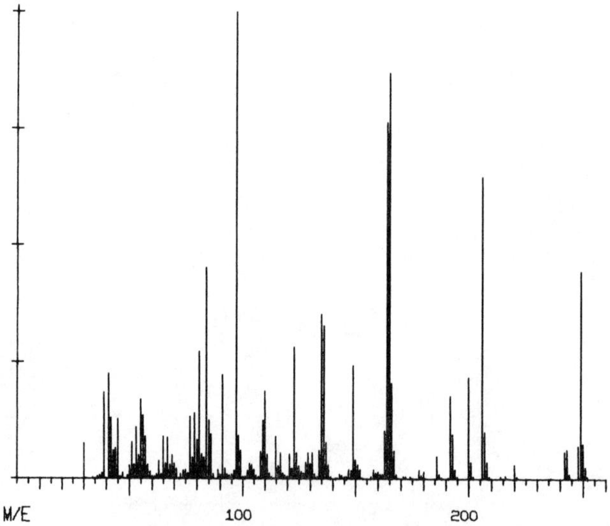

M/E
100
200

1-(1-THIOPHENYL CYCLOHEXYL) PIPERIDINE

$C_{15}H_{23}NS$ MW = 249

Base Peak = 97 33 Peaks

Mass	Int.	Mass	Int.	Mass	Int.	Mass	Int.	Mass	Int.	Mass	Int.	Mass	Int.	Mass	Int.	Mass	Int.
30	76	39	185	41	226	55	169	56	136	65	90	67	88	81	272	84	453
91	223	97	999	109	126	110	188	123	283	129	56	135	352	136	327	149	242
150	40	164	766	165	871	178	19	186	48	192	178	200	217	206	648	207	101
216	8	220	30	242	57	243	62	249	445	250	76						

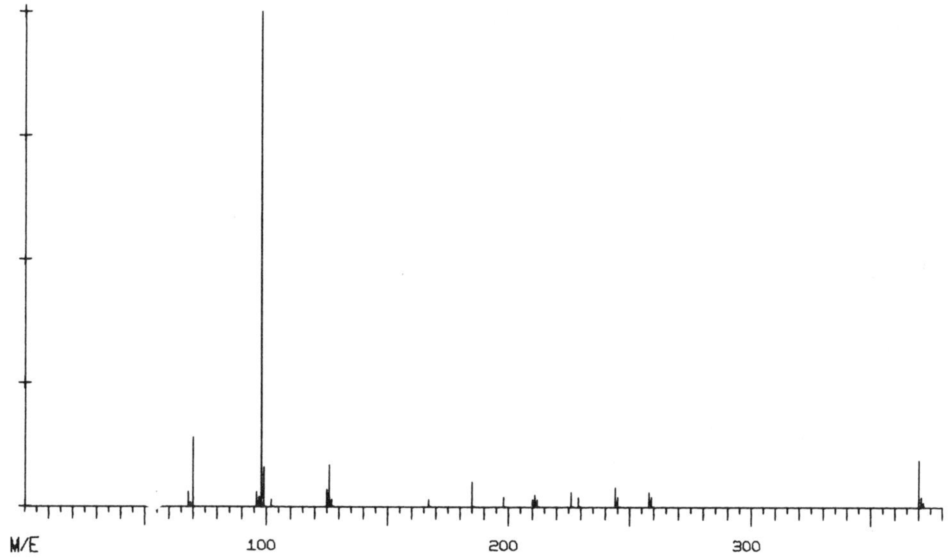
M/E
100
200
300

THIORIDAZINE

$C_{21}H_{26}N_2S_2$ MW = 370
Base Peak = 98 20 Peaks

Mass	Int.	Mass	Int.	Mass	Int.	Mass	Int.	Mass	Int.	Mass	Int.	Mass	Int.	Mass	Int.	Mass	Int.
55	29	68	29	70	139	98	999	99	79	125	34	126	84	167	14	185	49
198	19	210	14	211	24	226	29	229	19	244	39	245	19	258	29	259	19
370	94	371	19														

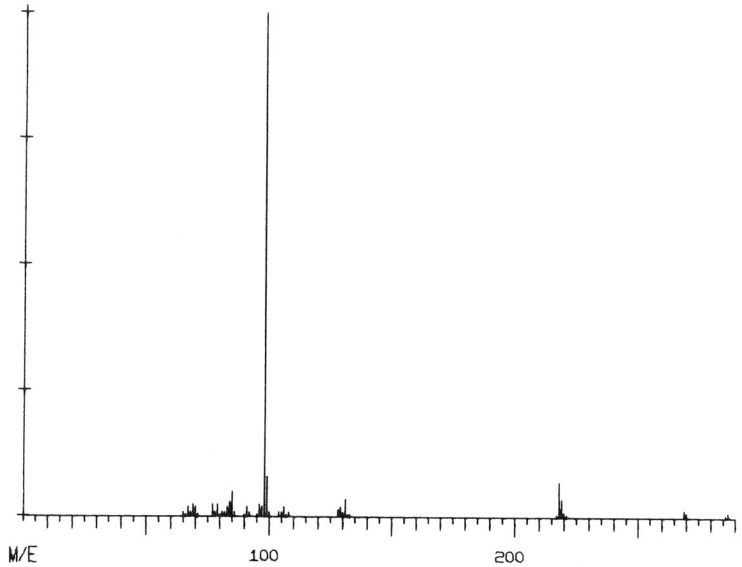
M/E
100
200

CYCRIMINE

$C_{19}H_{29}NO$ MW = 287
Base Peak = 98 27 Peaks

Mass	Int.	Mass	Int.	Mass	Int.	Mass	Int.	Mass	Int.	Mass	Int.	Mass	Int.	Mass	Int.	Mass	Int.
67	19	69	24	84	29	85	49	98	999	99	79	104	9	106	19	129	19
131	34	132	4	133	4	218	69	219	34	269	14	270	9	286	4	287	9

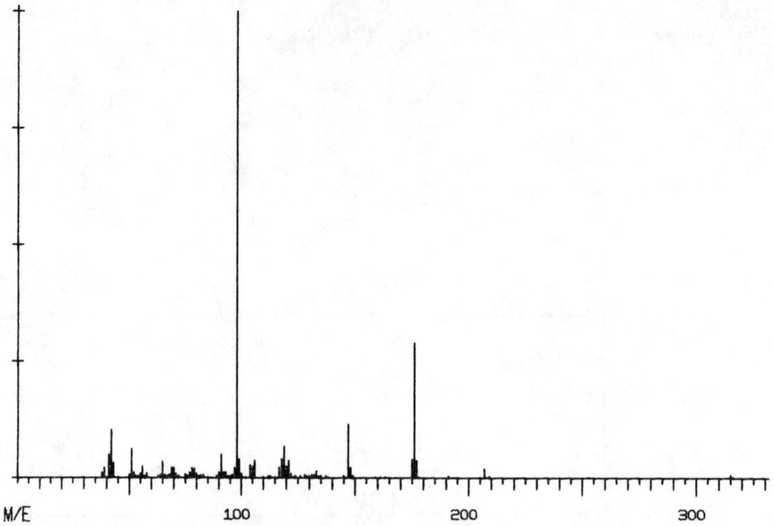
M/E
100
200
300

COTININE

$C_{10}H_{12}N_2O$ MW = 176

Base Peak = 98 30 Peaks

Mass	Int.	Mass	Int.	Mass	Int.	Mass	Int.	Mass	Int.	Mass	Int.	Mass	Int.	Mass	Int.	Mass	Int.
30	2	41	51	42	102	51	63	56	24	65	34	69	23	78	23	79	21
91	49	98	999	104	28	106	37	118	41	119	68	132	7	133	15	147	116
148	22	161	3	163	3	175	40	176	289	191	5	207	19	209	4	253	2
315	8	316	2	329	2												

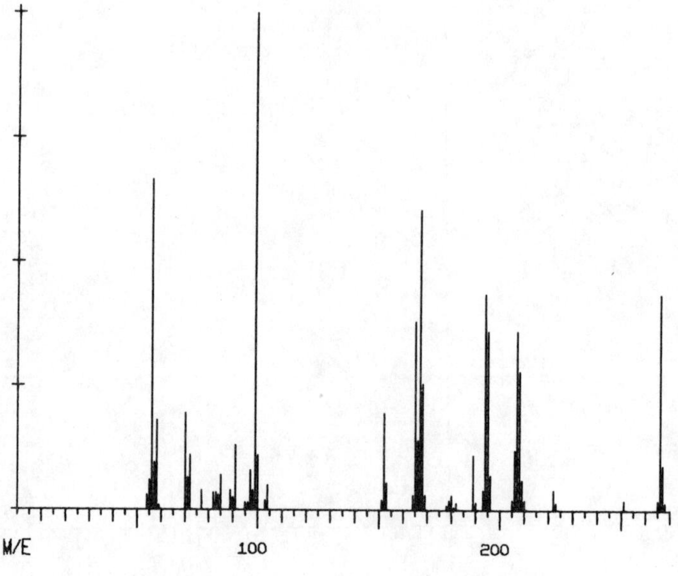
M/E
100
200

CYCLIZINE

$C_{18}H_{22}N_2$ MW = 266

Base Peak = 99 24 Peaks

Mass	Int.	Mass	Int.	Mass	Int.	Mass	Int.	Mass	Int.	Mass	Int.	Mass	Int.	Mass	Int.	Mass	Int.
56	664	58	179	70	194	72	109	77	39	85	69	91	129	99	999	104	49
152	194	153	54	165	379	167	604	179	19	180	29	194	434	195	359	207	359
208	279	222	39	223	14	251	19	266	434	267	89						

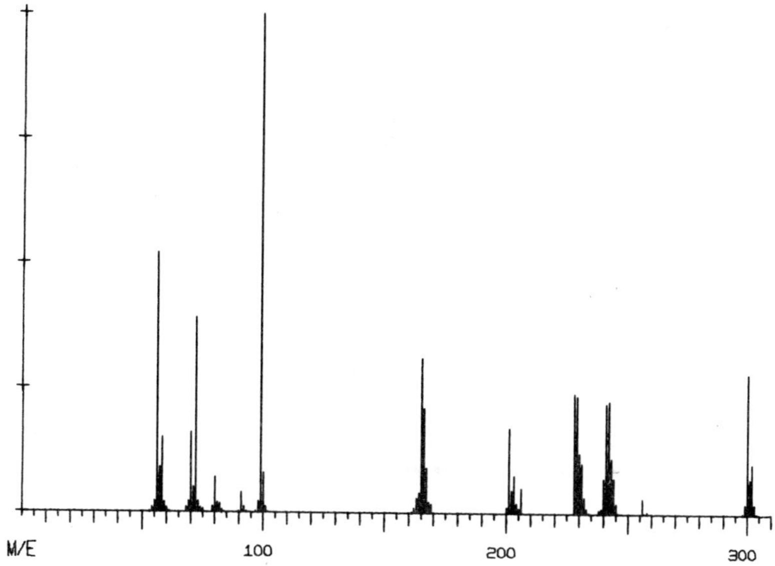
M/E
100
200
300

CHLORCYCLIZINE

$C_{18}H_{21}CLN_2$ MW = 300

Base Peak = 99 25 Peaks

Mass	Int.	Mass	Int.	Mass	Int.	Mass	Int.	Mass	Int.	Mass	Int.	Mass	Int.	Mass	Int.	Mass	Int.
56	519	58	149	70	159	72	389	80	69	81	20	99	999	100	79	165	309
166	209	200	13	201	169	203	74	206	49	228	239	229	234	241	219	242	224
244	69	256	29	258	4	298	3	299	20	300	279	302	99				

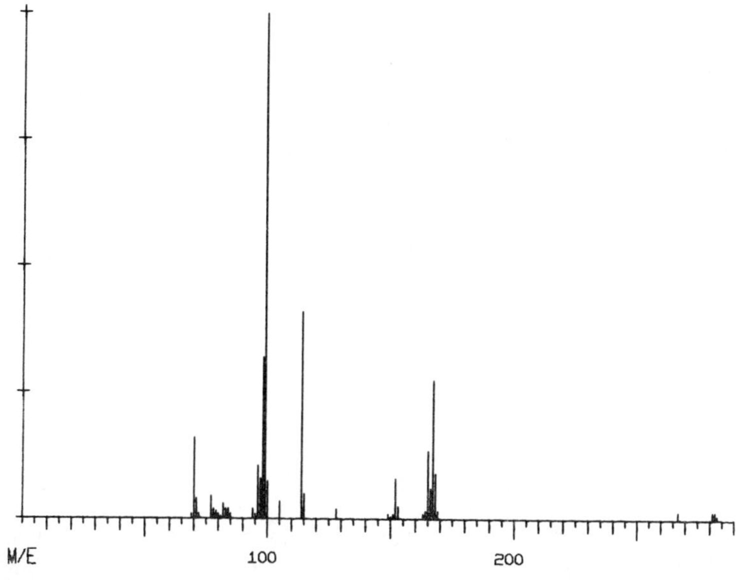
M/E
100
200

DIPHENYLPYRALINE

$C_{19}H_{23}NO$ MW = 281

Base Peak = 99 16 Peaks

Mass	Int.	Mass	Int.	Mass	Int.	Mass	Int.	Mass	Int.	Mass	Int.	Mass	Int.	Mass	Int.	Mass	Int.
70	159	71	39	77	44	82	29	98	319	99	999	114	409	115	49	128	19
152	79	153	24	165	134	167	274	267	14	281	14	282	14				

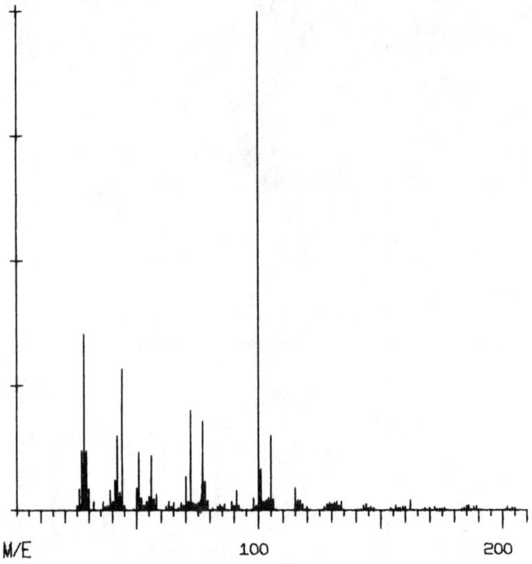

M/E
100
200

DIETHYLPROPION

$C_{13}H_{19}NO$ MW = 205

Base Peak = 100 28 Peaks

Mass	Int.	Mass	Int.	Mass	Int.	Mass	Int.	Mass	Int.	Mass	Int.	Mass	Int.	Mass	Int.	Mass	Int.
27	121	28	353	42	149	44	283	51	117	56	109	70	67	72	200	77	179
78	58	100	999	101	83	105	149	115	44	129	14	131	14	132	17	134	17
156	9	159	8	160	8	162	19	185	11	186	11	188	8	189	9	202	8
205	5																

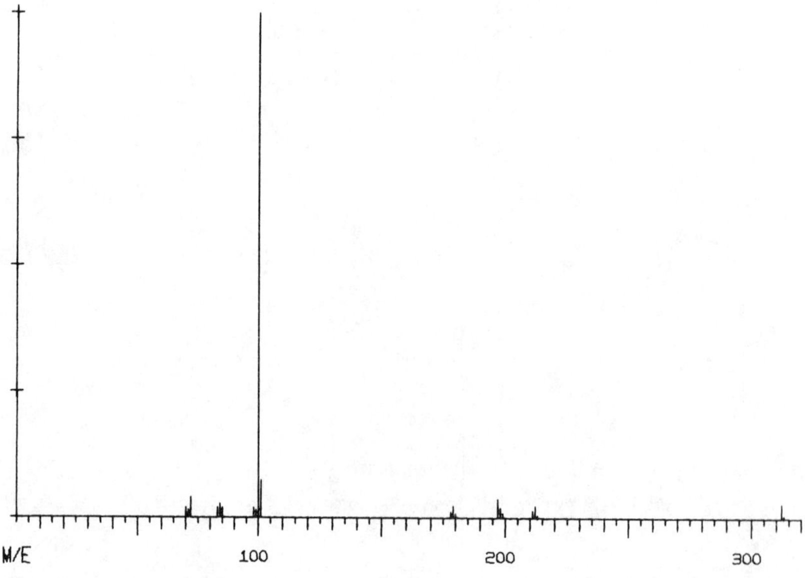

M/E
100
200
300

ETHOPROPAZINE

$C_{19}H_{24}N_2S$ MW = 312

Base Peak = 100 14 Peaks

Mass	Int.	Mass	Int.	Mass	Int.	Mass	Int.	Mass	Int.	Mass	Int.	Mass	Int.	Mass	Int.	Mass	Int.
70	19	72	39	83	21	84	27	100	999	101	74	178	13	179	24	197	38
198	19	211	14	212	24	312	30	313	4								

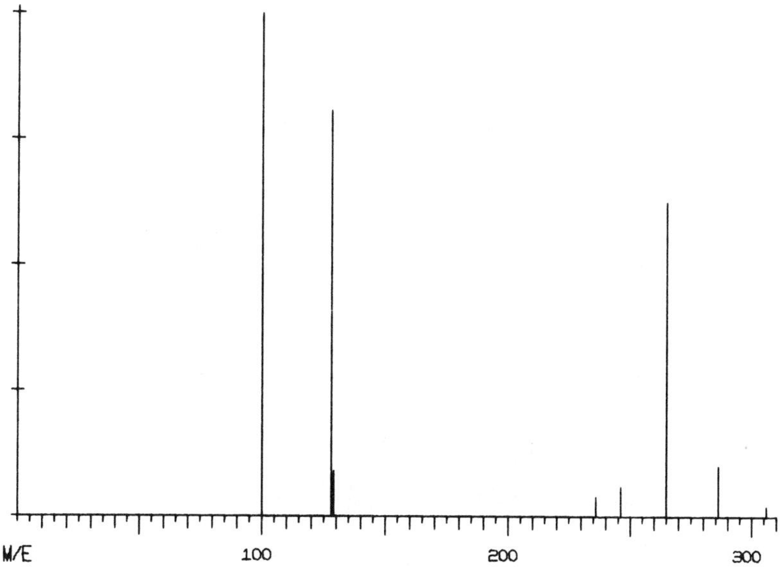
M/E
100
200
300

MORAMIDE

$C_{25}H_{32}N_2O_2$ MW = 392
Base Peak = 100 8 Peaks

Mass	Int.	Mass	Int.	Mass	Int.	Mass	Int.	Mass	Int.	Mass	Int.	Mass	Int.	Mass	Int.
100	999	128	808	129	90	236	40	246	60	265	626	286	101	306	20

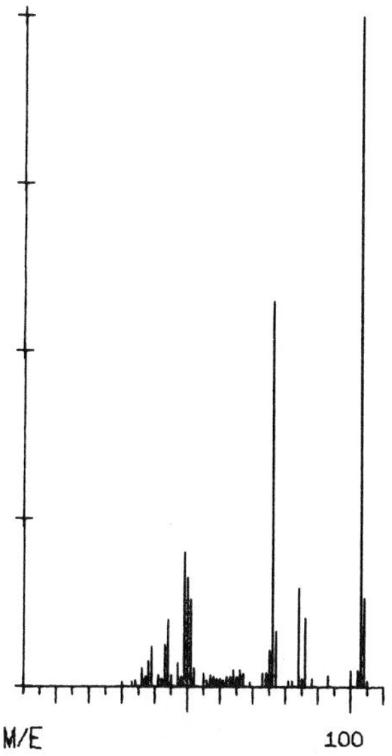
M/E
100

OXACILLIN MTB. 3

C_7H_5N MW = 103
Base Peak = 103 14 Peaks

Mass	Int.	Mass	Int.	Mass	Int.	Mass	Int.	Mass	Int.	Mass	Int.	Mass	Int.	Mass	Int.	Mass	Int.
30	7	33	7	43	62	44	101	49	201	50	163	64	24	75	54	76	76
84	147	102	225	103	999	104	133	105	11								

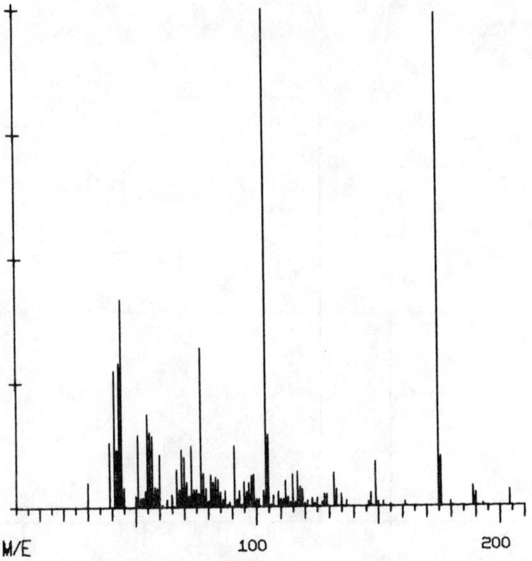
M/E
100
200

MEPHENYTOIN MTB. 1

$C_{11}H_{12}N_2O_2$ MW = 204

Base Peak = 104 25 Peaks

Mass	Int.	Mass	Int.	Mass	Int.	Mass	Int.	Mass	Int.	Mass	Int.	Mass	Int.	Mass	Int.	Mass	Int.
30	51	43	290	44	417	55	187	56	150	69	114	73	123	77	319	78	67
91	123	99	64	104	999	105	146	118	40	119	35	132	68	133	35	147	27
149	91	161	11	175	990	176	101	189	40	190	27	204	33				

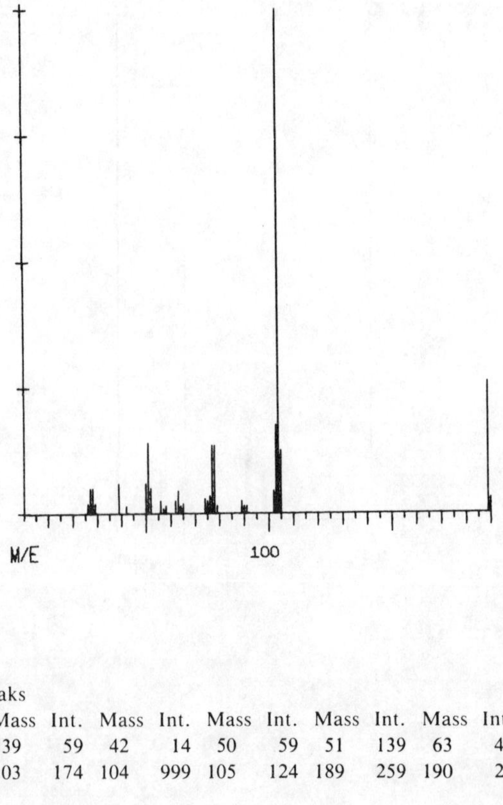
M/E
100

PHENSUXIMIDE

$C_{11}H_{11}NO_2$ MW = 189

Base Peak = 104 16 Peaks

Mass	Int.	Mass	Int.	Mass	Int.	Mass	Int.	Mass	Int.	Mass	Int.	Mass	Int.	Mass	Int.	Mass	Int.
27	49	28	49	39	59	42	14	50	59	51	139	63	44	74	29	77	134
78	134	102	44	103	174	104	999	105	124	189	259	190	29				

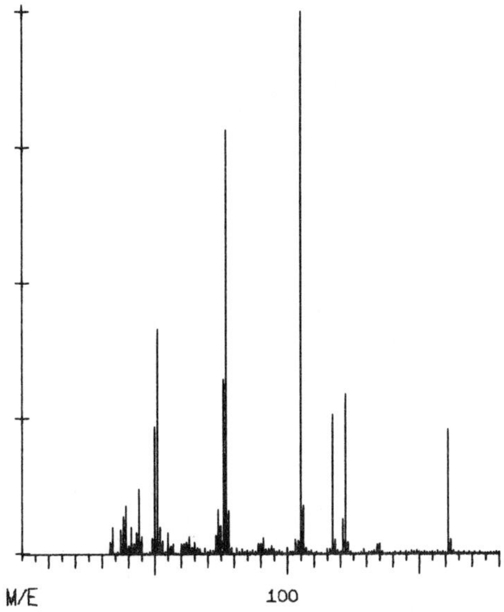
M/E
100

HIPPURIC ACID

$C_9H_9NO_3$ MW = 179

Base Peak = 105 23 Peaks

Mass	Int.	Mass	Int.	Mass	Int.	Mass	Int.	Mass	Int.	Mass	Int.	Mass	Int.	Mass	Int.	Mass	Int.
33	23	39	91	44	121	50	235	51	416	74	83	75	53	76	323	77	782
91	29	103	27	105	999	117	258	121	64	122	295	134	18	135	19	147	7
149	7	161	231	162	27	175	3	179	4								

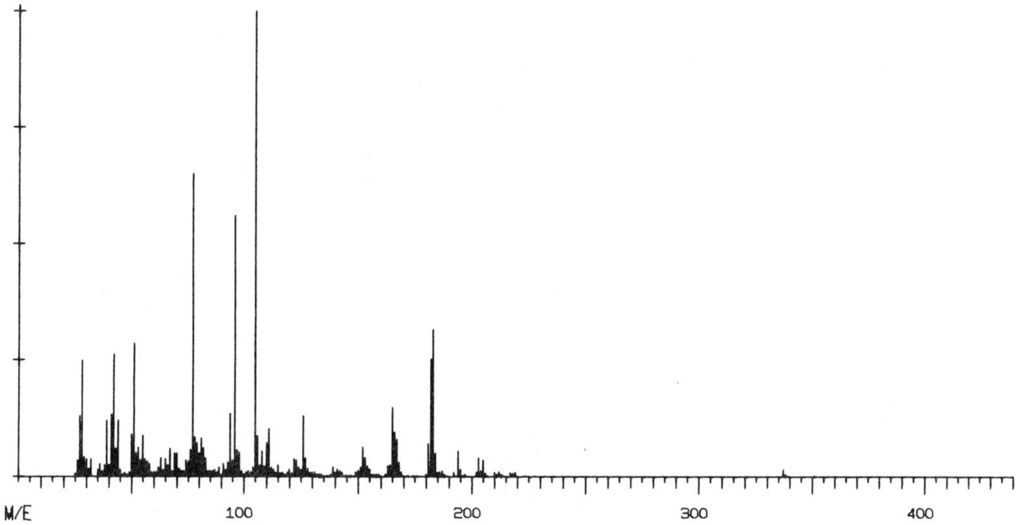
M/E
100
200
300
400

CLIDINIUM BROMIDE

$C_{22}H_{26}Br_1N_1O_3$ MW = 432

Base Peak = 105 35 Peaks

Mass	Int.	Mass	Int.	Mass	Int.	Mass	Int.	Mass	Int.	Mass	Int.	Mass	Int.	Mass	Int.	Mass	Int.
27	129	28	249	41	132	42	262	50	91	51	286	67	60	70	51	77	649
78	84	94	135	96	560	105	999	111	102	126	130	127	39	139	21	141	14
152	63	153	41	165	147	166	96	182	252	183	316	194	54	195	14	203	41
205	36	218	8	219	10	321	2	322	1	337	14	338	4	351	2		

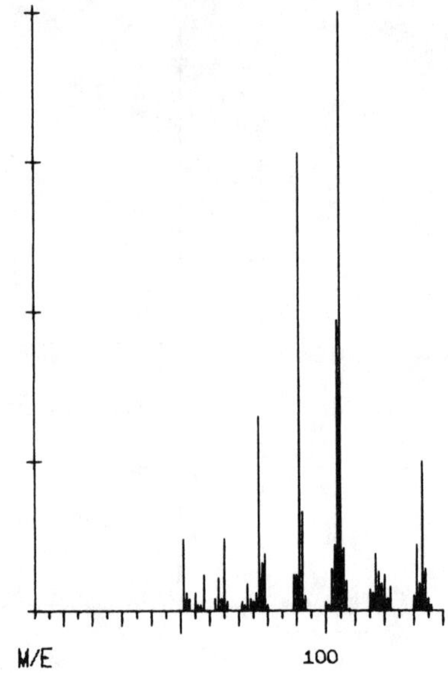
M/E
100

PHENELZINE

$C_8H_{12}N_2$ MW = 136

Base Peak = 105 14 Peaks

Mass	Int.	Mass	Int.	Mass	Int.	Mass	Int.	Mass	Int.	Mass	Int.	Mass	Int.	Mass	Int.	Mass	Int.
51	119	58	59	63	54	65	119	77	324	79	94	91	764	92	164	104	484
105	999	118	64	131	109	133	249	134	69								

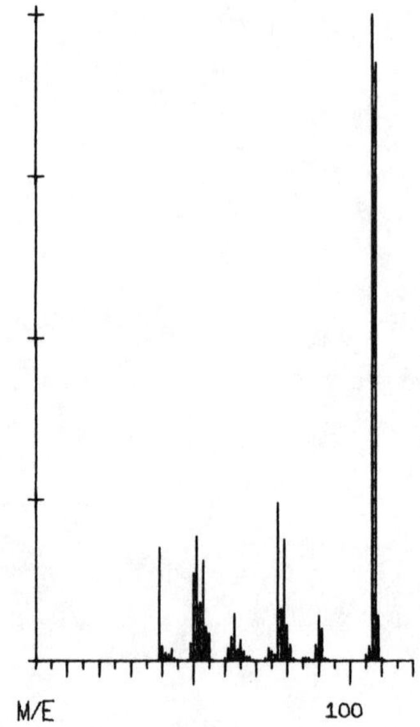
M/E
100

P-CRESOL

C_7H_8O MW = 108

Base Peak = 107 12 Peaks

Mass	Int.	Mass	Int.	Mass	Int.	Mass	Int.	Mass	Int.	Mass	Int.	Mass	Int.	Mass	Int.	Mass	Int.
39	174	40	22	51	192	53	154	62	38	63	73	77	244	79	188	90	71
91	49	107	999	108	926												

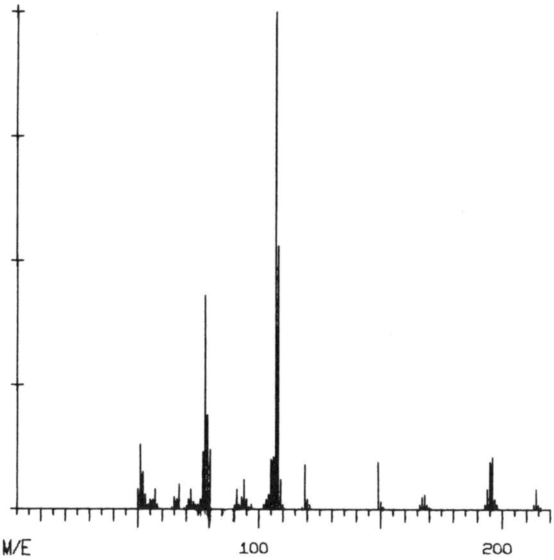
M/E
100
200

PHENYRAMIDOL

$C_{13}H_{14}N_2O$ MW = 214

Base Peak = 107 21 Peaks

Mass	Int.	Mass	Int.	Mass	Int.	Mass	Int.	Mass	Int.	Mass	Int.	Mass	Int.	Mass	Int.	Mass	Int.
51	129	52	74	67	49	72	39	78	429	79	189	91	39	94	59	107	999
108	529	119	89	120	19	149	94	150	14	167	24	168	29	195	94	196	104
213	9	214	39	216	4												

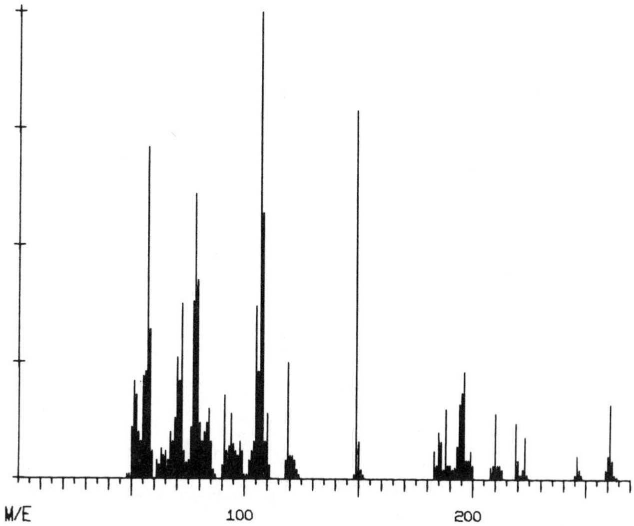
M/E
100
200

ETHOHEPTAZINE

$C_{16}H_{23}NO_2$ MW = 261

Base Peak = 107 26 Peaks

Mass	Int.	Mass	Int.	Mass	Int.	Mass	Int.	Mass	Int.	Mass	Int.	Mass	Int.	Mass	Int.	Mass	Int.
57	709	58	319	70	259	72	374	78	609	79	424	91	179	94	139	107	999
108	569	119	249	120	49	149	789	150	79	185	99	186	79	195	184	196	229
209	29	210	139	219	119	223	89	246	49	247	19	260	49	261	159		

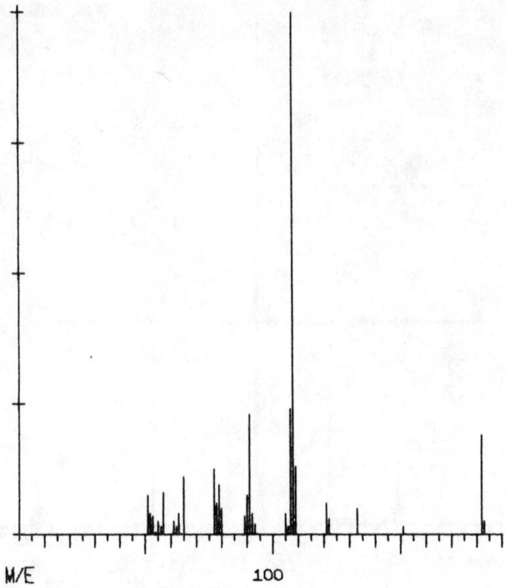
M/E
100

MEPHENESIN

$C_{10}H_{14}O_3$ MW = 182

Base Peak = 108 16 Peaks

Mass	Int.	Mass	Int.	Mass	Int.	Mass	Int.	Mass	Int.	Mass	Int.	Mass	Int.	Mass	Int.	Mass	Int.
51	74	57	79	63	39	65	109	77	124	79	94	90	74	91	229	107	239
108	999	121	59	122	29	133	49	151	14	182	189	183	24				

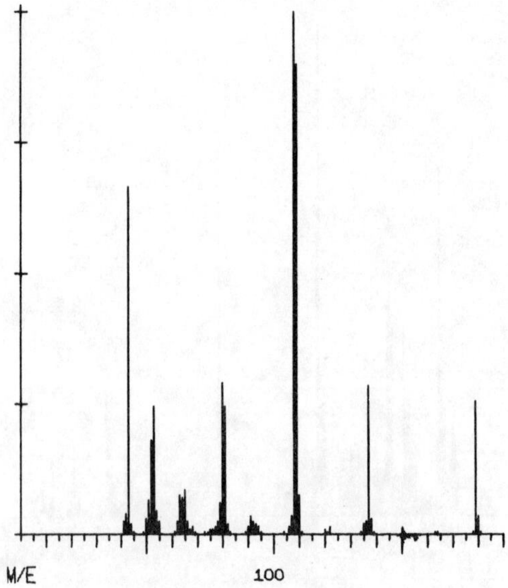
M/E
100

ACETOPHENETIDIN

$C_{10}H_{13}NO_2$ MW = 179

Base Peak = 108 22 Peaks

Mass	Int.	Mass	Int.	Mass	Int.	Mass	Int.	Mass	Int.	Mass	Int.	Mass	Int.	Mass	Int.	Mass	Int.
42	40	43	665	52	180	53	245	63	75	65	85	80	290	81	245	91	35
92	25	108	999	109	900	120	10	122	15	137	285	138	30	150	15	151	5
163	5	164	5	179	255	180	35										

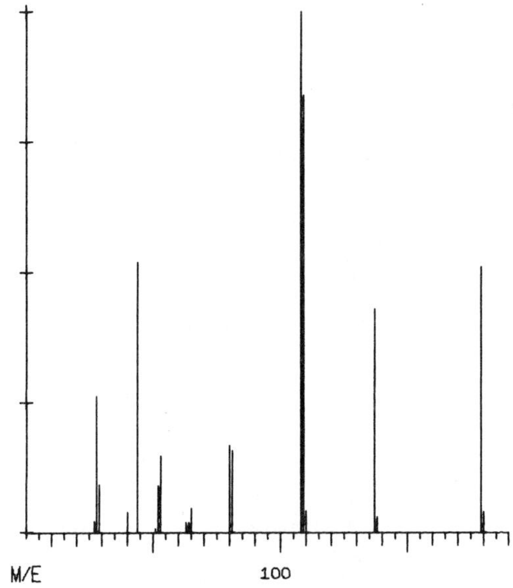
M/E
100

ACETOPHENETIDIN

$C_{10}H_{13}NO_2$ MW = 179

Base Peak = 108 16 Peaks

Mass	Int.	Mass	Int.	Mass	Int.	Mass	Int.	Mass	Int.	Mass	Int.	Mass	Int.	Mass	Int.	Mass	Int.
28	263	29	93	40	40	44	519	52	89	53	147	63	21	65	48	80	168
81	158	108	999	109	839	137	431	138	29	179	511	180	39				

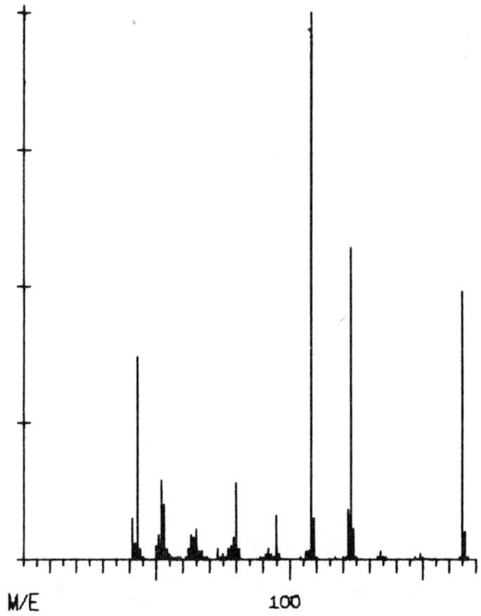
M/E
100

ACETAMINOPHEN METHYL ETHER

$C_9H_{11}O_2$ MW = 151

Base Peak = 108 20 Peaks

Mass	Int.	Mass	Int.	Mass	Int.	Mass	Int.	Mass	Int.	Mass	Int.	Mass	Int.	Mass	Int.	Mass	Int.
41	75	43	370	52	145	53	100	63	45	65	55	79	40	80	140	92	20
95	80	108	999	109	75	122	90	123	570	133	5	134	15	147	5	149	10
165	490	166	50														

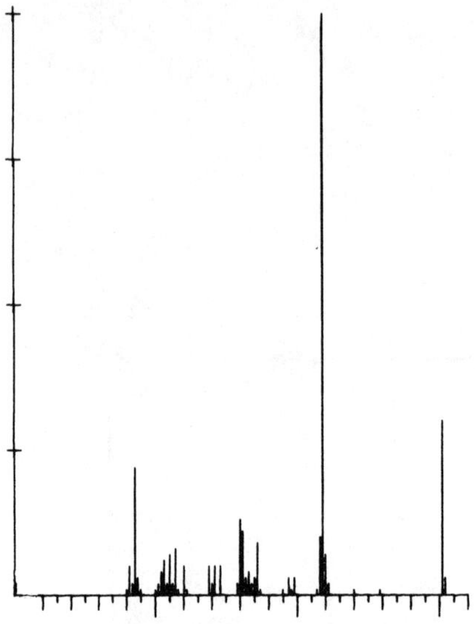

M/E
100

ACETAMINOPHEN

$C_8H_9NO_2$ MW = 151

Base Peak = 109 16 Peaks

Mass	Int.	Mass	Int.	Mass	Int.	Mass	Int.	Mass	Int.	Mass	Int.	Mass	Int.	Mass	Int.	Mass	Int.
41	50	43	220	55	70	57	80	69	50	71	50	80	130	81	110	97	30
99	30	108	100	109	999	120	10	129	10	151	300	152	30				

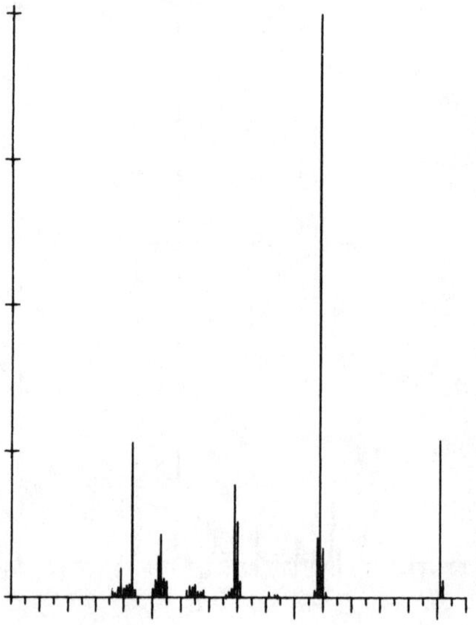

M/E
100

ACETAMINOPHEN

$C_8H_9NO_2$ MW = 151

Base Peak = 109 14 Peaks

Mass	Int.	Mass	Int.	Mass	Int.	Mass	Int.	Mass	Int.	Mass	Int.	Mass	Int.	Mass	Int.	Mass	Int.
39	49	43	264	52	71	53	107	63	21	65	22	79	193	80	129	91	9
93	6	108	103	109	999	151	269	152	30								

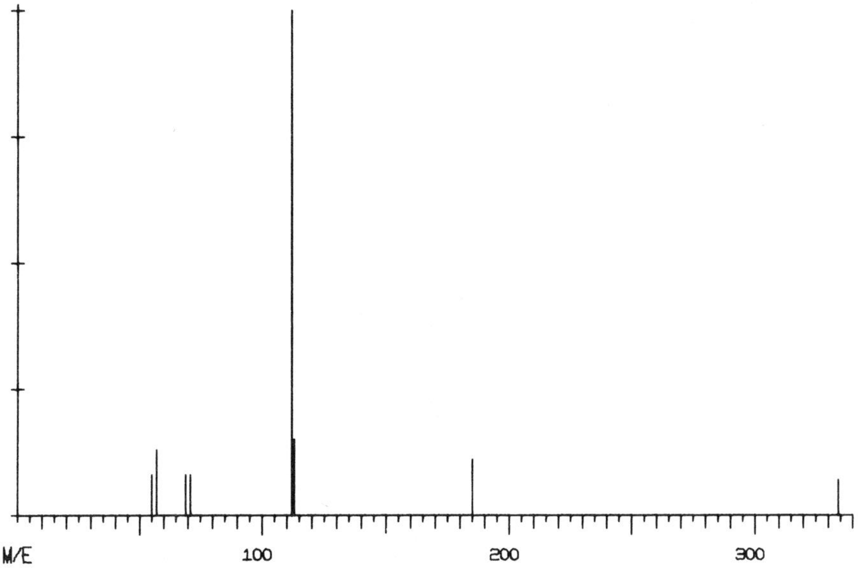
M/E
100
200
300

DIPIPANONE

$C_{24}H_{31}NO$ MW = 349

Base Peak = 112 8 Peaks

Mass	Int.	Mass	Int.	Mass	Int.	Mass	Int.	Mass	Int.	Mass	Int.	Mass	Int.	Mass	Int.
55	80	57	131	69	80	71	80	112	999	113	151	185	111	334	70

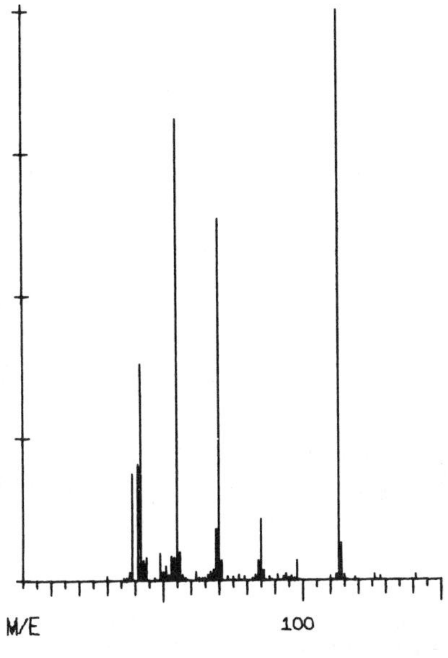
M/E
100

ETHOSUXIMIDE

$C_7H_{11}NO_2$ MW = 141

Base Peak = 113 16 Peaks

Mass	Int.	Mass	Int.	Mass	Int.	Mass	Int.	Mass	Int.	Mass	Int.	Mass	Int.	Mass	Int.	Mass	Int.
30	8	41	206	42	379	55	810	56	50	69	89	70	636	84	34	85	108
94	12	98	36	113	999	114	64	126	10	128	8	141	10				

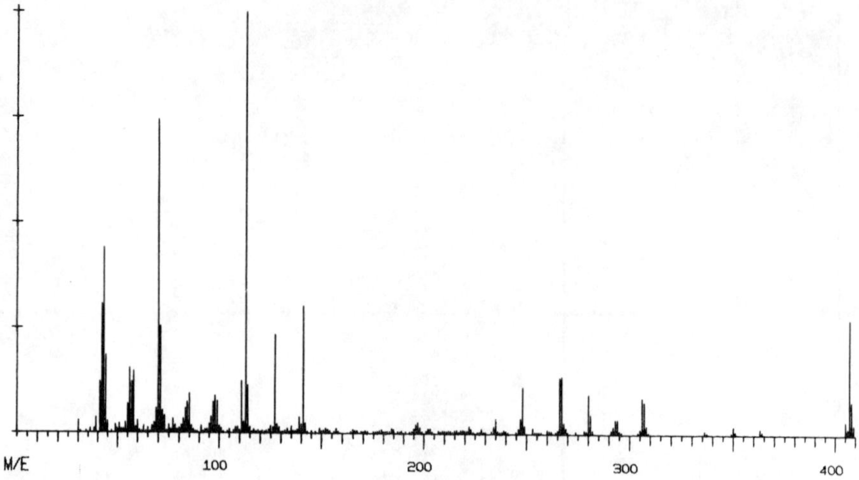
M/E
100
200
300
400

TRIFLUORPERAZINE

$C_{21}H_{24}F_3N_3S$ MW = 407

Base Peak = 113 50 Peaks

Mass	Int.	Mass	Int.	Mass	Int.	Mass	Int.	Mass	Int.	Mass	Int.	Mass	Int.	Mass	Int.	Mass	Int.
30	29	31	5	42	307	43	441	56	156	58	148	70	746	71	255	84	76
85	96	98	89	99	78	111	124	113	999	127	234	128	23	139	41	141	302
149	13	152	12	165	11	166	8	179	9	184	13	196	23	197	28	202	13
203	12	222	18	223	12	234	17	235	37	247	38	248	113	266	135	267	137
280	94	281	47	293	35	294	36	306	87	307	77	336	10	337	4	350	21
351	7	363	16	364	7	407	276	408	80								

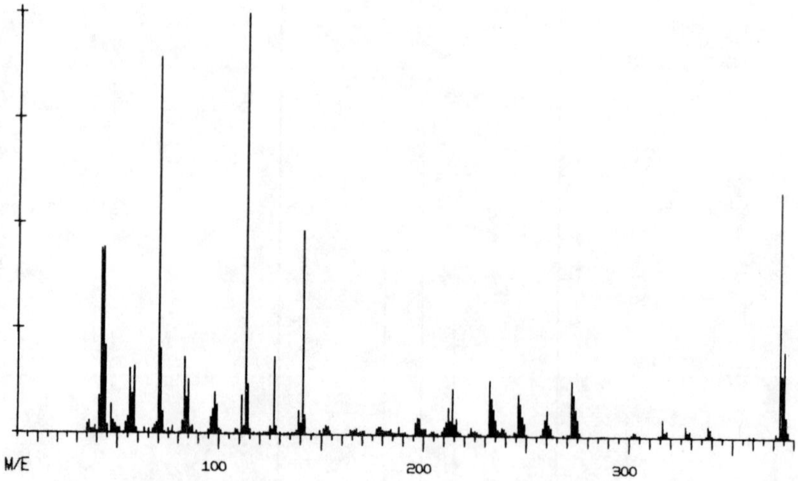
M/E
100
200
300

PROCHLORPERAZINE

$C_{20}H_{24}CLN_3S$ MW = 373

Base Peak = 113 50 Peaks

Mass	Int.	Mass	Int.	Mass	Int.	Mass	Int.	Mass	Int.	Mass	Int.	Mass	Int.	Mass	Int.	Mass	Int.
42	441	43	443	56	156	58	161	70	896	71	203	83	183	85	130	98	99
99	69	113	999	114	119	125	19	127	186	139	57	141	486	152	23	153	21
164	13	167	16	178	20	179	21	197	43	198	39	212	68	214	113	216	41
223	19	232	133	233	91	246	101	247	78	259	41	260	63	272	133	273	101
286	3	287	3	302	12	303	9	316	42	327	27	338	27	339	21	343	2
344	2	358	6	360	4	373	585	375	208								

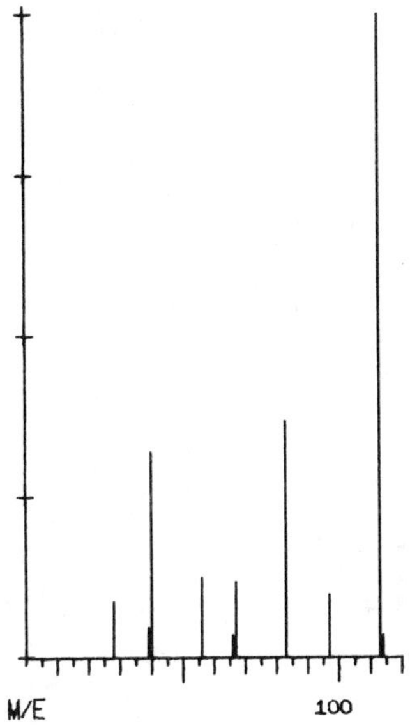
M/E
100

2-NITRO IMIDAZOLE

$C_3H_3N_3O_2$ MW = 113

Base Peak = 113 10 Peaks

Mass	Int.	Mass	Int.	Mass	Int.	Mass	Int.	Mass	Int.	Mass	Int.	Mass	Int.	Mass	Int.	Mass	Int.
28	87	39	47	40	319	56	126	66	35	67	117	83	367	97	98	113	999
114	34																

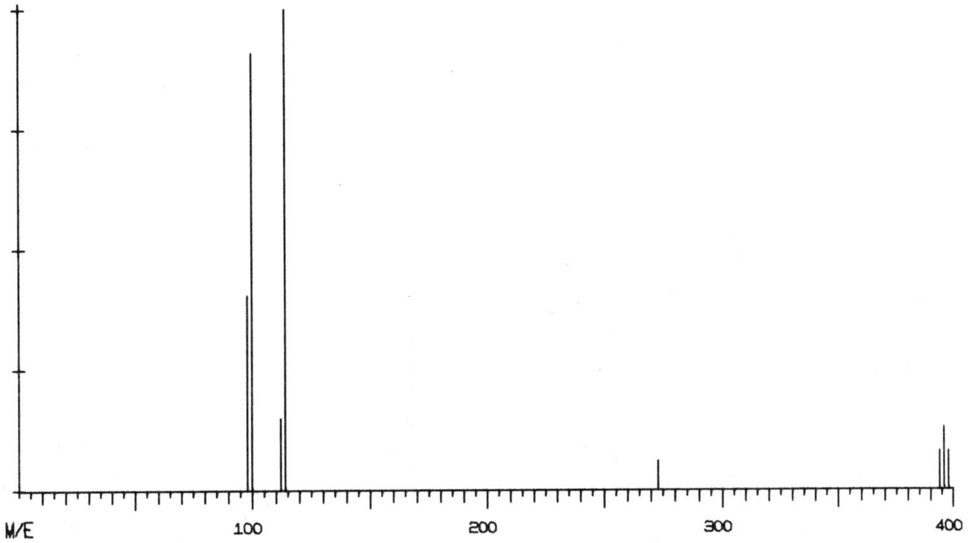
M/E
100
200
300
400

PHOLCODINE

$C_{23}H_{30}N_2O_4$ MW = 398

Base Peak = 114 8 Peaks

Mass	Int.	Mass	Int.	Mass	Int.	Mass	Int.	Mass	Int.	Mass	Int.	Mass	Int.	Mass	Int.
98	404	100	909	112	151	114	999	273	60	394	80	396	131	398	80

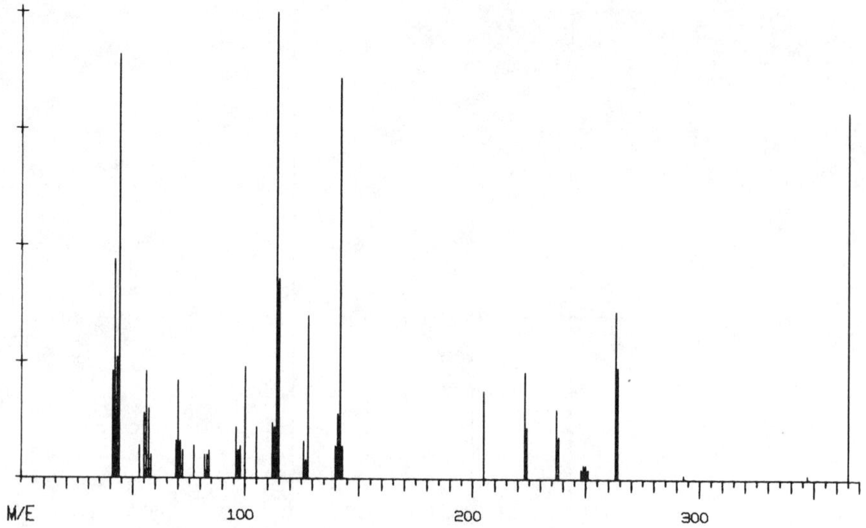
M/E
100
200
300

PERICYAZINE

$C_{21}H_{23}N_3O_1S_1$ MW = 365

Base Peak = 114 28 Peaks

Mass	Int.	Mass	Int.	Mass	Int.	Mass	Int.	Mass	Int.	Mass	Int.	Mass	Int.	Mass	Int.	Mass	Int.
42	469	44	909	56	229	57	149	69	79	70	209	77	69	84	59	96	109
100	239	114	999	115	429	126	79	128	349	141	139	142	859	205	189	223	229
224	109	237	149	238	89	249	29	250	29	263	359	264	239	293	9	347	9
365	789																

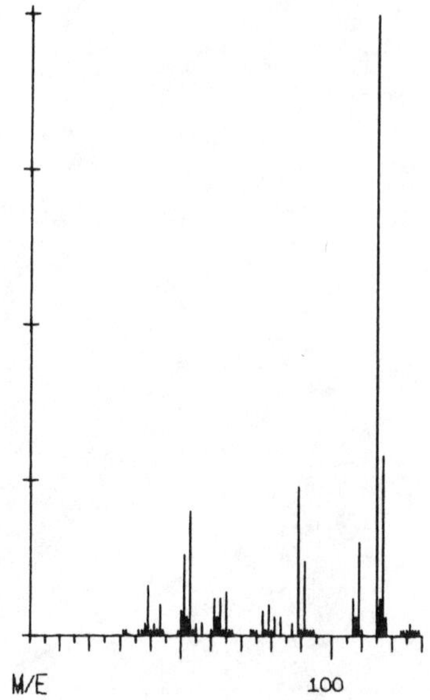
M/E
100

ETHCHLORVYNOL

C_7H_9CLO MW = 144

Base Peak = 115 16 Peaks

Mass	Int.	Mass	Int.	Mass	Int.	Mass	Int.	Mass	Int.	Mass	Int.	Mass	Int.	Mass	Int.	Mass	Int.
31	10	32	10	39	80	43	50	51	130	53	200	63	60	65	70	79	50
89	240	90	10	91	120	115	999	117	290	118	30	126	20				

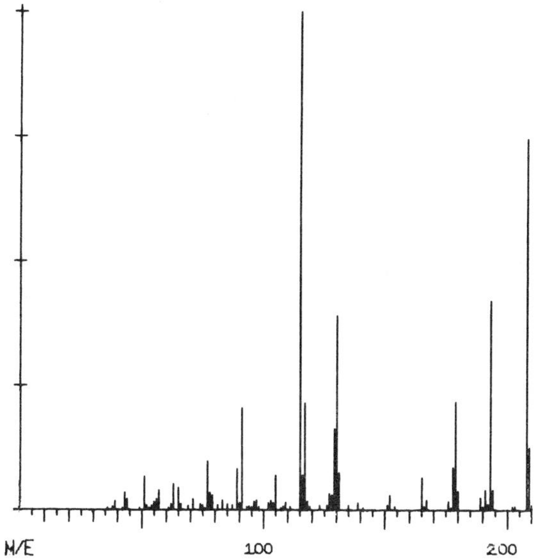
M/E
100
200

PROPOXYPHENE DECOMPOSITION PRODUCT, PEAK 1

$C_{16}H_{16}$ MW = 208

Base Peak = 115 26 Peaks

Mass	Int.	Mass	Int.	Mass	Int.	Mass	Int.	Mass	Int.	Mass	Int.	Mass	Int.	Mass	Int.	Mass	Int.
43	34	44	22	51	67	57	39	63	53	65	45	77	98	89	82	91	206
103	20	115	999	117	214	129	163	130	389	135	9	139	15	151	11	152	29
165	64	167	21	178	85	179	217	191	39	193	420	208	745	209	124		

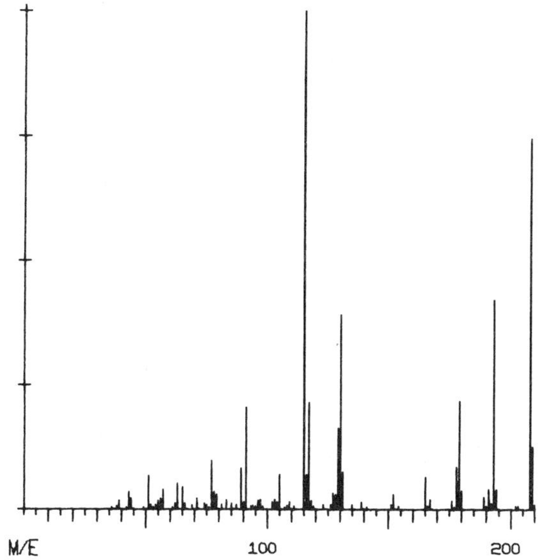
M/E
100
200

PROPOXYPHENE DECOMPOSITION PEAK 2

$C_{16}H_{16}$ MW = 208

Base Peak = 115 26 Peaks

Mass	Int.	Mass	Int.	Mass	Int.	Mass	Int.	Mass	Int.	Mass	Int.	Mass	Int.	Mass	Int.	Mass	Int.
43	34	44	22	51	67	57	39	63	53	65	45	77	98	89	82	91	206
103	20	115	999	117	214	129	163	130	389	135	9	139	15	151	11	152	29
165	64	167	21	178	85	179	217	191	39	193	420	208	745	209	124		

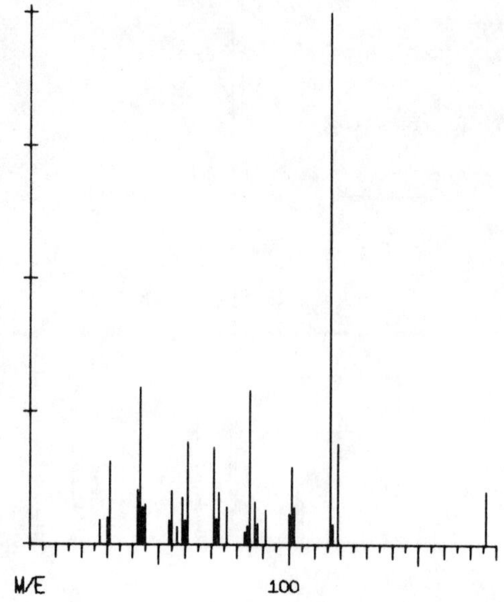
M/E
100

ASCORBIC ACID

$C_6H_8O_6$ MW = 176

Base Peak = 116 16 Peaks

Mass	Int.	Mass	Int.	Mass	Int.	Mass	Int.	Mass	Int.	Mass	Int.	Mass	Int.	Mass	Int.	Mass	Int.
30	51	31	156	42	102	43	294	55	100	61	192	71	183	73	97	85	290
87	80	101	145	102	70	116	999	117	37	119	189	176	100				

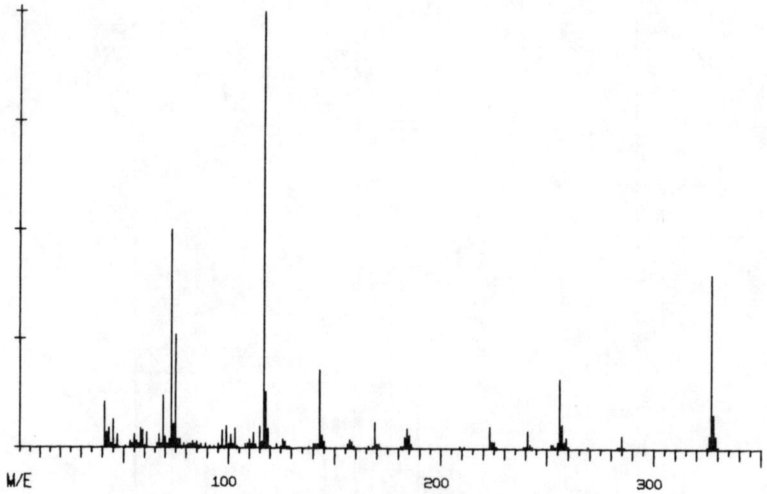
M/E
100
200
300

HYDROXYPENTOBARBITAL 1,3-DIMETHYL DERIVATIVE TMS ETHER

$C_{16}H_{30}N_2O_4Si$ MW = 342

Base Peak = 117 40 Peaks

Mass	Int.	Mass	Int.	Mass	Int.	Mass	Int.	Mass	Int.	Mass	Int.	Mass	Int.	Mass	Int.	Mass	Int.
41	105	45	65	58	45	59	40	73	500	75	260	76	20	77	20	99	50
103	45	115	50	117	999	118	130	119	45	143	180	144	30	157	20	158	15
169	60	170	10	184	45	185	30	209	5	211	5	223	50	224	15	241	40
242	10	256	160	257	55	258	15	259	25	283	5	285	30	286	5	326	30
327	400	328	80	329	30	342	1										

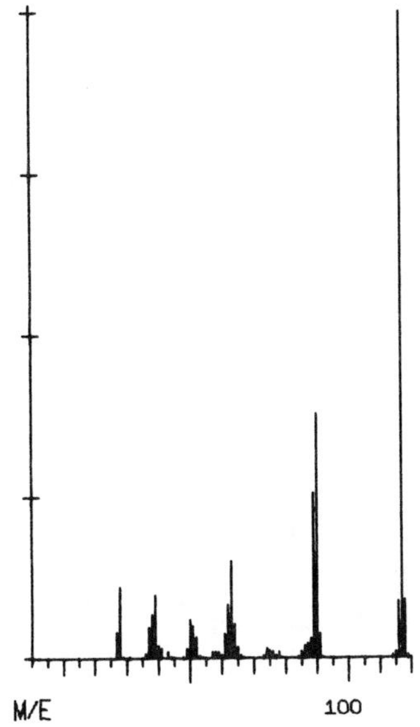
M/E
100

INDOLE

C_8H_7N MW = 117

Base Peak = 117 16 Peaks

Mass	Int.	Mass	Int.	Mass	Int.	Mass	Int.	Mass	Int.	Mass	Int.	Mass	Int.	Mass	Int.	Mass	Int.
27	39	28	110	38	67	39	98	50	60	51	51	62	82	63	150	88	29
89	255	90	377	91	37	116	87	117	999	118	91	119	4				

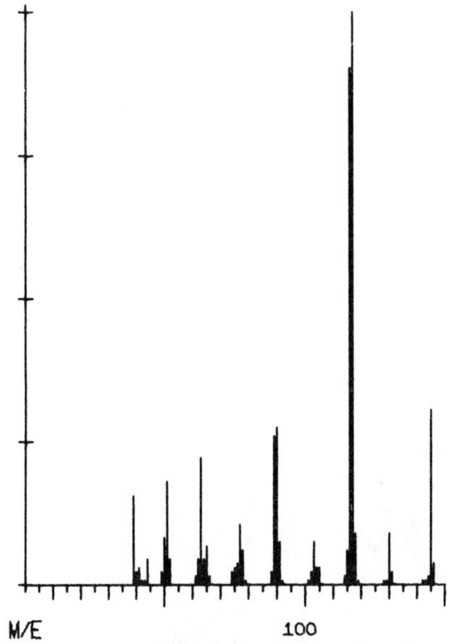
M/E
100

PHENOBARBITAL MTB. 1

$C_{10}H_{11}N$ MW = 117

Base Peak = 117 17 Peaks

Mass	Int.	Mass	Int.	Mass	Int.	Mass	Int.	Mass	Int.	Mass	Int.	Mass	Int.	Mass	Int.	Mass	Int.
39	156	44	44	50	82	51	179	63	223	65	67	77	104	89	261	90	276
91	74	116	902	117	999	118	89	130	89	144	14	145	305	146	37		

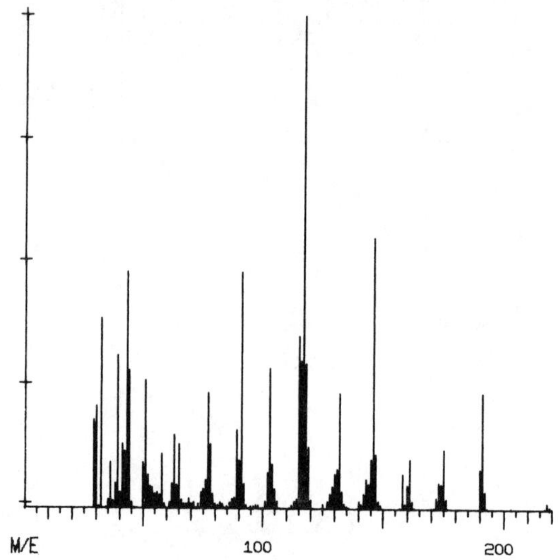
M/E
100
200

PRIMACLONE

$C_{12}H_{14}N_2O_2$ MW = 218

Base Peak = 117 30 Peaks

Mass	Int.	Mass	Int.	Mass	Int.	Mass	Int.	Mass	Int.	Mass	Int.	Mass	Int.	Mass	Int.	Mass	Int.
30	207	32	384	39	309	43	480	51	259	58	109	63	149	65	130	77	236
89	159	91	479	103	284	115	349	117	999	118	294	119	124	132	235	145	99
146	549	147	110	161	101	173	52	174	49	175	120	190	80	191	236	203	3
210	1	218	13	219	4												

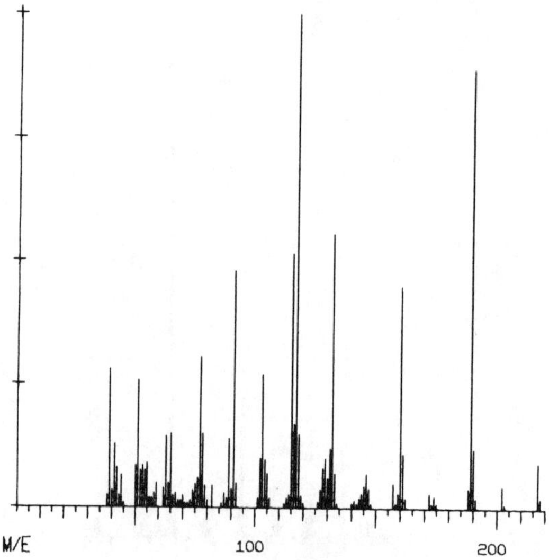
M/E
100
200

GLUTETHIMIDE

$C_{13}H_{15}NO_2$ MW = 217

Base Peak = 117 28 Peaks

Mass	Int.	Mass	Int.	Mass	Int.	Mass	Int.	Mass	Int.	Mass	Int.	Mass	Int.	Mass	Int.	Mass	Int.
39	281	41	127	51	258	55	89	63	144	65	149	77	304	78	151	91	479
103	269	115	514	117	999	118	149	131	119	132	555	133	69	146	69	157	49
160	449	161	109	174	24	175	11	189	889	190	119	202	44	203	9	217	92
218	19																

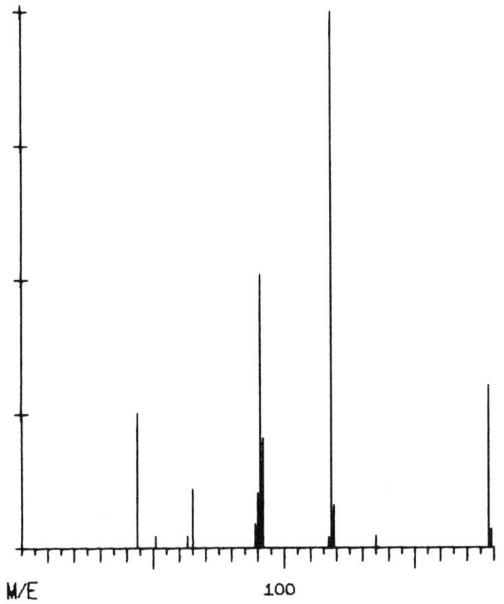
M/E
100

PHENACETYL UREA

$C_9H_{10}N_2O_2$ MW = 178
Base Peak = 118 13 Peaks

Mass	Int.	Mass	Int.	Mass	Int.	Mass	Int.	Mass	Int.	Mass	Int.	Mass	Int.	Mass	Int.	Mass	Int.
44	253	51	22	63	23	65	109	89	44	91	510	92	206	117	20	118	999
119	80	135	22	178	303	179	34										

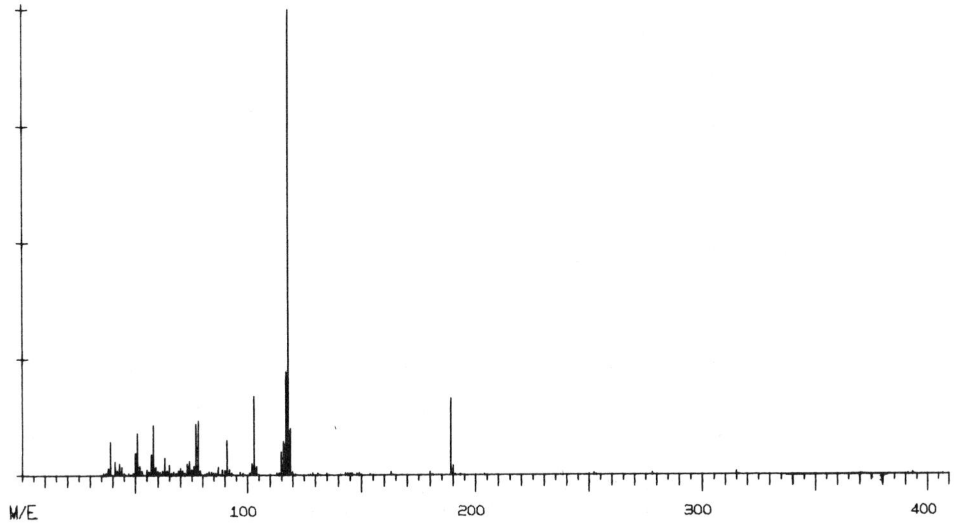
M/E
100
200
300
400

METHSUXIMIDE MTB 1

$C_{11}H_{11}NO_2$ MW = 189
Base Peak = 118 31 Peaks

Mass	Int.	Mass	Int.	Mass	Int.	Mass	Int.	Mass	Int.	Mass	Int.	Mass	Int.	Mass	Int.	Mass	Int.
39	73	41	31	51	91	58	108	63	37	74	30	77	111	78	117	91	74
103	169	116	72	117	223	118	999	119	101	144	5	145	5	146	6	148	5
163	7	180	7	181	3	189	166	190	21	204	3	250	3	252	4	278	4
315	7	391	3	393	5	407	3										

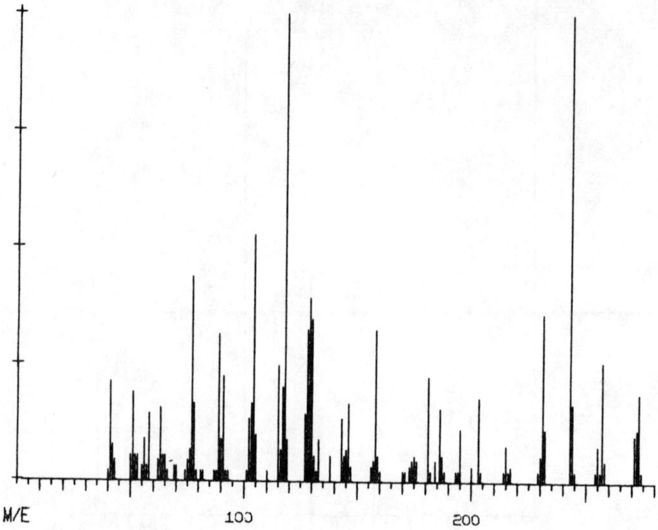
M/E
100
200

ALPHENAL 1,3-DIMETHYL DERIVATIVE

$C_{15}H_{16}N_2O_3$ MW = 272

Base Peak = 118 36 Peaks

Mass	Int.	Mass	Int.	Mass	Int.	Mass	Int.	Mass	Int.	Mass	Int.	Mass	Int.	Mass	Int.	Mass	Int.
41	213	42	78	51	191	58	146	63	157	64	56	77	438	89	314	91	224
103	168	104	528	115	247	118	999	129	393	133	89	143	134	146	168	158	325
160	22	173	33	181	224	186	157	195	112	200	33	203	179	215	78	216	22
217	33	231	359	243	999	244	168	257	258	258	44	271	101	272	112	273	191

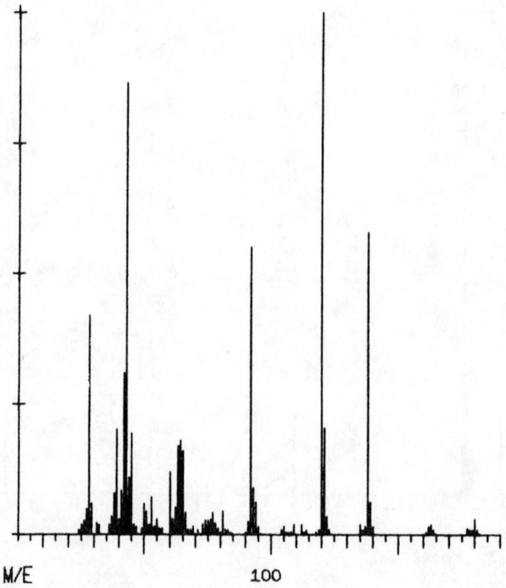
M/E
100

ACETYLSALICYLIC ACID

$C_9H_8O_4$ MW = 180

Base Peak = 120 22 Peaks

Mass	Int.	Mass	Int.	Mass	Int.	Mass	Int.	Mass	Int.	Mass	Int.	Mass	Int.	Mass	Int.	Mass	Int.
28	419	29	60	42	310	43	864	53	72	60	121	63	170	64	180	77	42
81	44	92	552	93	90	109	19	112	20	120	999	121	204	138	580	139	62
162	14	163	21	177	12	180	30										

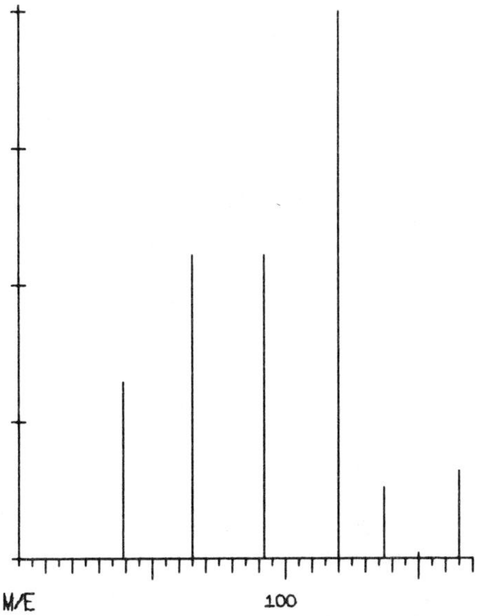
M/E
100

BENZOCAINE

$C_9H_{11}NO_2$ MW = 165

Base Peak = 120 7 Peaks

Mass	Int.	Mass	Int.	Mass	Int.	Mass	Int.	Mass	Int.	Mass	Int.	Mass	Int.
39	323	65	555	92	555	120	999	137	131	150	10	165	161

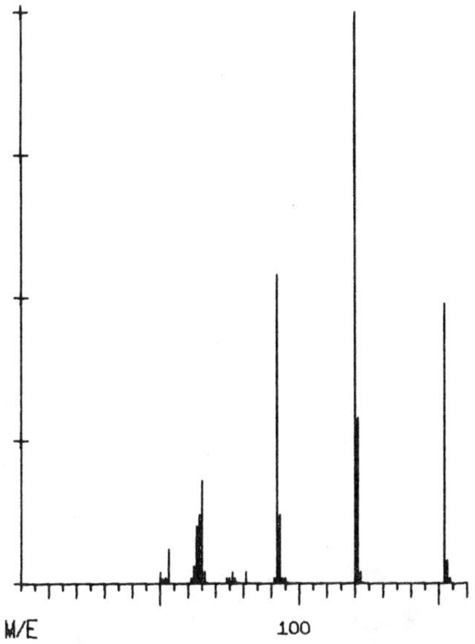
M/E
100

METHYL SALICYLATE

$C_8H_8O_3$ MW = 152

Base Peak = 120 26 Peaks

Mass	Int.	Mass	Int.	Mass	Int.	Mass	Int.	Mass	Int.	Mass	Int.	Mass	Int.	Mass	Int.	Mass	Int.
50	20	51	10	52	10	53	60	61	10	62	30	63	100	64	120	65	180
66	20	74	10	75	10	76	20	77	10	81	20	91	10	92	540	93	120
94	10	95	10	120	999	121	290	122	20	152	490	153	40	154	10		

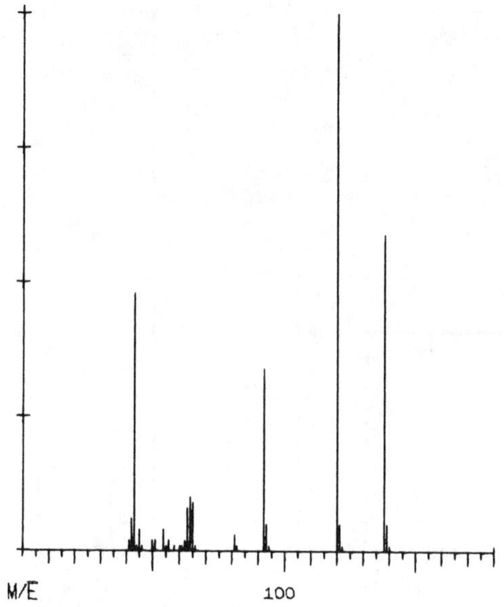
M/E
100

ACETYLSALICYLIC ACID

$C_9H_8O_4$ MW = 180

Base Peak = 120 15 Peaks

Mass	Int.	Mass	Int.	Mass	Int.	Mass	Int.	Mass	Int.	Mass	Int.	Mass	Int.	Mass	Int.	Mass	Int.
42	60	43	480	50	20	54	40	64	100	65	90	81	30	82	10	92	340
93	50	120	999	121	50	138	590	139	50	180	10						

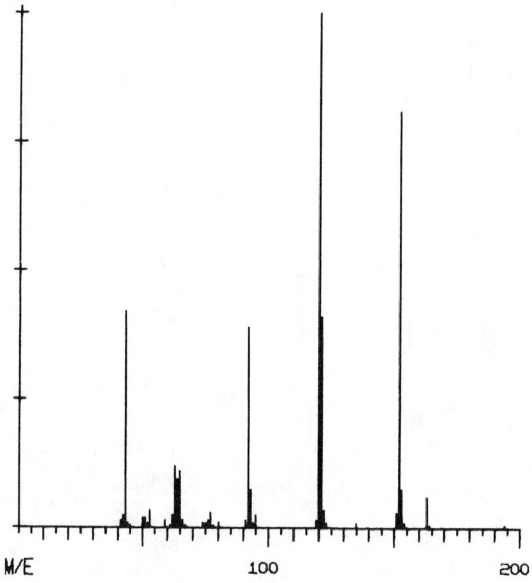
M/E
100
200

ACETYLSALICYLIC ACID METHYL ESTER

$C_{10}H_{10}O_4$ MW = 194

Base Peak = 120 18 Peaks

Mass	Int.	Mass	Int.	Mass	Int.	Mass	Int.	Mass	Int.	Mass	Int.	Mass	Int.	Mass	Int.	Mass	Int.
42	25	43	420	50	20	53	35	63	120	65	110	76	15	77	30	92	390
93	75	120	999	121	410	135	10	152	810	153	75	163	60	164	5	194	5

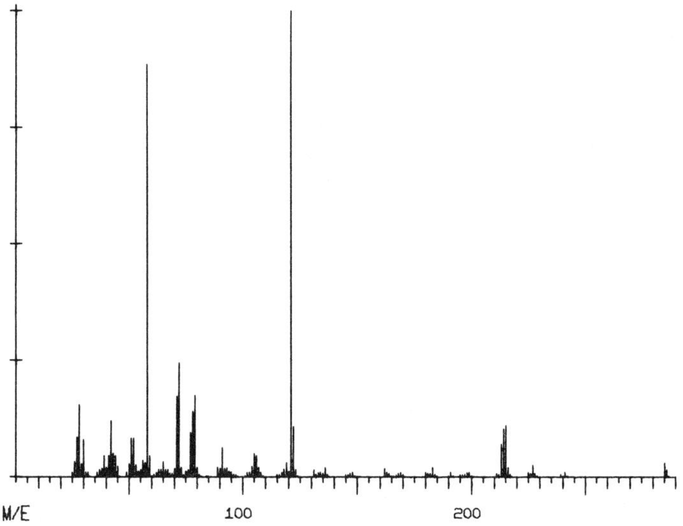
M/E
100
200

PYRILAMINE

$C_{17}H_{23}N_3O$ MW = 285

Base Peak = 121 35 Peaks

Mass	Int.	Mass	Int.	Mass	Int.	Mass	Int.	Mass	Int.	Mass	Int.	Mass	Int.	Mass	Int.	Mass	Int.
27	84	28	154	42	119	43	51	51	82	58	885	71	172	72	246	78	140
79	174	91	62	93	20	105	49	106	46	121	999	122	108	133	9	136	19
147	8	148	10	162	17	169	11	180	10	183	19	191	10	199	10	214	102
215	109	216	21	227	24	239	4	241	10	285	30	286	14	287	3		

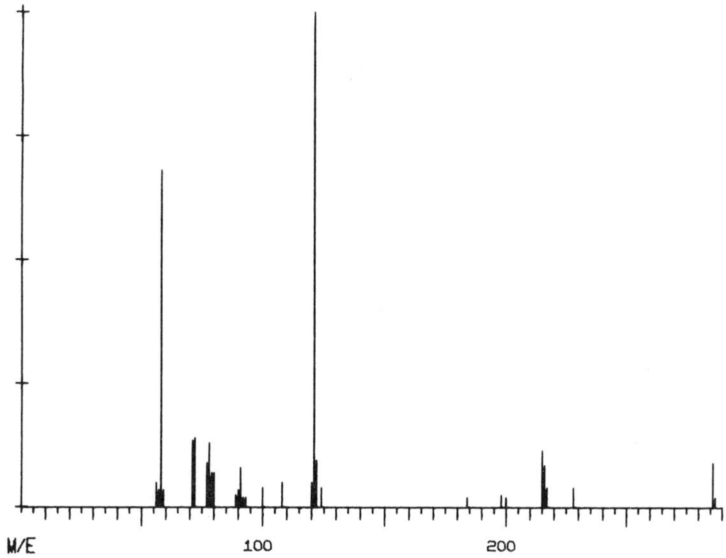
M/E
100
200

THONZYLAMINE

$C_{16}H_{22}N_4O$ MW = 286

Base Peak = 121 19 Peaks

Mass	Int.	Mass	Int.	Mass	Int.	Mass	Int.	Mass	Int.	Mass	Int.	Mass	Int.	Mass	Int.	Mass	Int.
56	49	58	679	71	134	72	139	77	89	78	129	91	79	100	39	108	49
121	999	122	94	184	19	198	24	200	19	215	114	216	84	217	39	286	89
287	21																

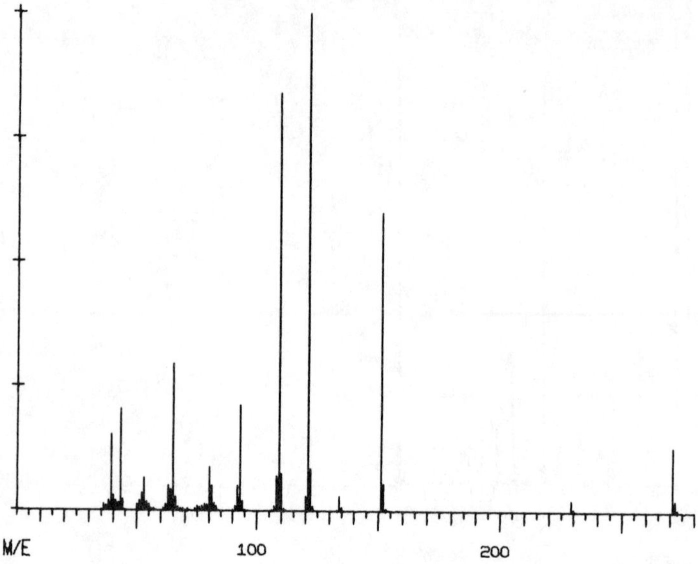
M/E
100
200

PHENETSAL

$C_{15}H_{13}N_1O_4$ MW = 271

Base Peak = 121 23 Peaks

Mass	Int.	Mass	Int.	Mass	Int.	Mass	Int.	Mass	Int.	Mass	Int.	Mass	Int.	Mass	Int.	Mass	Int.
39	152	43	204	52	35	53	64	63	49	65	296	80	88	81	49	92	49
93	213	109	839	110	74	121	999	122	84	134	31	135	7	151	599	152	55
229	23	230	5	271	129	272	23	273	6								

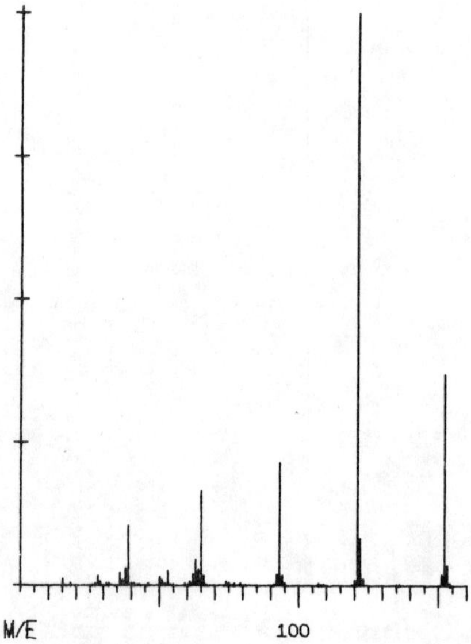
M/E
100

METHYL P-HYDROXYBENZOATE

$C_8H_8O_3$ MW = 152

Base Peak = 121 20 Peaks

Mass	Int.	Mass	Int.	Mass	Int.	Mass	Int.	Mass	Int.	Mass	Int.	Mass	Int.	Mass	Int.	Mass	Int.
28	17	29	8	38	29	39	106	50	14	53	25	63	45	65	166	77	4
79	4	92	21	93	215	107	5	108	3	121	999	122	82	135	2	136	2
152	369	153	34														

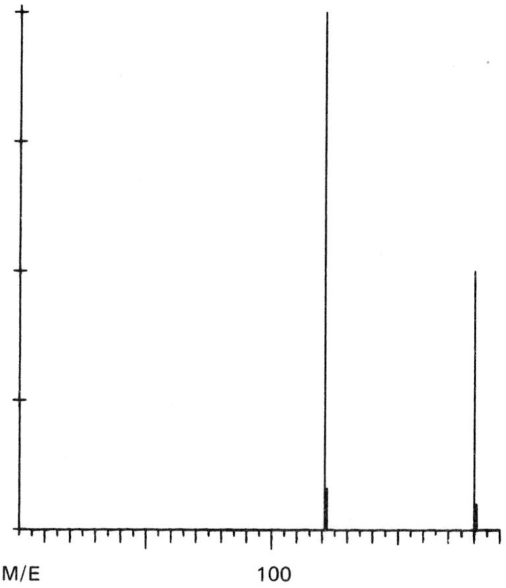
M/E
100

P-METHOXYPHENYLACETIC ACID METHYL ESTER

$C_{10}H_{12}O_3$ MW = 180

Base Peak = 121   4 Peaks

Mass	Int.	Mass	Int.	Mass	Int.	Mass	Int.
121	999	122	79	180	499	181	49

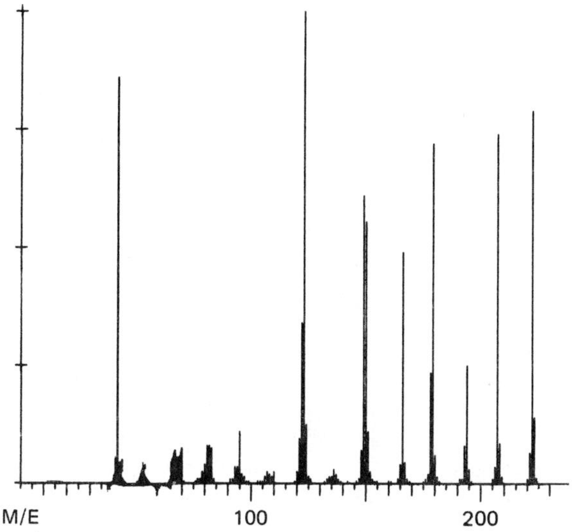
M/E
100
200

1,3-DI-ETHYL-9-METHYLXANTHINE

$C_{10}H_{14}N_4O_2$ MW = 222

Base Peak = 123   28 Peaks

Mass	Int.	Mass	Int.	Mass	Int.	Mass	Int.	Mass	Int.	Mass	Int.	Mass	Int.	Mass	Int.	Mass	Int.
41	55	42	860	53	45	54	40	67	70	70	75	81	80	82	80	93	35
95	110	107	25	110	25	122	340	123	999	136	30	137	20	149	610	150	555
166	490	167	45	178	235	179	720	193	80	194	250	207	740	208	85	222	790
223	140																

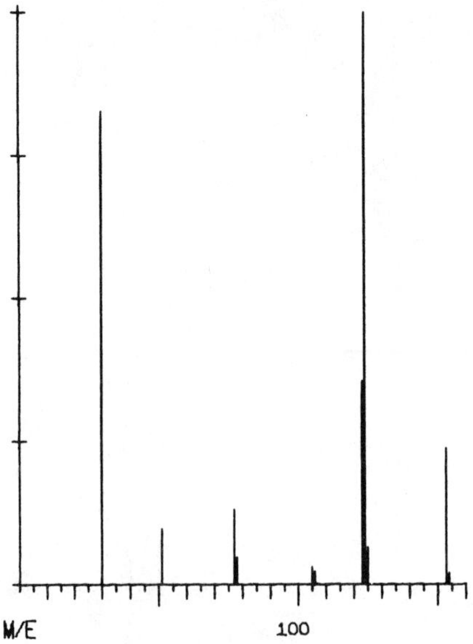
M/E
100

4-(2-AMINOETHYL) PYRO-CATECHOL

$C_8H_{11}NO_2$ MW 153

Base Peak = 124 10 Peaks

Mass	Int.	Mass	Int.	Mass	Int.	Mass	Int.	Mass	Int.	Mass	Int.	Mass	Int.	Mass	Int.	Mass	Int.
30	828	51	97	77	129	78	47	105	31	106	23	123	355	124	999	153	237
154	20																

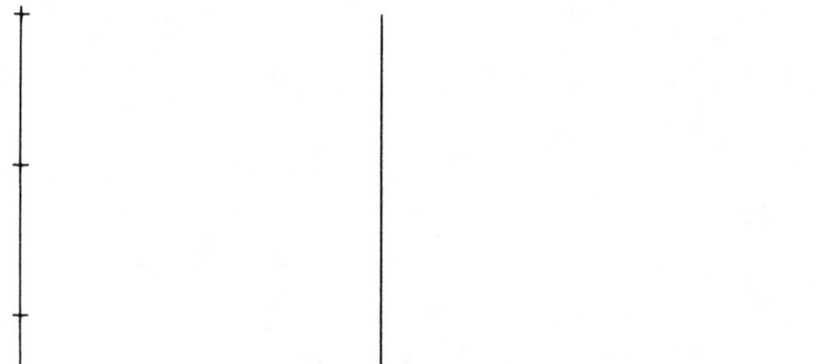
M/E
100
200

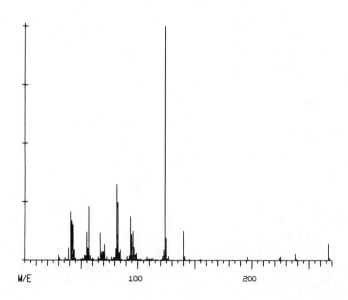

ANISOTROPINE

$C_{16}H_{29}NO_2$ MW = 267

Base Peak = 124 30 Peaks

Mass	Int.	Mass	Int.	Mass	Int.	Mass	Int.	Mass	Int.	Mass	Int.	Mass	Int.	Mass	Int.	Mass	Int.
30	22	31	9	41	208	42	169	55	120	57	230	67	118	71	67	82	324
83	248	94	188	96	126	108	14	113	10	124	999	125	94	140	125	141	17
154	3	155	4	169	3	183	3	196	16	197	3	224	11	225	15	238	27
239	5	267	71	268	11												

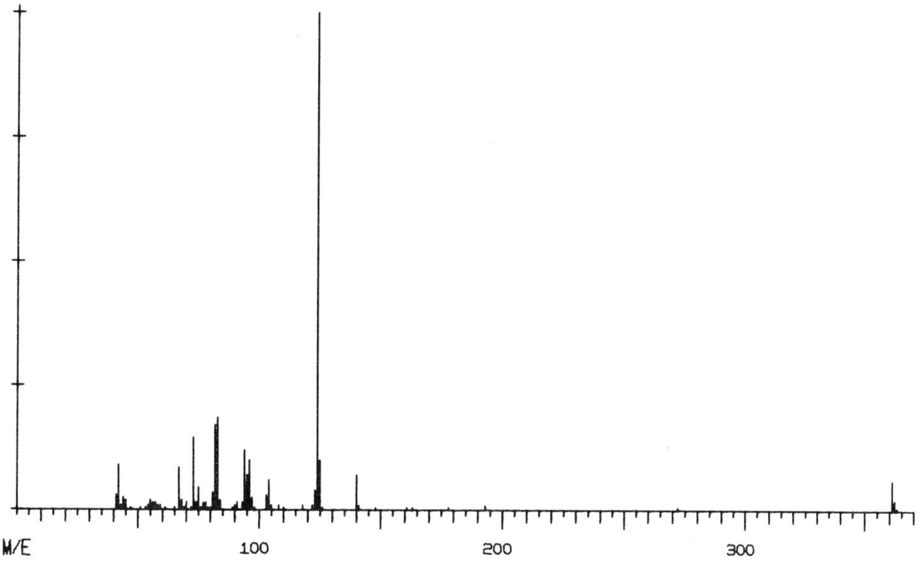
M/E
100
200
300

ATROPINE TMS ETHER

$C_{20}H_{31}NO_3Si$ MW = 361

Base Peak = 124   16 Peaks

Mass	Int.	Mass	Int.	Mass	Int.	Mass	Int.	Mass	Int.	Mass	Int.	Mass	Int.	Mass	Int.	Mass	Int.
41	30	42	90	55	20	56	15	67	85	73	145	82	170	83	185	94	120
96	100	104	60	105	10	124	999	125	100	140	70	141	10	148	5	161	5
163	5	178	5	193	10	272	5	361	60	362	20						

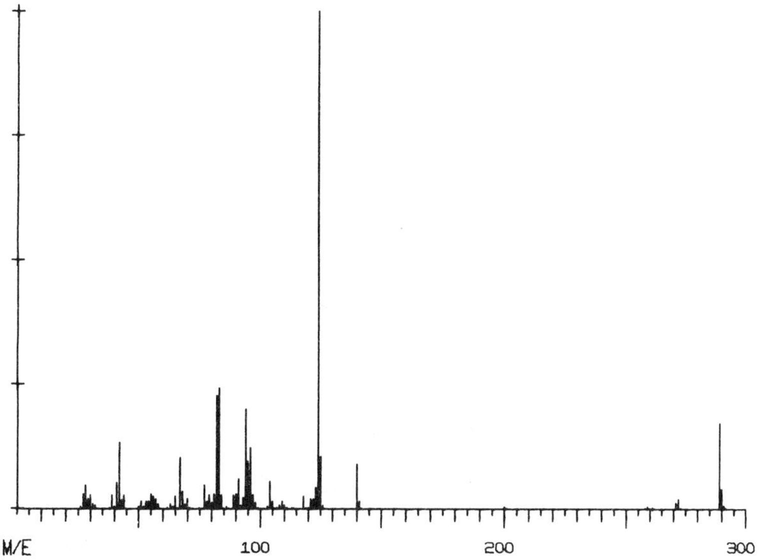
M/E
100
200
300

HYOSCYAMINE (ATROPINE)

$C_{17}H_{23}NO_3$ MW = 289

Base Peak = 124   25 Peaks

Mass	Int.	Mass	Int.	Mass	Int.	Mass	Int.	Mass	Int.	Mass	Int.	Mass	Int.	Mass	Int.	Mass	Int.
27	29	28	48	41	53	42	132	55	29	56	24	67	103	68	34	82	228
83	243	94	201	96	122	104	55	109	15	124	999	125	105	140	91	141	15
200	4	259	5	271	13	272	21	273	3	289	172	290	41				

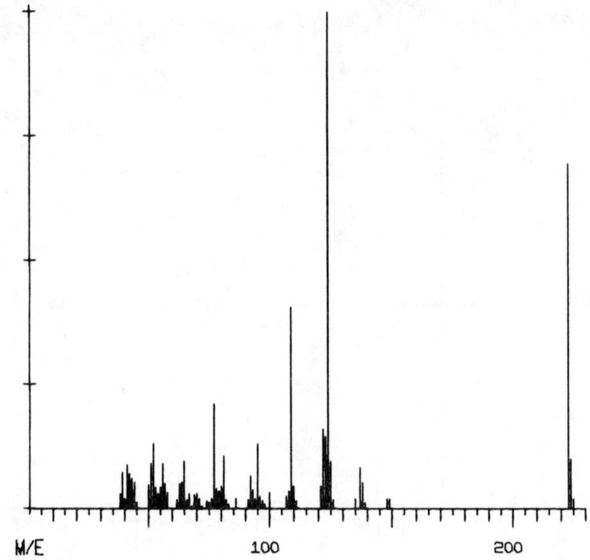
M/E
100
200

MEPHENOXALONE

$C_{11}H_{13}NO_4$ MW = 223

Base Peak = 124 20 Peaks

Mass	Int.	Mass	Int.	Mass	Int.	Mass	Int.	Mass	Int.	Mass	Int.	Mass	Int.	Mass	Int.	Mass	Int.
39	73	41	88	51	89	52	129	64	53	65	95	77	209	81	104	92	66
95	129	109	406	110	45	122	159	124	999	137	83	138	52	148	20	149	19
223	694	224	99														

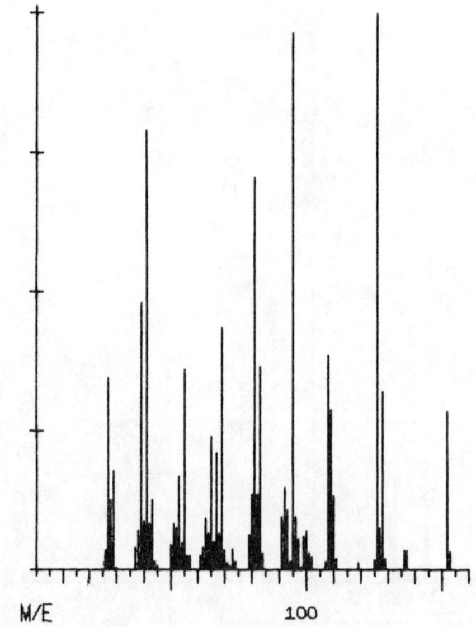
M/E
100

CAMPHOR

$C_{10}H_{16}O$ MW = 152

Base Peak = 126 20 Peaks

Mass	Int.	Mass	Int.	Mass	Int.	Mass	Int.	Mass	Int.	Mass	Int.	Mass	Int.	Mass	Int.	Mass	Int.
27	345	29	177	39	480	41	791	53	167	55	360	65	239	69	434	81	705
83	365	92	147	95	965	108	385	109	288	126	999	128	320	136	35	137	36
152	285	153	32														

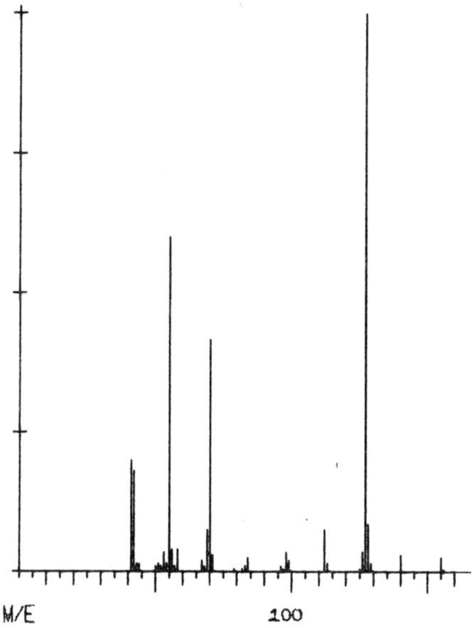
M/E
100

ETHOSUXIMIDE N-METHYL DERIVATIVE

$C_8H_{13}NO_2$ MW = 155

Base Peak = 127   17 Peaks

Mass	Int.	Mass	Int.	Mass	Int.	Mass	Int.	Mass	Int.	Mass	Int.	Mass	Int.	Mass	Int.	Mass	Int.
41	200	42	180	55	600	56	40	69	75	70	415	83	10	84	25	98	35
99	20	112	75	113	15	127	999	128	85	140	30	155	25	156	5		

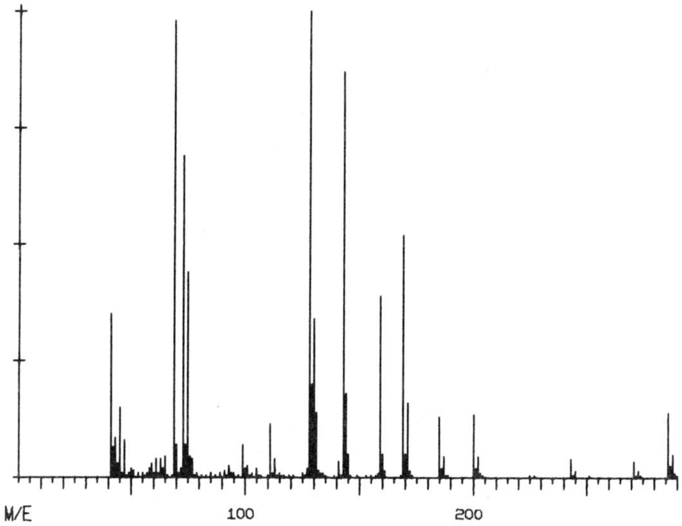
M/E
100
200

2-(4-CHLOROPHENOXY)-2-METHYLPROPIONIC ACID TMS ESTER (FROM CLOFIBRATE)

$C_{13}H_{19}CLO_3SI$ MW = 286

Base Peak = 128   36 Peaks

Mass	Int.	Mass	Int.	Mass	Int.	Mass	Int.	Mass	Int.	Mass	Int.	Mass	Int.	Mass	Int.	Mass	Int.
41	350	45	150	59	30	61	40	69	980	73	690	76	45	77	40	93	25
99	70	111	115	113	40	128	999	130	340	143	870	144	180	158	10	159	390
169	520	171	160	185	130	187	45	200	135	201	20	202	45	203	10	225	5
227	5	243	40	244	5	245	15	271	35	272	5	273	15	286	140	288	50

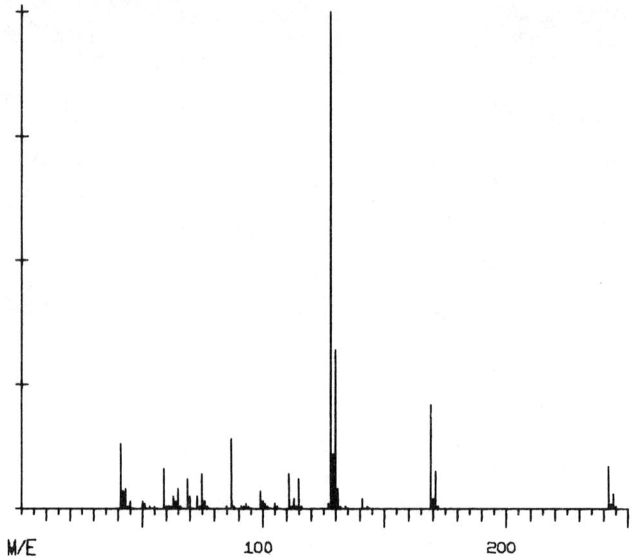

M/E
100
200

CLOFIBRATE

$C_{12}H_{15}CLO_3$ MW = 242

Base Peak = 128 22 Peaks

Mass	Int.	Mass	Int.	Mass	Int.	Mass	Int.	Mass	Int.	Mass	Int.	Mass	Int.	Mass	Int.	Mass	Int.
41	130	43	40	50	15	59	80	69	60	75	70	76	15	87	140	99	35
100	15	111	70	115	60	128	999	130	320	132	5	141	20	169	210	171	75
242	85	243	10	244	30	245	5										

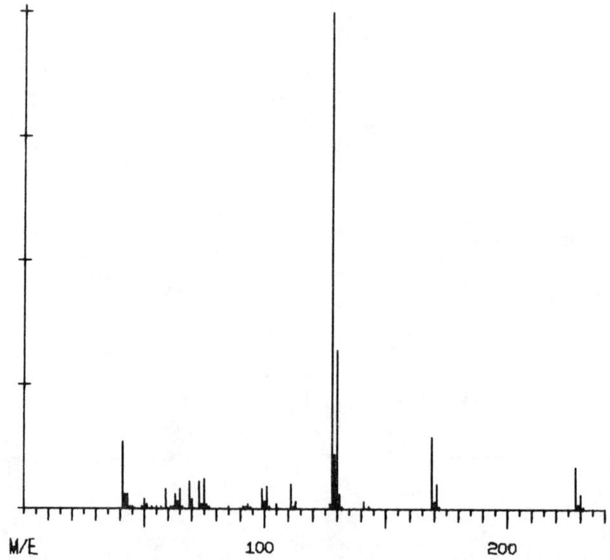

M/E
100
200

2-(4-CHLOROPHENOXY)-2-METHYLPROPIONIC ACID ME ESTER (FROM CLOFIBRATE)

$C_{11}H_{13}CO_3$ MW = 228

Base Peak = 128 22 Peaks

Mass	Int.	Mass	Int.	Mass	Int.	Mass	Int.	Mass	Int.	Mass	Int.	Mass	Int.	Mass	Int.	Mass	Int.
41	135	42	30	50	20	59	40	69	55	75	60	76	10	77	5	99	40
101	45	111	50	113	15	128	999	130	320	132	5	141	15	169	145	171	50
228	85	229	10	230	30	231	5										

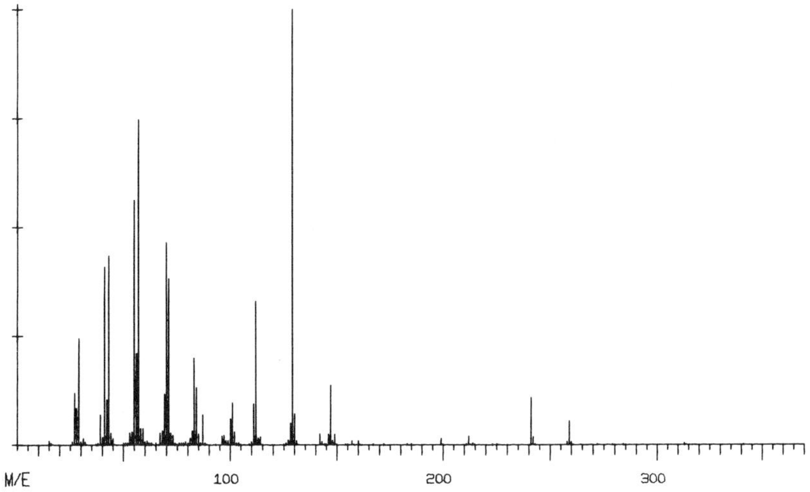
M/E
100
200
300

DIOCTYL ADIPATE

$C_{22}H_{42}O_4$ MW = 370

Base Peak = 129 40 Peaks

Mass	Int.	Mass	Int.	Mass	Int.	Mass	Int.	Mass	Int.	Mass	Int.	Mass	Int.	Mass	Int.	Mass	Int.
27	119	29	246	41	411	43	436	55	563	57	748	70	465	71	382	83	200
84	133	100	61	101	98	111	94	112	331	129	999	130	72	142	25	143	8
146	26	147	137	160	10	167	2	183	2	185	3	199	14	200	2	212	20
214	6	223	2	225	3	241	109	242	17	259	54	260	8	272	2	279	2
313	4	327	1	341	2	370	1										

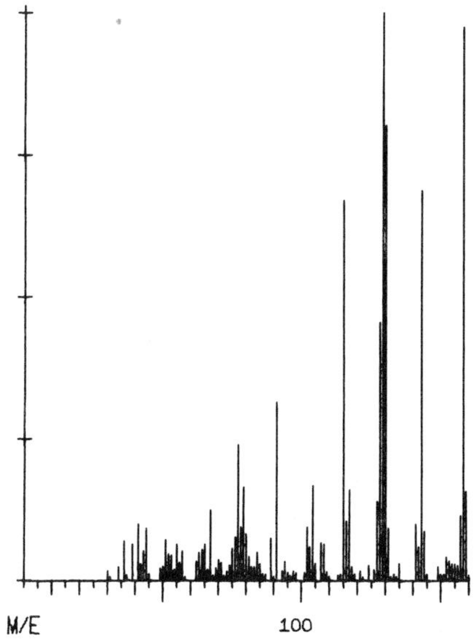
M/E
100

1-PHENYL CYCLOHEXENE

$C_{12}H_{14}$ MW = 158

Base Peak = 129 21 Peaks

Mass	Int.	Mass	Int.	Mass	Int.	Mass	Int.	Mass	Int.	Mass	Int.	Mass	Int.	Mass	Int.	Mass	Int.
30	17	31	8	41	101	44	93	51	72	55	65	65	64	67	124	77	240
79	164	91	314	102	96	104	167	115	669	129	999	130	803	141	100	143	688
158	974	159	158	160	11												

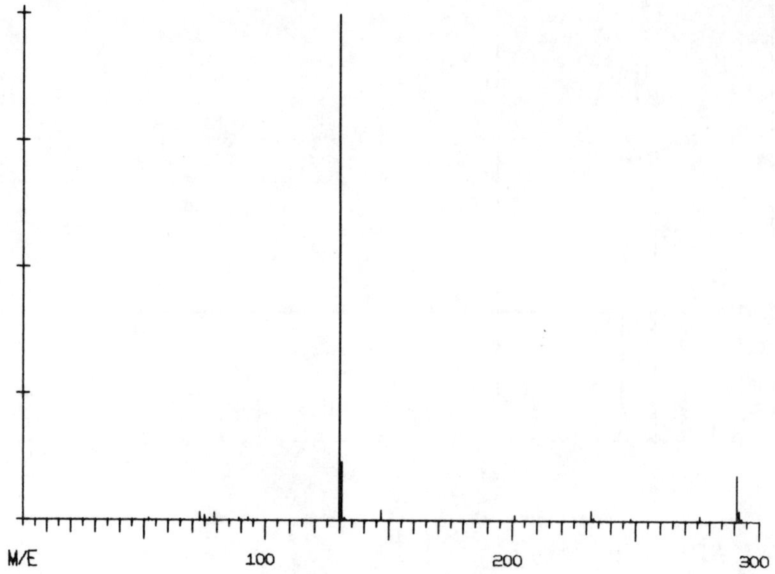
M/E
100
200
300

INDOLEACETIC ACID METHYL ESTER TRIMETHYLSILYL ETHER

$C_{14}H_{19}NO_2SI$ MW = 261
Base Peak = 130 17 Peaks

Mass	Int.	Mass	Int.	Mass	Int.	Mass	Int.	Mass	Int.	Mass	Int.	Mass	Int.	Mass	Int.	Mass	Int.
52	4	73	14	75	9	77	4	79	14	93	4	130	999	131	114	132	4
147	19	201	9	232	19	233	4	248	4	276	9	291	89	292	19		

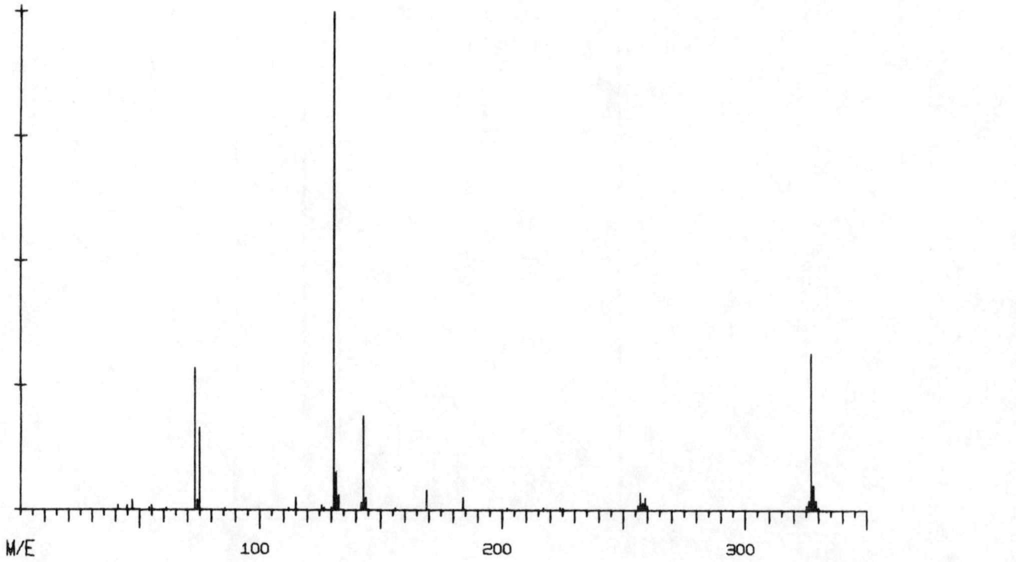
M/E
100
200
300

HYDROXYAMOBARBITAL 1,3-DIMETHYL DERIVATIVE TMS ETHER

$C_{16}H_{30}N_2O_4Si$ MW = 342
Base Peak = 131 27 Peaks

Mass	Int.	Mass	Int.	Mass	Int.	Mass	Int.	Mass	Int.	Mass	Int.	Mass	Int.	Mass	Int.	Mass	Int.
41	10	47	20	54	5	55	10	73	285	75	165	112	5	115	25	126	10
131	999	132	75	143	190	156	5	169	40	184	25	202	5	217	5	224	5
256	10	257	35	258	15	259	25	326	20	327	315	328	50	329	20	342	1

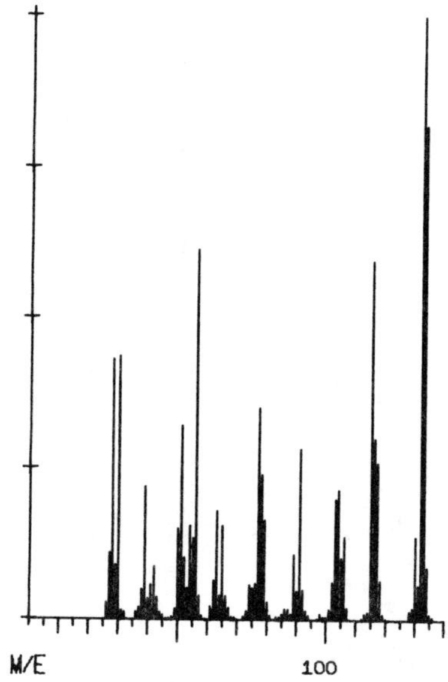
M/E
100

TRANYLCYPROMINE

$C_9H_{11}N$ MW = 263

Base Peak = 132 18 Peaks

Mass	Int.	Mass	Int.	Mass	Int.	Mass	Int.	Mass	Int.	Mass	Int.	Mass	Int.	Mass	Int.	Mass	Int.
28	431	30	434	39	219	42	88	51	319	56	612	63	181	65	154	77	349
78	240	91	283	103	199	115	594	116	300	118	65	130	138	132	999	133	820

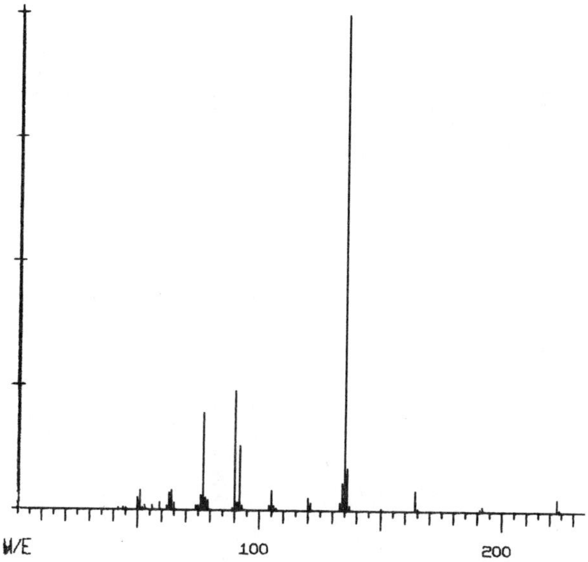
M/E
100
200

2-HYDROXYBENZOYLGLYCINE ACID METHYL ESTER METHYL ETHER

$C_{11}H_{13}NO_4$ MW = 223

Base Peak = 135 25 Peaks

Mass	Int.	Mass	Int.	Mass	Int.	Mass	Int.	Mass	Int.	Mass	Int.	Mass	Int.	Mass	Int.	Mass	Int.
42	20	45	20	51	100	59	60	63	80	64	90	76	55	77	370	90	355
92	220	104	15	105	55	120	40	121	20	135	999	136	90	148	10	150	10
164	35	165	5	191	5	192	15	206	5	223	50	224	20				

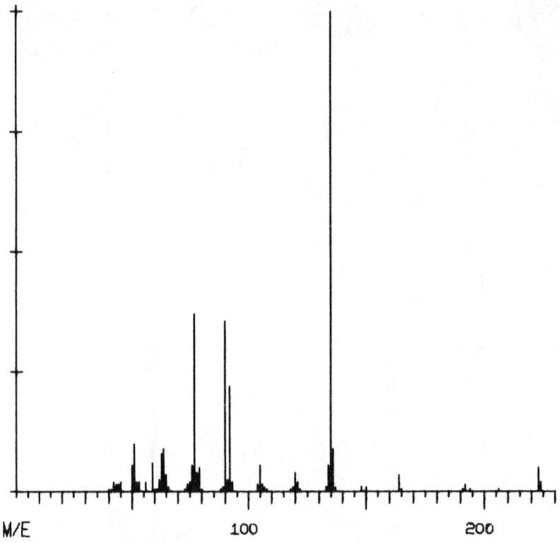
M/E
100
200

2-HYDROXY BENZOYLGLYCINE ACID METHYL ESTER METHYL ETHER

$C_{11}H_{13}NO_4$ MW = 223

Base Peak = 135 42 Peaks

Mass	Int.	Mass	Int.	Mass	Int.	Mass	Int.	Mass	Int.	Mass	Int.	Mass	Int.	Mass	Int.	Mass	Int.
42	5	44	5	45	5	50	25	51	40	52	5	53	10	56	10	59	15
62	10	63	35	64	40	65	15	74	10	75	10	76	30	77	195	78	25
79	20	89	5	90	240	91	15	92	130	93	10	104	10	105	40	106	10
107	5	120	25	121	15	133	15	134	55	135	999	136	85	137	10	150	5
164	40	165	5	191	5	192	10	223	25	224	5						

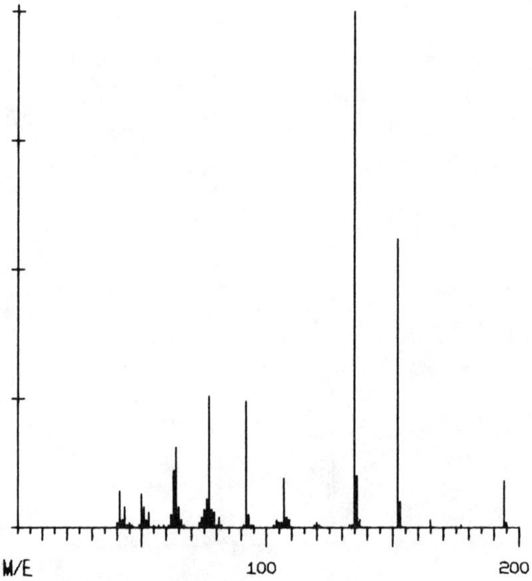
M/E
100
200

PROPYLPARABEN METHYL ETHER

$C_{11}H_{14}O_3$ MW = 194

Base Peak = 135 22 Peaks

Mass	Int.	Mass	Int.	Mass	Int.	Mass	Int.	Mass	Int.	Mass	Int.	Mass	Int.	Mass	Int.	Mass	Int.
41	70	43	40	50	65	51	40	63	110	64	155	76	55	77	255	92	245
93	25	107	95	108	20	119	5	120	10	135	999	136	100	152	560	153	50
165	15	177	5	194	90	195	10										

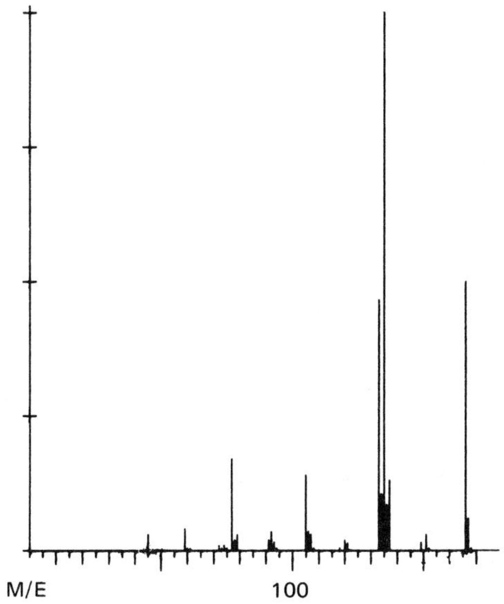
M/E
100

O-METHOXYBENZOIC ACID METHYL ESTER

$C_9H_{10}O_3$ MW = 166

Base Peak = 135 19 Peaks

Mass	Int.	Mass	Int.	Mass	Int.	Mass	Int.	Mass	Int.	Mass	Int.	Mass	Int.	Mass	Int.	Mass	Int.
45	29	59	39	60	4	72	9	74	9	77	169	79	29	91	19	92	34
105	139	106	34	120	19	121	14	133	464	135	999	149	14	151	29	166	499
167	59																

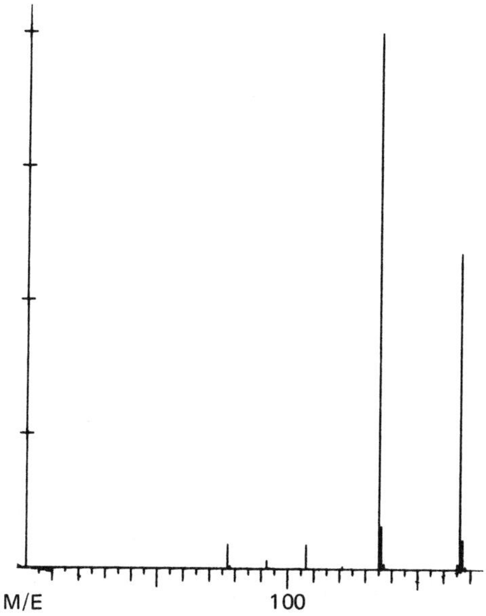
M/E
100

P-METHOXYBENZOIC ACID METHYL ESTER

$C_9H_{10}O_3$ MW = 166

Base Peak = 135 9 Peaks

Mass	Int.	Mass	Int.	Mass	Int.	Mass	Int.	Mass	Int.	Mass	Int.	Mass	Int.	Mass	Int.	Mass	Int.
77	44	78	4	92	14	107	44	121	4	135	999	136	79	166	589	167	54

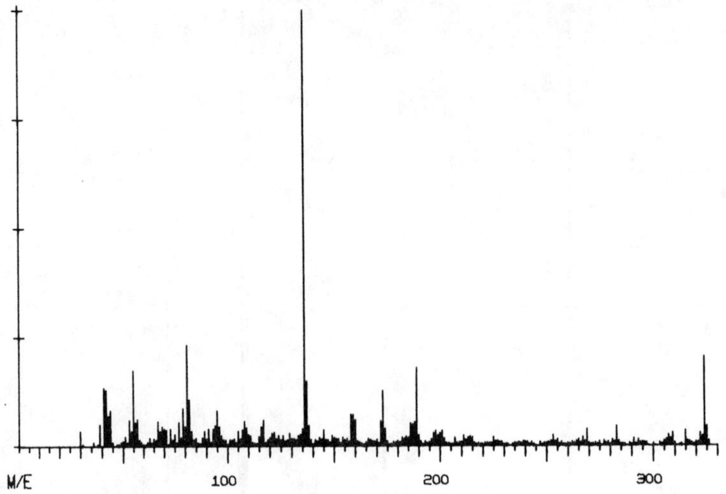

M/E
100
200
300

QUINIDINE

$C_{20}H_{24}N_2O_2$ MW = 324

Base Peak = 136 44 Peaks

Mass	Int.	Mass	Int.	Mass	Int.	Mass	Int.	Mass	Int.	Mass	Int.	Mass	Int.	Mass	Int.	Mass	Int.
30	36	31	5	41	134	42	131	55	175	57	63	67	57	69	44	81	232
82	107	94	48	95	83	108	57	117	60	121	30	122	32	136	999	137	150
158	72	159	73	160	59	173	128	186	53	187	49	188	56	189	179	211	24
214	23	225	20	226	13	237	11	239	13	253	25	255	14	267	19	269	37
281	16	283	44	291	18	293	16	307	24	309	31	324	204	325	46		

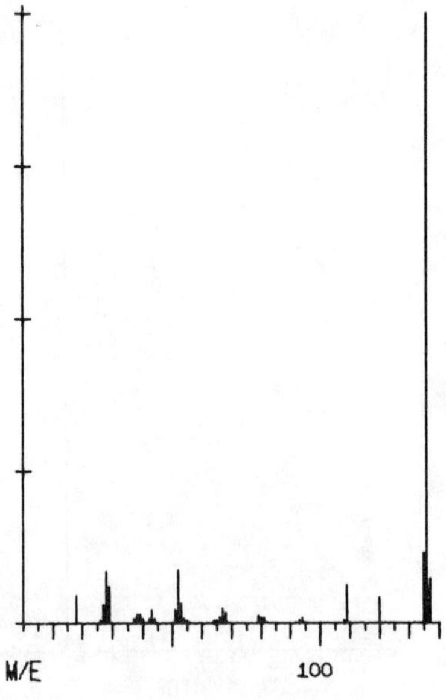

M/E
100

ALLOPURINOL

$C_5H_4N_4O$ MW = 136

Base Peak = 136 17 Peaks

Mass	Int.	Mass	Int.	Mass	Int.	Mass	Int.	Mass	Int.	Mass	Int.	Mass	Int.	Mass	Int.	Mass	Int.
28	84	29	60	39	15	43	22	52	87	53	33	67	25	68	18	79	13
80	9	93	5	94	10	108	6	109	62	120	42	135	114	136	999		

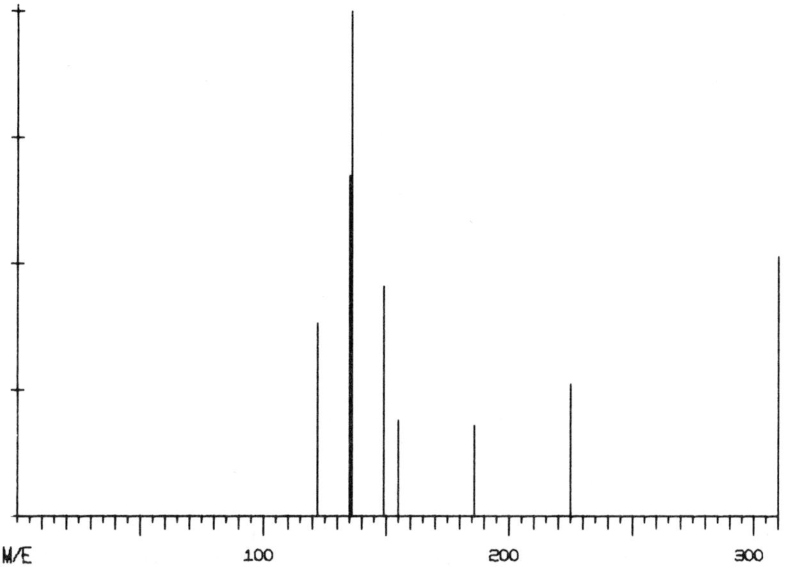
M/E
100
200
300

IBOGAINE

$C_{20}H_{26}N_2O$ MW = 310

Base Peak = 136 8 Peaks

Mass	Int.	Mass	Int.	Mass	Int.	Mass	Int.	Mass	Int.	Mass	Int.	Mass	Int.	Mass	Int.
122	383	135	676	136	999	149	454	155	191	186	181	225	262	310	515

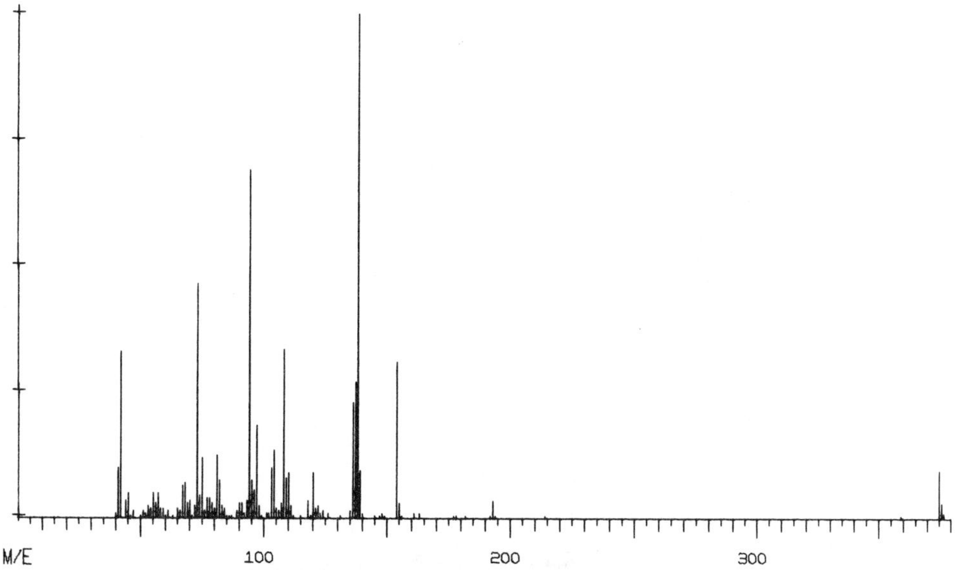
M/E
100
200
300

SCOPOLAMINE TMS ETHER

$C_{20}H_{29}NO_4SI$ MW = 375

Base Peak = 138 28 Peaks

Mass	Int.	Mass	Int.	Mass	Int.	Mass	Int.	Mass	Int.	Mass	Int.	Mass	Int.	Mass	Int.	Mass	Int.
41	100	42	330	55	50	57	50	73	465	75	120	81	125	82	75	94	690
97	185	104	135	108	335	118	35	120	90	137	270	138	999	154	310	155	30
161	10	163	10	177	5	178	5	192	5	193	35	214	5	359	5	375	95
376	30																

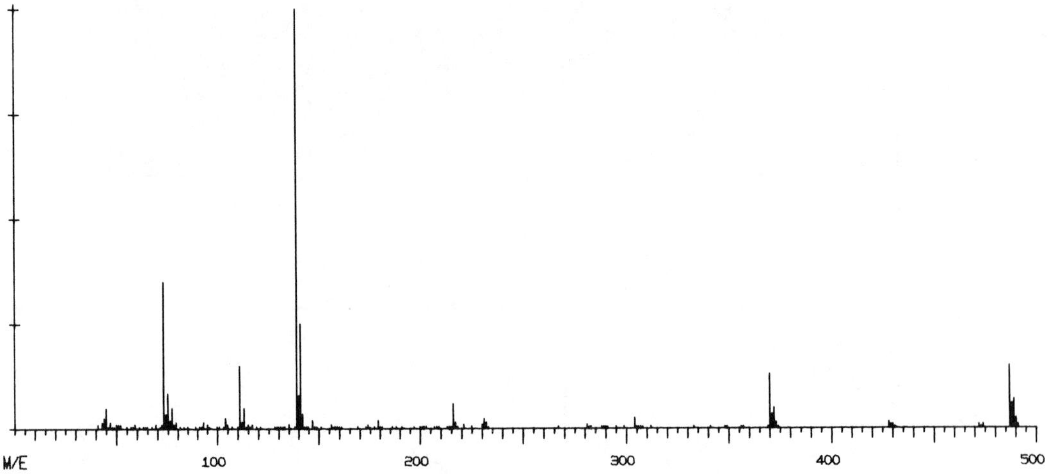
M/E
100
200
300
400
500

1-(4-CHLOROBENZOYL)-2-METHYL-5-TRIMETHYLSILYLOXY-INDOLE-3-ACETIC ACID TRIMETHYLSILYOXYESTER (FROM INDOMETHACIN)

$C_{24}H_{30}NO_4Si_2Cl$ MW = 487

Base Peak = 139 51 Peaks

Mass	Int.	Mass	Int.	Mass	Int.	Mass	Int.	Mass	Int.	Mass	Int.	Mass	Int.	Mass	Int.	Mass	Int.
44	25	45	50	50	10	51	10	73	350	75	85	76	20	77	50	93	15
95	10	111	150	113	50	119	5	121	5	139	999	141	250	147	20	156	10
160	5	161	5	174	10	179	20	188	5	191	5	202	5	203	5	216	60
217	15	231	25	232	15	267	5	281	10	282	5	288	5	289	5	304	25
305	5	333	5	341	5	348	5	349	5	356	5	357	5	370	130	372	50
428	15	429	10	472	10	474	10	387	150	489	70						

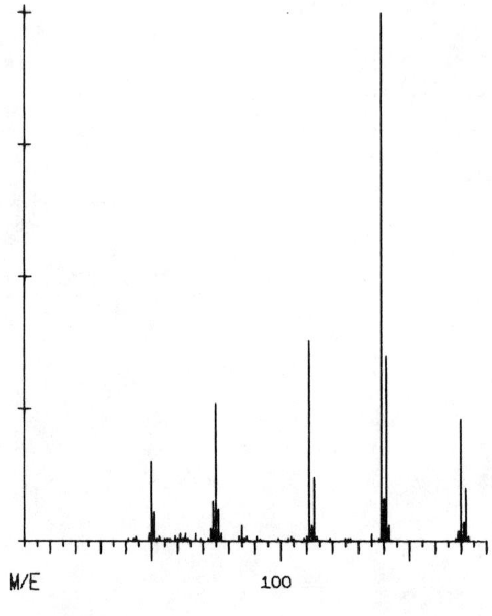
M/E
100

4-CHLOROBENZOIC ACID METHYL ESTER (FROM INDOMETHACIN)

$C_8II_7CLO_2$ MW – 170

Base Peak = 139 18 Peaks

Mass	Int.	Mass	Int.	Mass	Int.	Mass	Int.	Mass	Int.	Mass	Int.	Mass	Int.	Mass	Int.	Mass	Int.
41	5	44	10	50	150	51	55	74	75	75	260	76	60	85	30	91	10
92	5	111	380	113	120	119	5	125	5	139	999	141	350	170	230	172	100

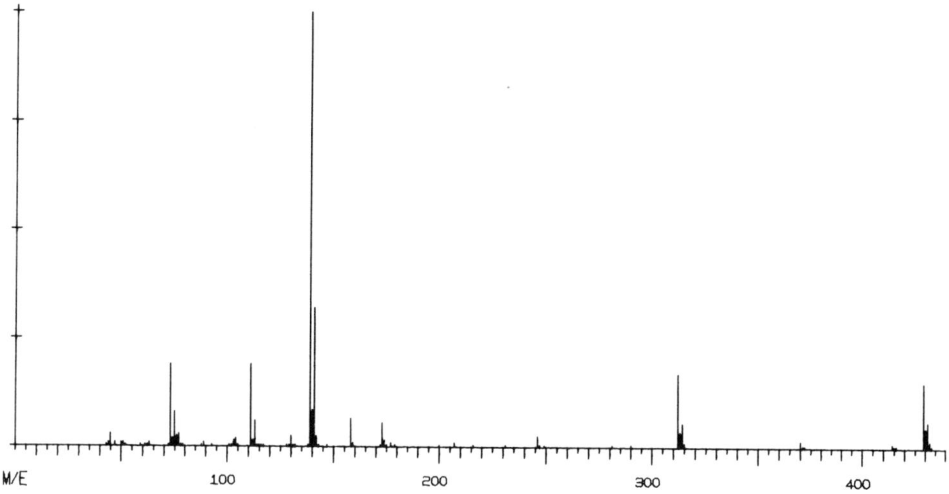
M/E
100
200
300
400

INDOMETHACIN TMS ESTER

$C_{22}H_{24}NO_4SiCl$ MW = 429

Base Peak = 139 40 Peaks

Mass	Int.	Mass	Int.	Mass	Int.	Mass	Int.	Mass	Int.	Mass	Int.	Mass	Int.	Mass	Int.	Mass	Int.
44	10	45	30	50	10	51	10	73	190	75	80	76	25	77	30	93	5
103	15	111	190	113	60	128	5	130	25	139	999	141	320	158	65	159	10
172	5	173	55	174	15	177	10	200	5	207	10	216	5	231	5	246	25
247	5	281	5	290	5	312	170	313	35	314	55	315	10	370	15	371	5
414	10	415	5	429	150	431	60										

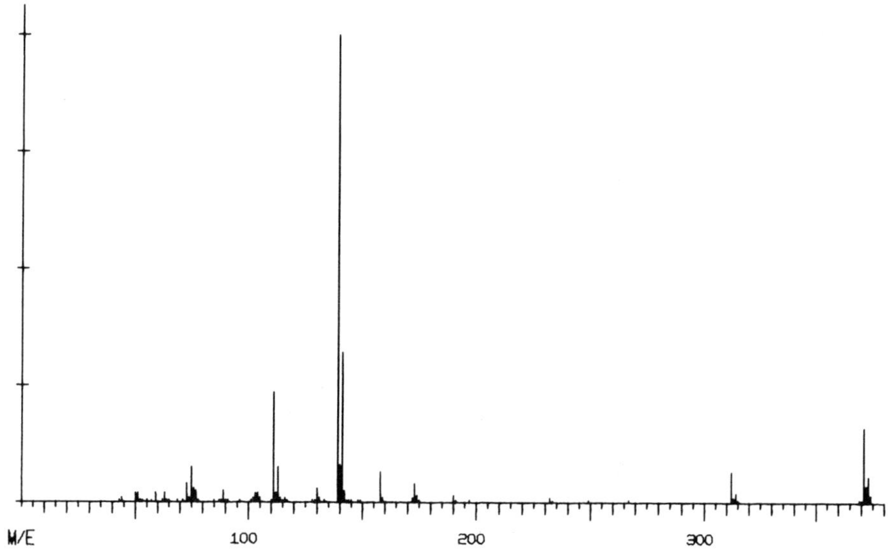
M/E
100
200
300

INDOMETHACIN METHYL ESTER

$C_{20}H_{18}NO_4Cl$ MW = 371

Base Peak = 139 35 Peaks

Mass	Int.	Mass	Int.	Mass	Int.	Mass	Int.	Mass	Int.	Mass	Int.	Mass	Int.	Mass	Int.	Mass	Int.
43	5	44	10	50	20	51	20	73	40	75	75	76	30	77	25	102	10
103	20	111	235	113	75	130	30	131	10	139	999	141	320	158	65	159	10
172	10	173	40	174	15	175	5	190	15	191	5	232	10	233	5	249	5
267	5	312	65	313	10	314	20	315	5	369	5	371	160	373	55		

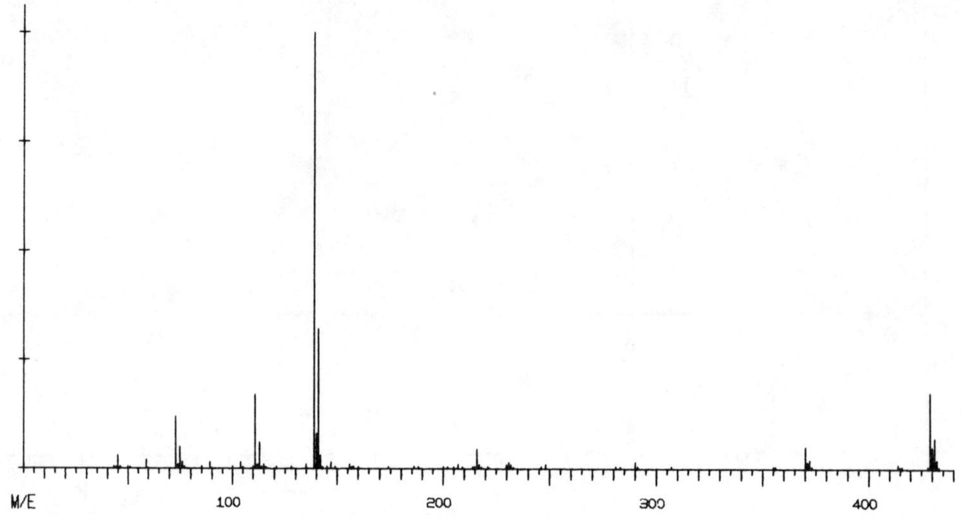
M/E
100
200
300
400

1-(4-CHLOROBENZOYL)-2-METHYL-5-TRIMETHYLSILYLOXYINDOLE-3-ACETIC ACID ME ESTER (FROM INDOMETHACIN)

$C_{22}H_{24}NO_4SiCl$ MW = 429
Base Peak = 139 43 Peaks

Mass	Int.	Mass	Int.	Mass	Int.	Mass	Int.	Mass	Int.	Mass	Int.	Mass	Int.	Mass	Int.	Mass	Int.
43	5	45	30	50	5	59	20	73	120	75	50	76	15	89	15	100	5
111	170	113	60	121	5	128	5	139	999	141	320	147	15	156	10	160	5
174	5	186	5	188	5	200	5	202	5	207	10	216	45	217	10	230	10
231	15	246	5	248	10	281	5	283	5	290	15	291	5	307	5	355	5
356	5	370	50	372	20	414	10	415	5	429	175	431	70				

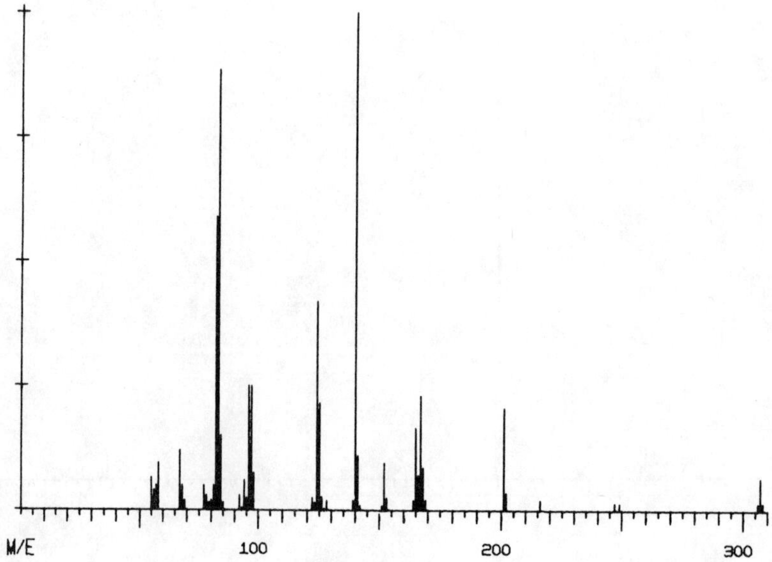
M/E
100
200
300

BENZTROPINE

$C_{21}H_{25}NO$ MW = 307
Base Peak = 140 23 Peaks

Mass	Int.	Mass	Int.	Mass	Int.	Mass	Int.	Mass	Int.	Mass	Int.	Mass	Int.	Mass	Int.	Mass	Int.
55	64	58	94	67	119	68	49	82	591	83	884	96	249	97	249	124	419
125	214	140	999	141	109	152	94	153	24	165	164	167	229	201	204	202	34
216	19	247	14	249	14	307	64	308	16								

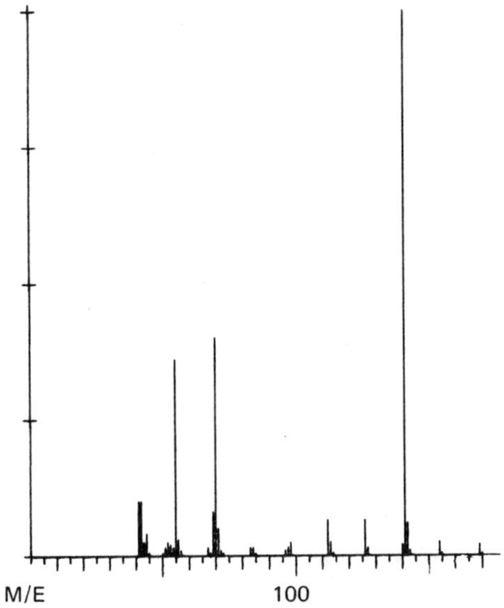
M/E
100

ETHOSUXIMIDE N-ETHYL DERIVATIVE

$C_9H_{15}NO_2$ MW = 169

Base Peak = 141 20 Peaks

Mass	Int.	Mass	Int.	Mass	Int.	Mass	Int.	Mass	Int.	Mass	Int.	Mass	Int.	Mass	Int.	Mass	Int.
41	100	42	100	55	360	56	30	69	80	70	400	83	15	84	15	97	15
98	25	112	65	113	25	126	65	127	15	141	999	142	60	154	25	155	5
169	20	170	5														

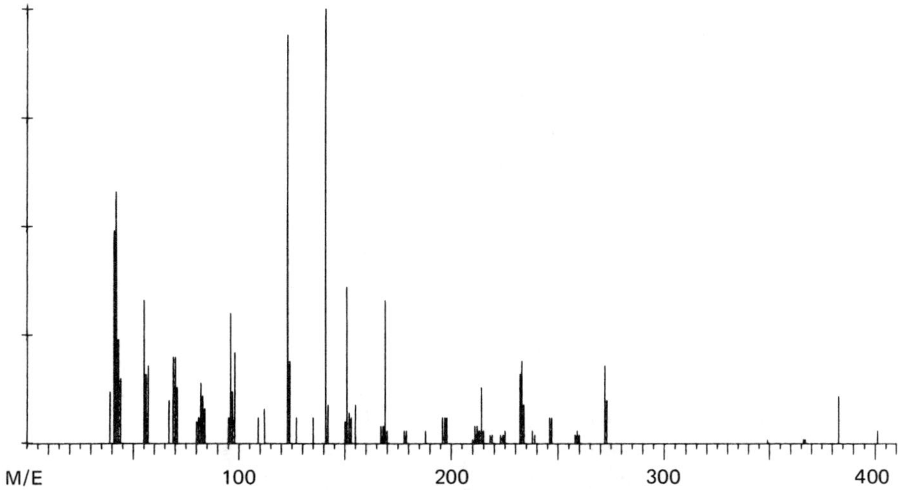
M/E
100
200
300
400

PIPAMAZINE

$C_{21}H_{24}CLN_3OS$ MW = 401

Base Peak = 141 41 Peaks

Mass	Int.	Mass	Int.	Mass	Int.	Mass	Int.	Mass	Int.	Mass	Int.	Mass	Int.	Mass	Int.	Mass	Int.
41	489	42	579	55	329	57	179	69	199	70	199	82	139	83	109	96	299
98	209	109	59	112	79	123	939	124	189	141	999	142	89	151	359	155	89
167	39	169	329	178	29	179	29	196	59	197	59	211	39	214	129	218	19
225	29	232	159	233	189	246	59	247	59	258	19	259	29	272	179	273	99
349	9	366	9	367	9	383	109	401	29								

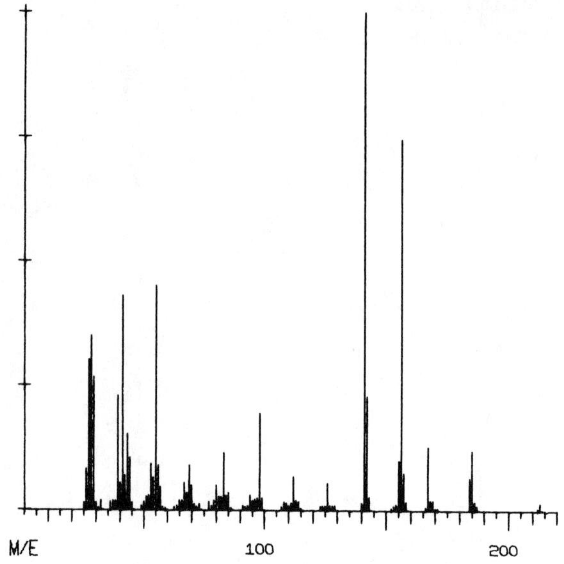
M/E
100
200

BUTETHAL

$C_{10}H_{16}N_2O_3$ MW = 212

Base Peak = 141 26 Peaks

Mass	Int.	Mass	Int.	Mass	Int.	Mass	Int.	Mass	Int.	Mass	Int.	Mass	Int.	Mass	Int.	Mass	Int.
27	302	28	350	39	230	41	431	53	92	55	451	67	54	69	90	80	49
83	115	94	29	98	194	112	68	113	19	124	11	126	54	141	999	142	231
155	101	156	745	167	127	168	19	184	64	185	120	212	4	213	15		

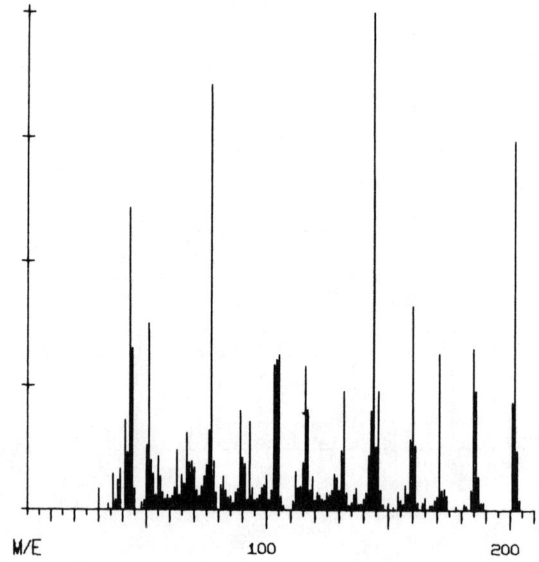
M/E
100
200

OXACILLIN MTB.2

$C_{11}H_{10}N_2O_2$ MW = 202

Base Peak = 144 27 Peaks

Mass	Int.	Mass	Int.	Mass	Int.	Mass	Int.	Mass	Int.	Mass	Int.	Mass	Int.	Mass	Int.	Mass	Int.
30	43	43	608	44	325	50	131	51	374	63	120	67	155	77	856	89	199
93	178	103	293	104	303	105	312	128	73	131	120	132	240	144	999	146	240
159	142	160	410	171	316	185	325	186	240	188	14	201	217	202	743	203	120

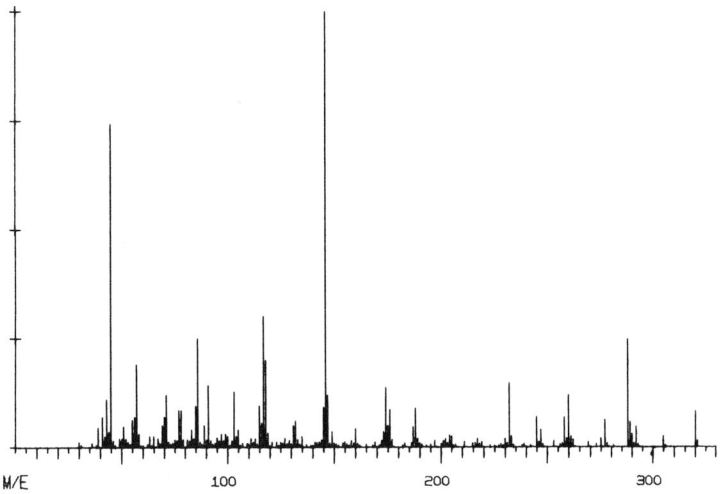
M/E
100
200
300

N,N′-DIMETHOXY METHYL PHENOBARBITAL

$C_{16}H_{20}N_2O_5$ MW = 320

Base Peak = 146 44 Peaks

Mass	Int.	Mass	Int.	Mass	Int.	Mass	Int.	Mass	Int.	Mass	Int.	Mass	Int.	Mass	Int.	Mass	Int.
30	12	31	6	43	110	45	743	56	69	57	190	70	69	71	120	85	95
86	250	91	143	103	127	115	94	117	301	118	199	131	49	132	60	145	93
146	999	147	121	160	43	173	35	174	137	176	87	188	90	189	21	204	27
205	24	217	20	219	12	232	148	233	26	245	71	247	39	258	71	260	121
275	19	277	63	288	248	289	58	305	24	306	5	320	82	321	15		

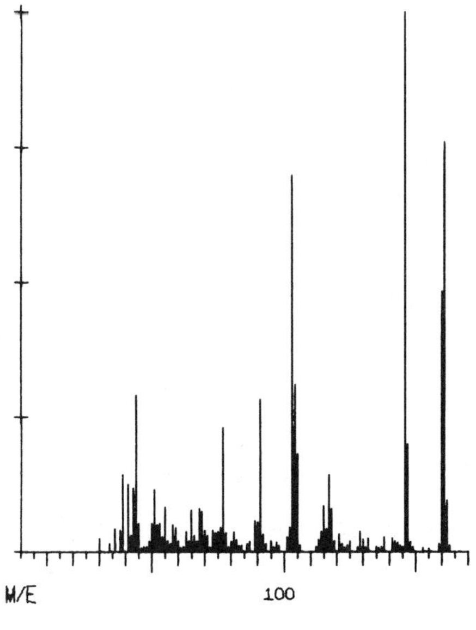
M/E
100

OXACILLIN MTB. 1

$C_{10}H_{11}NO$ MW = 161

Base Peak = 146 21 Peaks

Mass	Int.	Mass	Int.	Mass	Int.	Mass	Int.	Mass	Int.	Mass	Int.	Mass	Int.	Mass	Int.	Mass	Int.
30	24	39	143	44	298	51	114	55	83	65	78	68	81	77	229	89	57
91	283	103	697	104	309	105	182	118	80	129	37	138	28	141	25	146	999
147	199	160	482	161	761												

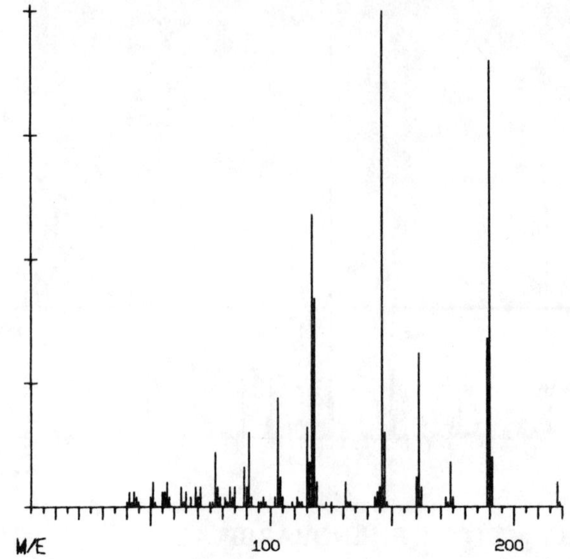
M/E
100
200

PRIMACLONE

$C_{12}H_{14}N_2O_2$ MW = 218

Base Peak = 146   26 Peaks

Mass	Int.	Mass	Int.	Mass	Int.	Mass	Int.	Mass	Int.	Mass	Int.	Mass	Int.	Mass	Int.	Mass	Int.
41	30	43	30	51	50	57	50	63	40	69	40	77	110	89	80	91	150
103	220	115	160	117	590	118	420	119	50	144	30	145	40	146	999	147	150
160	60	161	310	174	90	175	20	189	340	190	900	218	50	219	10		

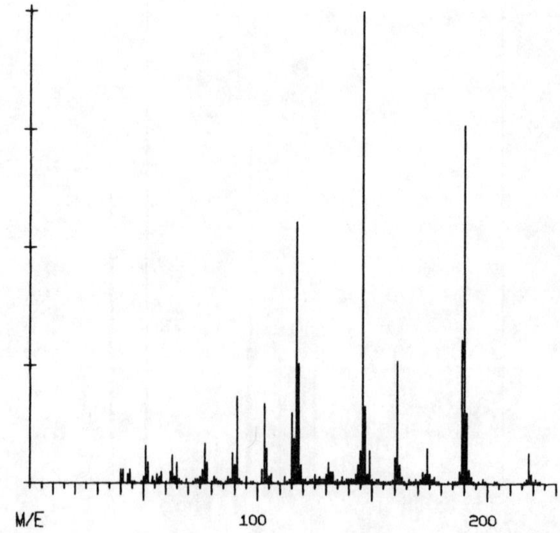
M/E
100
200

PRIMACLONE

$C_{12}H_{14}N_2O_2$ MW = 218

Base Peak = 146   28 Peaks

Mass	Int.	Mass	Int.	Mass	Int.	Mass	Int.	Mass	Int.	Mass	Int.	Mass	Int.	Mass	Int.	Mass	Int.
40	30	41	30	51	80	52	45	63	60	65	45	77	85	89	65	91	185
103	170	115	150	117	555	118	255	131	45	144	40	145	70	146	999	147	165
160	55	161	260	174	75	175	20	189	305	190	760	203	5	204	5	218	65
219	20																

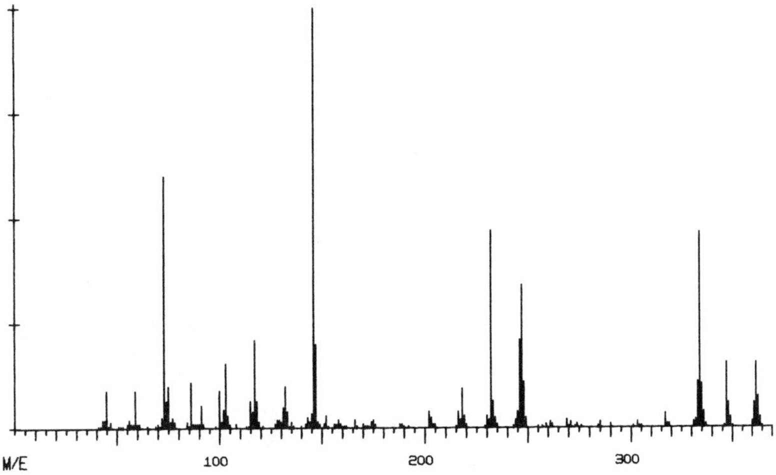
M/E
100
200
300

PRIMIDONE 1,3-DI-TMS DERIVATIVE

$C_{18}H_{30}N_2O_2Si_2$ MW = 362

Base Peak = 146 47 Peaks

Mass	Int.	Mass	Int.	Mass	Int.	Mass	Int.	Mass	Int.	Mass	Int.	Mass	Int.	Mass	Int.	Mass	Int.
43	19	45	89	56	19	59	89	73	599	75	99	77	24	86	109	100	89
103	154	115	64	117	209	118	64	131	49	132	99	133	39	146	999	147	199
166	19	170	9	174	14	175	19	188	9	189	9	202	39	203	24	216	39
218	94	232	469	233	64	246	209	247	339	261	14	269	19	274	9	285	14
290	9	301	4	303	14	317	34	318	9	333	109	334	464	347	154	348	59
362	154	363	74														

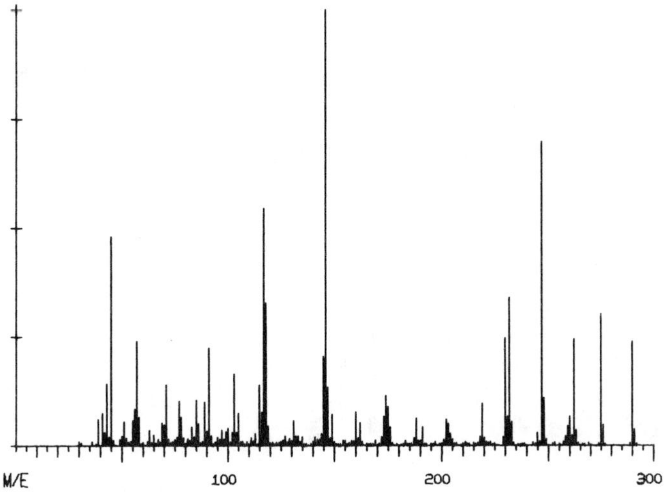
M/E
100
200
300

N-METHYL,N'-METHOXYMETHYL PHENOBARBITAL

$C_{15}H_{18}N_2O_4$ MW = 290

Base Peak = 146 40 Peaks

Mass	Int.	Mass	Int.	Mass	Int.	Mass	Int.	Mass	Int.	Mass	Int.	Mass	Int.	Mass	Int.	Mass	Int.
30	9	31	7	43	142	45	480	56	84	57	240	69	53	71	139	77	102
85	104	91	225	103	166	115	140	117	546	118	328	131	57	144	28	145	205
146	999	147	134	160	77	173	67	174	114	175	89	188	62	191	42	202	59
203	50	218	19	219	98	230	248	232	339	247	697	248	111	260	67	262	246
275	303	276	47	290	239	291	38										

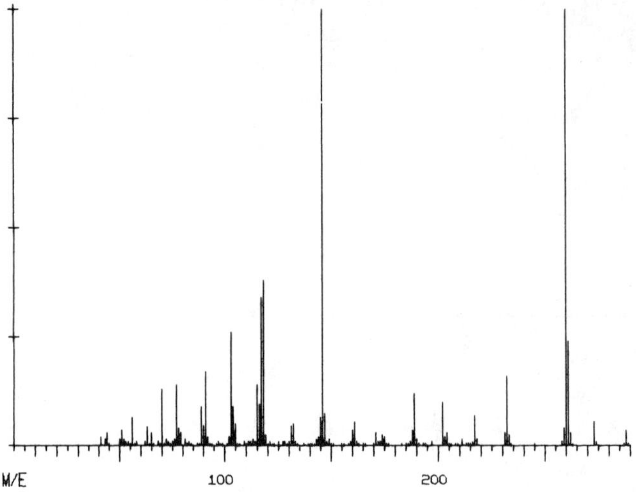
M/E
100
200

PHENOBARBITAL 1,3-DIETHYL DERIVATIVE

$C_{16}H_{20}N_2O_3$ MW = 288

Base Peak = 146 37 Peaks

Mass	Int.	Mass	Int.	Mass	Int.	Mass	Int.	Mass	Int.	Mass	Int.	Mass	Int.	Mass	Int.	Mass	Int.
41	20	44	30	51	35	56	65	63	45	70	130	77	140	89	90	91	170
103	260	115	140	117	340	118	380	131	45	132	50	145	65	146	999	147	75
160	35	161	55	174	25	175	20	188	35	189	120	202	100	204	30	217	70
218	15	231	30	232	160	245	5	260	999	261	240	273	55	274	10	287	5
288	35																

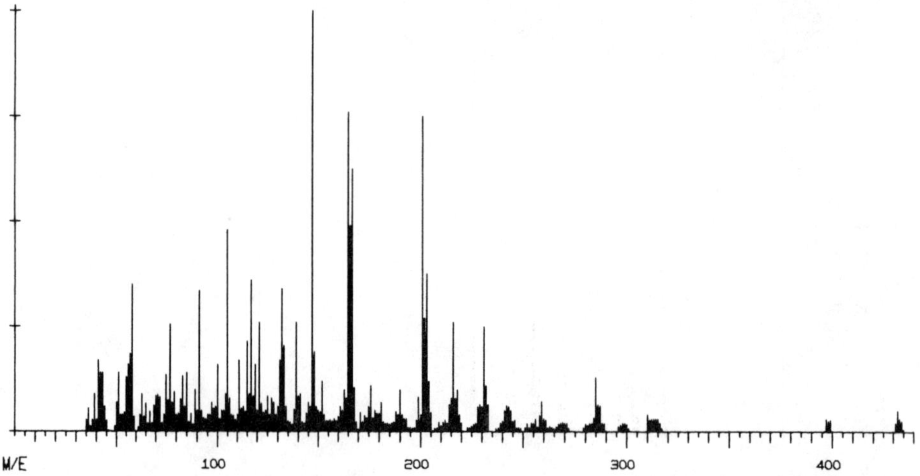
M/E
100
200
300
400

BUCLIZINE

$C_{28}H_{33}CLN_2$ MW = 432

Base Peak = 147 48 Peaks

Mass	Int.	Mass	Int.	Mass	Int.	Mass	Int.	Mass	Int.	Mass	Int.	Mass	Int.	Mass	Int.	Mass	Int.
41	171	42	140	57	184	58	349	63	89	75	134	77	254	85	139	91	334
100	159	105	479	117	359	121	259	131	169	132	339	139	259	147	999	148	189
165	759	167	624	176	109	181	69	190	99	201	749	202	269	203	374	216	259
218	101	231	249	232	109	244	49	256	30	258	38	259	73	284	33	285	129
286	64	287	62	310	41	313	31	314	31	315	30	396	3	397	30	398	25
399	27	432	49	433	29												

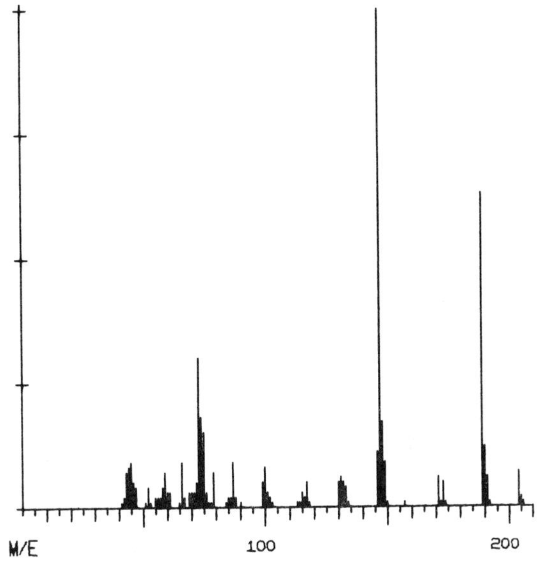
M/E
100
200

UREA DI-TMS DERIVATIVE

$C_7H_{20}N_2OSI_2$ MW = 204

Base Peak = 147   25 Peaks

Mass	Int.	Mass	Int.	Mass	Int.	Mass	Int.	Mass	Int.	Mass	Int.	Mass	Int.	Mass	Int.	Mass	Int.
44	79	45	89	52	39	59	69	73	299	74	179	79	69	87	89	99	49
100	79	115	29	117	49	130	49	131	59	132	49	133	39	147	999	148	169
171	59	173	49	174	9	189	629	190	119	204	69	205	19				

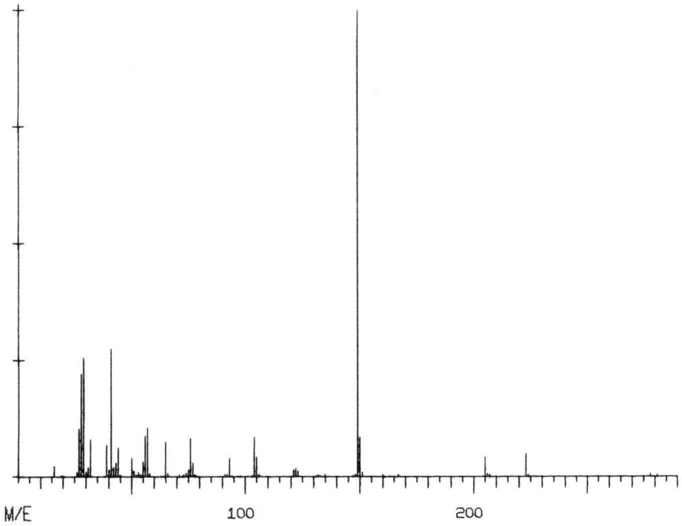
M/E
100
200

DIBUTYL PHTHALATE

$C_{16}H_{22}O_4$ MW = 278

Base Peak = 149   32 Peaks

Mass	Int.	Mass	Int.	Mass	Int.	Mass	Int.	Mass	Int.	Mass	Int.	Mass	Int.	Mass	Int.	Mass	Int.
28	220	29	255	39	68	41	275	56	87	57	106	65	76	75	14	76	82
77	31	91	4	93	39	104	85	105	43	121	16	122	17	132	6	135	4
149	999	150	85	160	4	167	5	176	1	177	1	189	1	191	1	205	42
206	7	223	49	224	6	278	7	281	4								

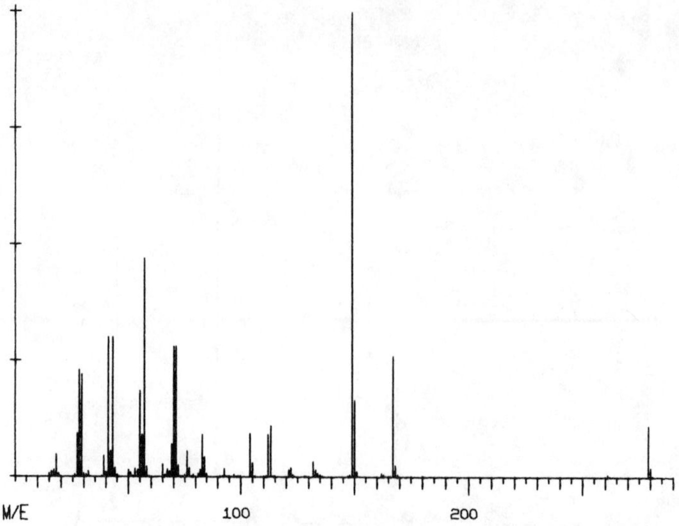
M/E
100
200

DIOCTYL PHTHALATE

$C_{24}H_{38}O_4$ MW = 390

Base Peak = 149 30 Peaks

Mass	Int.	Mass	Int.	Mass	Int.	Mass	Int.	Mass	Int.	Mass	Int.	Mass	Int.	Mass	Int.	Mass	Int.
28	230	29	220	41	300	43	300	55	185	57	470	70	280	71	280	76	54
83	90	93	17	97	6	104	92	113	110	121	15	122	19	132	33	133	15
149	999	150	165	167	260	168	25	175	2	180	2	191	1	192	1	261	5
262	2	279	110	280	20												

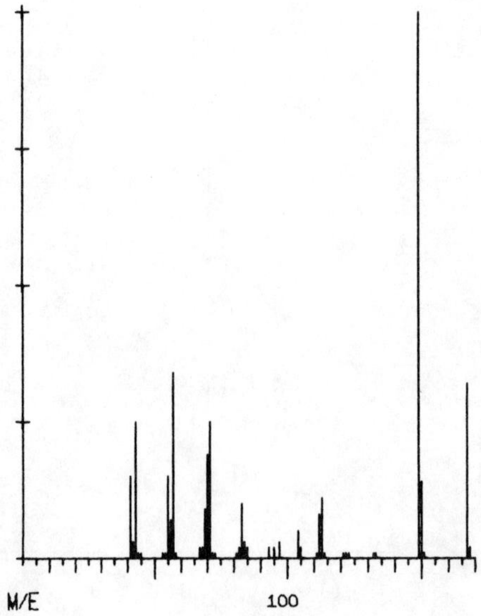
M/E
100

DIOCTYL PHTHALATE

$C_{24}H_{38}O_4$ MW = 390

Base Peak = 149 20 Peaks

Mass	Int.	Mass	Int.	Mass	Int.	Mass	Int.	Mass	Int.	Mass	Int.	Mass	Int.	Mass	Int.	Mass	Int.
41	150	43	250	55	150	57	340	70	190	71	250	83	100	84	30	93	20
97	30	112	80	113	110	121	10	122	10	132	10	133	10	149	999	150	140
167	320	168	20														

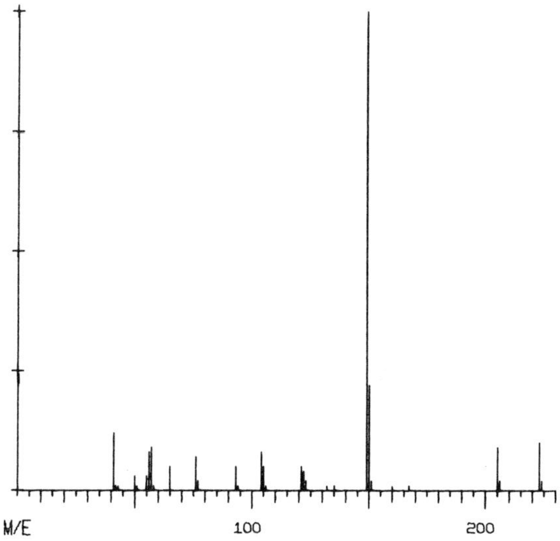
M/E
100
200

DI-N-BUTYL PHTHALATE

$C_{16}H_{22}O_4$ MW = 278

Base Peak = 149 32 Peaks

Mass	Int.	Mass	Int.	Mass	Int.	Mass	Int.	Mass	Int.	Mass	Int.	Mass	Int.	Mass	Int.	Mass	Int.
28	220	29	255	39	68	41	275	56	87	57	106	65	76	75	14	76	82
77	31	91	4	93	39	104	85	105	43	121	16	122	17	132	6	135	4
149	999	150	85	160	4	167	5	176	1	177	1	189	1	191	1	205	42
206	7	223	49	224	6	278	7	281	4								

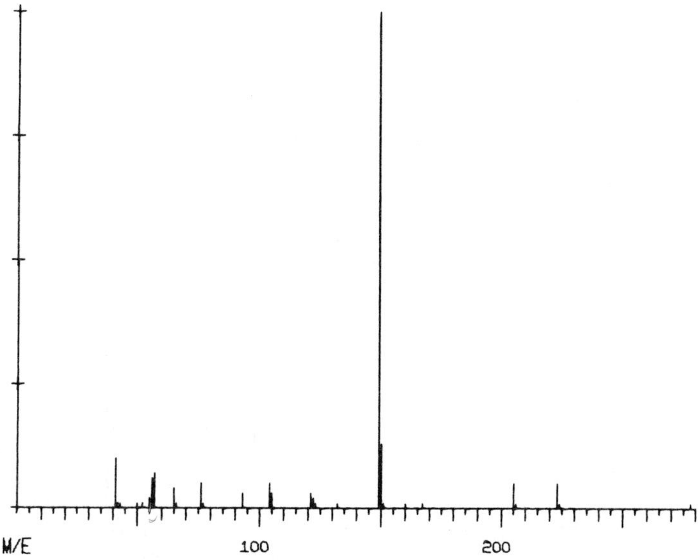
M/E
100
200

BUTYL CARBOBUTOXYMETHYL PHTHALATE

$C_{18}H_{24}O_6$ MW = 336

Base Peak = 149 23 Peaks

Mass	Int.	Mass	Int.	Mass	Int.	Mass	Int.	Mass	Int.	Mass	Int.	Mass	Int.	Mass	Int.	Mass	Int.
41	100	42	10	56	60	57	70	65	40	66	10	76	50	77	10	93	30
104	50	105	30	121	30	122	20	132	10	149	999	150	130	160	10	167	10
205	50	206	10	223	50	224	10	278	10								

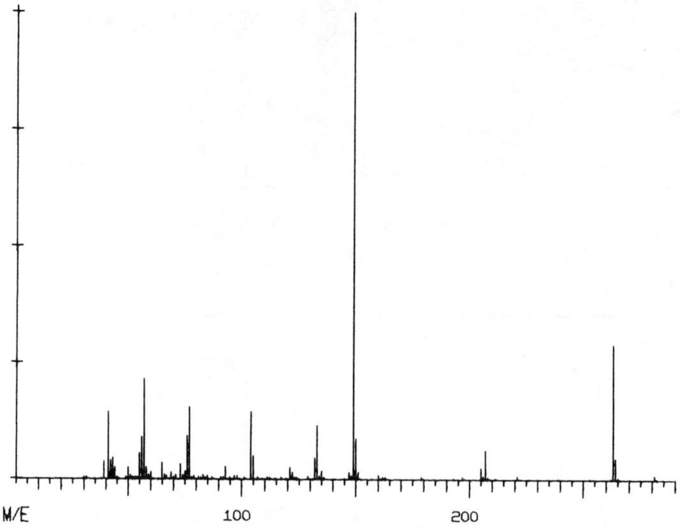
M/E
100
200

BUTYL BUTOXYETHYL PHTHALATE

$C_{18}H_{26}O_5$ MW = 322

Base Peak = 149 34 Peaks

Mass	Int.	Mass	Int.	Mass	Int.	Mass	Int.	Mass	Int.	Mass	Int.	Mass	Int.	Mass	Int.	Mass	Int.
30	4	31	6	41	144	43	46	56	90	57	215	65	35	73	32	76	92
77	154	93	28	97	8	104	145	105	51	121	25	122	14	132	46	133	114
149	999	150	88	160	10	162	5	169	4	193	2	197	4	205	25	207	63
221	7	239	3	253	2	263	291	264	44	281	10	282	2				

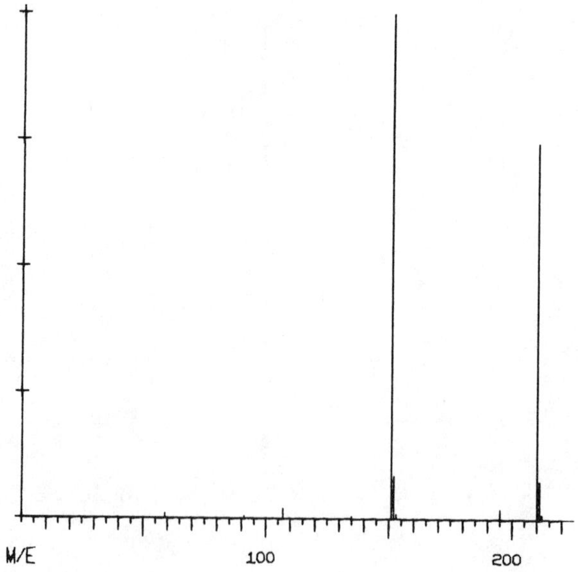
M/E
100
200

3,4-DIMETHOXYPHENYLACETIC ACID METHYL ESTER

$C_{11}H_{14}O_4$ MW = 210

Base Peak = 151 9 Peaks

Mass	Int.	Mass	Int.	Mass	Int.	Mass	Int.	Mass	Int.	Mass	Int.	Mass	Int.	Mass	Int.	Mass	Int.
59	9	91	4	107	19	135	4	151	999	152	84	195	19	210	744	211	74

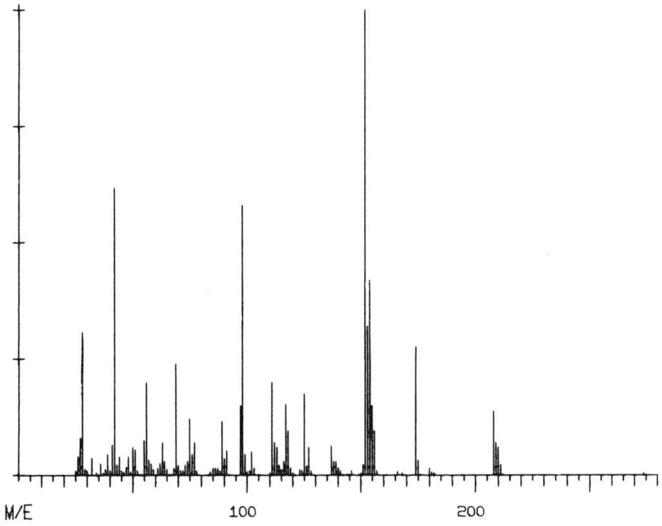
M/E
100
200

CHLORMEZANONE

$C_{11}H_{12}CLNO_3S$ MW = 273
Base Peak = 152 28 Peaks

Mass	Int.	Mass	Int.	Mass	Int.	Mass	Int.	Mass	Int.	Mass	Int.	Mass	Int.	Mass	Int.	Mass	Int.
27	81	28	307	41	65	42	618	55	74	56	201	69	240	75	122	77	70
89	115	97	150	98	580	111	200	117	153	118	95	125	174	137	63	139	31
152	999	154	421	166	7	168	4	174	277	175	33	194	3	208	137	209	69
274	4																

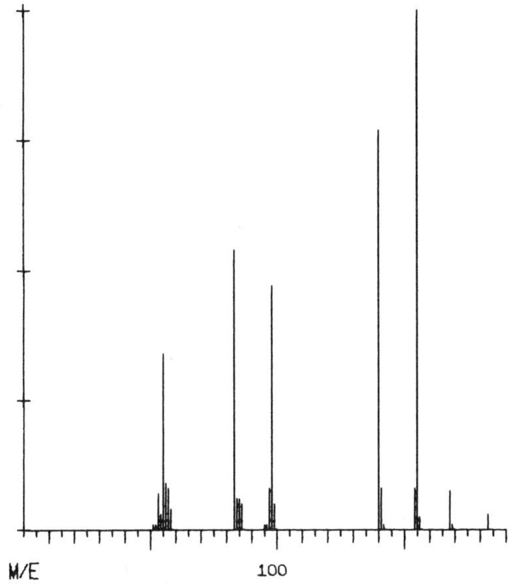
M/E
100

METHYPRYLON

$C_{10}H_{17}NO_2$ MW = 183
Base Peak = 155 13 Peaks

Mass	Int.	Mass	Int.	Mass	Int.	Mass	Int.	Mass	Int.	Mass	Int.	Mass	Int.	Mass	Int.	Mass	Int.
55	339	56	89	83	539	84	59	97	79	98	469	140	769	141	79	154	79
155	999	168	74	169	9	183	29										

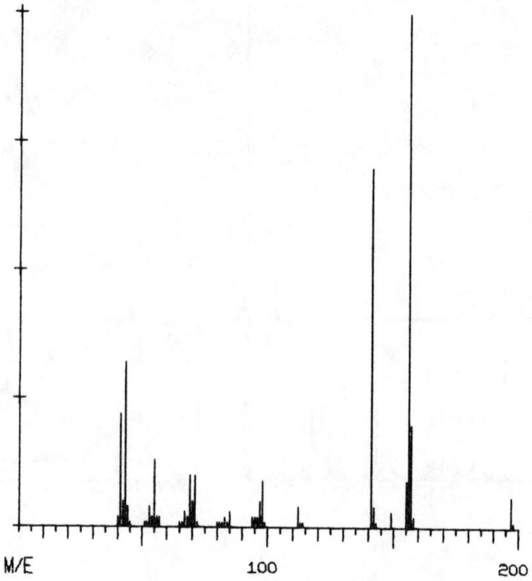
M/E
100
200

PENTOBARBITAL

$C_{11}H_{18}N_2O_3$ MW = 226

Base Peak = 156 18 Peaks

Mass	Int.	Mass	Int.	Mass	Int.	Mass	Int.	Mass	Int.	Mass	Int.	Mass	Int.	Mass	Int.	Mass	Int.
41	220	43	320	53	40	55	130	69	100	71	100	83	20	85	30	97	50
98	90	112	40	113	10	141	700	142	40	156	999	157	200	197	60	198	10

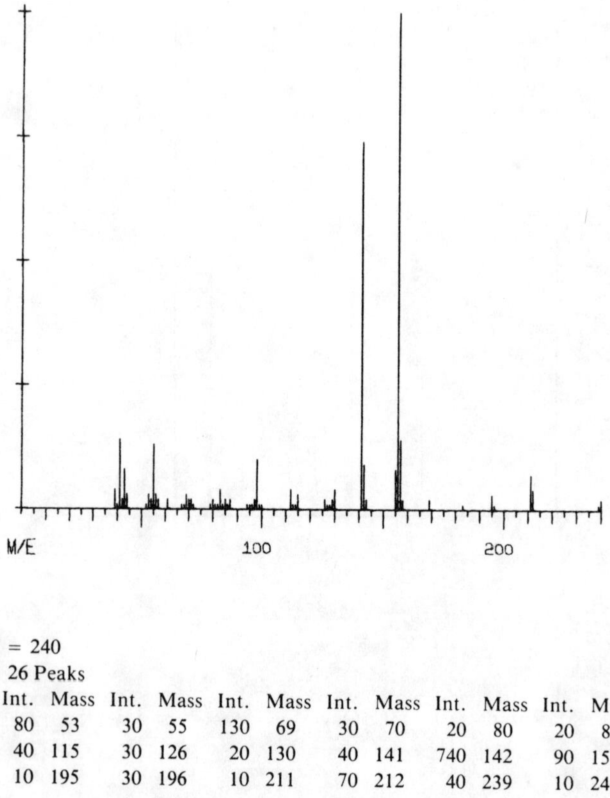
M/E
100
200

HEXETHAL

$C_{12}H_{20}N_2O_3$ MW = 240

Base Peak = 156 26 Peaks

Mass	Int.	Mass	Int.	Mass	Int.	Mass	Int.	Mass	Int.	Mass	Int.	Mass	Int.	Mass	Int.	Mass	Int.
41	140	43	80	53	30	55	130	69	30	70	20	80	20	83	40	97	20
98	100	112	40	115	30	126	20	130	40	141	740	142	90	156	999	157	140
169	20	183	10	195	30	196	10	211	70	212	40	239	10	240	20		

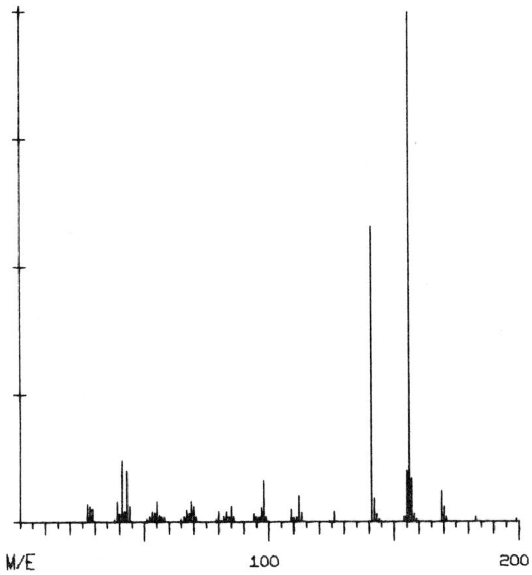
M/E
100
200

PROBARBITAL

$C_9H_{14}N_2O_3$ MW = 198

Base Peak = 156 26 Peaks

Mass	Int.	Mass	Int.	Mass	Int.	Mass	Int.	Mass	Int.	Mass	Int.	Mass	Int.	Mass	Int.	Mass	Int.
27	34	28	29	41	119	43	99	53	19	55	39	69	39	70	31	80	19
85	29	97	27	98	79	109	24	112	49	124	2	126	19	141	579	142	44
155	99	156	999	169	59	170	29	182	1	183	9	198	1	199	6		

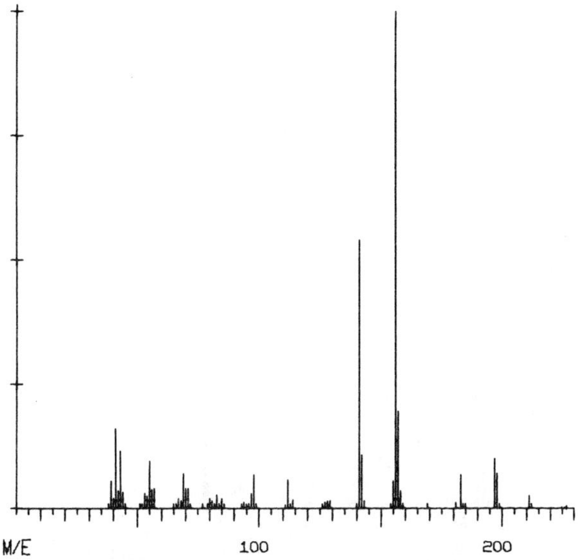
M/E
100
200

AMOBARBITAL

$C_{11}H_{18}N_2O_3$ MW = 226

Base Peak = 156 27 Peaks

Mass	Int.	Mass	Int.	Mass	Int.	Mass	Int.	Mass	Int.	Mass	Int.	Mass	Int.	Mass	Int.	Mass	Int.
41	159	43	114	55	94	57	39	69	69	70	39	80	19	83	28	97	31
98	67	112	57	114	17	128	14	129	14	141	539	142	107	156	999	157	195
169	9	181	12	183	67	197	99	198	69	211	24	212	9	225	3	227	4

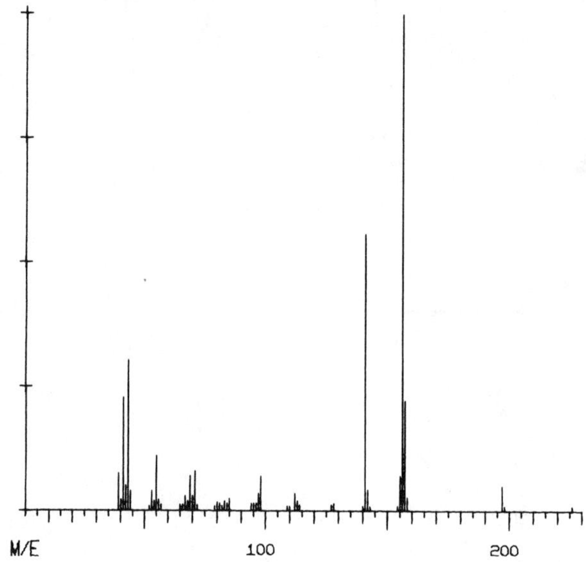
M/E
100
200

PENTOBARBITAL

$C_{11}H_{18}N_2O_3$ MW = 226

Base Peak = 156 21 Peaks

Mass	Int.	Mass	Int.	Mass	Int.	Mass	Int.	Mass	Int.	Mass	Int.	Mass	Int.	Mass	Int.	Mass	Int.
41	227	43	303	53	39	55	109	69	69	71	79	83	19	85	24	97	36
98	69	112	35	113	19	127	13	128	14	141	558	142	42	156	999	157	222
197	49	198	11	226	11												

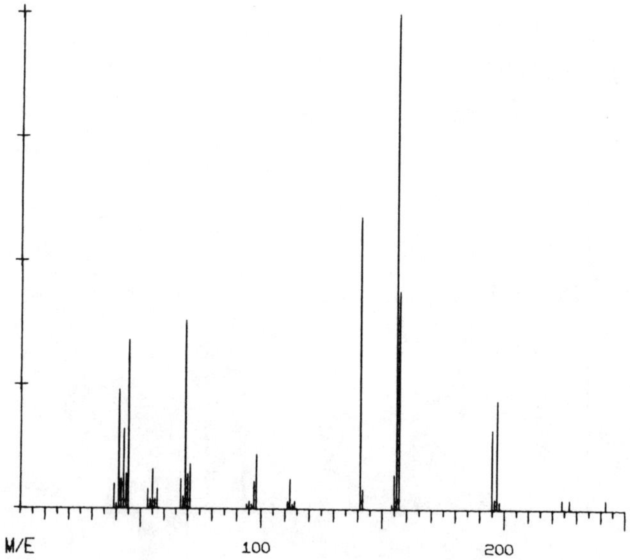
M/E
100
200

PENTOBARBITAL MTB. 1

$C_{11}H_{18}N_2O_4$ MW = 242

Base Peak = 156 19 Peaks

Mass	Int.	Mass	Int.	Mass	Int.	Mass	Int.	Mass	Int.	Mass	Int.	Mass	Int.	Mass	Int.	Mass	Int.
41	239	45	339	53	39	55	79	69	379	71	89	97	54	98	109	111	14
112	59	141	589	142	39	156	999	157	439	195	159	197	219	224	19	227	19
242	19																

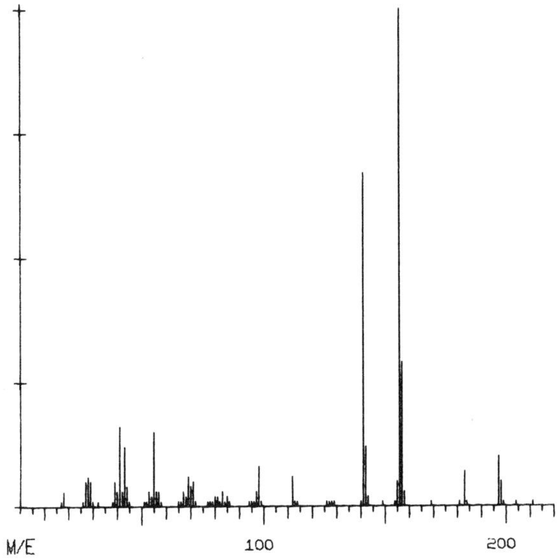
M/E
100
200

AMOBARBITAL

$C_{11}H_{18}N_2O_3$ MW = 226

Base Peak = 156   27 Peaks

Mass	Int.	Mass	Int.	Mass	Int.	Mass	Int.	Mass	Int.	Mass	Int.	Mass	Int.	Mass	Int.	Mass	Int.
41	159	43	114	55	94	57	39	69	69	70	39	80	19	83	28	97	31
98	67	112	57	114	17	128	14	129	14	141	539	142	107	156	999	157	195
169	9	181	12	183	67	197	99	198	69	211	24	212	9	225	3	227	4

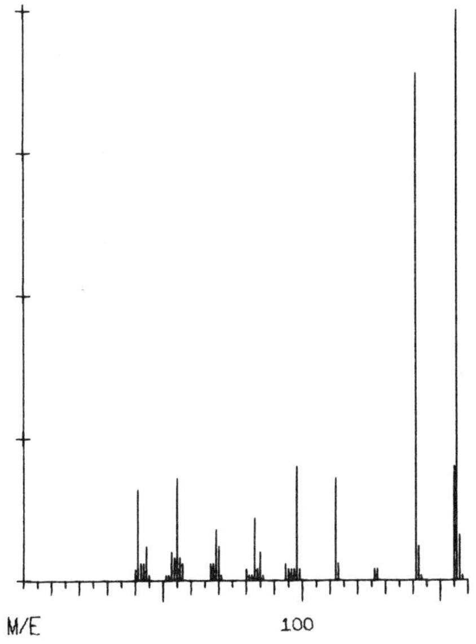
M/E
100

BARBITAL

$C_8H_{12}N_2O_3$ MW = 184

Base Peak = 156   18 Peaks

Mass	Int.	Mass	Int.	Mass	Int.	Mass	Int.	Mass	Int.	Mass	Int.	Mass	Int.	Mass	Int.	Mass	Int.
41	160	44	60	53	50	55	180	69	90	70	60	83	110	85	50	94	30
98	200	112	180	113	30	126	20	127	20	141	890	142	60	155	200	156	999

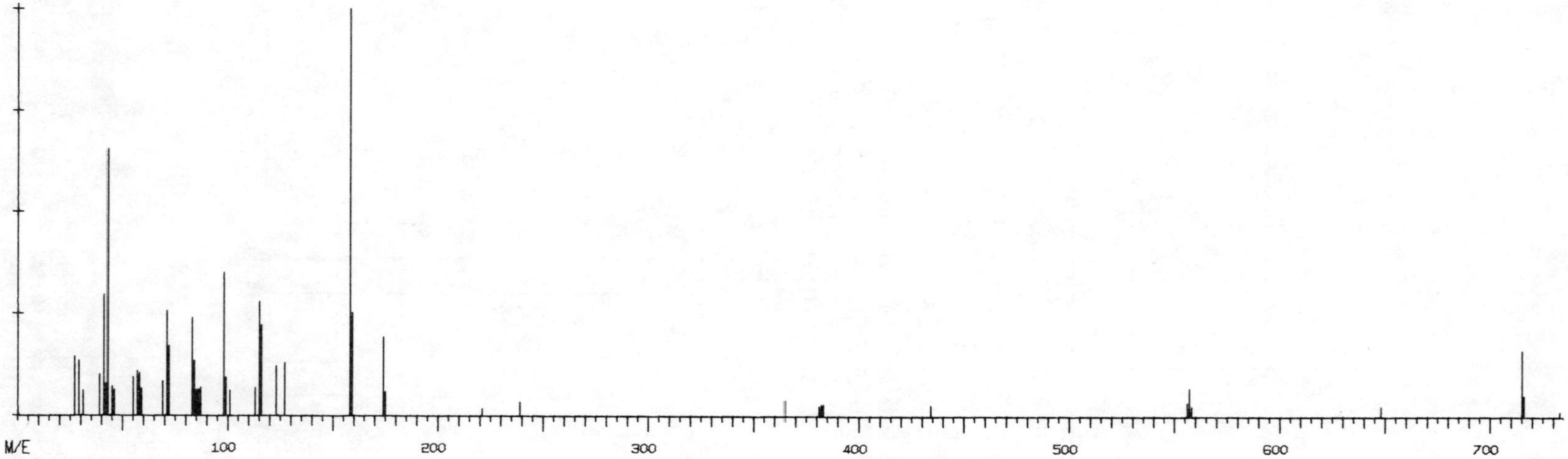
M/E
100
200
300
400
500
600
700

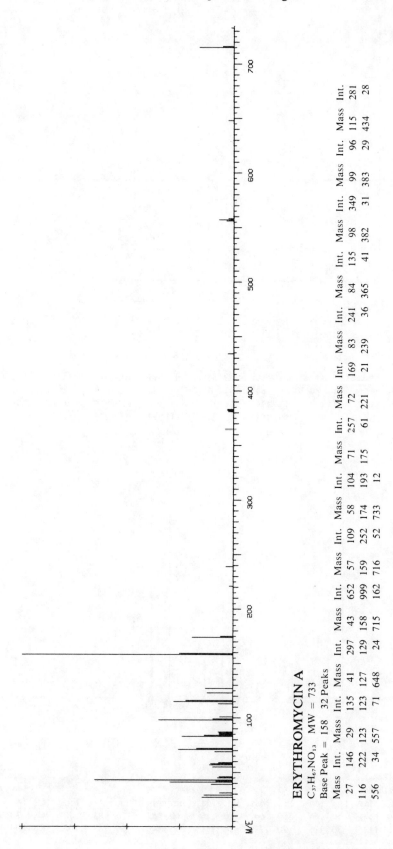

ERYTHROMYCIN A

$C_{37}H_{67}NO_{13}$ MW = 733

Base Peak = 158 32 Peaks

Mass	Int.	Mass	Int.	Mass	Int.	Mass	Int.	Mass	Int.	Mass	Int.	Mass	Int.	Mass	Int.	Mass	Int.	Mass	Int.	Mass	Int.	Mass	Int.	Mass	Int.
27	146	29	135	41	297	43	652	57	109	58	104	71	257	72	169	83	241	84	135	98	349	99	96	115	281
116	222	123	123	127	129	158	999	159	252	174	193	175	61	221	21	239	36	365	41	382	31	383	29	434	28
556	34	557	71	648	24	715	162	716	52	733	12														

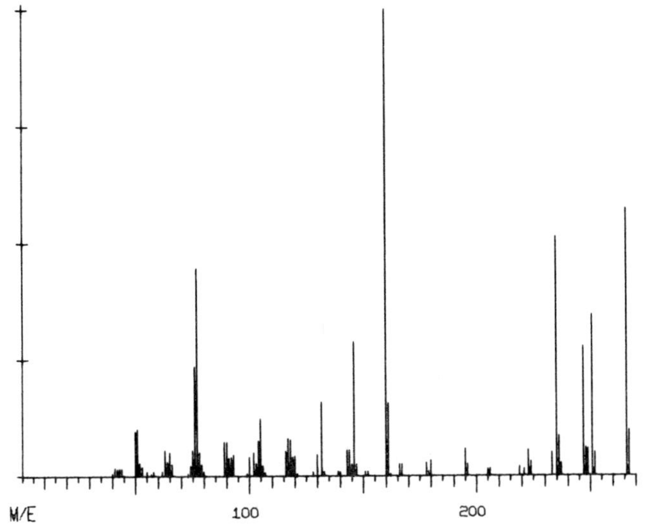
M/E
100
200

METHAQUALONE MTB. 1

$C_{16}H_{14}N_2O_2$ MW = 266

Base Peak = 160 36 Peaks

Mass	Int.	Mass	Int.	Mass	Int.	Mass	Int.	Mass	Int.	Mass	Int.	Mass	Int.	Mass	Int.	Mass	Int.
30	5	31	28	39	192	41	196	50	209	51	295	69	506	75	155	76	473
77	731	90	157	91	151	105	283	117	200	118	131	130	132	132	341	143	139
146	598	147	93	160	999	161	236	178	52	180	83	195	100	196	83	205	39
206	43	223	121	224	109	235	936	236	203	247	365	251	718	266	891	267	192

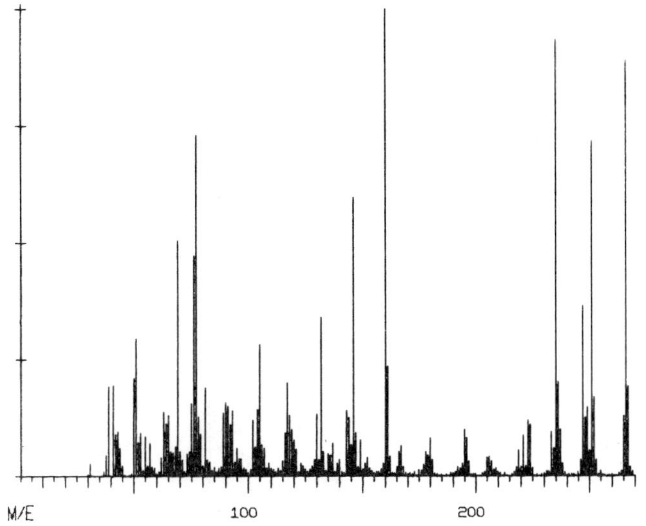
M/E
100
200

METHAQUALONE MTB. 1

$C_{16}H_{14}N_2O_2$ MW = 266

Base Peak = 160 34 Peaks

Mass	Int.	Mass	Int.	Mass	Int.	Mass	Int.	Mass	Int.	Mass	Int.	Mass	Int.	Mass	Int.	Mass	Int.
41	18	42	16	50	96	51	100	63	55	75	54	76	234	17	446	90	72
102	49	105	123	117	81	118	78	130	46	132	157	143	55	146	287	147	25
160	999	161	156	178	28	180	33	195	58	196	25	205	15	206	16	223	55
224	31	235	513	236	85	247	278	251	346	266	572	267	97				

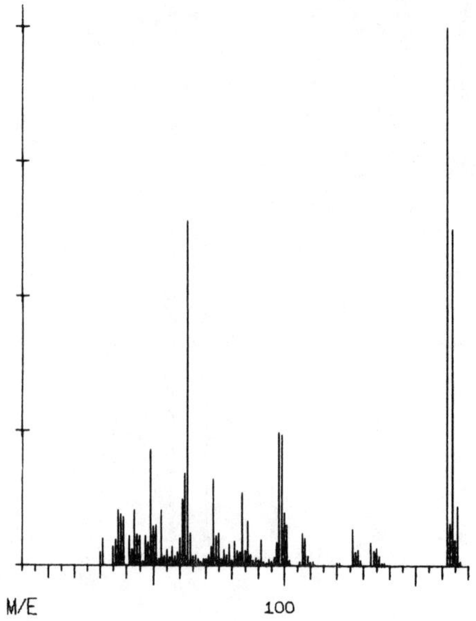
M/E
100

2,4-DICHLOROPHENOL

$C_6H_4CL_2O$ MW = 162
Base Peak = 162 20 Peaks

Mass	Int.	Mass	Int.	Mass	Int.	Mass	Int.	Mass	Int.	Mass	Int.	Mass	Int.	Mass	Int.	Mass	Int.
30	26	31	50	37	103	43	102	49	215	61	122	62	170	63	640	84	135
86	83	98	247	99	242	107	60	108	51	126	68	128	29	133	42	135	32
162	999	164	626														

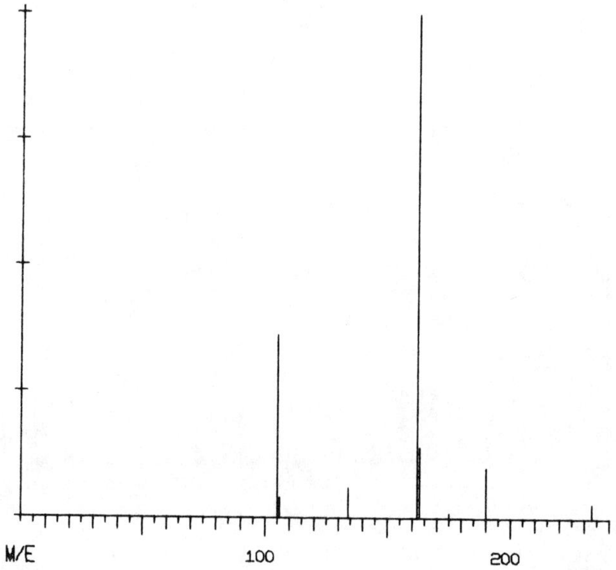
M/E
100
200

DIAMPROMIDE

$C_{21}H_{28}N_2O$ MW = 324
Base Peak = 162 8 Peaks

Mass	Int.	Mass	Int.	Mass	Int.	Mass	Int.	Mass	Int.	Mass	Int.	Mass	Int.	Mass	Int.
105	363	106	40	134	60	162	999	163	141	175	10	190	101	233	30

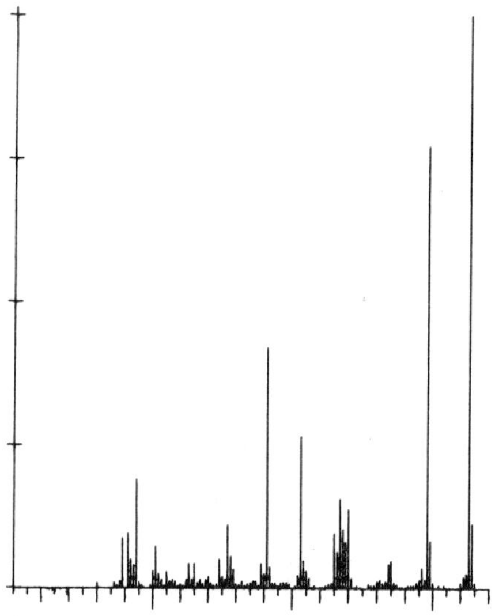
M/E
100

PRIMACLONE MTB. 1

$C_{11}H_{14}N_2O_2$ MW = 206

Base Peak = 163 21 Peaks

Mass	Int.	Mass	Int.	Mass	Int.	Mass	Int.	Mass	Int.	Mass	Int.	Mass	Int.	Mass	Int.	Mass	Int.
30	10	41	95	44	190	50	30	51	73	65	43	74	51	77	111	78	56
91	419	103	264	115	96	117	154	118	102	120	137	134	43	135	48	148	773
149	82	163	999	164	113												

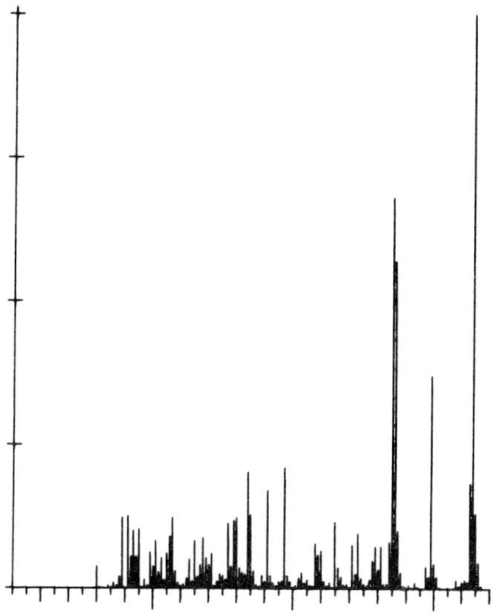
M/E
100

1-THIOPHENYL CYCLOHEXENE

$C_{10}H_{12}S$ MW = 164

Base Peak = 164 22 Peaks

Mass	Int.	Mass	Int.	Mass	Int.	Mass	Int.	Mass	Int.	Mass	Int.	Mass	Int.	Mass	Int.	Mass	Int.
30	37	31	2	39	123	41	126	56	91	57	122	65	83	68	87	84	202
85	128	91	170	97	210	108	77	115	114	121	76	123	96	135	679	136	569
149	371	150	42	163	182	164	999										

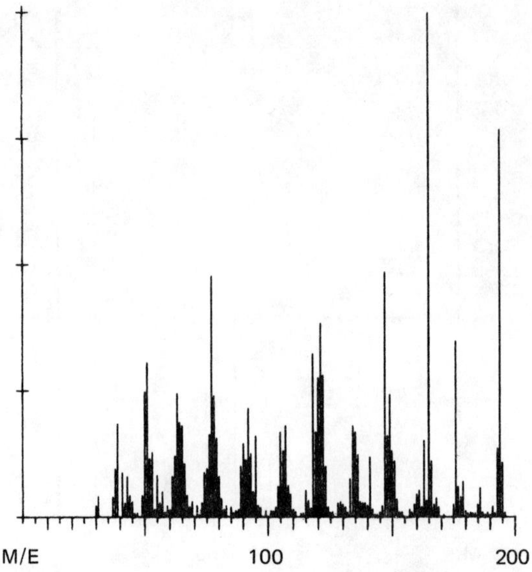
M/E
100
200

MECONIN

$C_{10}H_{10}O_4$ MW = 194

Base Peak = 165 26 Peaks

Mass	Int.	Mass	Int.	Mass	Int.	Mass	Int.	Mass	Int.	Mass	Int.	Mass	Int.	Mass	Int.	Mass	Int.
30	23	31	43	38	94	39	184	50	248	51	305	63	245	64	188	77	478
78	241	92	216	95	159	105	168	107	179	118	323	121	382	134	180	135	167
147	486	149	242	163	149	165	999	176	347	179	70	193	134	194	768		

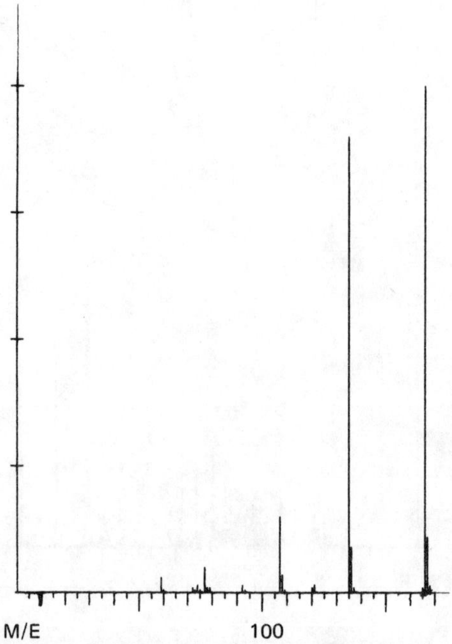
M/E
100

M-METHOXYBENZOIC ACID METHYL ESTER

$C_9H_{10}O_3$ MW = 166

Base Peak = 166 16 Peaks

Mass	Int.	Mass	Int.	Mass	Int.	Mass	Int.	Mass	Int.	Mass	Int.	Mass	Int.	Mass	Int.	Mass	Int.
59	29	60	4	72	9	74	14	77	49	78	9	92	14	93	4	107	149
108	34	120	9	121	14	135	899	136	89	166	999	167	109				

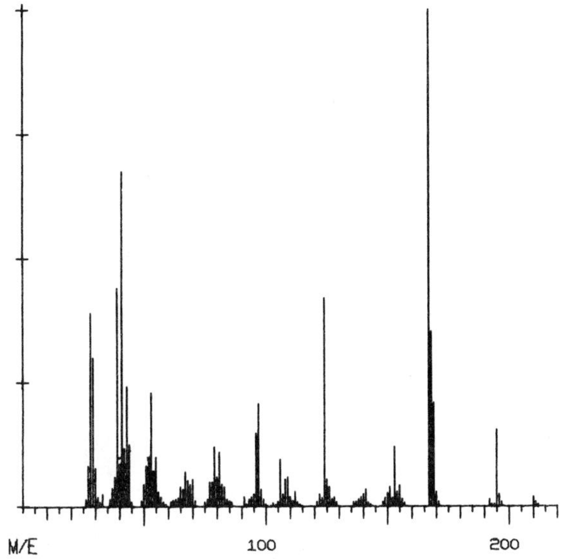
M/E
100
200

APROBARBITAL

$C_{10}H_{14}N_2O_3$ MW = 210

Base Peak = 167 26 Peaks

Mass	Int.	Mass	Int.	Mass	Int.	Mass	Int.	Mass	Int.	Mass	Int.	Mass	Int.	Mass	Int.	Mass	Int.
28	391	29	300	39	441	41	675	52	100	53	231	67	70	70	54	79	120
81	110	96	147	97	207	106	94	109	60	124	419	125	54	140	24	141	34
153	120	155	42	167	999	168	352	195	154	196	24	210	20	211	11		

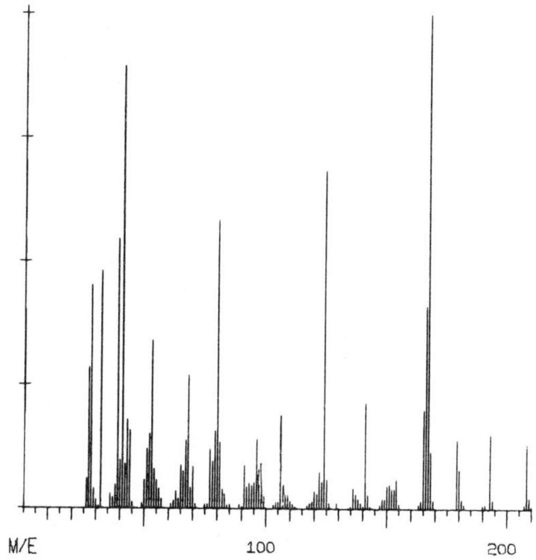
M/E
100
200

ALLOBARBITAL

$C_{10}H_{12}N_2O_3$ MW = 208

Base Peak = 167 28 Peaks

Mass	Int.	Mass	Int.	Mass	Int.	Mass	Int.	Mass	Int.	Mass	Int.	Mass	Int.	Mass	Int.	Mass	Int.
28	450	32	480	39	545	41	895	52	150	53	340	67	138	68	271	79	157
80	585	96	141	98	93	106	190	107	47	122	74	124	685	136	42	141	215
151	50	154	60	166	410	167	999	179	140	180	80	193	150	194	20	208	132
209	23																

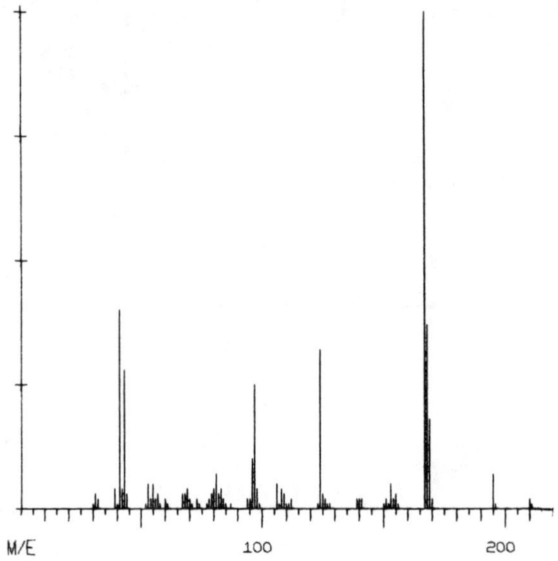
M/E
100
200

APROBARBITAL

$C_{10}H_{14}N_2O_3$ MW = 210
Base Peak = 167 26 Peaks

Mass	Int.	Mass	Int.	Mass	Int.	Mass	Int.	Mass	Int.	Mass	Int.	Mass	Int.	Mass	Int.	Mass	Int.
31	30	32	20	41	400	43	280	53	50	55	50	67	30	69	40	80	40
81	70	96	100	97	250	106	50	108	40	124	320	125	30	139	20	140	20
153	50	155	30	167	999	168	370	195	70	196	10	210	20	211	10		

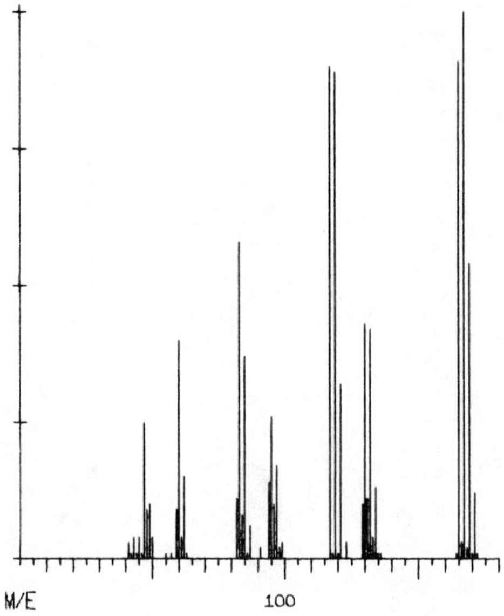
M/E
100

PENTACHLOROETHANE

$C_2H_1Cl_5$ MW = 202
Base Peak = 167 17 Peaks

Mass	Int.	Mass	Int.	Mass	Int.	Mass	Int.	Mass	Int.	Mass	Int.	Mass	Int.	Mass	Int.	Mass	Int.
43	40	47	250	49	100	60	400	62	150	63	10	83	580	85	370	95	260
97	170	117	900	119	890	130	430	132	420	134	130	165	910	167	999		

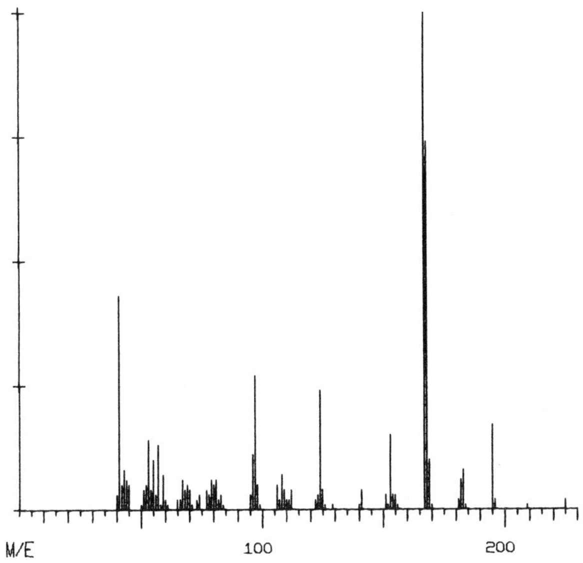
M/E
100
200

TALBUTAL

$C_{11}H_{16}N_2O_3$ MW = 224

Base Peak = 167 26 Peaks

Mass	Int.	Mass	Int.	Mass	Int.	Mass	Int.	Mass	Int.	Mass	Int.	Mass	Int.	Mass	Int.	Mass	Int.
41	430	43	80	53	140	57	130	67	60	69	50	79	60	81	60	96	110
97	270	106	50	108	70	124	240	125	40	140	10	141	40	151	30	153	150
167	999	168	740	182	60	183	80	195	170	196	20	209	10	225	20		

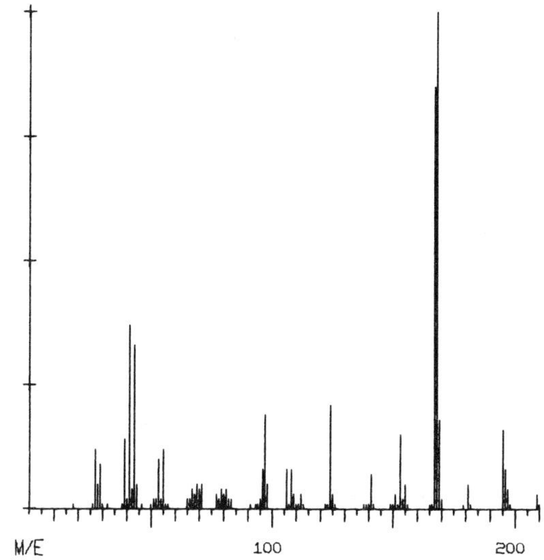
M/E
100
200

SECOBARBITAL

$C_{12}H_{18}N_2O_3$ MW = 238

Base Peak = 168 28 Peaks

Mass	Int.	Mass	Int.	Mass	Int.	Mass	Int.	Mass	Int.	Mass	Int.	Mass	Int.	Mass	Int.	Mass	Int.
27	120	29	90	41	370	43	330	53	100	55	120	69	50	71	50	79	40
81	40	96	80	97	190	106	80	108	80	124	210	125	30	138	10	141	70
153	150	155	50	167	850	168	999	179	10	181	50	195	160	196	80	209	30
210	10																

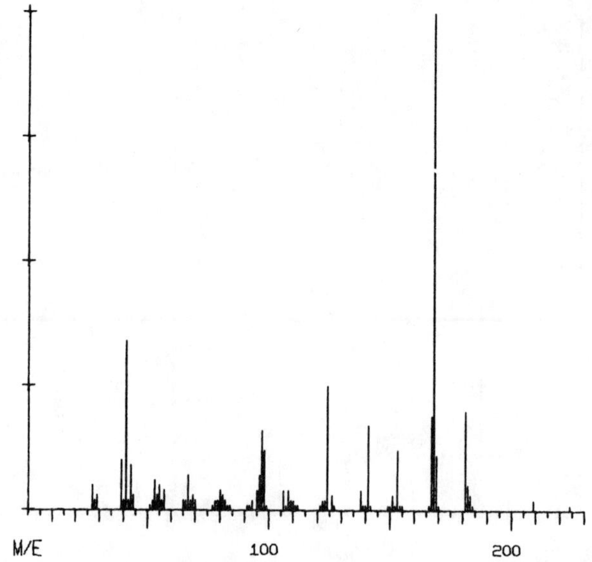
M/E
100
200

BUTALBITAL

$C_{11}H_{16}N_2O_3$ MW = 224

Base Peak = 168 26 Peaks

Mass	Int.	Mass	Int.	Mass	Int.	Mass	Int.	Mass	Int.	Mass	Int.	Mass	Int.	Mass	Int.	Mass	Int.
27	50	29	30	39	100	41	340	53	60	55	50	67	70	69	30	80	40
81	30	97	160	98	120	106	40	108	40	124	250	126	30	138	40	141	170
151	30	153	120	167	190	168	999	181	200	182	50	209	20	224	10		

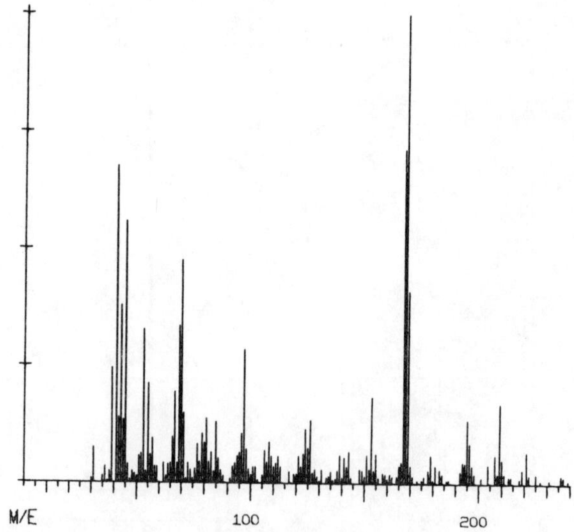
M/E
100
200

SECOBARBITAL MTB. 1

$C_{12}H_{18}N_2O_4$ MW = 254

Base Peak = 168 32 Peaks

Mass	Int.	Mass	Int.	Mass	Int.	Mass	Int.	Mass	Int.	Mass	Int.	Mass	Int.	Mass	Int.	Mass	Int.
30	11	31	75	41	676	45	557	53	328	55	212	69	335	70	475	81	137
85	129	96	105	97	285	106	71	108	88	124	116	126	134	139	60	143	67
153	185	155	63	167	712	168	999	179	60	181	38	195	135	196	85	207	59
209	170	219	30	221	68	236	18	237	15								

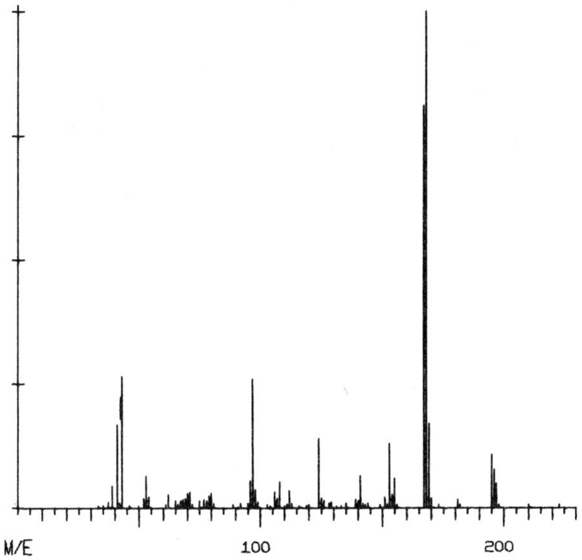
M/E
100
200

5-(2-BROMOALLYL) BARBITURIC ACID

$C_7H_7BrN_2O_3$ MW = 246

Base Peak = 168 27 Peaks

Mass	Int.	Mass	Int.	Mass	Int.	Mass	Int.	Mass	Int.	Mass	Int.	Mass	Int.	Mass	Int.	Mass	Int.
33	4	41	168	43	264	53	66	54	22	70	31	71	33	79	24	80	30
96	55	97	259	108	52	112	36	124	140	125	19	139	17	141	65	153	131
155	61	167	811	168	999	181	18	182	8	195	107	196	78	210	7	223	7

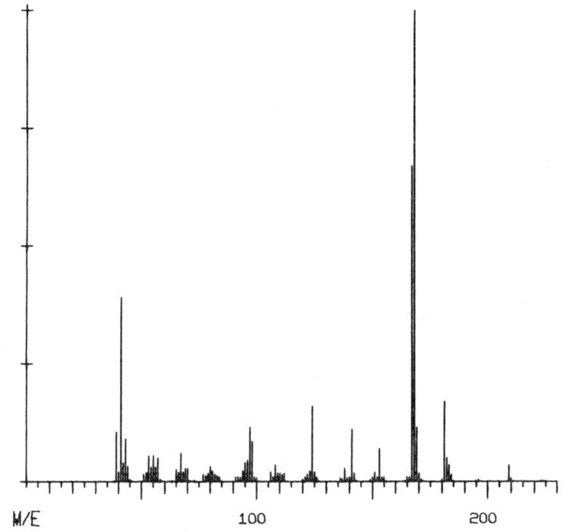
M/E
100
200

BUTALBITAL

$C_{11}H_{16}N_2O_3$ MW = 224

Base Peak = 168 28 Peaks

Mass	Int.	Mass	Int.	Mass	Int.	Mass	Int.	Mass	Int.	Mass	Int.	Mass	Int.	Mass	Int.	Mass	Int.
39	104	41	389	53	55	55	56	67	60	69	28	80	33	81	23	97	114
98	85	106	20	108	34	123	23	124	159	138	27	141	109	151	19	153	69
167	669	168	999	181	171	182	49	195	3	196	4	209	34	210	8	223	2
224	2																

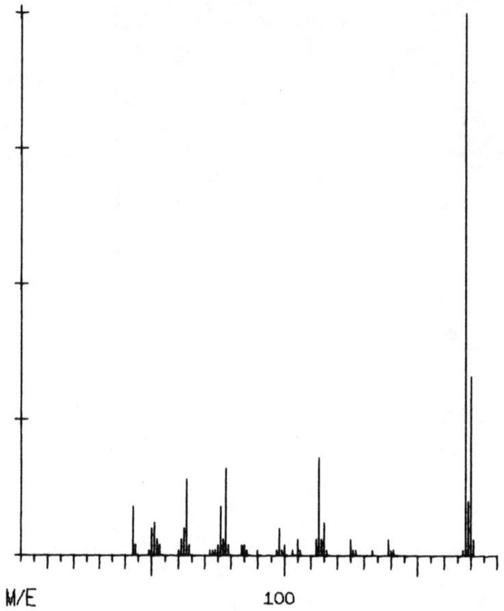
M/E
100

ZOXAZOLAMINE

$C_7H_5CLN_2O$ MW = 168

Base Peak = 168 18 Peaks

Mass	Int.	Mass	Int.	Mass	Int.	Mass	Int.	Mass	Int.	Mass	Int.	Mass	Int.	Mass	Int.	Mass	Int.
43	90	44	20	50	50	51	60	62	50	63	140	76	90	78	160	98	50
100	20	113	180	115	60	125	30	126	10	133	10	139	30	168	999	170	330

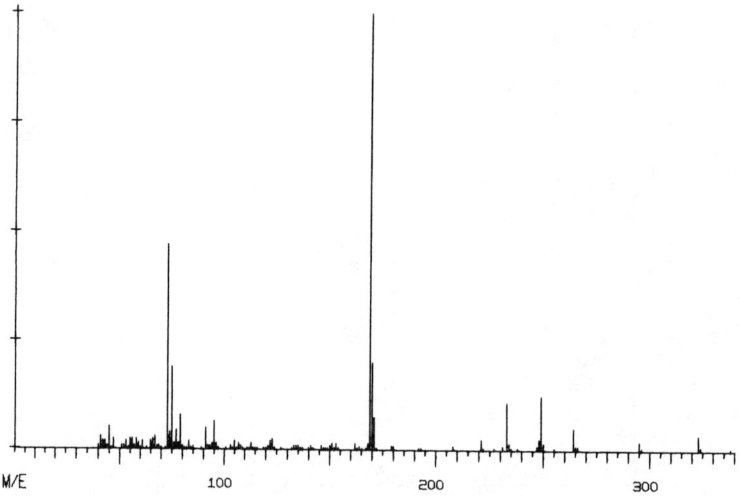
M/E
100
200
300

HYDROXYHEXOBARBITAL 3-METHYL DERIVATIVE TMS ETHER

$C_{16}H_{26}N_2O_4Si$ MW = 338

Base Peak = 169 38 Peaks

Mass	Int.	Mass	Int.	Mass	Int.	Mass	Int.	Mass	Int.	Mass	Int.	Mass	Int.	Mass	Int.	Mass	Int.
41	30	45	53	55	25	56	25	73	470	75	190	77	45	79	80	91	50
95	65	105	20	107	15	122	20	123	25	133	10	134	10	151	15	153	15
169	999	170	200	179	10	180	10	192	5	193	5	208	10	221	25	222	5
233	110	234	15	248	25	249	125	264	50	265	10	295	20	296	5	323	35
324	10	338	5														

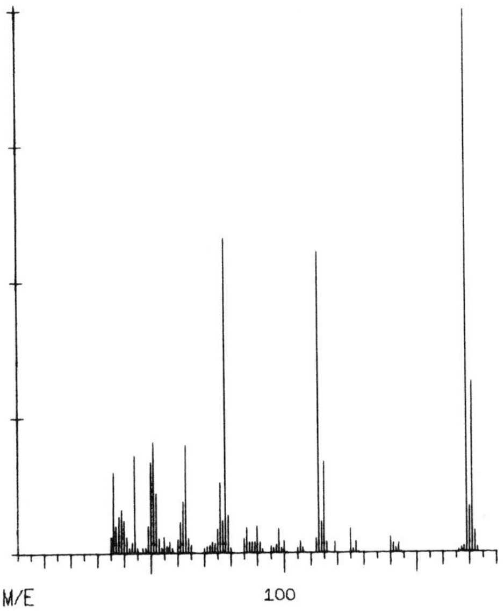
M/E
100

CHLORZOXAZONE

$C_7H_4CLNO_2$ MW = 169

Base Peak = 169 18 Peaks

Mass	Int.	Mass	Int.	Mass	Int.	Mass	Int.	Mass	Int.	Mass	Int.	Mass	Int.	Mass	Int.	Mass	Int.
36	149	44	179	50	168	51	204	62	94	63	199	76	129	78	579	90	49
98	44	113	554	115	167	119	19	125	44	140	29	141	18	169	999	171	314

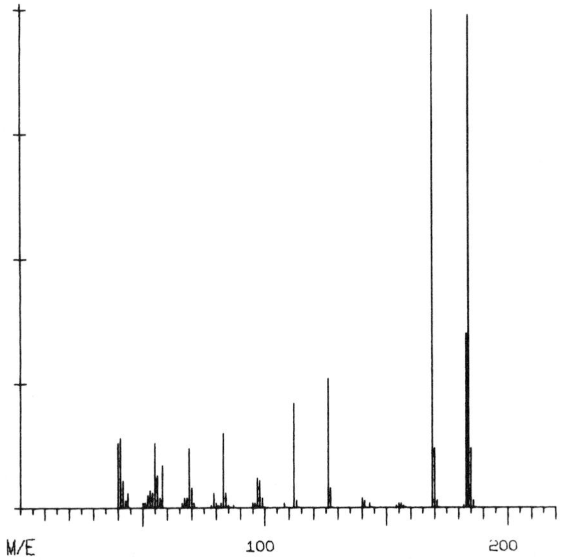
M/E
100
200

BARBITAL 1,3-DIMETHYL DERIVATIVE

$C_{10}H_{16}N_2O_3$ MW = 212

Base Peak = 169 23 Peaks

Mass	Int.	Mass	Int.	Mass	Int.	Mass	Int.	Mass	Int.	Mass	Int.	Mass	Int.	Mass	Int.	Mass	Int.
40	130	41	140	55	130	58	85	69	120	70	40	79	30	83	150	97	60
98	55	112	210	113	15	126	260	127	40	140	20	141	15	155	10	156	10
169	999	170	120	183	350	184	990	212	1								

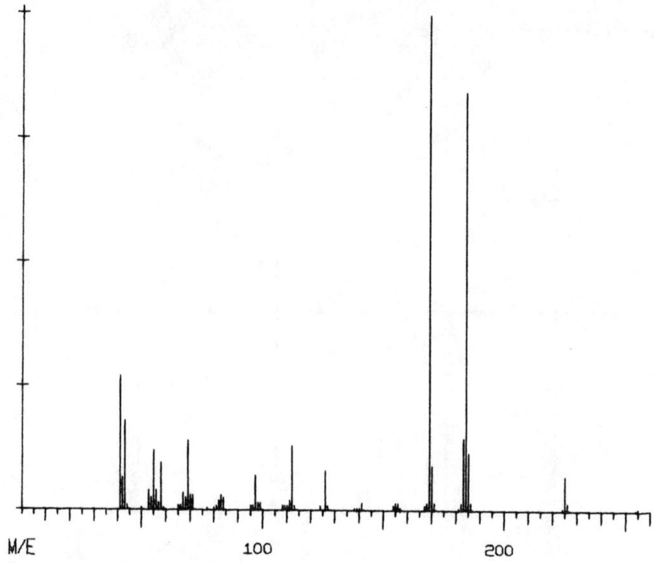
M/E
100
200

PENTOBARBITAL 1,3-DIMETHYL DERIVATIVE

$C_{13}H_{22}N_2O_3$ MW = 254

Base Peak = 169 26 Peaks

Mass	Int.	Mass	Int.	Mass	Int.	Mass	Int.	Mass	Int.	Mass	Int.	Mass	Int.	Mass	Int.	Mass	Int.
41	270	43	180	55	120	58	95	67	35	69	140	83	30	84	25	97	70
98	15	111	20	112	130	124	10	126	80	138	5	141	15	155	15	156	15
169	999	170	90	183	145	184	845	225	70	226	15	254	2	255	4		

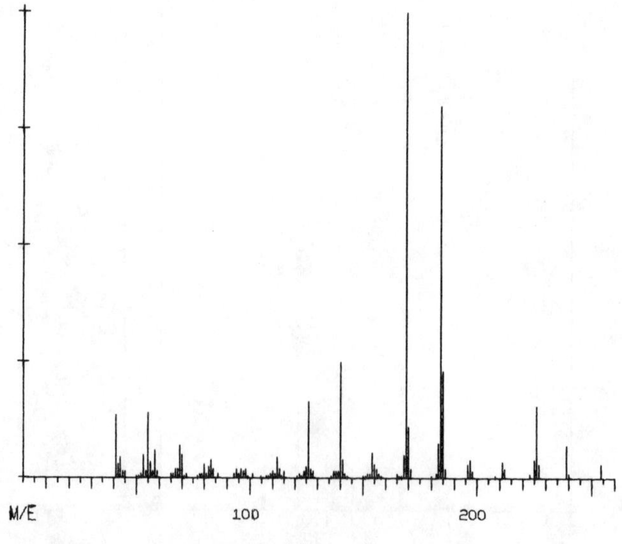
M/E
100
200

AMOBARBITAL 1,2(OR 4)-DIMETHYL DERIVATIVE

$C_{13}H_{22}N_2O_3$ MW = 254

Base Peak = 169 28 Peaks

Mass	Int.	Mass	Int.	Mass	Int.	Mass	Int.	Mass	Int.	Mass	Int.	Mass	Int.	Mass	Int.	Mass	Int.
41	135	43	45	55	140	58	60	69	70	70	50	80	30	83	40	94	20
96	20	112	45	113	20	125	25	126	165	140	250	141	40	154	55	155	30
169	999	170	110	184	800	185	230	196	30	197	40	211	35	212	20	225	40
226	155	239	70	240	10	254	30										

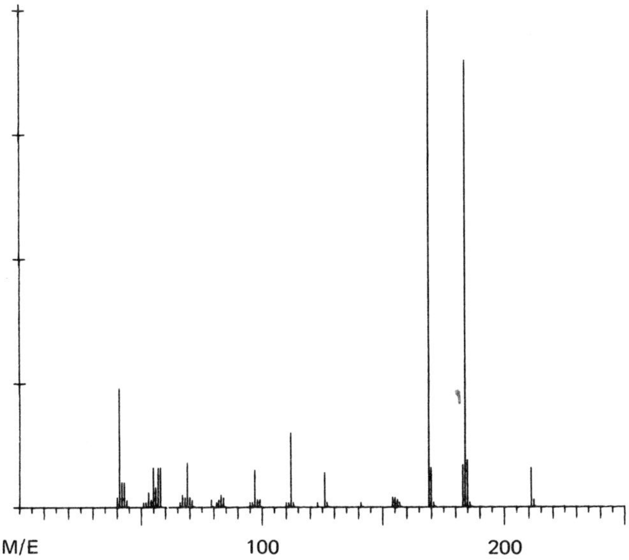
M/E
100
200

BUTABARBITAL 1,3-DIMETHYL DERIVATIVE

$C_{12}H_{20}N_2O_3$ MW = 240

Base Peak = 169 26 Peaks

Mass	Int.	Mass	Int.	Mass	Int.	Mass	Int.	Mass	Int.	Mass	Int.	Mass	Int.	Mass	Int.	Mass	Int.
41	240	42	50	55	80	57	80	67	25	69	90	83	25	84	20	97	75
98	15	110	10	112	150	123	10	126	70	141	10	154	20	155	20	169	999
170	80	184	900	185	95	211	80	212	15	225	2	239	1	240	1		

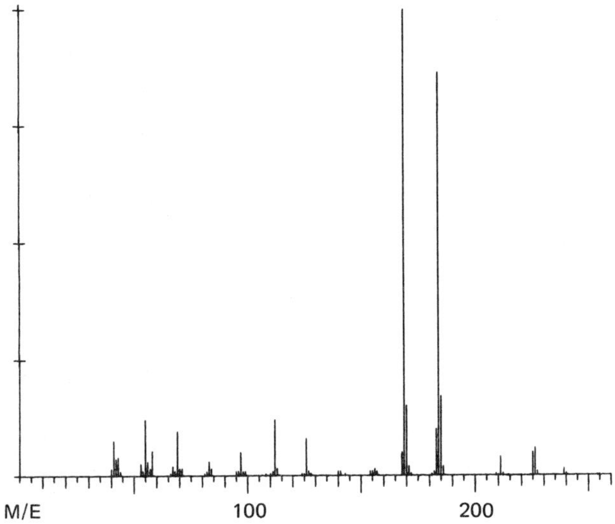
M/E
100
200

AMOBARBITAL 1,3-DIMETHYL DERIVATIVE

$C_{13}H_{22}N_2O_3$ MW = 254

Base Peak = 169 30 Peaks

Mass	Int.	Mass	Int.	Mass	Int.	Mass	Int.	Mass	Int.	Mass	Int.	Mass	Int.	Mass	Int.	Mass	Int.
41	75	43	40	55	120	58	53	67	20	69	95	83	30	84	15	112	120
113	15	126	80	127	10	140	10	141	10	154	10	156	15	184	865	185	170
209	5	211	40	225	50	226	60	239	15	240	4						

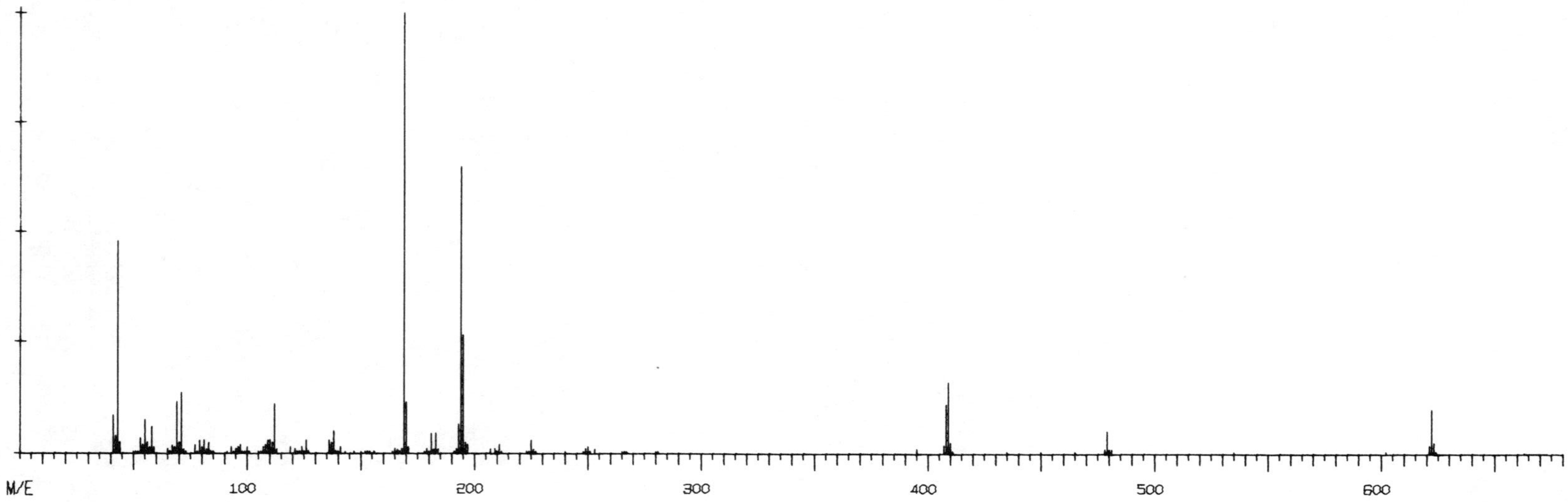
M/E
100
200
300
400
500
600

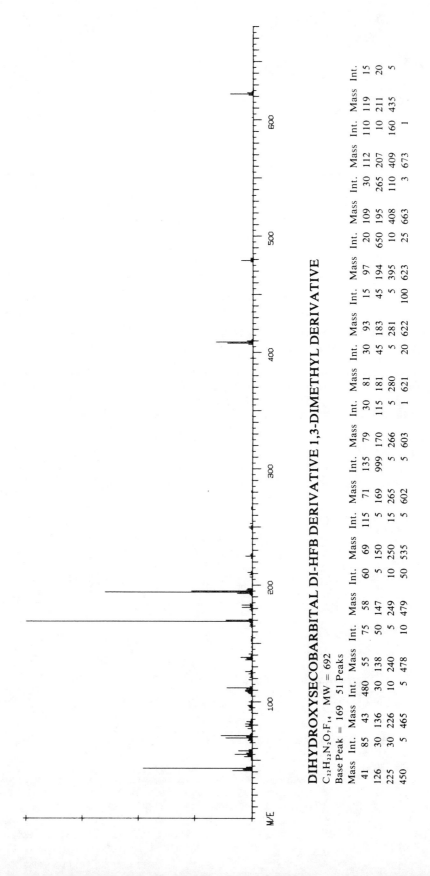

DIHYDROXYSECOBARBITAL DI-HFB DERIVATIVE 1,3-DIMETHYL DERIVATIVE

$C_{22}H_{22}N_2O_7F_{14}$ MW = 692

Base Peak = 169 51 Peaks

Mass	Int.	Mass	Int.	Mass	Int.	Mass	Int.	Mass	Int.	Mass	Int.	Mass	Int.	Mass	Int.	Mass	Int.	Mass	Int.	Mass	Int.	Mass	Int.	Mass	Int.
41	85	43	480	55	75	58	60	69	115	71	135	79	30	81	30	93	15	97	20	109	30	112	110	119	15
126	30	136	30	138	50	147	5	150	5	169	999	170	115	181	45	183	45	194	650	195	265	207	10	211	20
225	30	226	10	240	5	249	10	250	15	265	5	266	5	280	5	281	5	395	10	408	110	409	160	435	5
450	5	465	5	478	10	479	50	535	5	602	5	603	1	621	20	622	100	623	25	663	3	673	1		

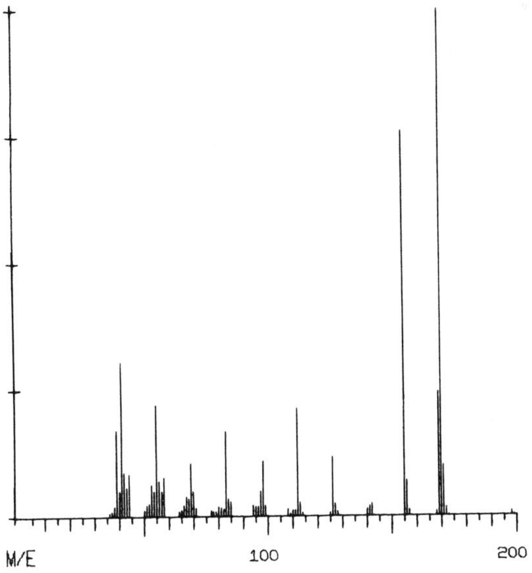
M/E
100
200

METHARBITAL

$C_9H_{14}N_2O_3$ MW = 198

Base Peak = 170 22 Peaks

Mass	Int.	Mass	Int.	Mass	Int.	Mass	Int.	Mass	Int.	Mass	Int.	Mass	Int.	Mass	Int.	Mass	Int.
39	171	41	304	55	220	58	78	69	105	70	50	83	168	84	34	97	51
98	110	112	212	113	28	126	118	127	24	141	20	142	24	155	760	156	71
169	246	170	999	198	11	199	2										

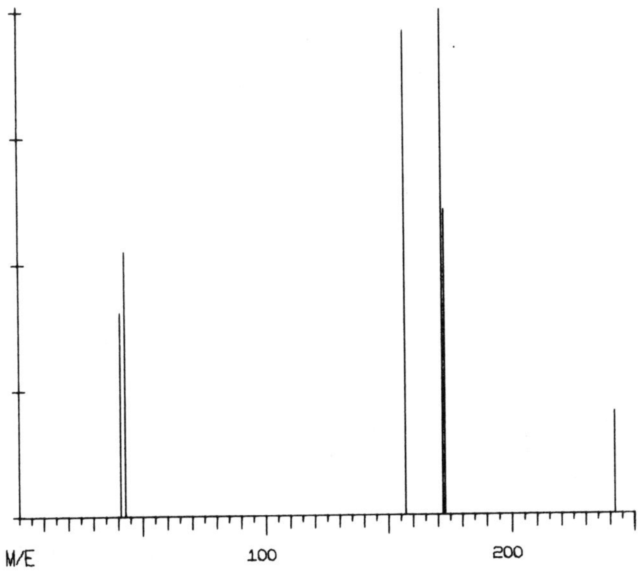
M/E
100
200

THIOPENTAL

$C_{11}H_{18}N_2O_2S$ MW = 242

Base Peak = 172 6 Peaks

Mass	Int.	Mass	Int.	Mass	Int.	Mass	Int.	Mass	Int.	Mass	Int.
41	404	43	525	157	959	172	999	173	606	242	202

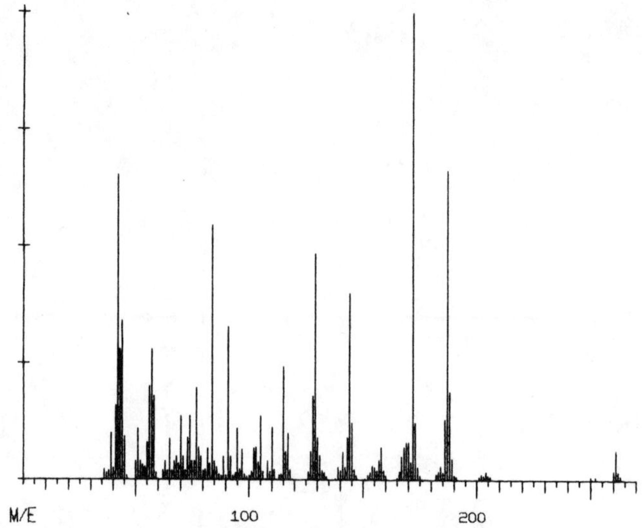
M/E
100
200

ALPHAPRODINE

$C_{16}H_{23}NO_2$ MW = 261

Base Peak = 172 30 Peaks

Mass	Int.	Mass	Int.	Mass	Int.	Mass	Int.	Mass	Int.	Mass	Int.	Mass	Int.	Mass	Int.	Mass	Int.
42	653	44	340	56	200	57	281	70	138	74	137	77	198	84	544	91	328
95	111	105	137	115	243	128	181	129	484	144	399	145	122	157	43	158	71
172	999	173	122	186	130	187	663	188	191	189	44	202	12	204	17	250	7
252	7	260	21	261	64												

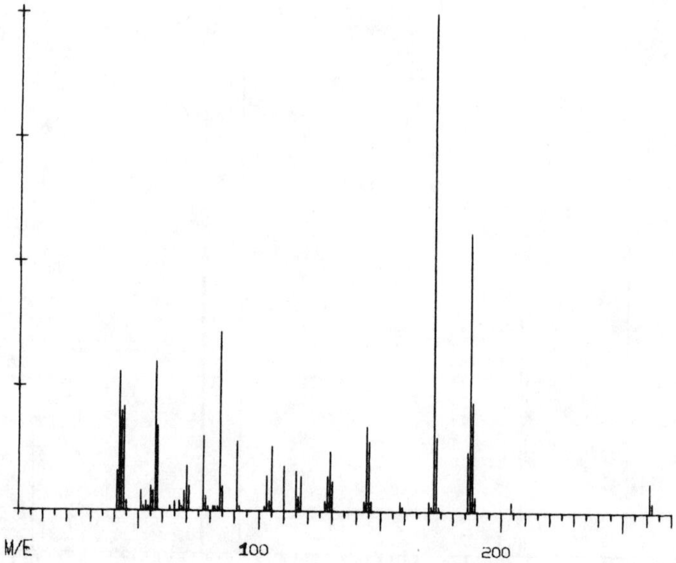
M/E
100
200

ALPHAPRODINE

$C_{16}H_{23}NO_2$ MW = 261

Base Peak = 172 27 Peaks

Mass	Int.	Mass	Int.	Mass	Int.	Mass	Int.	Mass	Int.	Mass	Int.	Mass	Int.	Mass	Int.	Mass	Int.
42	280	44	210	57	300	58	170	70	90	71	50	77	150	84	360	91	140
103	70	105	130	110	90	128	70	129	120	144	170	145	140	146	20	158	20
172	999	173	150	186	120	187	560	188	220	189	30	204	20	261	60	262	20

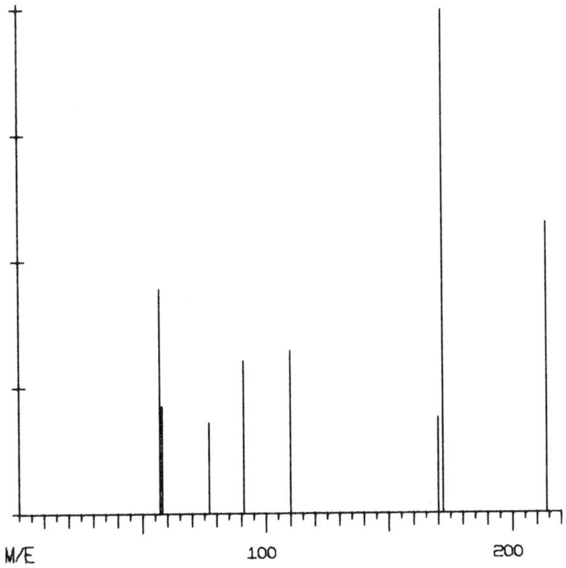

M/E
100
200

ALLYLPRODINE

$C_{18}H_{25}NO_2$ MW = 287

Base Peak = 172 8 Peaks

Mass	Int.	Mass	Int.	Mass	Int.	Mass	Int.	Mass	Int.	Mass	Int.	Mass	Int.	Mass	Int.
57	444	58	212	77	181	91	303	110	323	170	191	172	999	214	575

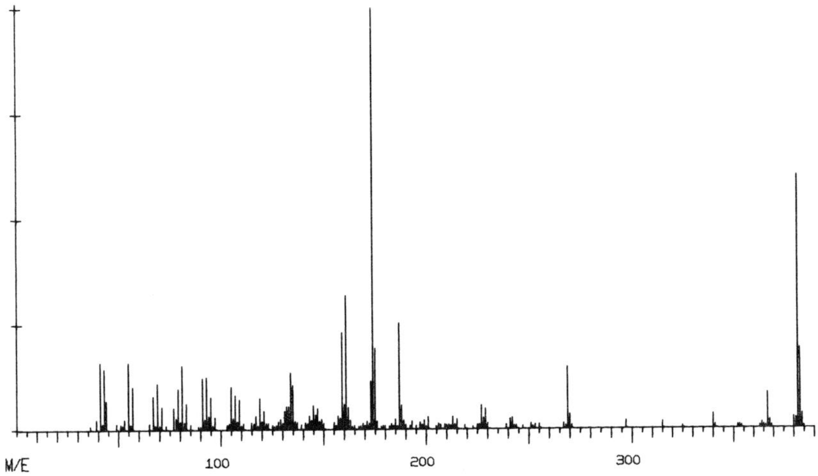

M/E
100
200
300

3,5-CHOLESTADIENE-7-ONE

$C_{27}H_{42}O$ MW = 382

Base Peak = 174 47 Peaks

Mass	Int.	Mass	Int.	Mass	Int.	Mass	Int.	Mass	Int.	Mass	Int.	Mass	Int.	Mass	Int.	Mass	Int.
41	159	43	146	55	159	57	102	67	79	69	109	79	97	81	152	91	122
93	125	105	103	107	83	119	75	131	45	134	136	135	106	147	51	159	229
161	318	173	116	174	999	187	252	188	57	201	29	213	31	215	24	227	58
229	51	241	24	242	27	2151	14	255	13	269	148	270	34	297	20	315	17
325	8	340	34	341	10	352	10	353	9	367	86	368	21	382	601	383	189
384	35	385	7														

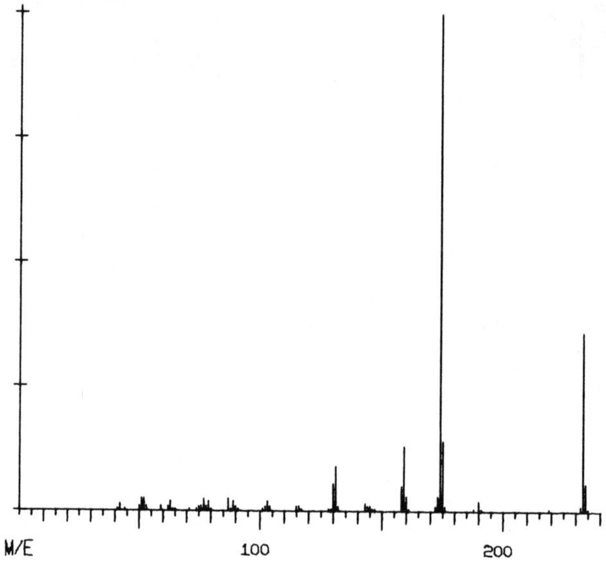
M/E
100
200

2-METHYL-5-METHOXYINDOLE-3-ACETIC ACID METHYL ESTER (FROM INDOMETHACIN)

$C_{13}H_{15}NO_3$ MW = 233

Base Peak = 174   27 Peaks

Mass	Int.	Mass	Int.	Mass	Int.	Mass	Int.	Mass	Int.	Mass	Int.	Mass	Int.	Mass	Int.	Mass	Int.
41	5	42	15	51	25	52	25	62	10	63	20	77	25	87	25	90	10
103	20	104	10	115	10	130	55	131	90	132	10	143	15	158	50	159	130
160	30	173	30	174	999	175	140	188	5	190	20	219	5	233	360	234	55

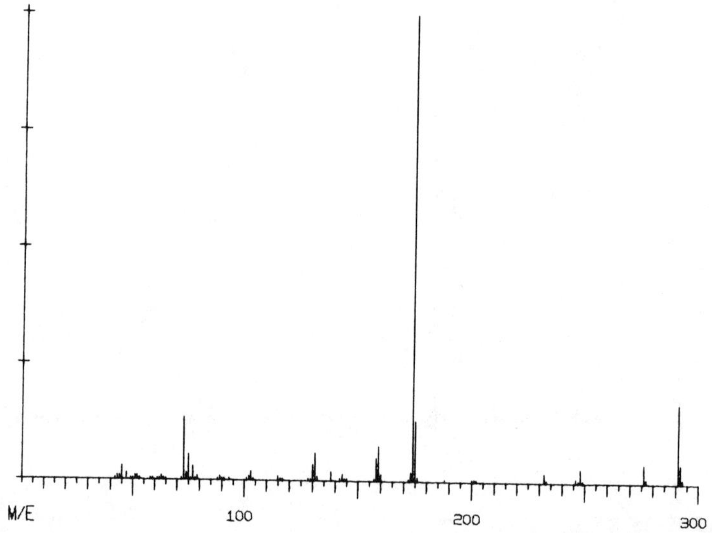
M/E
100
200
300

2-METHYL-5-METHOXYINDOLE-3-ACETIC ACID TMS ESTER (FROM INDOMETHACIN)

$C_{15}H_{21}NO_3Si$ MW = 291

Base Peak = 174   33 Peaks

Mass	Int.	Mass	Int.	Mass	Int.	Mass	Int.	Mass	Int.	Mass	Int.	Mass	Int.	Mass	Int.	Mass	Int.
45	30	47	15	51	10	52	10	73	135	75	55	77	30	79	10	102	10
103	20	104	5	115	10	130	35	131	60	138	20	143	15	158	50	159	75
160	15	173	20	174	999	175	130	188	5	200	5	202	5	232	20	233	5
246	10	248	30	276	40	277	10	291	170	292	40						

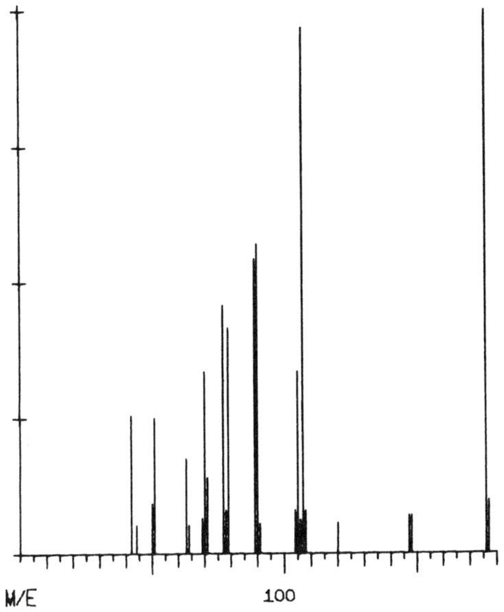
M/E
100

2-IMINO-5-PHENYL-4-OXAZOLIDINONE

$C_9H_8N_2O_2$ MW = 176

Base Peak = 176 17 Peaks

Mass	Int.	Mass	Int.	Mass	Int.	Mass	Int.	Mass	Int.	Mass	Int.	Mass	Int.	Mass	Int.	Mass	Int.
42	254	44	53	50	92	51	250	63	176	70	335	77	457	89	542	90	569
91	56	105	336	107	968	120	57	147	71	148	71	176	999	177	98		

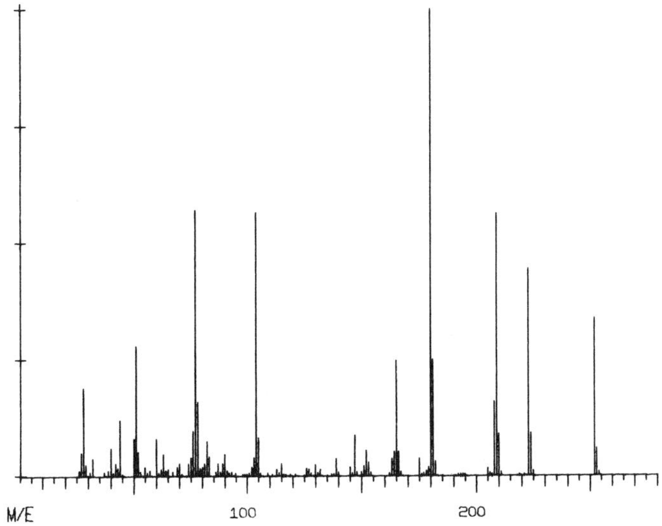
M/E
100
200

DIPHENYLHYDANTOIN

$C_{15}H_{12}N_2O_2$ MW = 252

Base Peak = 180 32 Peaks

Mass	Int.	Mass	Int.	Mass	Int.	Mass	Int.	Mass	Int.	Mass	Int.	Mass	Int.	Mass	Int.	Mass	Int.
27	49	28	189	40	61	44	119	50	79	51	279	63	48	75	39	77	569
78	159	90	48	103	39	104	564	105	82	126	17	130	24	139	37	145	19
147	87	152	56	164	53	165	248	180	999	181	251	192	4	193	5	208	161
209	563	223	444	224	93	252	337	253	61								

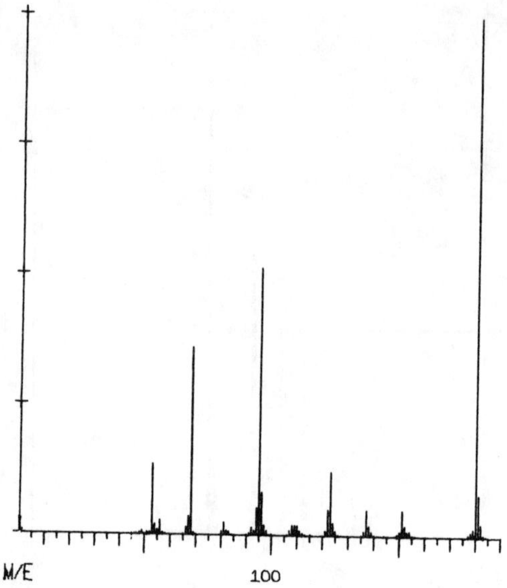
M/E
100

THEOPHYLLINE

$C_7H_8N_4O_2$ MW = 180

Base Peak = 180 20 Peaks

Mass	Int.	Mass	Int.	Mass	Int.	Mass	Int.	Mass	Int.	Mass	Int.	Mass	Int.	Mass	Int.	Mass	Int.
46	3	47	5	53	134	56	27	67	36	68	361	81	24	82	9	95	515
96	82	108	21	110	20	122	50	123	123	137	49	138	19	151	49	152	20
180	999	181	81														

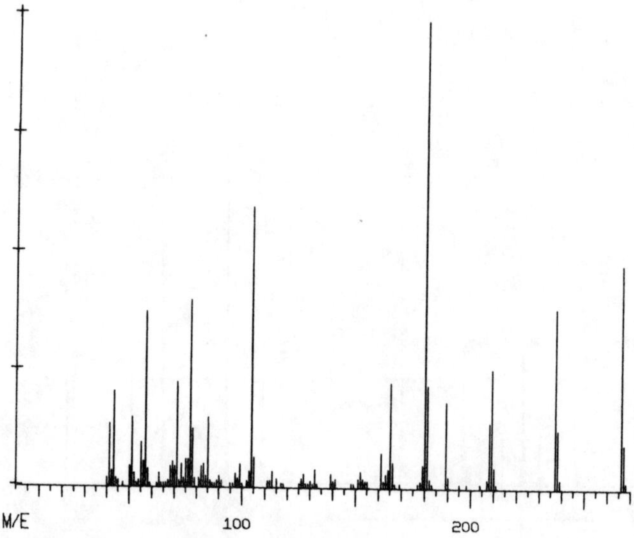
M/E
100
200

DIPHENYLHYDANTOIN 3-METHYL DERIVATIVE

$C_{16}H_{14}N_2O_2$ MW = 266

Base Peak = 180 30 Peaks

Mass	Int.	Mass	Int.	Mass	Int.	Mass	Int.	Mass	Int.	Mass	Int.	Mass	Int.	Mass	Int.	Mass	Int.
41	135	43	205	51	150	57	375	71	225	75	60	77	400	85	150	99	50
103	35	104	600	105	65	126	20	127	30	132	40	139	30	151	20	152	35
161	75	165	225	180	999	181	220	189	185	190	25	208	140	209	255	237	385
238	125	266	480	267	95												

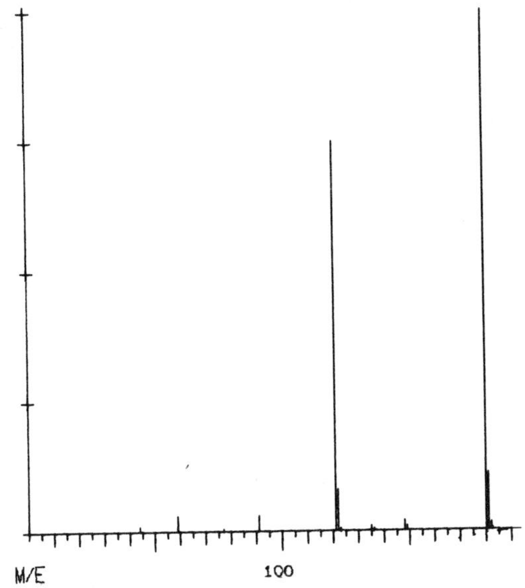
M/E
100

M-METHOXYPHENYLACETIC ACID METHYL ESTER

$C_{10}H_{12}O_3$ MW = 180

Base Peak = 180 12 Peaks

Mass	Int.	Mass	Int.	Mass	Int.	Mass	Int.	Mass	Int.	Mass	Int.	Mass	Int.	Mass	Int.	Mass	Int.
44	9	59	29	77	4	91	29	121	749	122	79	135	9	136	4	148	19
149	9	180	999	181	109												

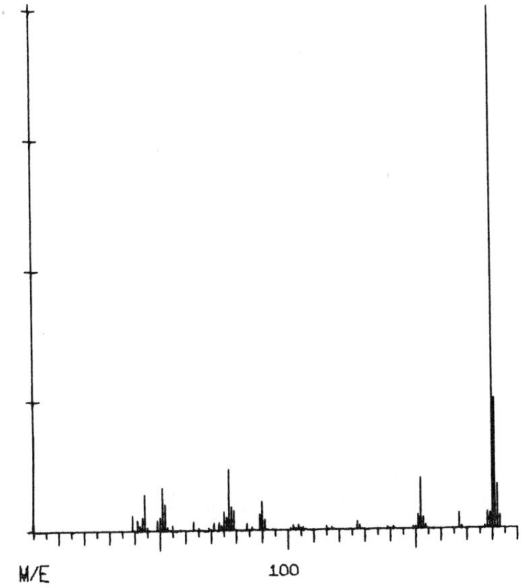
M/E
100

DOXYLAMINE MTB 2

$C_{13}H_{11}N$ MW = 181

Base Peak = 180 22 Peaks

Mass	Int.	Mass	Int.	Mass	Int.	Mass	Int.	Mass	Int.	Mass	Int.	Mass	Int.	Mass	Int.	Mass	Int.
39	31	44	69	51	83	52	49	63	17	75	35	77	118	78	45	90	55
91	21	104	9	115	7	127	14	128	11	139	4	141	4	151	28	152	98
167	29	168	5	180	999	181	250										

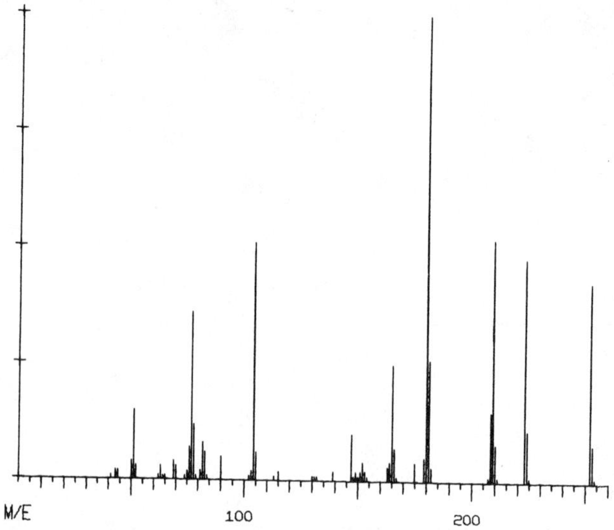
M/E
100
200

DIPHENYLHYDANTOIN

$C_{15}H_{12}N_2O_2$ MW = 252

Base Peak = 180 28 Peaks

Mass	Int.	Mass	Int.	Mass	Int.	Mass	Int.	Mass	Int.	Mass	Int.	Mass	Int.	Mass	Int.	Mass	Int.
43	20	44	20	50	40	51	150	63	30	69	40	77	360	78	120	90	50
103	20	104	510	105	60	130	10	131	10	132	10	139	20	147	100	152	40
165	250	166	70	180	999	181	260	208	150	209	520	223	480	224	110	252	430
253	80																

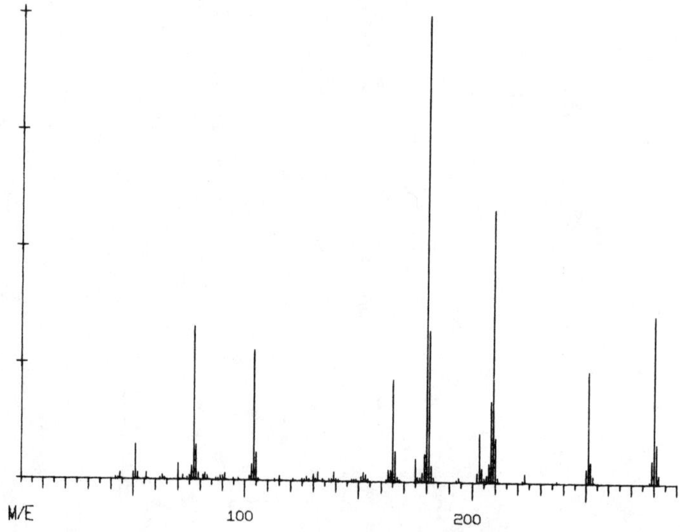
M/E
100
200

5,5-DIPHENYLHYDANTOIN 3-ETHYL DERIVATIVE

$C_{17}H_{16}N_2O_2$ MW = 280

Base Peak = 180 33 Peaks

Mass	Int.	Mass	Int.	Mass	Int.	Mass	Int.	Mass	Int.	Mass	Int.	Mass	Int.	Mass	Int.	Mass	Int.
42	5	44	15	50	15	51	75	63	10	70	35	77	330	78	75	91	15
103	35	104	280	105	60	127	10	130	15	132	20	139	20	152	20	153	15
165	220	166	65	180	999	181	325	193	5	194	10	208	175	209	585	222	5
223	20	237	5	251	240	252	45	280	360	281	85						

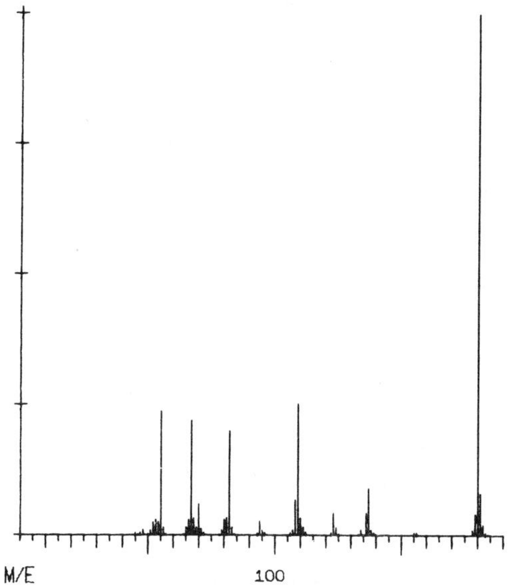
M/E
100

THEOBROMINE

$C_7H_8N_4O_2$ MW = 180

Base Peak = 180   20 Peaks

Mass	Int.	Mass	Int.	Mass	Int.	Mass	Int.	Mass	Int.	Mass	Int.	Mass	Int.	Mass	Int.	Mass	Int.
45	4	47	5	53	29	55	237	67	219	70	60	81	32	82	201	94	28
95	10	108	67	109	252	123	42	124	14	136	42	137	90	155	5	156	4
180	999	181	81														

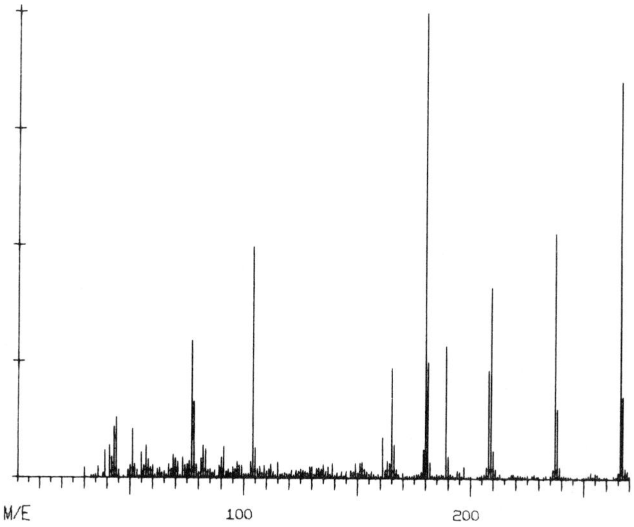
M/E
100
200

DIPHENYLHYDANTOIN MTB. 1

$C_{16}H_{14}N_2O_2$ MW = 266

Base Peak = 180   36 Peaks

Mass	Int.	Mass	Int.	Mass	Int.	Mass	Int.	Mass	Int.	Mass	Int.	Mass	Int.	Mass	Int.	Mass	Int.
30	23	33	4	43	110	44	130	51	105	57	71	69	50	73	44	77	295
78	166	90	44	91	67	104	498	105	64	129	24	130	24	135	30	139	33
149	33	152	34	161	88	165	237	180	999	181	250	189	286	190	147	208	233
209	409	218	9	228	10	237	528	238	151	253	16	255	13	266	856	267	178

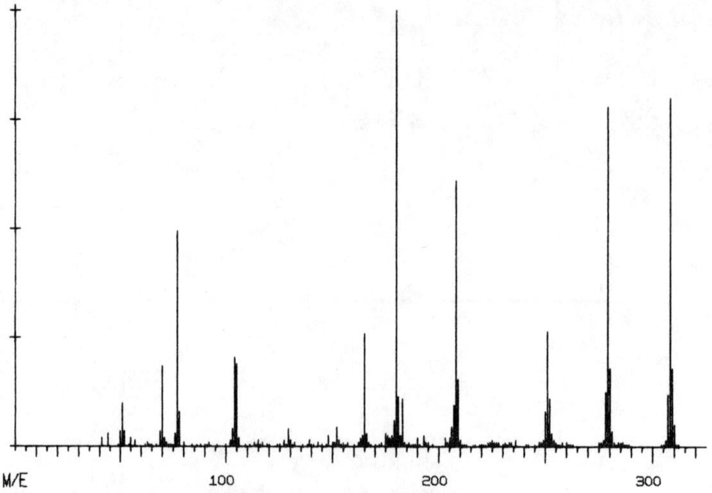
M/E
100
200
300

5,5-DIPHENYLHYDANTOIN 1,3-DIETHYL DERIVATIVE

$C_{19}H_{20}N_2O_2$ MW = 308

Base Peak = 180 40 Peaks

Mass	Int.	Mass	Int.	Mass	Int.	Mass	Int.	Mass	Int.	Mass	Int.	Mass	Int.	Mass	Int.	Mass	Int.
41	20	44	30	50	35	51	100	69	35	70	185	77	495	78	80	102	15
103	40	104	205	105	190	127	15	129	40	132	10	139	15	148	25	152	45
165	260	166	30	180	999	181	115	190	20	193	25	208	610	209	155	223	10
225	15	231	10	236	15	251	265	252	110	258	10	260	10	279	780	280	180
286	10	287	5	308	800	309	180										

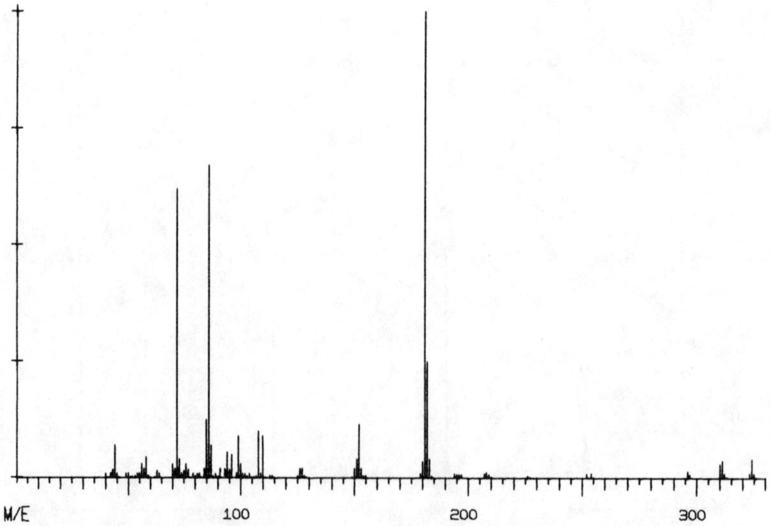
M/E
100
200
300

METHANTHELINE BROMIDE

$C_{21}H_{26}BRNO_3$ MW = 419

Base Peak = 181 32 Peaks

Mass	Int.	Mass	Int.	Mass	Int.	Mass	Int.	Mass	Int.	Mass	Int.	Mass	Int.	Mass	Int.	Mass	Int.
43	17	44	69	56	29	58	44	72	619	73	39	85	124	86	669	94	56
99	89	108	99	110	89	126	19	127	19	151	39	152	114	181	999	182	249
194	9	195	8	207	9	208	13	226	4	227	2	252	11	254	9	296	14
297	7	310	30	311	37	323	9	324	39								

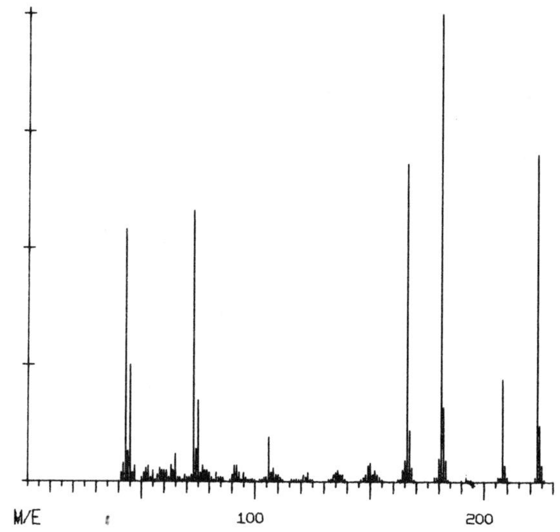
M/E
100
200

ACETAMINOPHEN TMS ETHER

$C_{11}H_{17}NO_2Si$ MW = 223

Base Peak = 181 26 Peaks

Mass	Int.	Mass	Int.	Mass	Int.	Mass	Int.	Mass	Int.	Mass	Int.	Mass	Int.	Mass	Int.	Mass	Int.
43	540	45	250	52	30	53	35	73	580	75	175	77	35	78	25	91	35
92	35	106	95	108	30	121	15	123	20	135	20	136	25	149	35	150	40
166	680	181	999	182	160	190	5	192	20	208	220	209	35	223	700	224	120

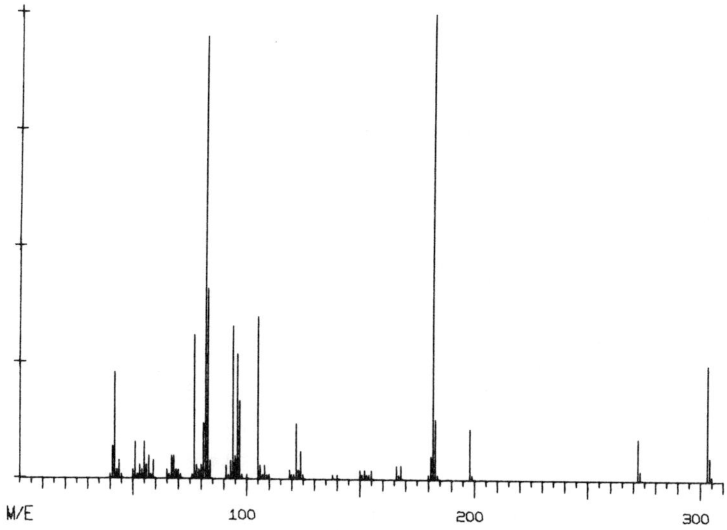
M/E
100
200
300

COCAINE

$C_{17}H_{21}NO_4$ MW = 303

Base Peak = 182 28 Peaks

Mass	Int.	Mass	Int.	Mass	Int.	Mass	Int.	Mass	Int.	Mass	Int.	Mass	Int.	Mass	Int.	Mass	Int.
41	70	42	230	51	80	55	80	67	50	68	50	82	950	83	410	94	330
96	270	105	350	106	30	122	120	124	60	138	10	140	10	150	20	152	20
166	30	168	30	182	999	183	130	198	110	199	10	272	90	273	20	303	250
304	50																

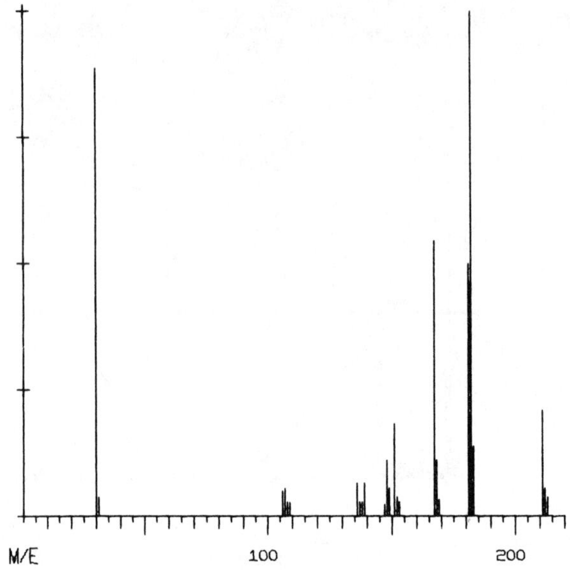
M/E
100
200

MESCALINE

$C_{11}H_{17}NO_3$ MW = 211

Base Peak = 182 14 Peaks

Mass	Int.	Mass	Int.	Mass	Int.	Mass	Int.	Mass	Int.	Mass	Int.	Mass	Int.	Mass	Int.	Mass	Int.
30	888	31	38	106	49	107	55	136	66	139	66	148	111	151	183	167	544
168	111	181	499	182	999	211	211	212	55								

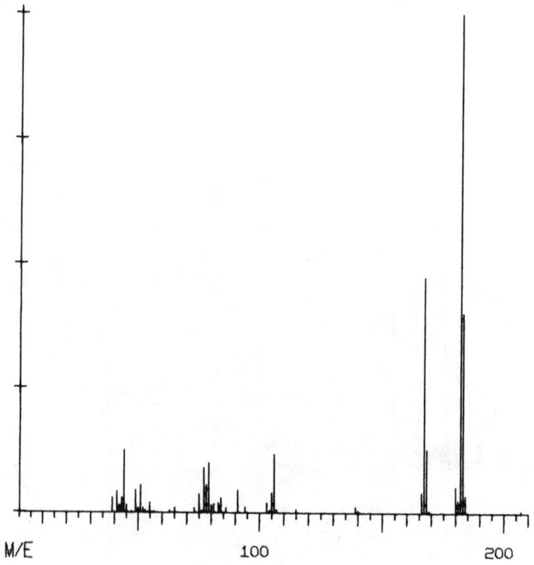
M/E
100
200

DOXYLAMINE MTB. 1

$C_{13}H_{13}NO$ MW = 199

Base Peak = 182 18 Peaks

Mass	Int.	Mass	Int.	Mass	Int.	Mass	Int.	Mass	Int.	Mass	Int.	Mass	Int.	Mass	Int.	Mass	Int.
41	43	44	124	49	46	51	55	65	10	75	37	77	90	79	99	91	44
103	20	105	39	106	118	139	12	140	5	167	473	168	128	182	999	183	400
207	5																

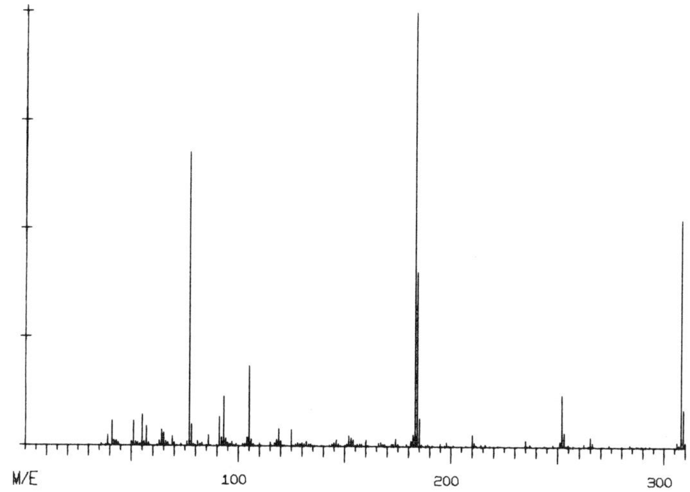
M/E
100
200
300

PHENYLBUTAZONE

$C_{19}H_{20}N_2O_2$ MW = 308

Base Peak = 183 41 Peaks

Mass	Int.	Mass	Int.	Mass	Int.	Mass	Int.	Mass	Int.	Mass	Int.	Mass	Int.	Mass	Int.	Mass	Int.
30	3	39	24	41	58	51	58	55	72	64	37	65	29	77	678	78	49
91	68	93	114	104	21	105	186	119	40	125	37	132	12	145	8	152	26
153	20	160	15	167	10	183	999	184	402	195	7	198	9	210	28	211	11
216	5	218	3	235	14	237	5	252	119	253	32	265	22	266	9	274	4
284	5	287	3	291	3	308	524	309	87								

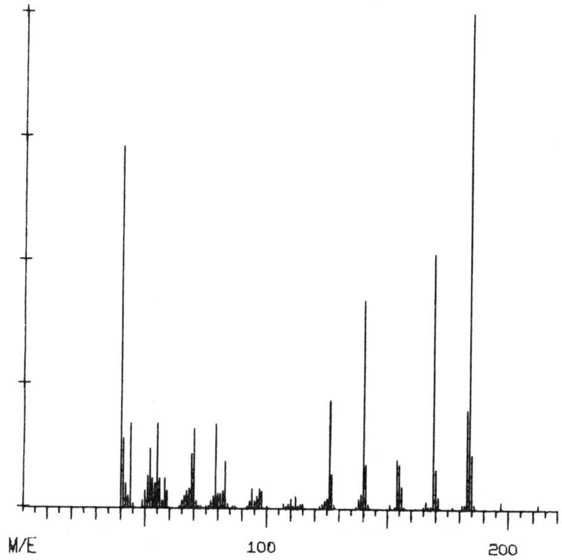
M/E
100
200

BARBITAL 1,2(OR 4)-DIMETHYL DERIVATIVE

$C_{10}H_{16}N_2O_3$ MW = 212

Base Peak = 184 24 Peaks

Mass	Int.	Mass	Int.	Mass	Int.	Mass	Int.	Mass	Int.	Mass	Int.	Mass	Int.	Mass	Int.	Mass	Int.
40	730	44	170	52	120	55	170	69	110	70	160	79	170	83	95	94	40
97	40	110	20	112	25	126	245	127	70	140	420	141	90	154	100	155	90
169	515	170	80	183	200	184	999	197	15	212	10						

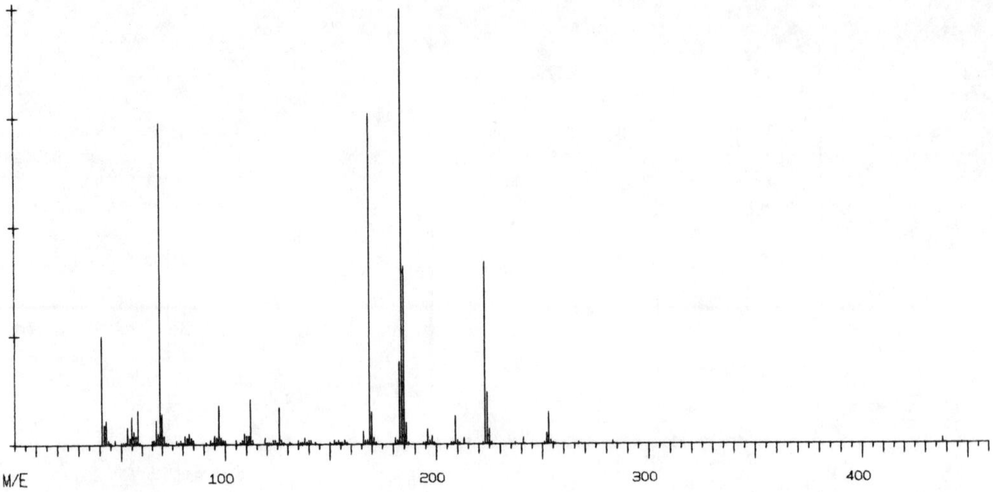

M/E
100
200
300
400

HYDROXYPENTO BARBITAL 1,3-DIMETHYL DERIVATIVE HFB DERIVATIVE

$C_{18}H_{21}F_7N_2O_5$ MW = 478
Base Peak = 184 41 Peaks

Mass	Int.	Mass	Int.	Mass	Int.	Mass	Int.	Mass	Int.	Mass	Int.	Mass	Int.	Mass	Int.	Mass	Int.
41	250	43	55	55	65	58	80	69	740	70	70	81	20	83	25	95	20
97	90	109	25	112	105	119	15	126	85	135	10	138	15	152	10	154	10
169	760	170	75	184	999	185	410	196	35	198	20	209	65	213	15	223	420
224	120	237	5	241	15	252	25	253	75	267	6	269	1	280	1	283	8
379	1	437	1	438	13	447	2	451	1								

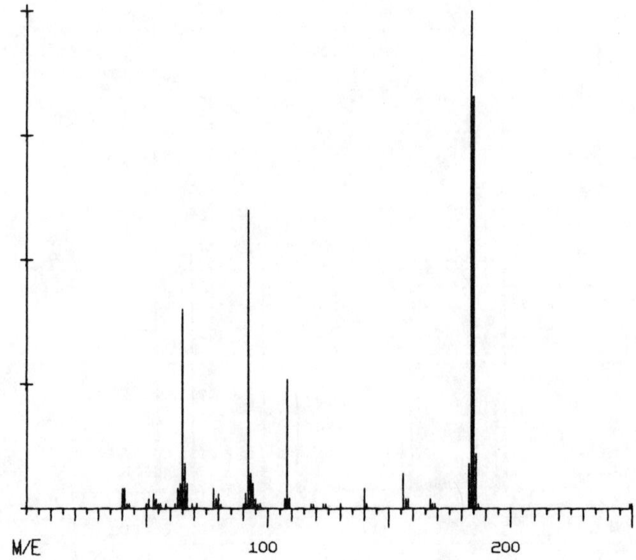

M/E
100
200

SULFAPYRIDINE

$C_{11}H_{11}N_3O_2S$ MW = 249
Base Peak = 184 24 Peaks

Mass	Int.	Mass	Int.	Mass	Int.	Mass	Int.	Mass	Int.	Mass	Int.	Mass	Int.	Mass	Int.	Mass	Int.
40	40	41	40	51	20	53	30	65	400	66	90	78	40	80	30	92	600
93	70	107	20	108	260	118	10	119	10	140	40	141	10	156	70	157	20
167	20	168	10	184	999	185	830	249	10	250	10						

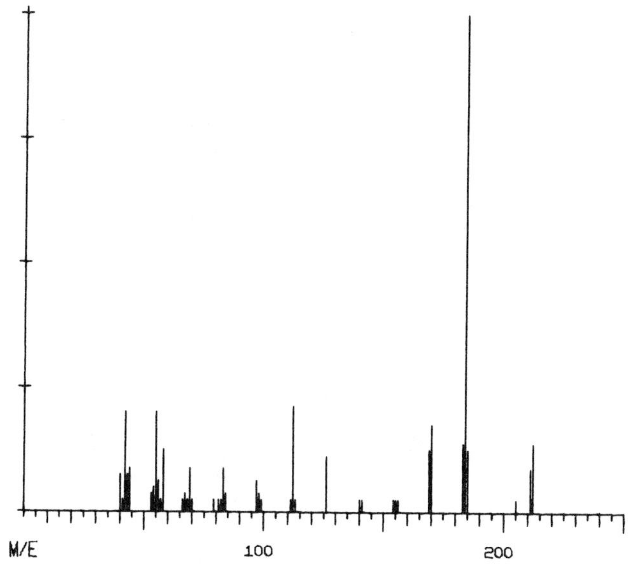
M/E
100
200

BUTETHAL 1,3-DIMETHYL DERIVATIVE

$C_{12}H_{20}N_2O_3$ MW = 240
Base Peak = 184 22 Peaks

Mass	Int.	Mass	Int.	Mass	Int.	Mass	Int.	Mass	Int.	Mass	Int.	Mass	Int.	Mass	Int.	Mass	Int.
42	199	44	87	55	199	58	124	67	37	69	87	83	87	84	37	97	62
98	37	111	24	112	212	126	112	140	24	141	24	154	24	155	24	169	124
170	174	183	137	184	999	211	87	212	137	225	2	239	2	240	2		

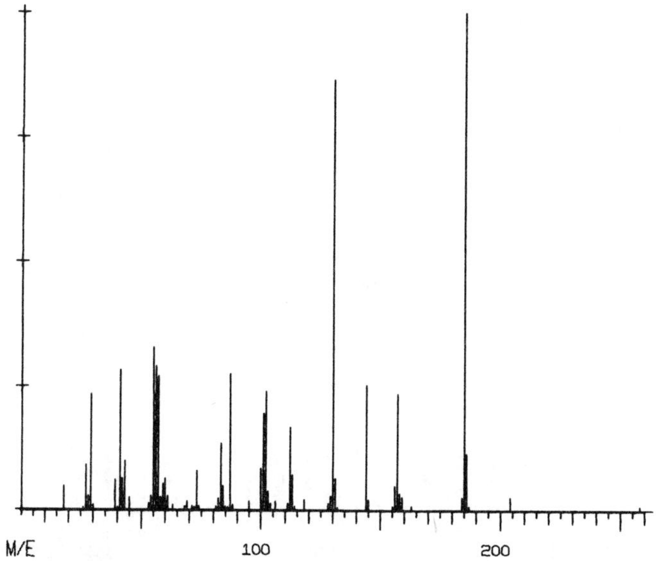
M/E
100
200

DIBUTYL ADIPATE

$C_{14}H_{26}O_4$ MW = 258
Base Peak = 185 25 Peaks

Mass	Int.	Mass	Int.	Mass	Int.	Mass	Int.	Mass	Int.	Mass	Int.	Mass	Int.	Mass	Int.	Mass	Int.
27	93	29	235	41	283	43	100	55	328	56	291	69	20	73	80	83	135
87	276	101	195	102	241	112	167	113	72	130	865	131	66	144	253	145	22
156	49	157	235	163	11	185	999	186	115	204	27	258	9				

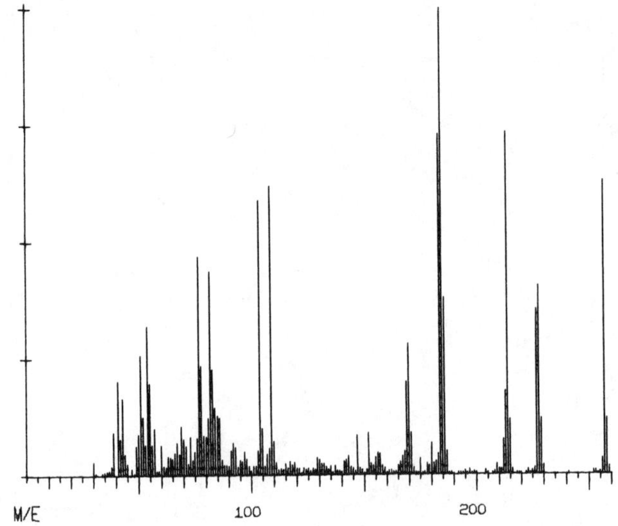
M/E
100
200

DIPHENYLHYDANTOIN (D5-PH)

$C_{11}H_7O_5N_2$ MW = 177

Base Peak = 185 36 Peaks

Mass	Int.	Mass	Int.	Mass	Int.	Mass	Int.	Mass	Int.	Mass	Int.	Mass	Int.	Mass	Int.	Mass	Int.
30	29	31	6	41	203	43	164	51	258	54	319	69	106	73	83	77	471
82	438	92	69	93	59	104	591	109	621	129	38	130	32	142	32	143	40
147	84	152	91	169	201	170	282	184	730	185	999	195	9	197	12	213	180
214	736	227	356	228	406	230	20	241	4	256	35	257	630	258	120	259	17

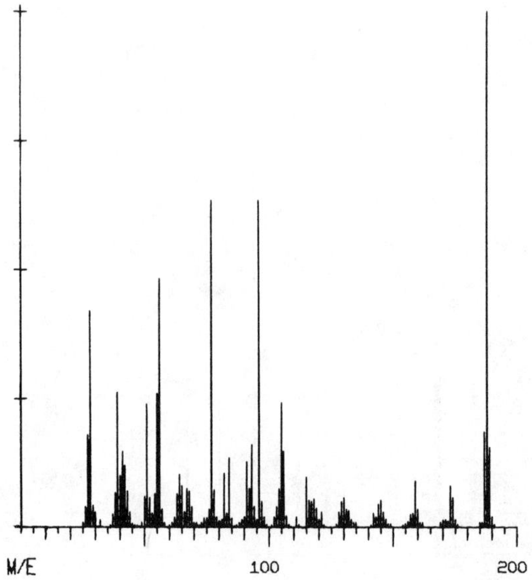
M/E
100
200

ANTIPYRINE

$C_{11}H_{12}N_2O$ MW = 188

Base Peak – 188 26 Peaks

Mass	Int.	Mass	Int.	Mass	Int.	Mass	Int.	Mass	Int.	Mass	Int.	Mass	Int.	Mass	Int.	Mass	Int.
27	180	28	420	39	263	41	147	55	260	56	482	64	103	65	81	77	634
84	136	93	160	96	635	105	242	106	147	118	54	130	58	144	44	145	52
146	31	159	89	160	34	173	80	174	58	187	186	188	999	189	155		

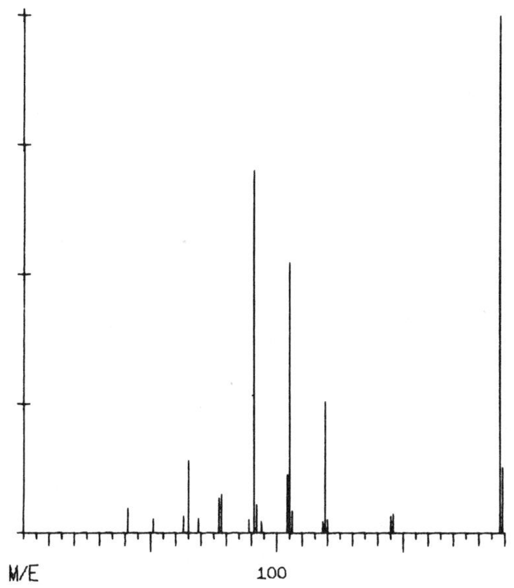
M/E
100

1-(P-TOLYL)-3-METHYL-PYRAZOL-5-ONE

$C_{11}H_{12}N_2O$ MW = 188

Base Peak = 188 16 Peaks

Mass	Int.	Mass	Int.	Mass	Int.	Mass	Int.	Mass	Int.	Mass	Int.	Mass	Int.	Mass	Int.	Mass	Int.
41	47	51	27	63	33	65	141	77	67	78	75	91	700	92	54	104	113
105	523	119	256	120	24	145	32	146	38	188	999	189	128				

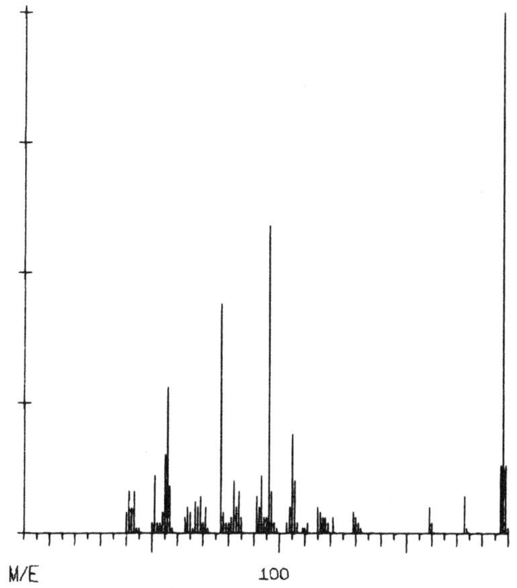
M/E
100

ANTIPYRENE

$C_{11}H_{12}N_2O$ MW = 188

Base Peak = 188 22 Peaks

Mass	Int.	Mass	Int.	Mass	Int.	Mass	Int.	Mass	Int.	Mass	Int.	Mass	Int.	Mass	Int.	Mass	Int.
41	80	43	80	55	150	56	280	67	60	69	70	77	440	82	100	93	110
96	590	105	190	106	100	118	30	129	40	132	10	159	50	160	20	173	70
174	10	187	130	188	999	189	130										

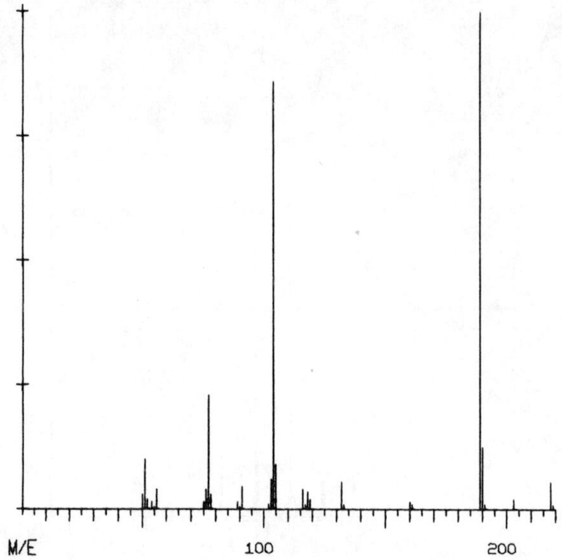
M/E
100
200

MEPHENYTOIN

$C_{12}H_{14}N_2O_2$ MW = 218

Base Peak = 189 20 Peaks

Mass	Int.	Mass	Int.	Mass	Int.	Mass	Int.	Mass	Int.	Mass	Int.	Mass	Int.	Mass	Int.	Mass	Int.
51	99	56	39	75	14	76	39	77	229	91	44	103	59	104	859	105	89
118	34	119	19	132	54	133	9	160	14	161	9	189	999	190	124	203	19
218	54	219	9														

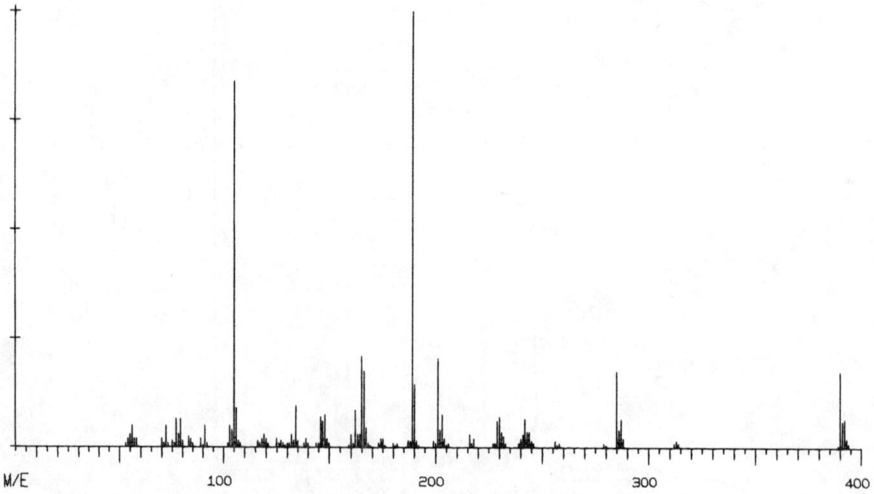
M/E
100
200
300
400

MECLIZINE

$C_{25}H_{27}CLN_2$ MW = 390

Base Peak = 189 40 Peaks

Mass	Int.	Mass	Int.	Mass	Int.	Mass	Int.	Mass	Int.	Mass	Int.	Mass	Int.	Mass	Int.	Mass	Int.
55	29	56	49	70	19	72	49	77	64	79	64	91	49	103	49	105	839
106	89	118	19	119	29	132	29	134	94	146	69	148	74	165	209	166	174
174	19	175	19	189	999	201	204	202	39	203	74	216	31	229	59	230	69
242	64	244	34	245	14	258	9	279	9	285	174	286	39	287	64	312	9
313	14	314	9	390	174	392	64										

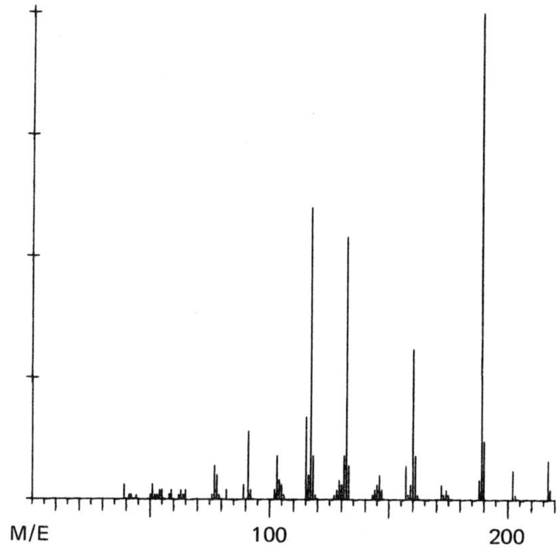
M/E
100
200

GLUTETHIMIDE

$C_{13}H_{15}NO_2$ MW = 217
Base Peak = 189 28 Peaks

Mass	Int.	Mass	Int.	Mass	Int.	Mass	Int.	Mass	Int.	Mass	Int.	Mass	Int.	Mass	Int.	Mass	Int.
39	30	41	10	51	30	54	20	63	20	65	20	77	70	78	50	91	140
103	90	115	170	117	600	118	90	131	90	132	540	133	70	146	50	157	70
160	310	161	90	174	20	175	10	189	999	190	120	202	60	203	10	217	80
218	20																

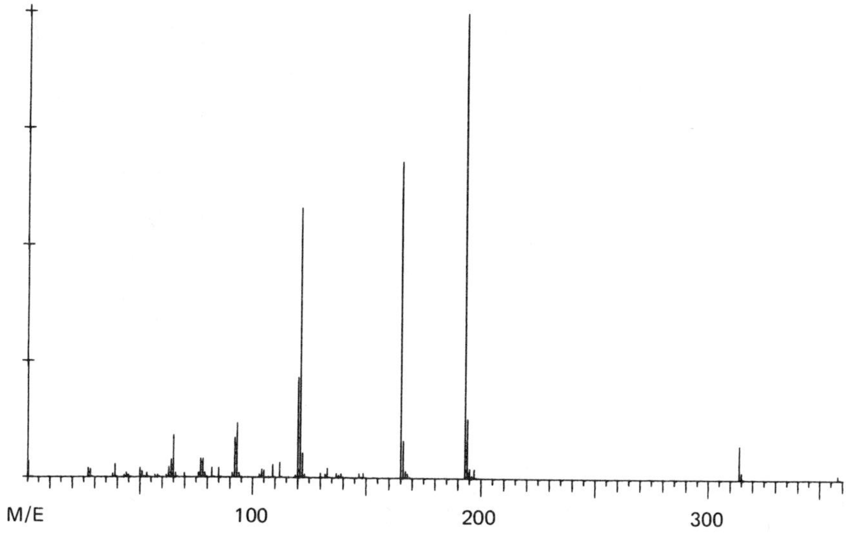
M/E
100
200
300

CARBETHYL SALICYLATE

$C_{19}H_{18}O_7$ MW = 358
Base Peak = 193 28 Peaks

Mass	Int.	Mass	Int.	Mass	Int.	Mass	Int.	Mass	Int.	Mass	Int.	Mass	Int.	Mass	Int.	Mass	Int.
27	19	28	17	39	27	44	9	50	19	51	13	64	37	65	89	77	39
78	39	92	84	93	117	109	27	112	33	120	215	121	579	133	19	137	9
147	10	149	10	165	679	166	79	193	999	194	128	314	73	315	16	358	9
359	1																

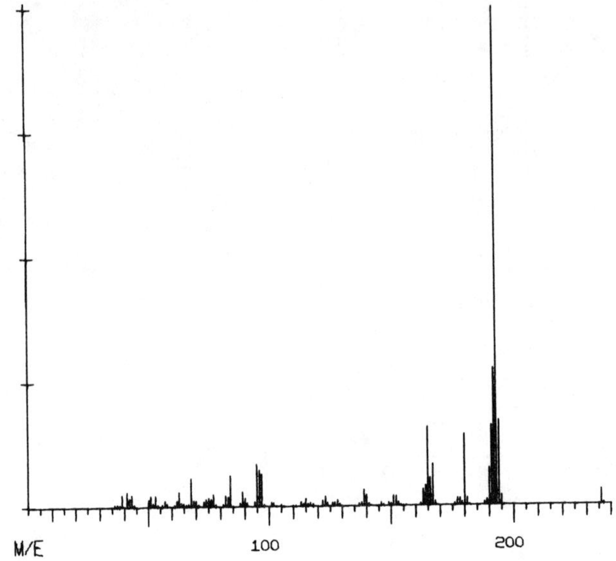
M/E
100
200

CARBAMAZEPINE

$C_{15}H_{12}N_2O$ MW = 236

Base Peak = 193 27 Peaks

Mass	Int.	Mass	Int.	Mass	Int.	Mass	Int.	Mass	Int.	Mass	Int.	Mass	Int.	Mass	Int.	Mass	Int.
41	31	43	26	51	23	53	22	63	29	68	57	84	62	89	29	95	86
96	72	113	9	115	16	122	12	123	21	139	32	140	22	151	19	152	19
165	158	167	82	177	15	180	142	192	274	193	999	221	3	236	29	237	6

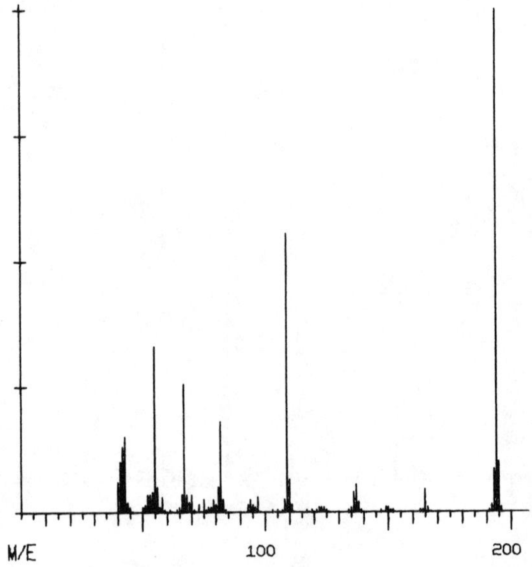
M/E
100
200

CAFFEINE

$C_8H_{10}N_4O_2$ MW = 194

Base Peak = 194 22 Peaks

Mass	Int.	Mass	Int.	Mass	Int.	Mass	Int.	Mass	Int.	Mass	Int.	Mass	Int.	Mass	Int.	Mass	Int.
42	130	43	150	55	330	56	50	66	35	67	255	81	50	82	180	94	25
97	30	109	555	110	65	122	10	123	10	136	40	137	55	149	10	150	10
165	45	166	10	194	999	195	100										

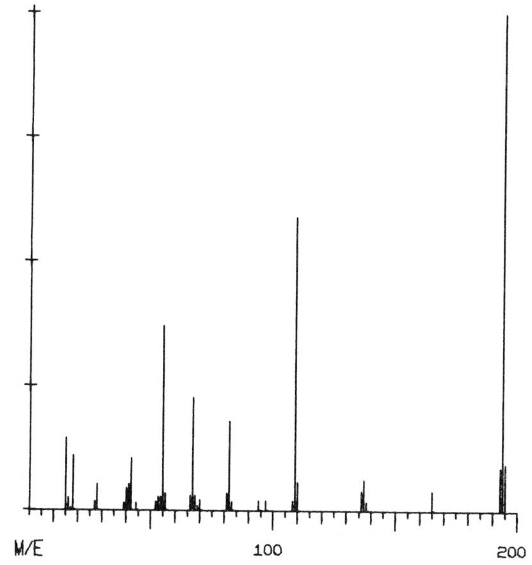
M/E
100
200

CAFFEINE

$C_8H_{10}N_4O_2$ MW = 194

Base Peak = 194   20 Peaks

Mass	Int.	Mass	Int.	Mass	Int.	Mass	Int.	Mass	Int.	Mass	Int.	Mass	Int.	Mass	Int.	Mass	Int.
27	17	28	52	41	53	42	105	55	370	56	34	67	228	68	31	81	36
82	179	94	21	97	20	109	591	110	58	136	39	137	63	165	40	179	1
194	999	195	94														

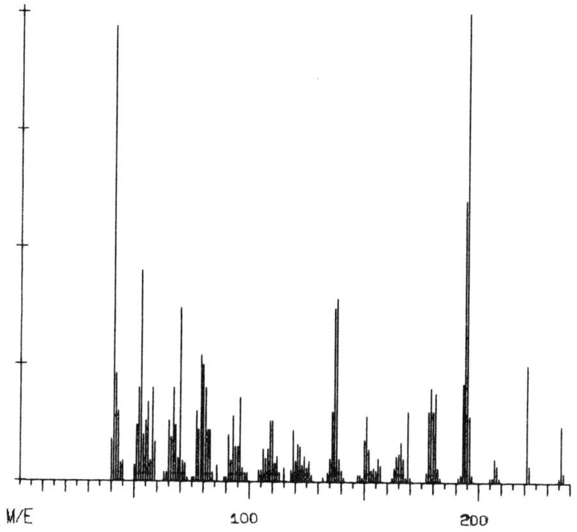
M/E
100
200

ALLOBARBITAL 1,2(OR 4)-DIMETHYL DERIVATIVE

$C_{12}H_{16}N_2O_3$ MW = 236

Base Peak = 195   30 Peaks

Mass	Int.	Mass	Int.	Mass	Int.	Mass	Int.	Mass	Int.	Mass	Int.	Mass	Int.	Mass	Int.	Mass	Int.
41	970	42	230	52	200	53	450	67	200	70	370	79	270	80	250	93	140
96	180	109	130	110	130	119	110	121	80	137	370	138	390	150	90	151	140
166	85	169	150	179	200	181	190	194	600	195	999	207	50	208	35	221	250
222	35	236	120	237	20												

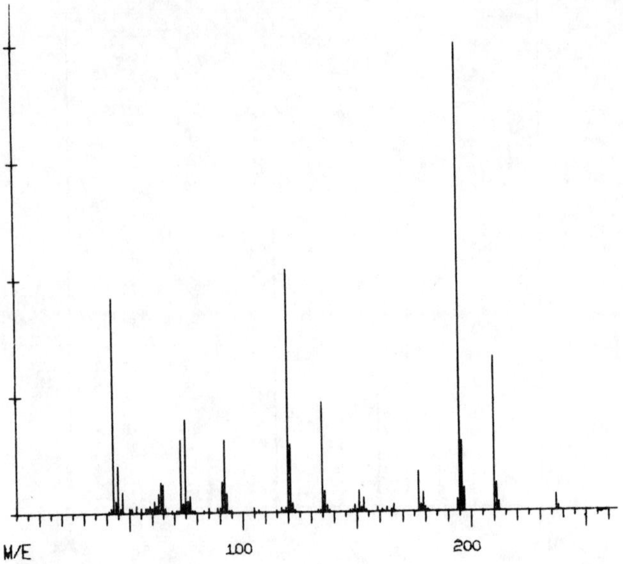
M/E
100
200

ASPIRIN TMS ESTER

$C_{12}H_{16}O_4Si$ MW = 252

Base Peak = 195 29 Peaks

Mass	Int.	Mass	Int.	Mass	Int.	Mass	Int.	Mass	Int.	Mass	Int.	Mass	Int.	Mass	Int.	Mass	Int.
43	460	45	100	53	15	61	25	73	155	75	200	76	25	77	35	91	65
92	155	105	10	117	10	120	520	121	145	135	235	136	45	151	45	153	30
163	10	166	15	177	85	179	40	195	999	196	150	210	330	211	60	237	35
238	10	252	1														

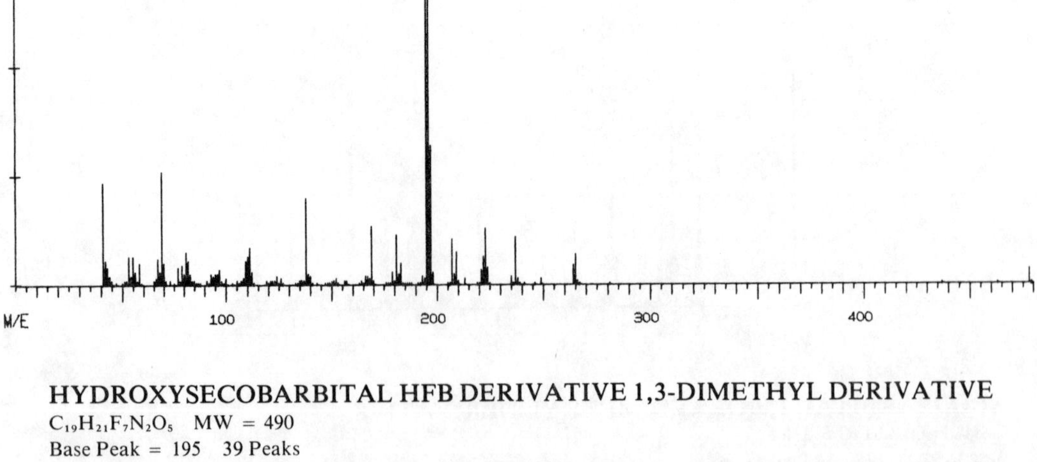
M/E
100
200
300
400

HYDROXYSECOBARBITAL HFB DERIVATIVE 1,3-DIMETHYL DERIVATIVE

$C_{19}H_{21}F_7N_2O_5$ MW = 490

Base Peak = 195 39 Peaks

Mass	Int.	Mass	Int.	Mass	Int.	Mass	Int.	Mass	Int.	Mass	Int.	Mass	Int.	Mass	Int.	Mass	Int.
41	235	42	55	53	65	55	65	67	60	69	260	81	75	82	55	93	25
97	35	110	65	111	85	124	20	126	15	138	200	139	25	150	10	152	15
166	20	169	135	181	115	183	50	195	999	196	970	207	105	209	75	222	65
223	130	235	20	237	110	246	5	249	15	264	45	265	70	281	5	437	5
463	5	478	35	479	5												

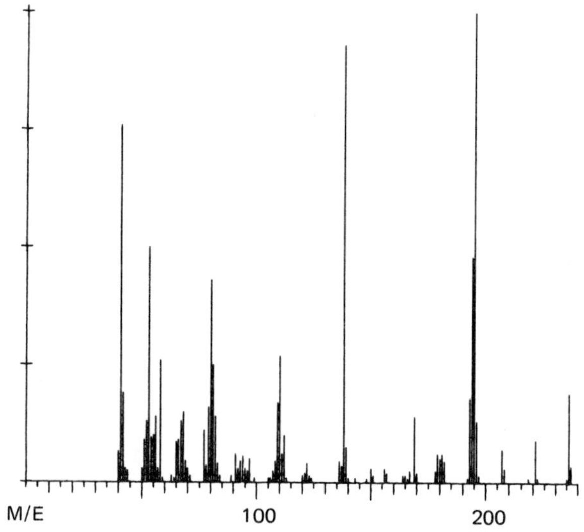
M/E
100
200

ALLOBARBITAL 1,3-DIMETHYL DERIVATIVE

$C_{12}H_{16}N_2O_3$ MW = 236
Base Peak = 195 27 Peaks

Mass	Int.	Mass	Int.	Mass	Int.	Mass	Int.	Mass	Int.	Mass	Int.	Mass	Int.	Mass	Int.	Mass	Int.
41	760	42	190	53	500	58	260	67	130	68	150	80	430	81	250	91	60
94	55	109	170	110	270	121	20	122	40	138	930	139	75	150	30	156	30
167	25	169	140	179	60	181	60	194	480	195	999	207	70	208	30	218	10
221	90	236	190	237	35												

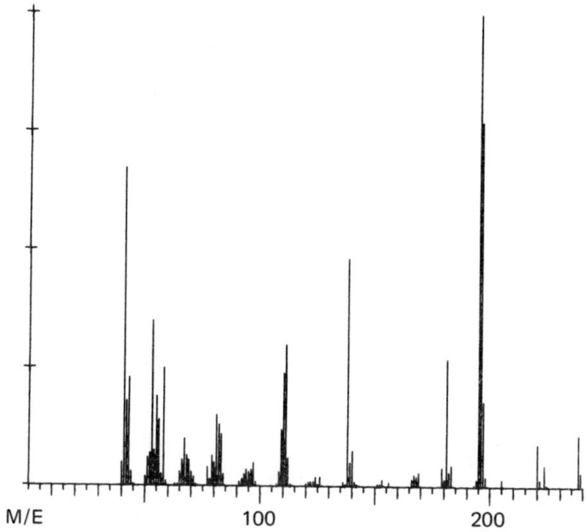
M/E
100
200

APROBARBITAL 1,3-DIMETHYL DERIVATIVE

$C_{12}H_{18}N_2O_3$ MW = 238
Base Peak = 195 29 Peaks

Mass	Int.	Mass	Int.	Mass	Int.	Mass	Int.	Mass	Int.	Mass	Int.	Mass	Int.	Mass	Int.	Mass	Int.
41	673	43	231	53	351	58	251	67	100	68	65	81	150	82	130	94	35
97	50	110	241	111	301	124	20	126	20	138	482	140	75	153	15	156	10
167	25	169	30	181	271	183	45	195	999	196	773	205	15	220	90	223	45
238	110	239	30														

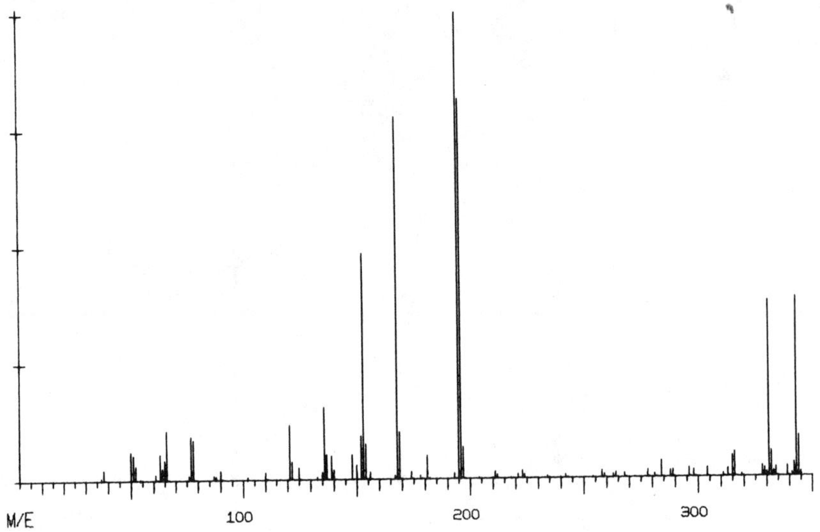
M/E
100
200
300

TRIMETHOBENZAMIDE MTB. 1

$C_{19}H_{21}NO_5$ MW = 343

Base Peak = 195 45 Peaks

Mass	Int.	Mass	Int.	Mass	Int.	Mass	Int.	Mass	Int.	Mass	Int.	Mass	Int.	Mass	Int.	Mass	Int.
37	4	38	23	50	60	51	52	63	56	66	105	77	92	78	85	90	21
102	4	110	14	121	117	122	37	136	156	137	52	152	92	153	484	168	777
169	101	174	15	181	50	195	999	196	814	211	15	212	7	221	11	223	18
234	4	242	7	254	2	257	1	258	15	264	13	278	14	284	36	288	15
296	20	304	20	313	17	315	44	316	53	331	378	332	56	343	386	344	88

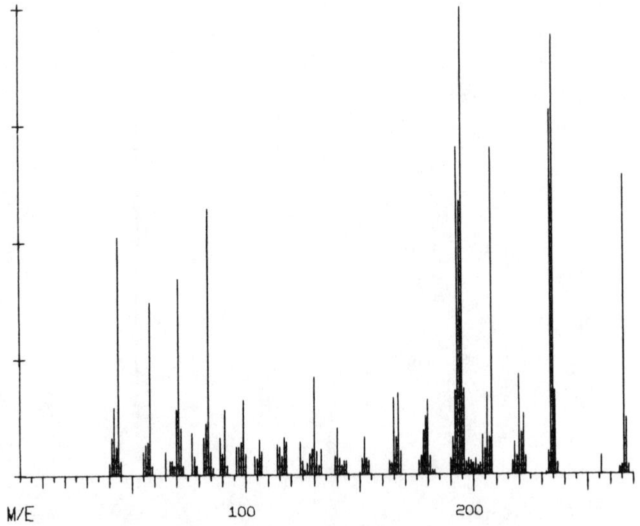
M/E
100
200

DESIPRAMINE

$C_{18}H_{22}N_2$ MW = 266

Base Peak = 195 33 Peaks

Mass	Int.	Mass	Int.	Mass	Int.	Mass	Int.	Mass	Int.	Mass	Int.	Mass	Int.	Mass	Int.	Mass	Int.
42	144	44	509	57	69	58	369	70	139	71	419	83	109	84	569	91	139
99	159	106	74	117	79	118	69	130	209	133	54	140	99	151	34	152	79
165	164	167	174	179	124	180	159	193	699	195	999	206	174	208	699	220	214
222	129	234	779	235	939	256	39	266	639	267	119						

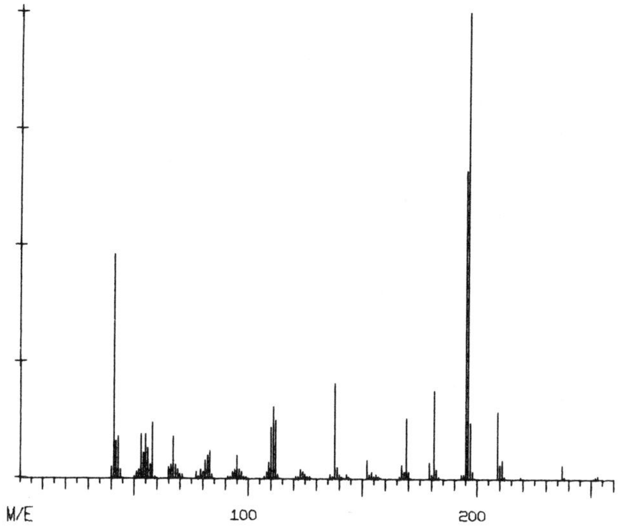
M/E
100
200

BUTALBITAL 1,3-DIMETHYL DERIVATIVE

$C_{13}H_{20}N_2O_3$ MW = 252
Base Peak = 196 31 Peaks

Mass	Int.	Mass	Int.	Mass	Int.	Mass	Int.	Mass	Int.	Mass	Int.	Mass	Int.	Mass	Int.	Mass	Int.
41	480	43	90	53	95	58	120	66	30	67	90	82	50	83	60	94	20
95	50	111	155	112	125	123	20	124	15	138	205	139	25	152	40	154	15
167	30	169	130	179	35	181	190	195	660	196	999	209	145	211	40	219	5
237	30	238	4	252	5	253	7										

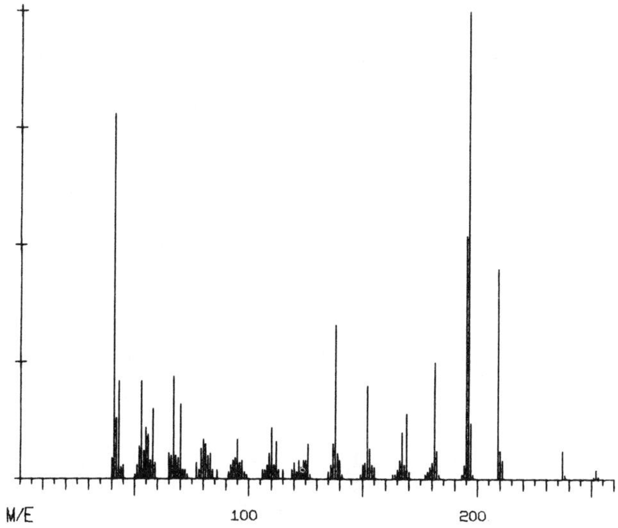
M/E
100
200

BUTALBITAL 1,2(OR 4)-DIMETHYL DERIVATIVE

$C_{13}H_{20}N_2O_3$ MW = 252
Base Peak = 196 30 Peaks

Mass	Int.	Mass	Int.	Mass	Int.	Mass	Int.	Mass	Int.	Mass	Int.	Mass	Int.	Mass	Int.	Mass	Int.
41	780	43	210	53	210	58	150	67	220	70	160	80	85	81	75	94	45
95	85	110	110	112	80	122	40	126	75	137	75	138	330	152	200	153	65
167	100	169	140	181	250	182	60	195	520	196	999	209	450	210	60	237	60
238	10	251	5	252	20												

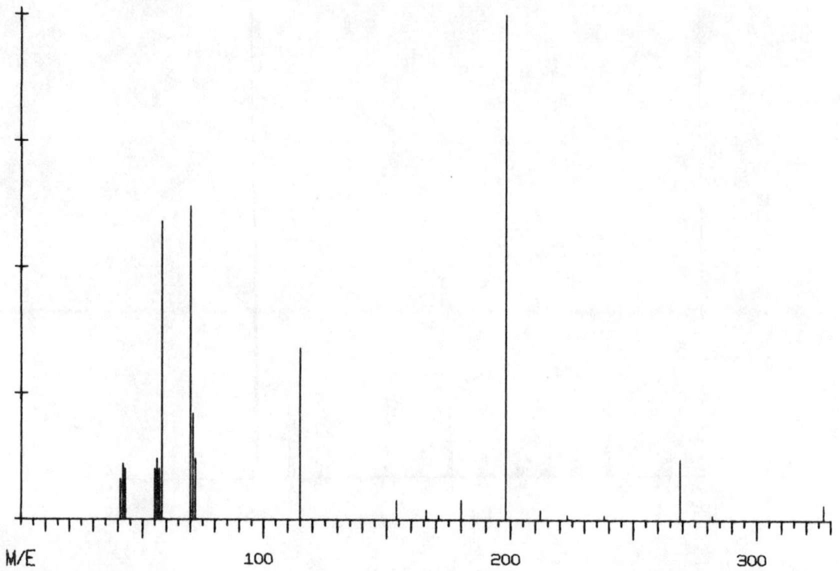
M/E
100
200
300

AMINOPROMAZINE

$C_{19}H_{25}N_3S$ MW = 327

Base Peak = 198 18 Peaks

Mass	Int.	Mass	Int.	Mass	Int.	Mass	Int.	Mass	Int.	Mass	Int.	Mass	Int.	Mass	Int.	Mass	Int.
42	109	43	99	56	119	58	589	70	619	71	209	115	339	154	39	166	19
171	9	180	39	198	999	212	19	223	9	238	9	269	119	282	9	327	29

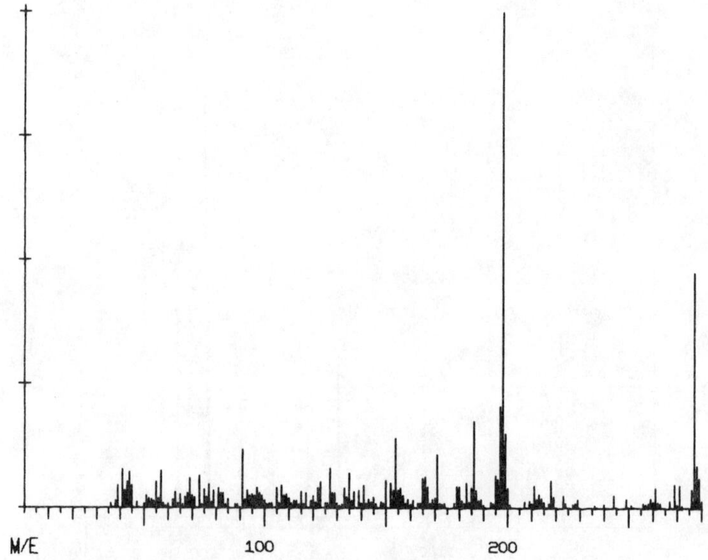
M/E
100
200

THIORIDAZINE MTB. 2

$C_{13}H_{11}NO_2S_2$ MW = 277

Base Peak = 198 36 Peaks

Mass	Int.	Mass	Int.	Mass	Int.	Mass	Int.	Mass	Int.	Mass	Int.	Mass	Int.	Mass	Int.	Mass	Int.
41	78	44	73	55	52	57	76	69	60	73	65	77	47	81	39	91	117
97	41	105	41	107	44	123	53	127	79	135	71	141	44	150	55	154	140
166	63	171	108	183	51	186	176	197	204	198	999	211	44	213	28	218	55
223	26	236	6	240	7	244	25	249	17	209	47	271	46	277	474	278	86

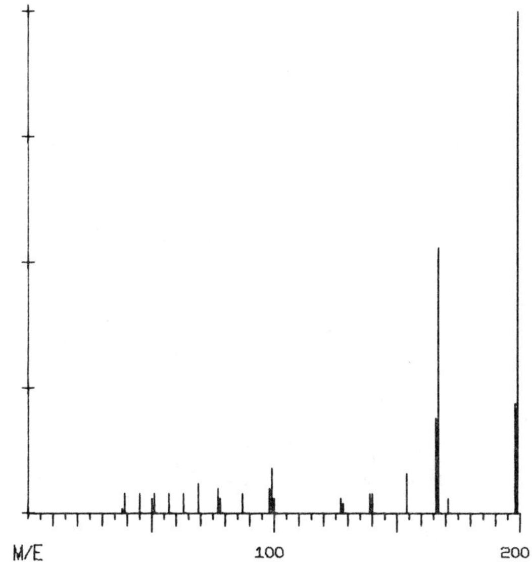
M/E
100
200

PHENOTHIAZINE

$C_{12}H_9NS$ MW = 199

Base Peak = 199 19 Peaks

Mass	Int.	Mass	Int.	Mass	Int.	Mass	Int.	Mass	Int.	Mass	Int.	Mass	Int.	Mass	Int.	Mass	Int.
39	39	45	39	51	39	57	39	63	39	69	59	77	49	87	39	98	49
99	89	127	29	128	19	139	39	140	39	154	79	166	189	167	529	198	219
199	999																

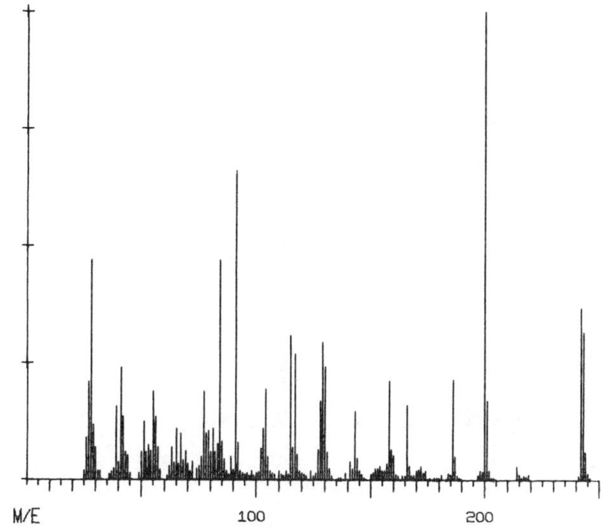
M/E
100
200

PHENCYCLIDINE

$C_{17}H_{25}N$ MW = 243

Base Peak = 200 34 Peaks

Mass	Int.	Mass	Int.	Mass	Int.	Mass	Int.	Mass	Int.	Mass	Int.	Mass	Int.	Mass	Int.	Mass	Int.
27	210	28	469	39	158	41	239	55	191	56	136	65	109	67	101	77	191
84	470	91	661	103	110	115	309	117	270	129	294	130	242	143	147	144	48
158	212	159	65	160	52	166	160	186	215	187	50	200	999	201	169	202	20
214	30	217	10	219	12	242	368	243	315	244	60	245	15				

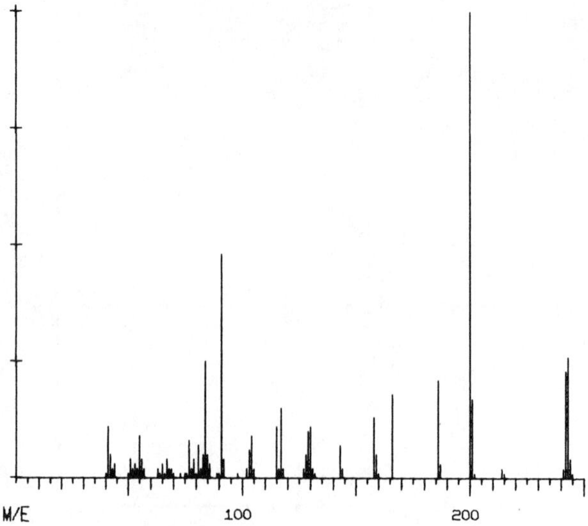
M/E
100
200

PHENCYCLIDINE

$C_{17}H_{25}N$ MW = 243

Base Peak = 200 30 Peaks

Mass	Int.	Mass	Int.	Mass	Int.	Mass	Int.	Mass	Int.	Mass	Int.	Mass	Int.	Mass	Int.	Mass	Int.
41	110	42	50	51	40	55	90	65	30	67	40	77	80	84	250	91	480
103	60	115	110	117	150	129	100	130	110	143	70	144	20	158	130	159	50
160	10	166	180	186	210	187	30	200	999	201	170	202	10	214	20	242	230
243	260	244	40	245	10												

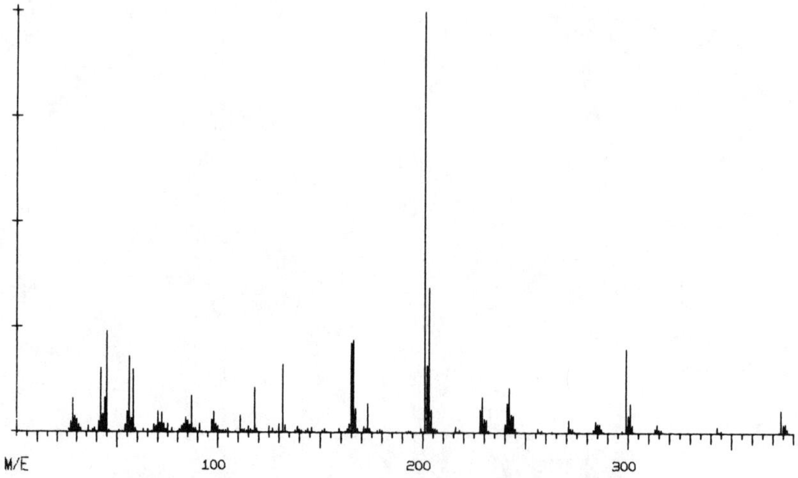
M/E
100
200
300

HYDROXYZINE

$C_{21}H_{27}CLN_2O_2$ MW = 374

Base Peak = 201 48 Peaks

Mass	Int.	Mass	Int.	Mass	Int.	Mass	Int.	Mass	Int.	Mass	Int.	Mass	Int.	Mass	Int.	Mass	Int.
28	80	29	37	42	153	45	240	56	179	58	149	70	49	72	48	84	36
87	88	97	29	98	49	111	41	115	14	118	107	130	19	132	163	133	18
146	12	159	10	165	213	166	220	174	10	179	8	199	9	201	999	202	161
203	345	228	55	229	85	241	71	242	108	244	39	245	9	258	4	271	29
284	27	285	20	286	19	299	200	300	41	301	70	314	20	315	8	343	14
345	4	374	56	376	23												

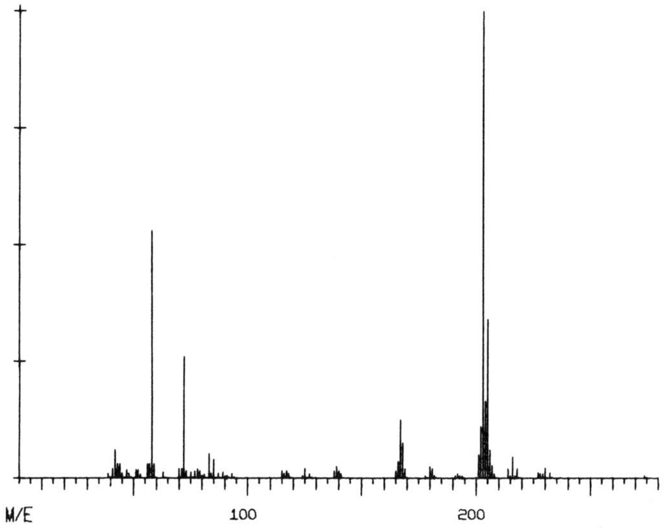
M/E
100
200

CHLORPHENIRAMINE

$C_{16}H_{19}CLN_2$ MW = 274
Base Peak = 203 29 Peaks

Mass	Int.	Mass	Int.	Mass	Int.	Mass	Int.	Mass	Int.	Mass	Int.	Mass	Int.	Mass	Int.	Mass	Int.
42	59	43	29	56	29	58	529	71	21	72	259	83	53	85	39	91	6
93	11	115	14	117	14	125	19	127	11	138	14	139	24	167	125	168	75
180	24	181	19	192	9	201	49	203	999	205	339	216	44	218	19	230	23
232	12	274	4														

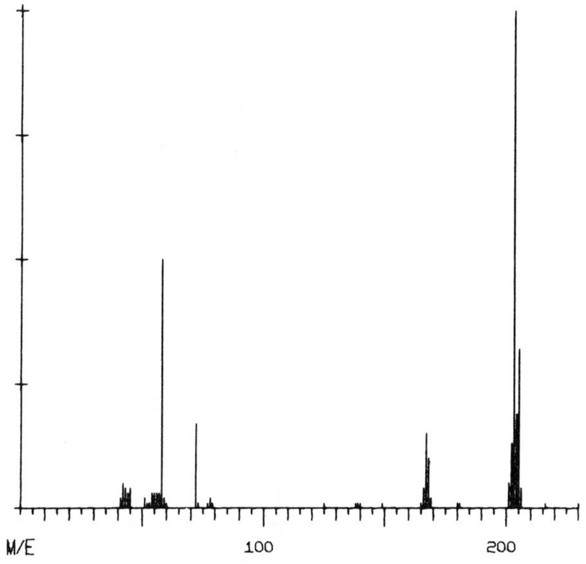
M/E
100
200

CHLORPHENIRAMINE

$C_{16}H_{19}CLN_2$ MW = 274
Base Peak = 203 21 Peaks

Mass	Int.	Mass	Int.	Mass	Int.	Mass	Int.	Mass	Int.	Mass	Int.	Mass	Int.	Mass	Int.	Mass	Int.
42	50	43	40	54	30	58	500	72	170	73	10	77	10	78	20	125	10
138	10	139	10	149	10	167	150	168	100	180	10	181	10	201	50	203	999
205	320	216	10	230	10												

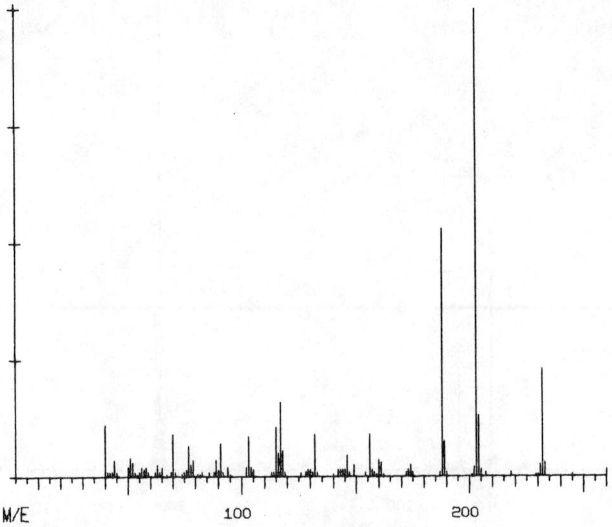
M/E
100
200

PHENOBARBITAL 1,2(OR 4)-DIMETHYL DERIVATIVE

$C_{14}H_{16}N_2O_3$ MW = 260
Base Peak = 203 30 Peaks

Mass	Int.	Mass	Int.	Mass	Int.	Mass	Int.	Mass	Int.	Mass	Int.	Mass	Int.	Mass	Int.	Mass	Int.
40	110	44	35	51	40	52	30	63	25	70	90	77	65	79	35	91	70
103	85	115	105	117	160	118	55	129	15	132	90	142	15	146	45	156	90
160	35	161	30	174	25	175	10	188	530	189	75	203	999	204	130	218	10
229	5	232	230	233	30	245	5	260	15								

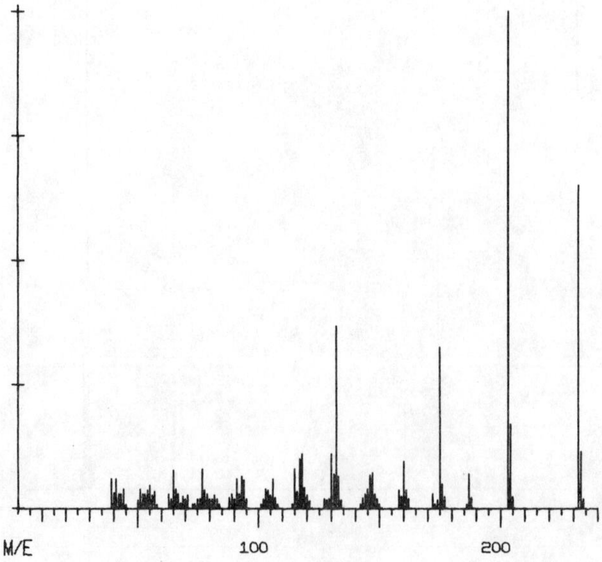
M/E
100
200

AMINOGLUTETHIMIDE

$C_{13}H_{16}N_2O$ MW = 216
Base Peak = 203 27 Peaks

Mass	Int.	Mass	Int.	Mass	Int.	Mass	Int.	Mass	Int.	Mass	Int.	Mass	Int.	Mass	Int.	Mass	Int.
39	60	41	59	51	40	55	48	65	78	66	39	77	79	78	37	91	59
93	66	115	80	117	99	118	109	130	110	132	367	133	64	146	67	147	73
160	95	161	37	175	325	187	69	188	23	203	999	204	169	232	650	233	114

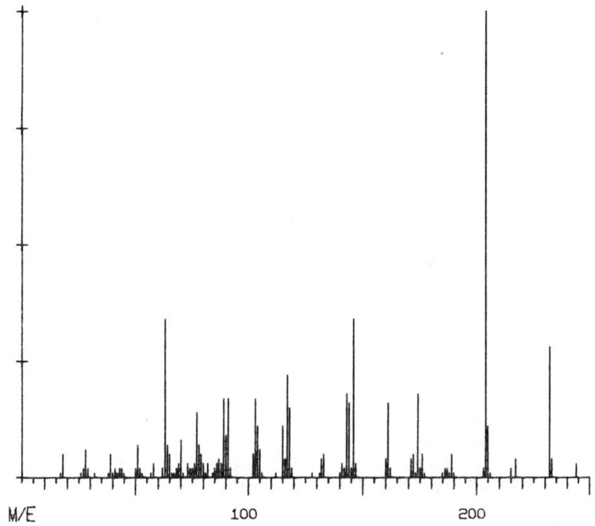
M/E
100
200

PHENOBARBITAL

$C_{12}H_{12}N_2O_3$ MW = 232

Base Peak = 204 32 Peaks

Mass	Int.	Mass	Int.	Mass	Int.	Mass	Int.	Mass	Int.	Mass	Int.	Mass	Int.	Mass	Int.	Mass	Int.
27	20	28	60	39	50	41	20	51	70	58	30	63	340	70	80	77	140
89	170	91	170	103	170	104	110	117	220	118	150	119	20	143	180	144	160
146	340	147	30	161	160	172	50	174	180	176	50	188	10	189	50	204	999
205	110	217	40	232	280	233	40	244	30								

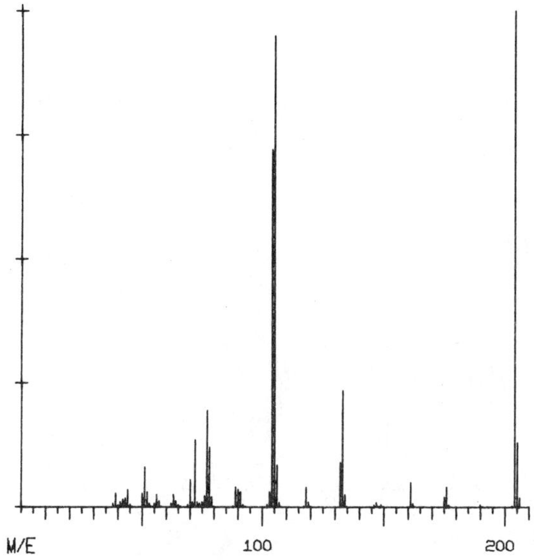
M/E
100
200

ETHOTOIN

$C_{11}H_{12}N_2O_2$ MW = 204

Base Peak = 204 26 Peaks

Mass	Int.	Mass	Int.	Mass	Int.	Mass	Int.	Mass	Int.	Mass	Int.	Mass	Int.	Mass	Int.	Mass	Int.
39	27	44	35	51	80	52	29	70	55	72	135	77	194	78	120	90	35
91	30	104	719	105	949	118	39	119	9	132	89	133	234	146	6	147	11
161	49	162	8	175	19	176	39	189	1	190	4	204	999	205	131		

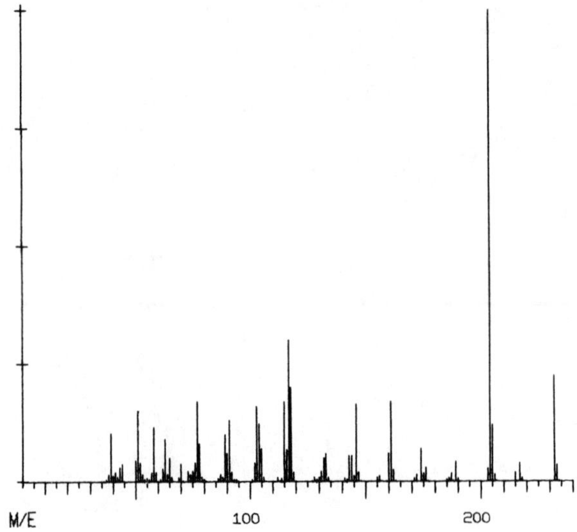
M/E
100
200

PHENOBARBITAL

$C_{12}H_{12}N_2O_3$ MW = 232
Base Peak = 204 30 Peaks

Mass	Int.	Mass	Int.	Mass	Int.	Mass	Int.	Mass	Int.	Mass	Int.	Mass	Int.	Mass	Int.	Mass	Int.
39	103	44	37	51	151	58	114	63	91	65	50	77	169	89	99	91	130
103	161	115	169	117	299	118	200	131	22	133	61	144	56	146	164	147	21
160	59	161	171	174	69	176	29	189	42	190	8	204	999	205	120	217	41
218	8	232	225	233	35												

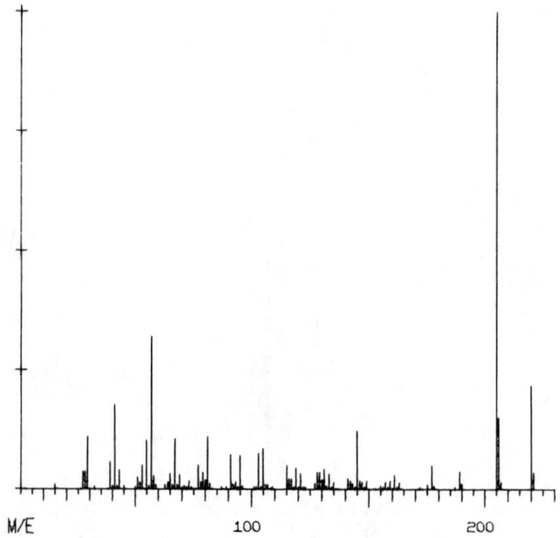
M/E
100
200

IONOL

$C_{15}H_{24}O$ MW = 220
Base Peak = 205 30 Peaks

Mass	Int.	Mass	Int.	Mass	Int.	Mass	Int.	Mass	Int.	Mass	Int.	Mass	Int.	Mass	Int.	Mass	Int.
27	37	29	109	39	57	41	177	55	102	57	320	65	32	67	106	77	51
81	109	91	73	103	74	105	84	115	49	119	45	131	42	133	33	145	123
146	17	149	18	161	30	163	16	175	11	177	51	189	37	190	12	205	999
206	151	220	217	221	35												

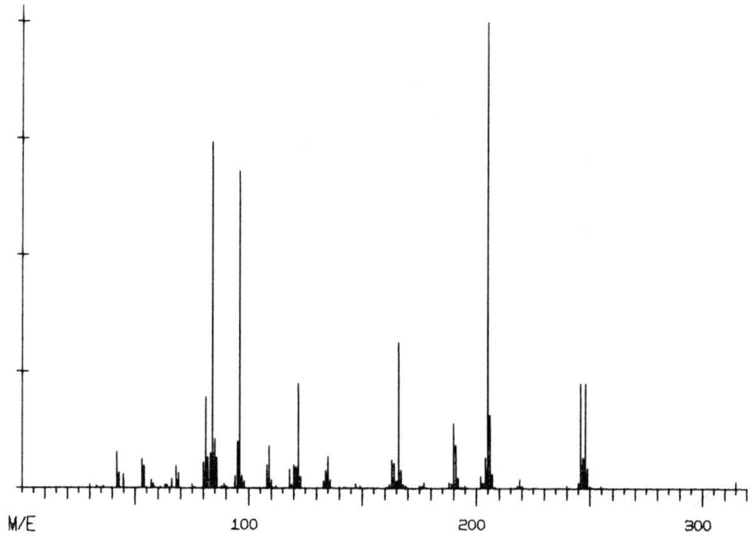
M/E
100
200
300

PHENCYCLIDINE (D5-PH)

$C_{17}H_{20}D_5N$ MW = 248

Base Peak = 205 34 Peaks

Mass	Int.	Mass	Int.	Mass	Int.	Mass	Int.	Mass	Int.	Mass	Int.	Mass	Int.	Mass	Int.	Mass	Int.
30	7	33	5	42	78	43	33	53	62	54	47	68	47	69	33	81	196
84	743	95	99	96	681	108	51	109	89	120	49	122	225	134	37	135	68
147	9	149	6	163	61	166	312	175	5	177	13	190	141	191	92	205	999
206	157	218	5	219	21	240	4	246	224	248	226	315	14				

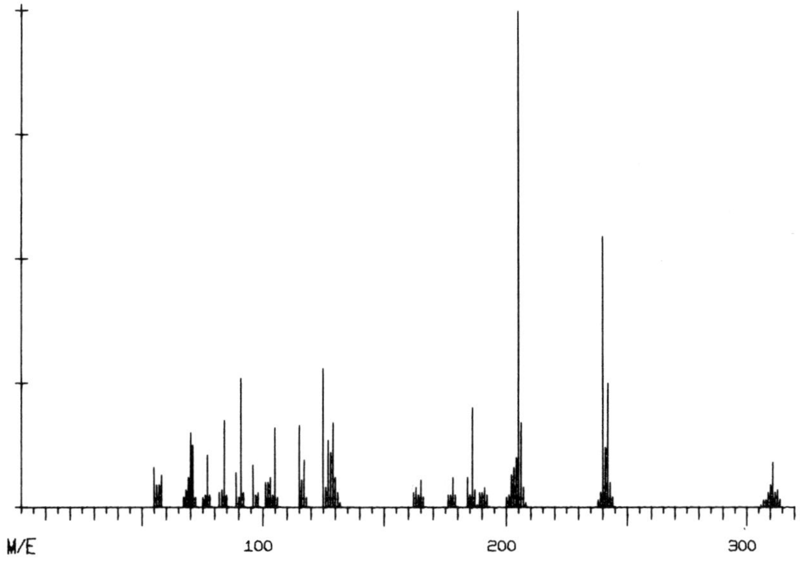
M/E
100
200
300

PYRROBUTAMINE

$C_{20}H_{22}CLN$ MW = 311

Base Peak = 205 27 Peaks

Mass	Int.	Mass	Int.	Mass	Int.	Mass	Int.	Mass	Int.	Mass	Int.	Mass	Int.	Mass	Int.	Mass	Int.
55	79	58	66	70	149	71	124	77	104	84	174	91	259	96	84	105	159
115	164	125	279	129	169	132	9	163	39	165	54	178	59	186	199	189	29
191	39	205	999	206	169	240	544	242	249	244	21	310	44	311	89	314	14

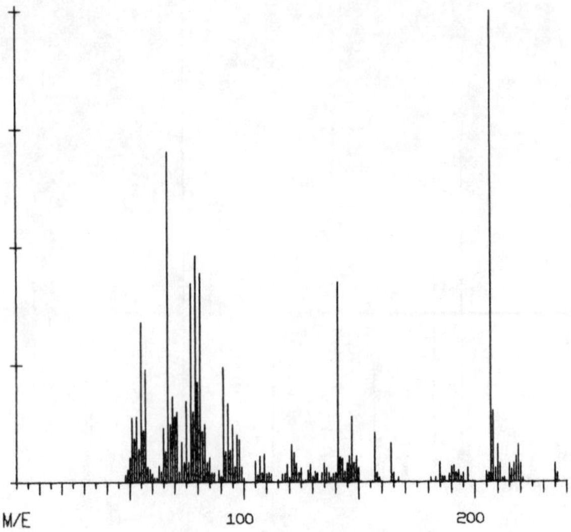
M/E
100
200

CYCLOBARBITAL

$C_{12}H_{16}N_2O_3$ MW = 236
Base Peak = 207 28 Peaks

Mass	Int.	Mass	Int.	Mass	Int.	Mass	Int.	Mass	Int.	Mass	Int.	Mass	Int.	Mass	Int.	Mass	Int.
55	340	57	240	67	703	69	183	79	483	81	444	91	245	93	168	107	55
109	60	121	80	122	63	141	424	142	53	147	140	157	105	164	83	165	20
185	43	186	13	190	33	191	34	207	999	208	153	218	55	219	81	235	41
236	21																

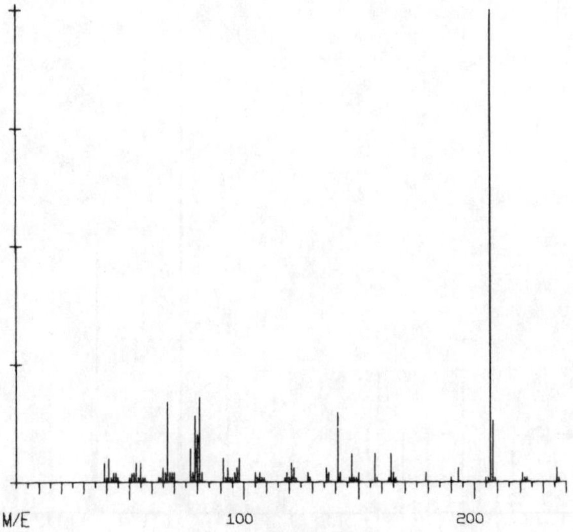
M/E
100
200

CYCLOBARBITAL

$C_{12}H_{16}N_2O_3$ MW = 236
Base Peak = 207 29 Peaks

Mass	Int.	Mass	Int.	Mass	Int.	Mass	Int.	Mass	Int.	Mass	Int.	Mass	Int.	Mass	Int.	Mass	Int.
39	40	41	50	53	40	55	40	65	30	67	170	79	140	81	180	91	50
98	50	105	20	107	20	121	40	122	30	136	30	141	280	147	60	157	60
164	60	165	20	179	20	190	10	193	30	207	999	208	130	221	20	222	10
236	30	237	10														

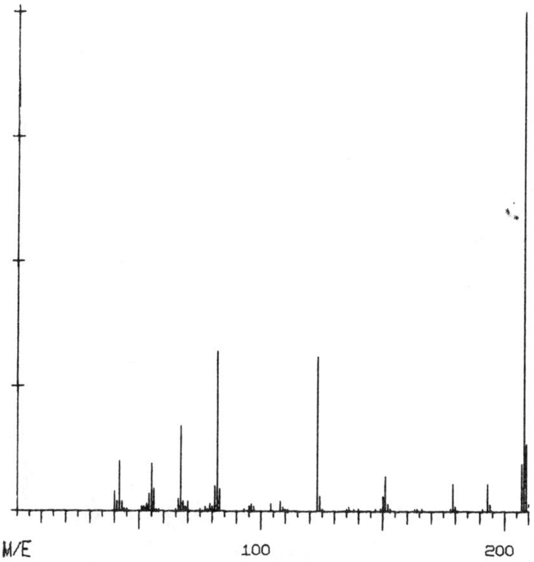
M/E
100
200

1,3,7(OR 9),8-TETRAMETHYLXANTHINE

$C_9H_{12}N_4O_2$ MW = 208

Base Peak = 208 26 Peaks

Mass	Int.	Mass	Int.	Mass	Int.	Mass	Int.	Mass	Int.	Mass	Int.	Mass	Int.	Mass	Int.	Mass	Int.
40	40	42	100	55	95	56	45	66	25	67	170	81	50	82	320	95	10
96	15	104	15	108	20	123	310	124	30	135	5	136	10	150	30	151	70
163	5	164	5	179	55	180	10	193	55	194	15	208	999	209	135		

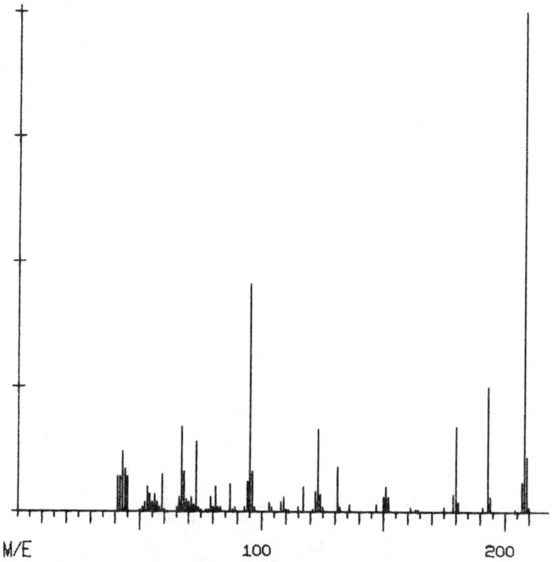
M/E
100
200

THEOPHYLLINE 7-ETHYL DERIVATIVE

$C_9H_{12}N_4O_2$ MW = 208

Base Peak = 208 26 Peaks

Mass	Int.	Mass	Int.	Mass	Int.	Mass	Int.	Mass	Int.	Mass	Int.	Mass	Int.	Mass	Int.	Mass	Int.
43	120	44	85	53	50	59	75	67	170	73	140	81	50	87	55	95	455
96	80	109	30	117	50	123	165	131	90	132	10	136	15	150	30	151	50
161	10	163	5	179	35	180	170	193	250	194	30	208	999	209	110		

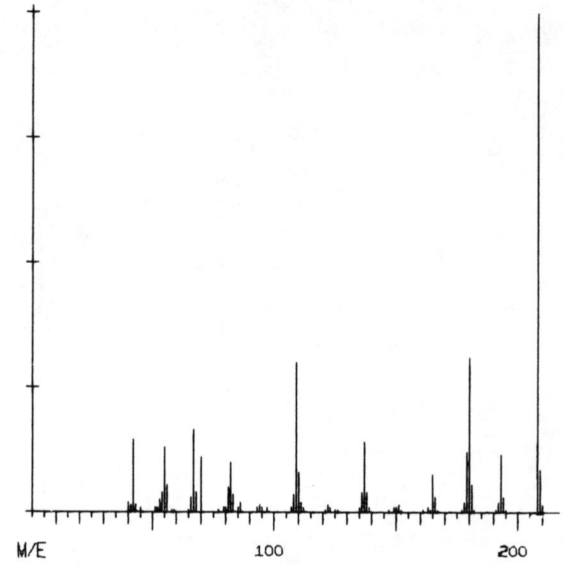
M/E
100
200

THEOBROMINE 1-ETHYL DERIVATIVE

$C_9H_{12}N_4O_2$ MW = 208

Base Peak = 208 26 Peaks

Mass	Int.	Mass	Int.	Mass	Int.	Mass	Int.	Mass	Int.	Mass	Int.	Mass	Int.	Mass	Int.	Mass	Int.
40	19	42	144	55	129	56	54	67	164	70	109	81	49	82	99	93	9
94	14	109	299	110	79	122	14	123	9	136	39	137	139	149	9	151	14
165	74	166	29	119	18	180	309	193	114	194	29	208	999	209	84		

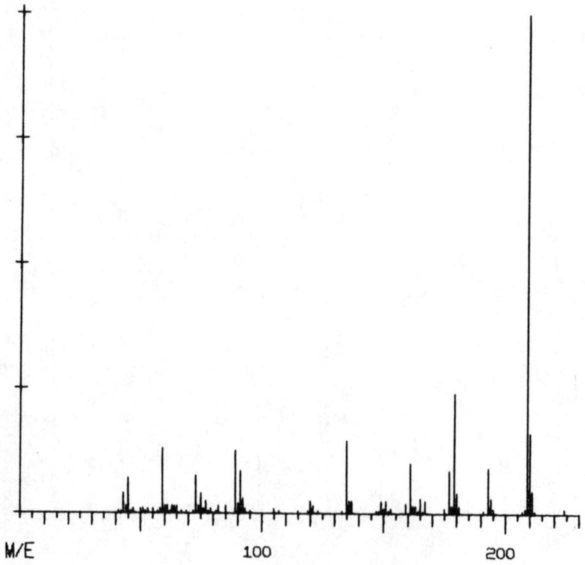
M/E
100
200

SALICYCLIC ACID METHYL ESTER TMS ETHER

$C_{11}H_{16}O_3SI$ MW = 224

Base Peak = 209 27 Peaks

Mass	Int.	Mass	Int.	Mass	Int.	Mass	Int.	Mass	Int.	Mass	Int.	Mass	Int.	Mass	Int.	Mass	Int.
43	40	45	70	59	130	60	15	73	75	75	40	77	25	89	125	91	85
92	30	105	10	107	5	120	25	121	15	135	145	136	25	149	25	151	25
161	100	165	30	177	85	179	240	193	90	194	30	209	999	210	160	224	10

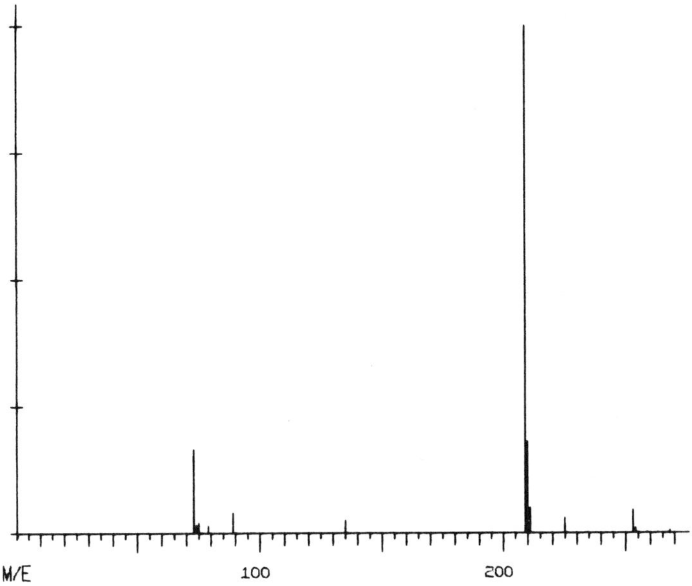
M/E
100
200

P-METHOXYMANDELIC ACID METHYL ESTER TRIMETHYLSILYL ETHER

$C_{13}H_{20}O_4SI$ MW = 268

Base Peak = 209 11 Peaks

Mass	Int.	Mass	Int.	Mass	Int.	Mass	Int.	Mass	Int.	Mass	Int.	Mass	Int.	Mass	Int.	Mass	Int.
73	164	75	19	79	14	89	39	135	24	209	999	210	179	225	29	253	44
254	9	268	4														

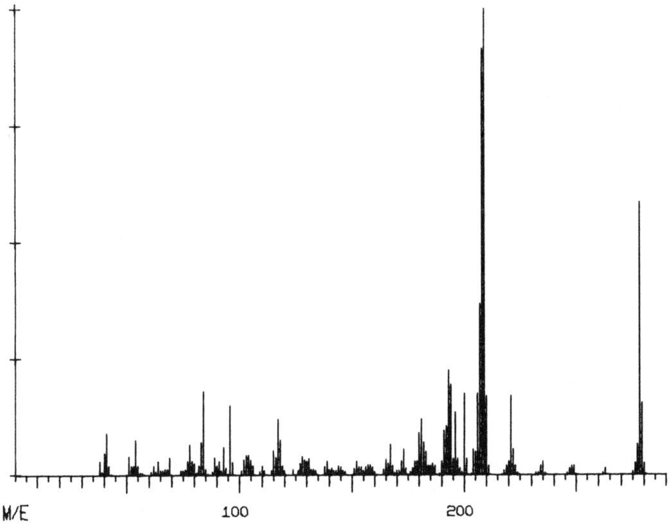
M/E
100
200

TRIPROLIDINE

$C_{19}H_{22}N_2$ MW = 278

Base Peak = 209 36 Peaks

Mass	Int.	Mass	Int.	Mass	Int.	Mass	Int.	Mass	Int.	Mass	Int.	Mass	Int.	Mass	Int.	Mass	Int.
40	48	41	90	51	39	54	74	64	29	69	38	83	69	84	179	93	60
96	149	115	53	117	119	118	74	128	40	139	29	144	20	152	29	157	22
167	64	173	56	180	89	181	119	193	224	194	194	208	914	209	999	221	169
222	54	234	19	235	29	248	19	249	19	262	4	263	14	278	585	279	156

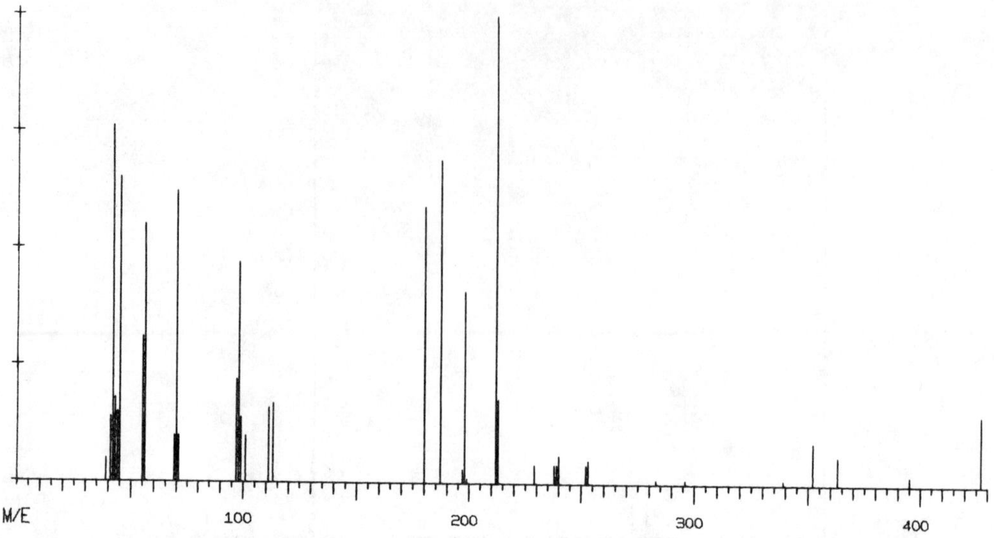
M/E
100
200
300
400

DIXYRAZINE

$C_{24}H_{33}N_3O_2S$ MW = 427

Base Peak = 212 28 Peaks

Mass	Int.	Mass	Int.	Mass	Int.	Mass	Int.	Mass	Int.	Mass	Int.	Mass	Int.	Mass	Int.	Mass	Int.
42	759	45	649	55	309	56	549	69	99	70	619	97	219	98	469	111	159
113	169	180	589	187	689	197	29	198	409	212	999	213	179	229	39	238	39
240	59	252	39	253	49	283	9	296	9	339	9	352	89	363	59	395	19
427	149																

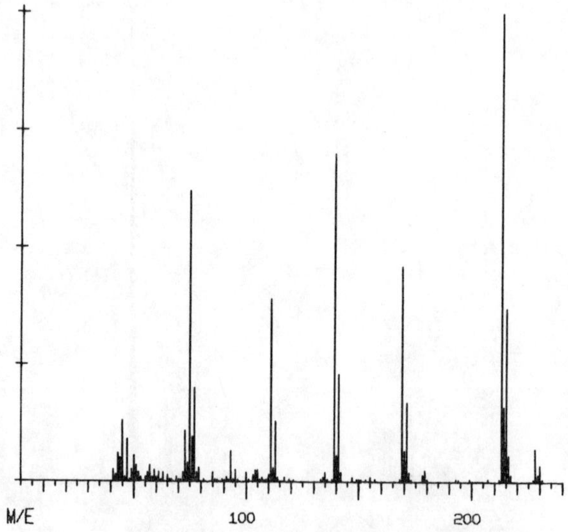
M/E
100
200

4-CHLOROBENZOIC ACID TMS ESTER (FROM INDOMETHACIN)

$C_{10}H_{13}ClO_2Si$ MW = 228

Base Peak = 213 30 Peaks

Mass	Int.	Mass	Int.	Mass	Int.	Mass	Int.	Mass	Int.	Mass	Int.	Mass	Int.	Mass	Int.	Mass	Int.
45	130	47	90	50	55	51	35	73	110	75	620	76	95	77	200	93	65
95	25	111	390	113	130	119	5	121	5	139	700	141	230	147	10	155	10
169	460	171	170	178	15	179	25	193	5	194	5	213	999	215	370	216	55
228	70	230	35	231	5												

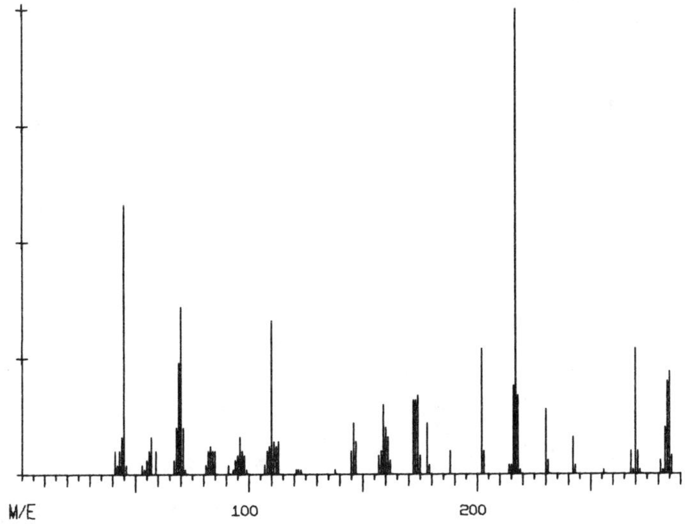
M/E
100
200

PENTAZOCINE

$C_{19}H_{27}NO$ MW = 285

Base Peak = 217 35 Peaks

Mass	Int.	Mass	Int.	Mass	Int.	Mass	Int.	Mass	Int.	Mass	Int.	Mass	Int.	Mass	Int.	Mass	Int.
44	80	45	580	56	50	57	80	69	240	70	360	82	50	83	60	96	80
97	50	110	330	111	70	121	10	122	10	138	10	145	50	146	110	159	150
172	160	173	160	174	170	178	110	188	50	202	270	203	50	216	190	217	999
230	140	242	80	256	10	268	50	270	270	284	200	285	220	286	40		

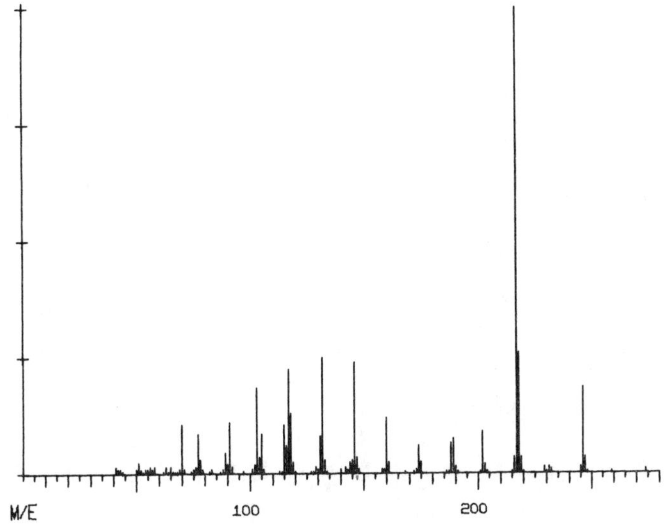
M/E
100
200

METHOBARBITAL 4-ETHYL DERIVATIVE

$C_{15}H_{18}N_2O_3$ MW = 274

Base Peak = 217 34 Peaks

Mass	Int.	Mass	Int.	Mass	Int.	Mass	Int.	Mass	Int.	Mass	Int.	Mass	Int.	Mass	Int.	Mass	Int.
41	15	42	10	51	25	56	15	63	15	70	105	77	85	89	45	91	110
103	185	115	105	117	225	118	130	131	80	132	250	133	30	146	240	147	35
160	120	161	25	174	60	175	25	188	65	189	75	202	90	203	20	217	999
218	260	231	15	232	10	246	185	247	35	259	5	274	10				

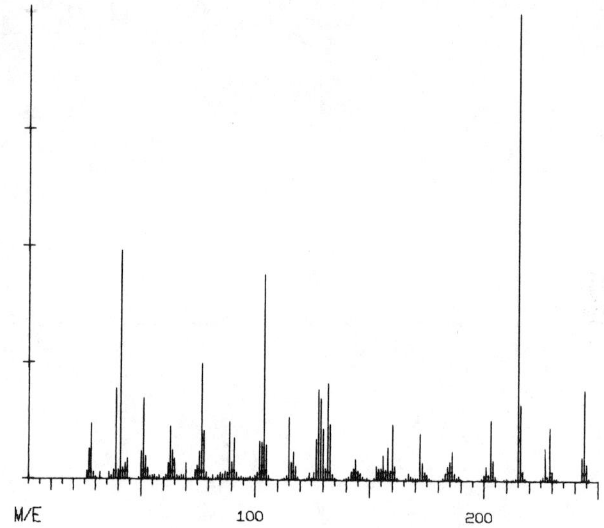
M/E
100
200

ALPHENAL

$C_{13}H_{12}N_2O_3$ MW = 244

Base Peak = 215 34 Peaks

Mass	Int.	Mass	Int.	Mass	Int.	Mass	Int.	Mass	Int.	Mass	Int.	Mass	Int.	Mass	Int.	Mass	Int.
27	64	28	120	39	196	41	490	50	61	51	176	63	114	64	62	77	249
89	126	91	90	102	82	104	440	115	134	128	195	129	174	132	208	133	121
156	53	158	70	160	119	172	101	185	40	186	63	200	13	201	29	203	131
215	999	216	162	229	116	230	19	243	49	244	194	245	34				

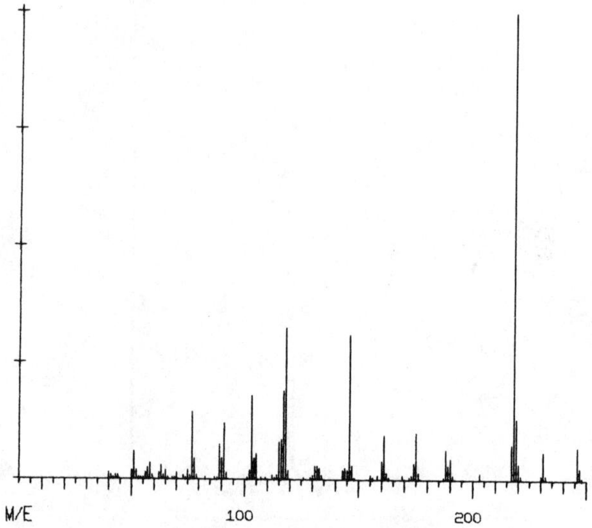
M/E
100
200

MEPHOBARBITAL

$C_{13}H_{14}N_2O_3$ MW = 246

Base Peak = 218 31 Peaks

Mass	Int.	Mass	Int.	Mass	Int.	Mass	Int.	Mass	Int.	Mass	Int.	Mass	Int.	Mass	Int.	Mass	Int.
40	15	41	10	51	60	58	35	63	30	65	20	77	145	89	75	91	120
103	180	116	85	117	190	118	325	131	30	132	30	133	25	146	310	147	30
160	40	161	95	174	35	175	100	188	65	190	45	203	15	218	999	219	130
230	10	231	60	246	70	247	25										

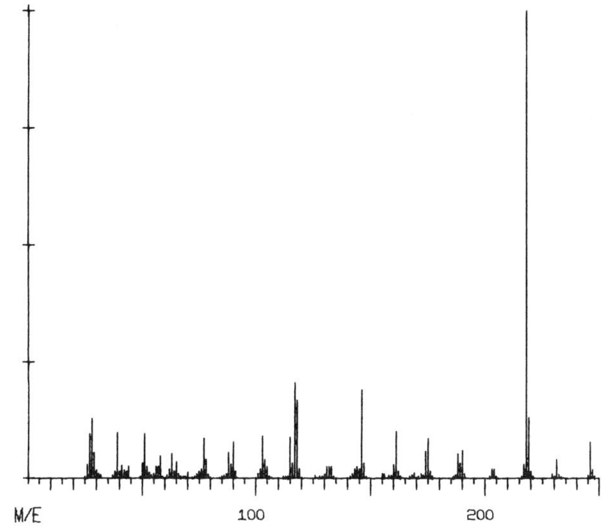
M/E
100
200

MEPHOBARBITAL

$C_{13}H_{14}N_2O_3$ MW = 246

Base Peak = 218 34 Peaks

Mass	Int.	Mass	Int.	Mass	Int.	Mass	Int.	Mass	Int.	Mass	Int.	Mass	Int.	Mass	Int.	Mass	Int.
27	96	28	127	39	98	41	28	51	96	58	48	63	53	65	34	77	85
88	54	90	78	103	90	115	88	117	204	118	168	131	26	133	26	144	26
146	189	147	33	160	29	161	99	174	57	175	84	188	53	190	61	203	20
204	19	218	999	219	130	231	41	232	10	246	77	247	18				

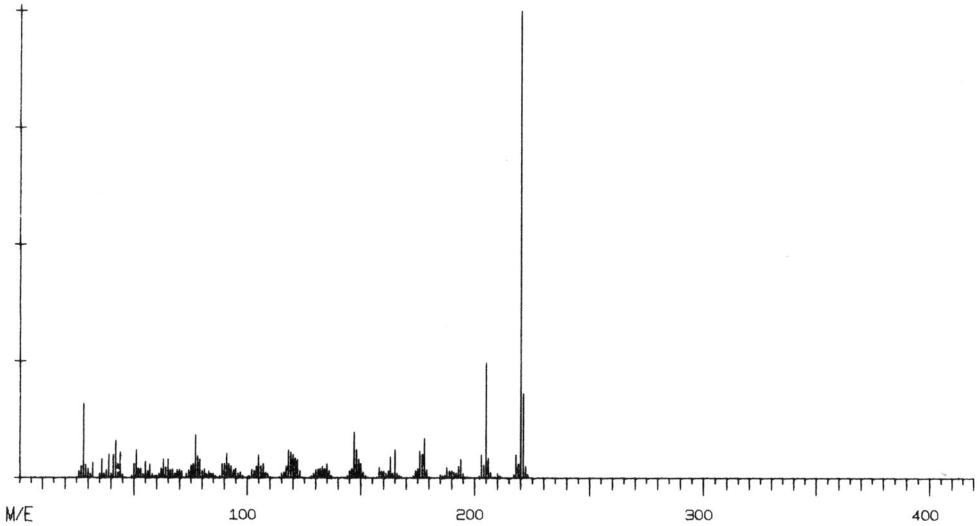
M/E
100
200
300
400

NOSCAPINE

$C_{22}H_{23}NO_7$ MW = 413

Base Peak = 220 31 Peaks

Mass	Int.	Mass	Int.	Mass	Int.	Mass	Int.	Mass	Int.	Mass	Int.	Mass	Int.	Mass	Int.	Mass	Int.
28	161	32	32	42	80	44	54	51	61	55	34	63	41	65	41	77	92
78	47	91	52	92	30	105	50	107	29	118	59	119	54	133	26	135	29
147	97	148	60	163	46	165	59	176	57	178	84	193	26	194	41	203	49
205	247	220	999	221	182	413	2										

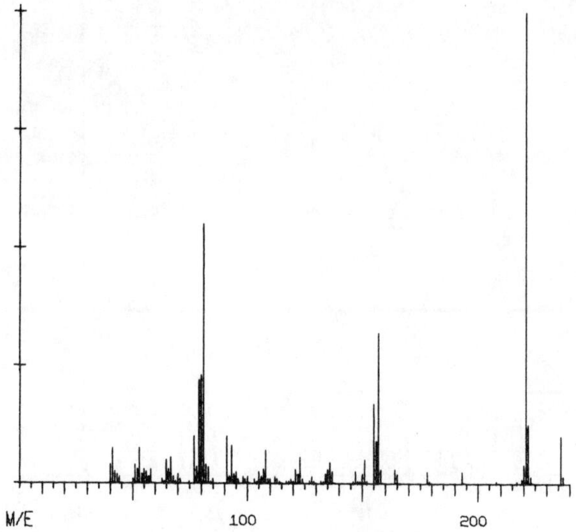

M/E
100
200

HEXOBARBITAL

$C_{12}H_{16}N_2O_3$ MW = 236

Base Peak = 221 28 Peaks

Mass	Int.	Mass	Int.	Mass	Int.	Mass	Int.	Mass	Int.	Mass	Int.	Mass	Int.	Mass	Int.	Mass	Int.
40	40	41	75	51	40	53	75	65	50	67	55	80	230	81	550	91	100
93	80	107	30	108	70	121	30	123	55	135	30	136	45	155	170	157	320
164	30	165	20	178	25	179	5	193	25	208	5	221	999	222	125	236	100
237	15																

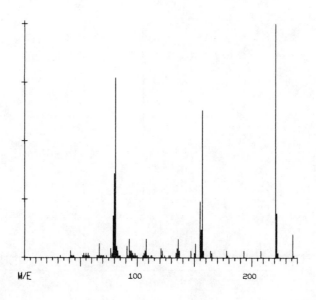

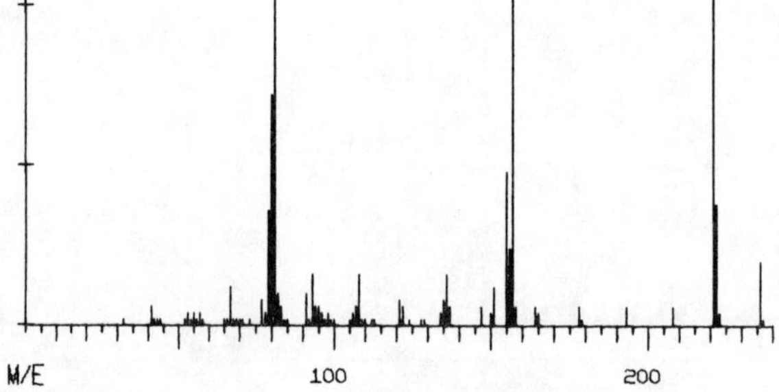

M/E
100
200

HEXOBARBITAL

$C_{12}H_{15}N_2O_3$ MW = 235

Base Peak = 221 29 Peaks

Mass	Int.	Mass	Int.	Mass	Int.	Mass	Int.	Mass	Int.	Mass	Int.	Mass	Int.	Mass	Int.	Mass	Int.
32	10	41	30	42	10	53	20	55	20	65	10	67	60	80	360	81	770
91	50	93	80	107	30	108	80	121	40	122	30	135	40	136	80	155	240
157	630	164	30	165	20	178	30	179	10	193	30	208	30	221	999	222	190
236	100	237	10														

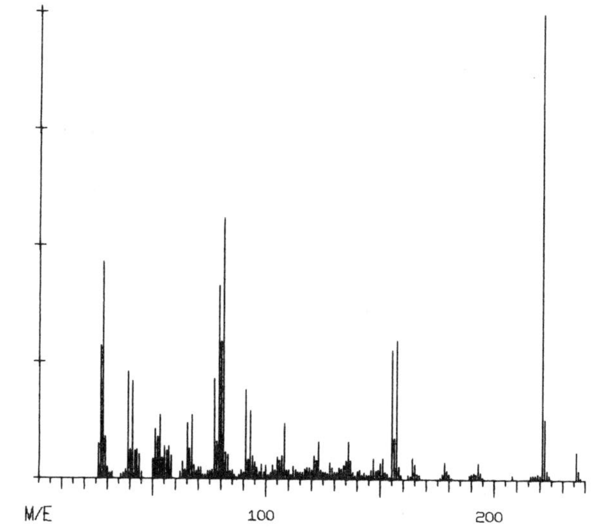
M/E
100
200

HEXOBARBITAL

$C_{12}H_{16}N_2O_3$ MW = 236

Base Peak = 221 31 Peaks

Mass	Int.	Mass	Int.	Mass	Int.	Mass	Int.	Mass	Int.	Mass	Int.	Mass	Int.	Mass	Int.	Mass	Int.
27	284	28	464	39	229	41	209	51	108	53	138	65	120	67	138	79	414
81	560	91	192	93	147	107	50	108	121	121	50	123	79	135	40	136	81
155	278	157	298	164	48	165	32	178	38	179	20	191	14	193	34	208	10
221	999	222	131	236	61	237	20										

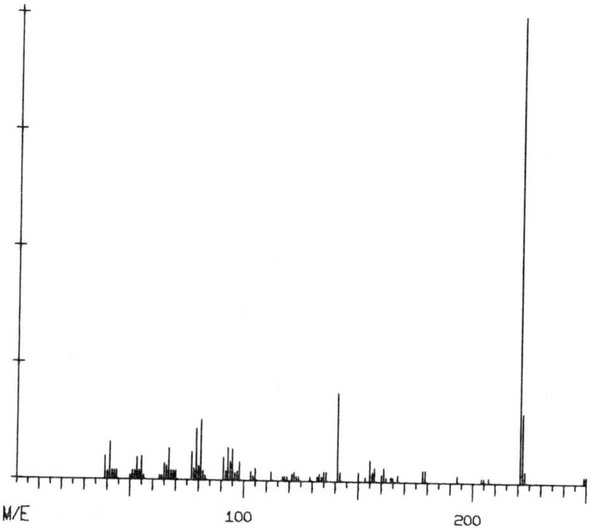
M/E
100
200

HEPTABARBITAL

$C_{13}H_{18}N_2O_3$ MW = 250

Base Peak = 221 29 Peaks

Mass	Int.	Mass	Int.	Mass	Int.	Mass	Int.	Mass	Int.	Mass	Int.	Mass	Int.	Mass	Int.	Mass	Int.
39	49	41	79	53	47	55	49	65	35	67	67	79	109	81	129	93	71
95	68	105	27	112	19	121	14	122	19	135	19	141	189	155	47	157	29
160	16	161	29	178	26	179	24	193	14	204	9	205	9	221	999	222	149
249	14	250	14														

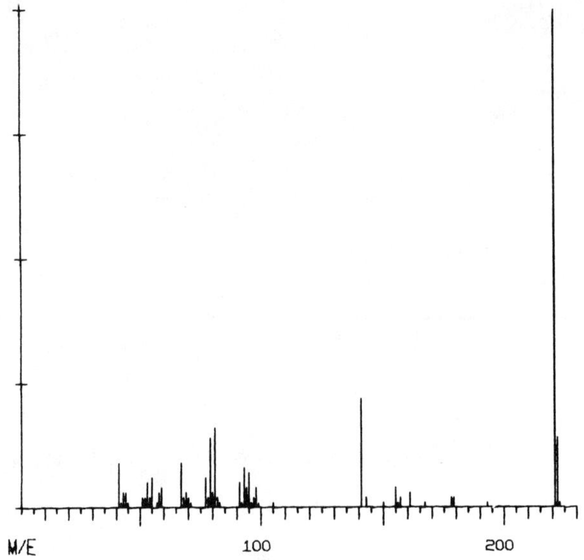
M/E
100
200

HEPTABARBITAL

$C_{13}H_{18}N_2O_3$ MW = 250

Base Peak = 221 22 Peaks

Mass	Int.	Mass	Int.	Mass	Int.	Mass	Int.	Mass	Int.	Mass	Int.	Mass	Int.	Mass	Int.	Mass	Int.
41	90	43	30	53	50	55	60	67	90	69	30	79	140	81	160	93	80
95	70	105	10	141	220	143	20	155	40	157	20	161	30	167	10	178	20
179	20	193	10	221	999	222	140										

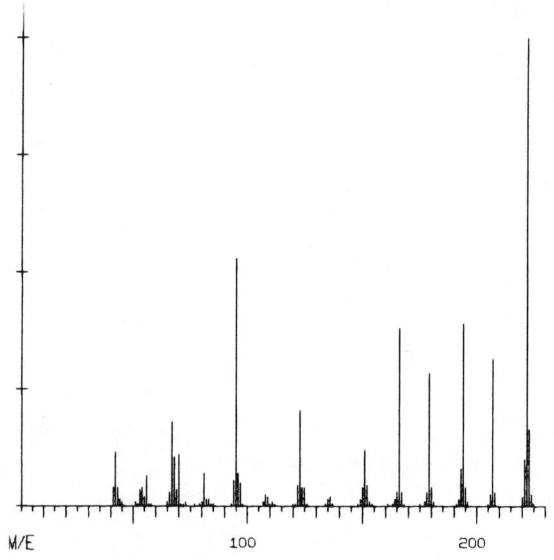
M/E
100
200

1,7-DI-ETHYL-3-METHYLXANTHINE

$C_{10}H_{14}N_4O_2$ MW = 222

Base Peak = 222 28 Peaks

Mass	Int.	Mass	Int.	Mass	Int.	Mass	Int.	Mass	Int.	Mass	Int.	Mass	Int.	Mass	Int.	Mass	Int.
41	40	42	115	54	40	56	65	67	180	70	110	81	70	82	15	95	530
96	70	108	25	109	20	122	45	123	205	135	15	136	20	151	120	152	45
165	30	166	380	179	285	180	40	193	80	194	390	207	315	208	30	222	999
223	165																

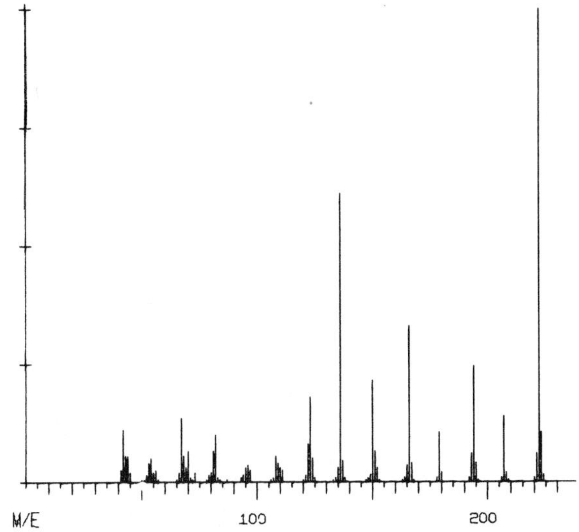
M/E
100
200

1,3-DI-ETHYL-7-METHYLXANTHINE

$C_{10}H_{14}N_4O_2$ MW = 222
Base Peak = 222 28 Peaks

Mass	Int.	Mass	Int.	Mass	Int.	Mass	Int.	Mass	Int.	Mass	Int.	Mass	Int.	Mass	Int.	Mass	Int.
42	110	43	55	53	40	54	50	67	135	70	65	81	65	82	100	95	30
96	35	108	55	109	40	122	80	123	180	136	610	137	45	150	215	151	65
166	330	167	40	179	105	180	20	193	60	194	245	207	140	208	20	222	999
223	105																

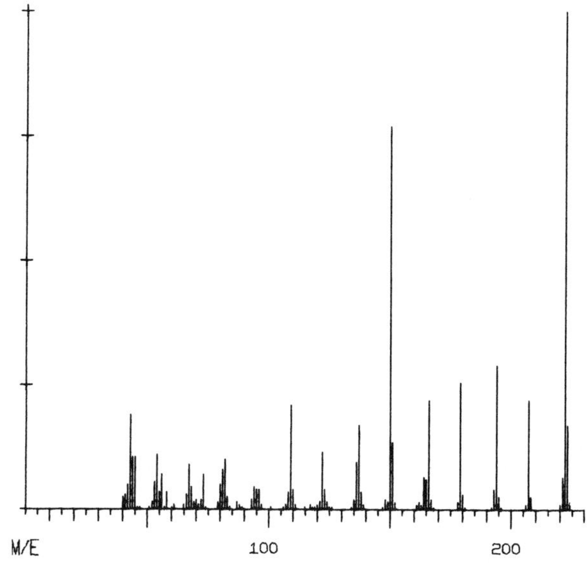
M/E
100
200

3,7-DIETHYL-1-METHYLXANTHINE

$C_{10}H_{14}N_4O_2$ MW = 222
Base Peak = 222 28 Peaks

Mass	Int.	Mass	Int.	Mass	Int.	Mass	Int.	Mass	Int.	Mass	Int.	Mass	Int.	Mass	Int.	Mass	Int.
43	190	44	105	54	110	56	70	67	90	73	70	81	80	82	100	94	45
95	40	109	210	110	40	122	115	123	40	136	95	137	170	150	770	151	135
164	65	166	220	179	255	180	30	193	40	194	290	207	220	208	25	222	999
223	170																

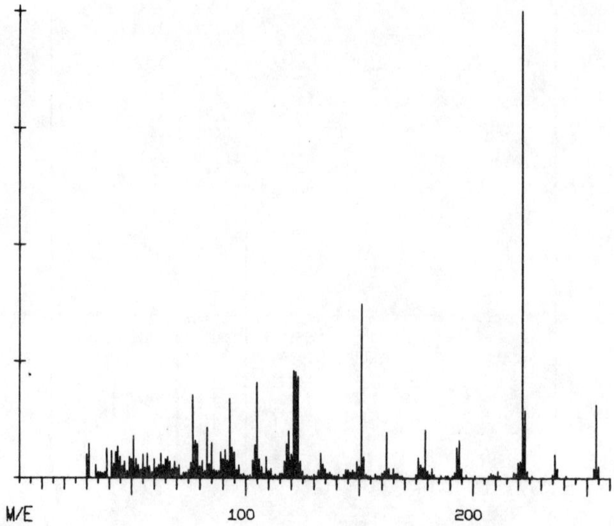
M/E
100
200

N-D3-Me-PHENOBARBITAL (D5-ET)

$C_{13}H_6D_8N_2O_3$ MW = 254

Base Peak = 222 34 Peaks

Mass	Int.	Mass	Int.	Mass	Int.	Mass	Int.	Mass	Int.	Mass	Int.	Mass	Int.	Mass	Int.	Mass	Int.
30	50	31	72	39	62	44	68	51	89	57	53	63	52	66	46	77	178
83	108	93	169	94	65	104	69	105	205	121	230	122	228	133	53	134	30
151	372	152	46	162	97	165	20	176	43	179	103	193	65	194	79	208	11
211	16	222	999	223	145	236	50	237	19	254	157	255	26				

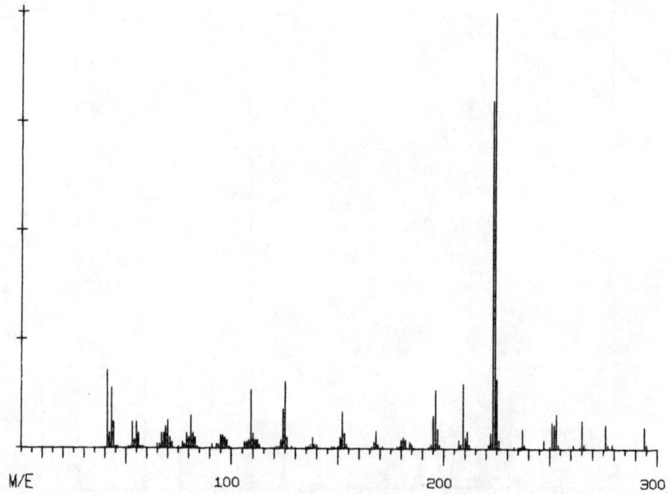
M/E
100
200
300

SECOBARBITAL 1,3-DI-ETHYL DERIVATIVE

$C_{16}H_{26}N_2O_3$ MW = 294

Base Peak = 224 38 Peaks

Mass	Int.	Mass	Int.	Mass	Int.	Mass	Int.	Mass	Int.	Mass	Int.	Mass	Int.	Mass	Int.	Mass	Int.
41	180	43	140	53	60	55	60	69	50	70	65	79	35	81	75	95	30
96	30	109	135	110	35	124	90	125	155	137	10	138	25	152	85	153	35
167	15	168	40	180	20	181	25	195	75	196	135	209	150	211	40	223	800
224	999	237	45	238	10	251	60	253	80	261	10	265	65	276	55	277	10
294	50	295	10														

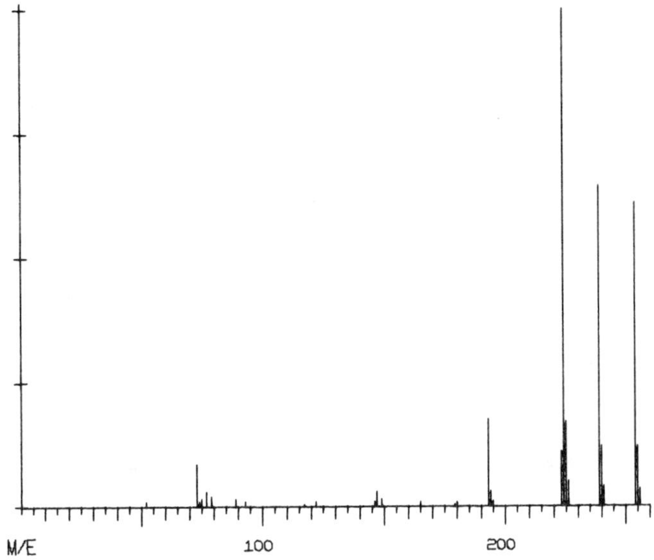
M/E
100
200

3-METHOXY-4-HYDROXYBENZOIC ACID METHYL ESTER TMS ETHER

$C_{12}H_{18}O_4SI$ MW = 254

Base Peak = 224 21 Peaks

Mass	Int.	Mass	Int.	Mass	Int.	Mass	Int.	Mass	Int.	Mass	Int.	Mass	Int.	Mass	Int.	Mass	Int.
52	9	73	84	75	14	77	29	79	19	93	9	117	4	122	9	147	29
149	14	165	9	179	4	180	9	193	174	194	29	224	999	225	169	239	644
240	119	254	609	255	119												

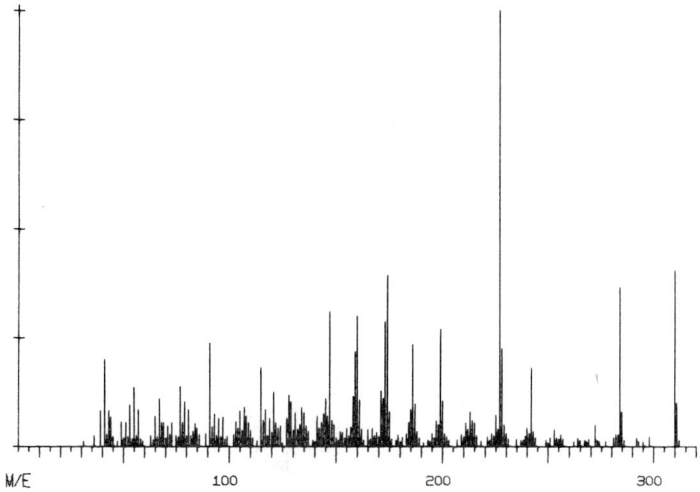
M/E
100
200
300

MESTRANOL

$C_{21}H_{26}O_2$ MW = 310

Base Peak = 227 41 Peaks

Mass	Int.	Mass	Int.	Mass	Int.	Mass	Int.	Mass	Int.	Mass	Int.	Mass	Int.	Mass	Int.	Mass	Int.
31	12	39	83	41	199	53	95	55	136	65	70	67	109	77	138	69	103
91	238	93	75	107	89	115	182	121	125	128	117	134	91	145	109	147	309
159	218	160	299	173	288	174	392	186	234	199	270	200	104	213	79	214	60
227	999	228	225	240	42	241	182	253	41	256	28	264	20	269	18	284	364
285	79	292	19	298	22	310	403	311	99								

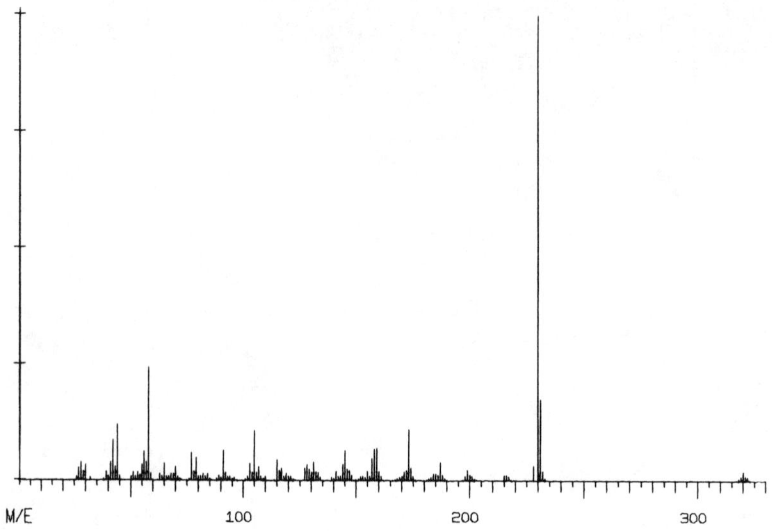
M/E
100
200
300

PHENAZOCINE

$C_{22}H_{27}NO$ MW = 321

Base Peak = 230 34 Peaks

Mass	Int.	Mass	Int.	Mass	Int.	Mass	Int.	Mass	Int.	Mass	Int.	Mass	Int.	Mass	Int.	Mass	Int.
28	41	30	34	42	87	44	119	56	62	58	242	65	37	70	29	77	60
79	50	91	65	103	37	105	107	115	46	128	34	131	40	144	34	145	64
158	67	159	69	172	23	173	110	174	27	187	40	188	12	199	23	202	4
215	13	216	12	228	33	230	999	231	176	319	10	320	19				

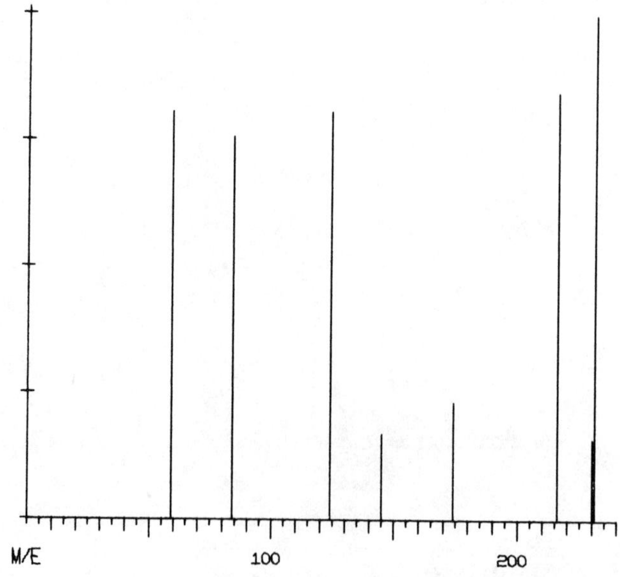
M/E
100
200

METAZOCINE

$C_{15}H_{21}NO$ MW = 231

Base Peak = 231 8 Peaks

Mass	Int.	Mass	Int.	Mass	Int.	Mass	Int.	Mass	Int.	Mass	Int.	Mass	Int.	Mass	Int.	Mass	Int.
59	808	84	757	124	808	145	171	174	232	216	848	230	161	231	999		

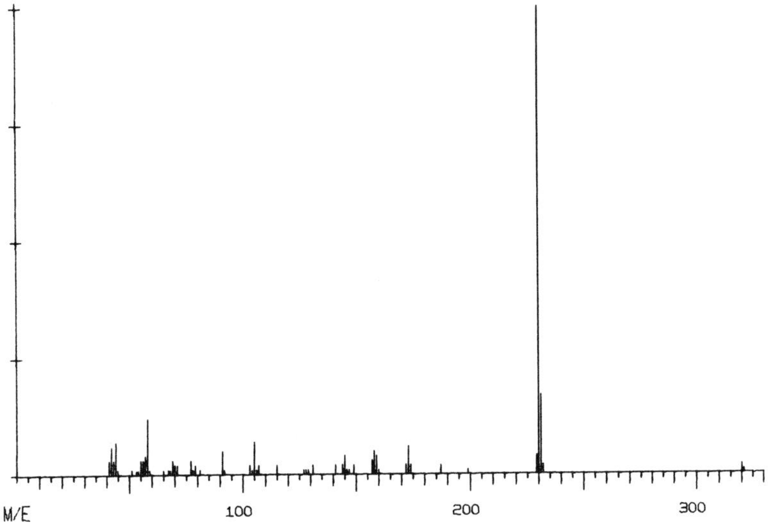
M/E
100
200
300

PHENAZOCINE

$C_{22}H_{27}NO$ MW = 321

Base Peak = 230 28 Peaks

Mass	Int.	Mass	Int.	Mass	Int.	Mass	Int.	Mass	Int.	Mass	Int.	Mass	Int.	Mass	Int.	Mass	Int.
42	60	44	70	57	40	58	120	69	30	70	20	77	30	79	20	91	50
103	20	105	70	107	20	127	10	131	20	141	20	145	40	158	50	159	40
172	20	173	60	174	20	187	20	199	10	229	40	230	999	231	170	320	20
321	10																

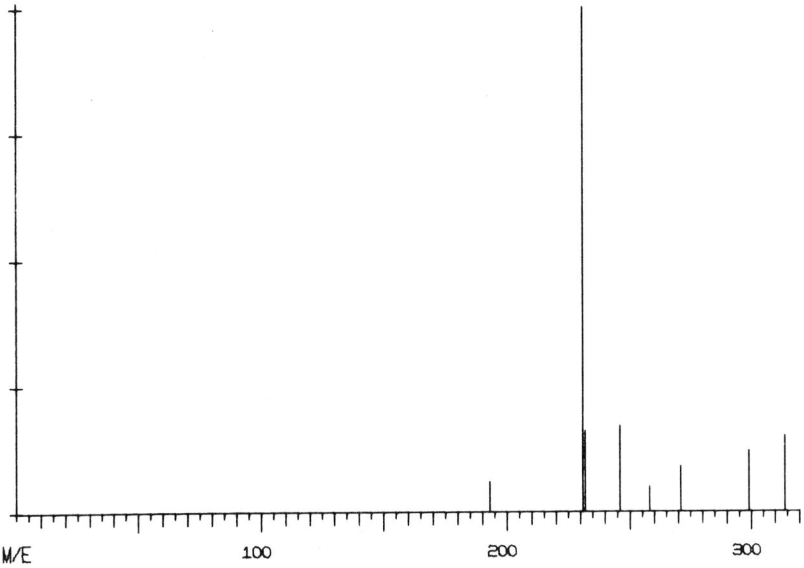
M/E
100
200
300

CANNABIDIOL

$C_{21}H_{30}O_2$ MW = 314

Base Peak = 231 8 Peaks

Mass	Int.	Mass	Int.	Mass	Int.	Mass	Int.	Mass	Int.	Mass	Int.	Mass	Int.	Mass	Int.
193	60	231	999	232	161	246	171	258	50	271	90	299	121	314	151

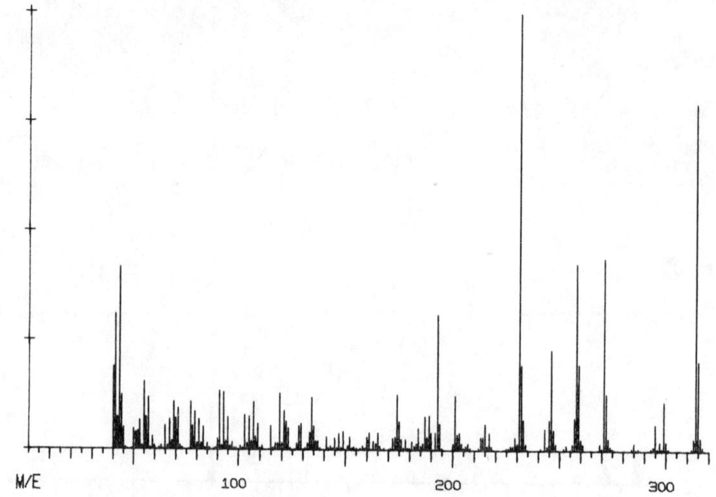
M/E
100
200
300

TETRAHYDROCANNABINOL-DELTA-8

$C_{21}H_{30}O_2$ MW = 314

Base Peak = 231 42 Peaks

Mass	Int.	Mass	Int.	Mass	Int.	Mass	Int.	Mass	Int.	Mass	Int.	Mass	Int.	Mass	Int.	Mass	Int.
41	310	43	417	55	154	57	120	69	110	71	94	77	110	79	87	91	135
93	132	105	78	107	110	119	131	121	91	134	121	135	54	147	37	149	42
161	41	165	41	174	127	187	77	193	311	201	124	203	41	215	59	217	39
229	29	231	999	232	194	246	229	257	74	258	427	271	440	272	131	273	27
295	59	299	112	300	21	313	27	314	795	315	204						

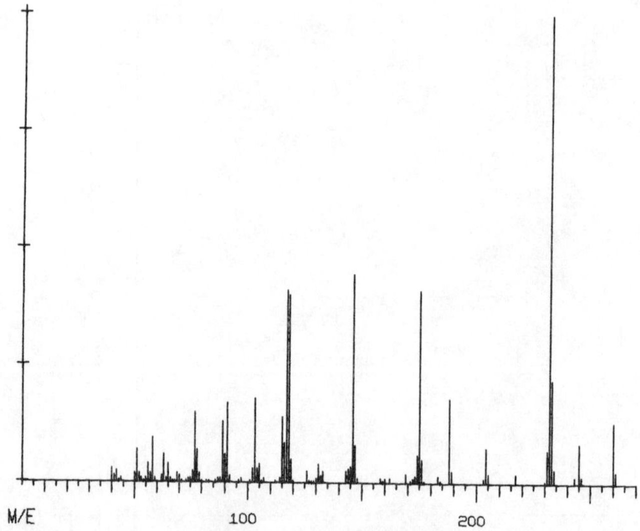
M/E
100
200

PHENOBARBITAL 1,3-DIMETHYL DERIVATIVE

$C_{14}H_{16}N_2O_3$ MW = 260

Base Peak = 232 33 Peaks

Mass	Int.	Mass	Int.	Mass	Int.	Mass	Int.	Mass	Int.	Mass	Int.	Mass	Int.	Mass	Int.	Mass	Int.
40	30	42	25	51	70	58	95	63	60	65	40	77	150	89	130	91	170
103	180	115	140	117	410	118	400	119	50	144	30	145	35	146	445	147	80
169	20	173	15	174	60	175	410	188	180	189	25	204	75	205	20	217	20
232	999	233	220	245	85	246	15	260	130	261	25						

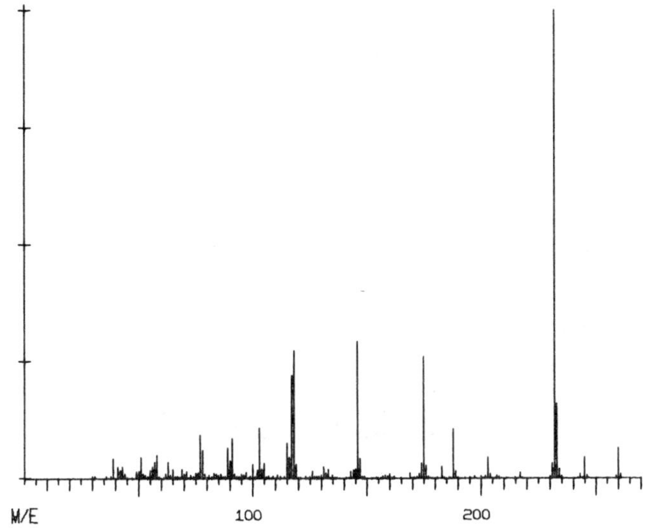
M/E
100
200

N,N-DIMETHYL PHENOBARBITAL

$C_{14}H_{16}N_2O_3$ MW = 260

Base Peak = 232 30 Peaks

Mass	Int.	Mass	Int.	Mass	Int.	Mass	Int.	Mass	Int.	Mass	Int.	Mass	Int.	Mass	Int.	Mass	Int.
30	4	31	4	39	42	41	26	51	46	58	50	63	36	69	21	77	93
89	65	91	84	103	108	115	75	117	219	118	273	119	31	133	19	145	19
146	292	147	43	160	9	169	13	174	33	175	261	188	104	189	15	203	46
204	11	217	13	218	3	232	999	233	159	245	46	246	7	260	65	261	11

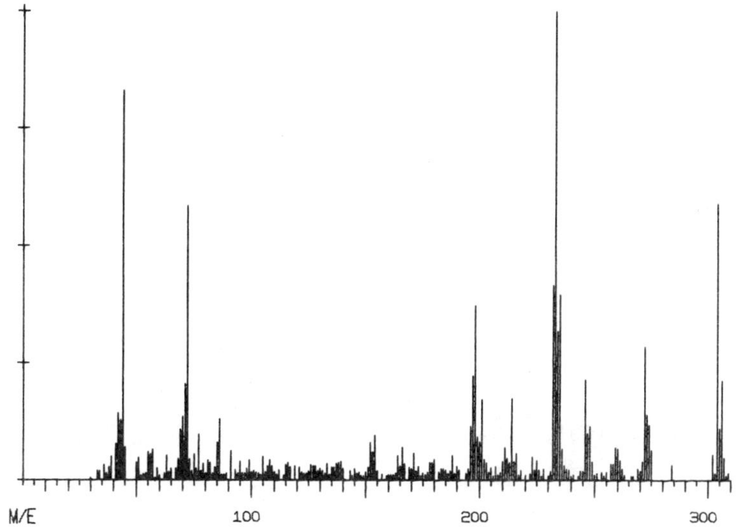
M/E
100
200
300

CHLORPROMAZINE MTB. 1

$C_{12}H_8CLNS$ MW = 233

Base Peak = 233 31 Peaks

Mass	Int.	Mass	Int.	Mass	Int.	Mass	Int.	Mass	Int.	Mass	Int.	Mass	Int.	Mass	Int.	Mass	Int.
30	6	39	25	45	50	50	23	58	26	63	52	69	57	77	46	85	35
95	26	99	106	108	17	117	103	118	32	127	30	139	27	140	25	153	45
154	143	166	94	171	92	174	15	180	16	198	626	201	262	202	54	203	76
223	3	227	4	233	999	235	384										

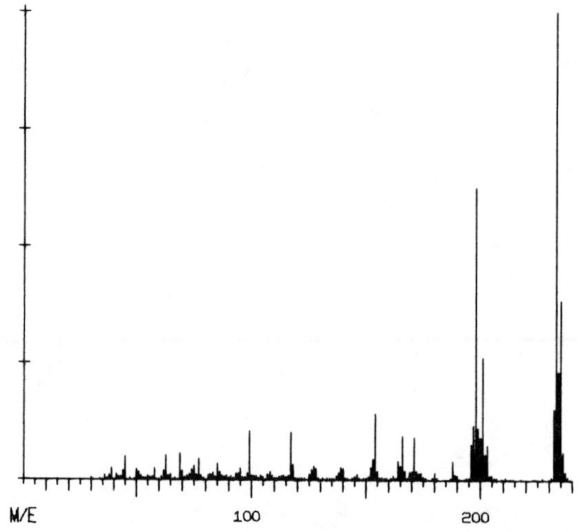

M/E
100
200

CHLORPROMAZINE MTB. 2

$C_{16}H_{17}CLN_2S$ MW = 304

Base Peak = 233 40 Peaks

Mass	Int.	Mass	Int.	Mass	Int.	Mass	Int.	Mass	Int.	Mass	Int.	Mass	Int.	Mass	Int.	Mass	Int.
30	6	33	19	42	142	44	829	55	60	57	64	71	205	72	586	77	97
86	130	91	62	99	42	105	50	108	42	119	30	126	32	138	38	139	41
152	79	154	95	166	69	171	57	179	38	180	44	197	222	198	372	211	71
214	175	216	58	223	50	232	415	233	999	246	214	248	115	259	71	260	67
272	286	273	139	304	591	306	213										

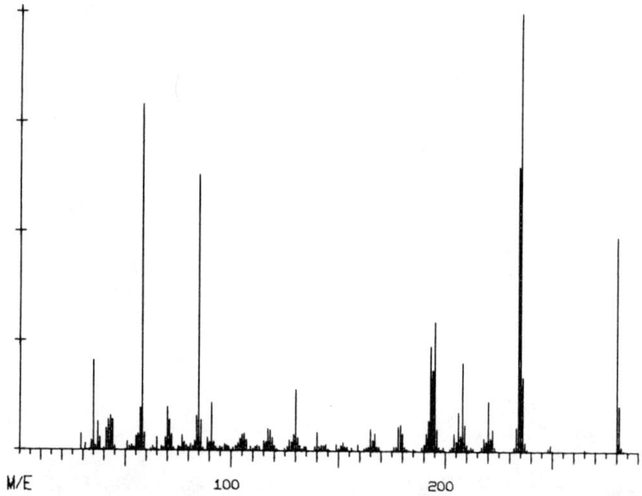

M/E
100
200

IMIPRAMINE

$C_{19}H_{24}N_2$ MW = 280

Base Peak = 235 38 Peaks

Mass	Int.	Mass	Int.	Mass	Int.	Mass	Int.	Mass	Int.	Mass	Int.	Mass	Int.	Mass	Int.	Mass	Int.
29	38	31	14	35	204	43	81	57	97	58	789	70	99	71	69	84	79
85	629	90	20	91	109	106	39	117	49	118	47	130	141	140	43	144	14
149	16	152	19	165	49	167	39	178	54	179	59	193	239	195	294	206	91
208	203	220	114	222	49	234	649	235	999	248	7	249	14	264	1	265	6
280	489	281	104														

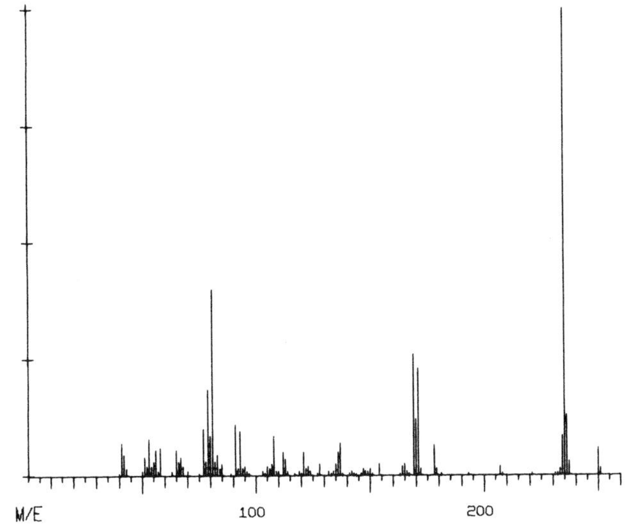
M/E
100
200

HEXOBARBITAL-3-METHYL DERIVATIVE

$C_{13}H_{18}N_2O_3$ MW = 250

Base Peak = 235 30 Peaks

Mass	Int.	Mass	Int.	Mass	Int.	Mass	Int.	Mass	Int.	Mass	Int.	Mass	Int.	Mass	Int.	Mass	Int.
41	70	42	45	53	80	58	60	65	55	67	40	79	185	81	400	91	110
93	95	108	85	112	50	121	50	128	25	136	50	137	70	147	15	154	25
169	260	171	230	178	65	179	15	193	5	207	20	208	5	221	5	235	999
236	130	250	60	251	15												

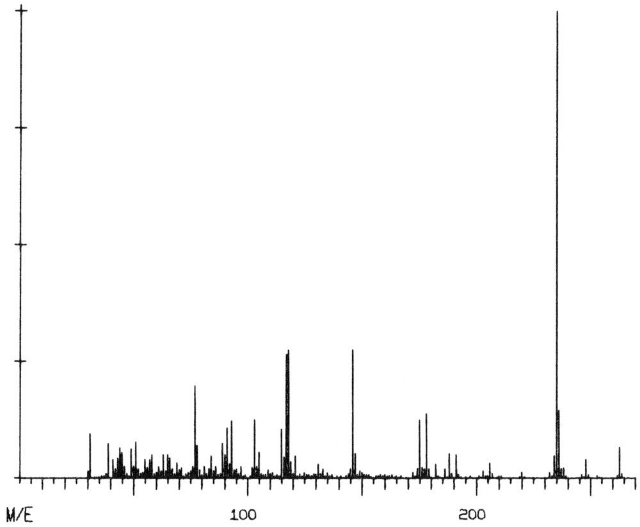
M/E
100
200

N-Me,N-D3-Me PHENOBARBITAL

$C_{14}H_{13}D_3N_2O_3$ MW = 263

Base Peak = 235 36 Peaks

Mass	Int.	Mass	Int.	Mass	Int.	Mass	Int.	Mass	Int.	Mass	Int.	Mass	Int.	Mass	Int.	Mass	Int.
30	15	31	95	39	76	44	66	49	62	51	77	63	51	65	50	77	197
89	75	93	123	103	124	115	104	117	266	118	274	121	47	133	20	145	19
146	276	147	53	160	7	172	13	175	124	178	137	188	52	191	51	203	16
206	33	219	3	220	12	235	999	236	145	248	41	249	8	263	65	264	9

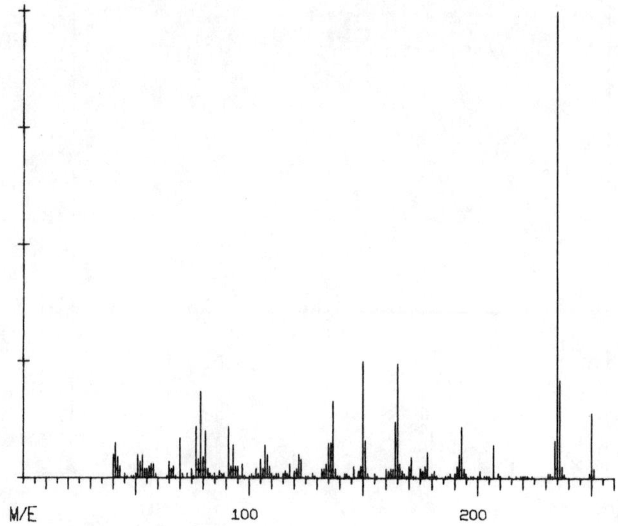
M/E
100
200

HEXOBARBITAL-2(OR 4)-METHYL DERIVATIVE

$C_{13}H_{18}N_2O_3$ MW = 250

Base Peak = 235 32 Peaks

Mass	Int.	Mass	Int.	Mass	Int.	Mass	Int.	Mass	Int.	Mass	Int.	Mass	Int.	Mass	Int.	Mass	Int.
40	50	41	75	51	50	53	50	65	35	70	85	77	110	79	185	91	110
93	70	107	70	108	50	122	50	123	40	135	75	137	165	150	250	151	80
164	120	165	245	177	25	178	55	192	50	193	110	207	70	209	10	217	5
219	5	235	999	236	210	250	140	251	20								

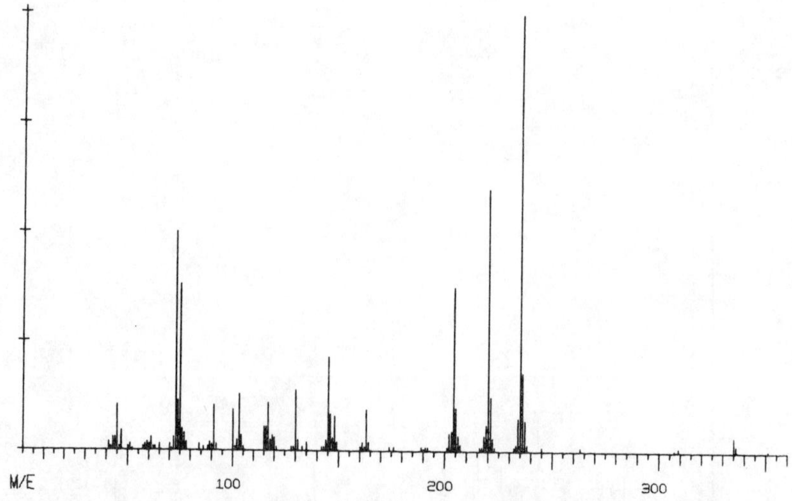
M/E
100
200
300

ETHYLPHENYLMALONDIAMIDE DI-TMS DERIVATIVE (FROM PRIMIDONE)

$C_{17}H_{30}N_2O_2Si_2$ MW = 350

Base Peak = 235 37 Peaks

Mass	Int.	Mass	Int.	Mass	Int.	Mass	Int.	Mass	Int.	Mass	Int.	Mass	Int.	Mass	Int.	Mass	Int.
45	104	47	44	59	19	61	29	73	499	75	379	76	49	77	39	91	104
103	129	115	54	117	109	119	34	130	139	144	24	145	214	146	84	148	79
161	19	163	94	174	9	176	9	189	9	191	9	204	374	205	99	220	599
221	124	235	999	236	179	245	9	263	9	307	4	309	9	335	34	336	14
351	4																

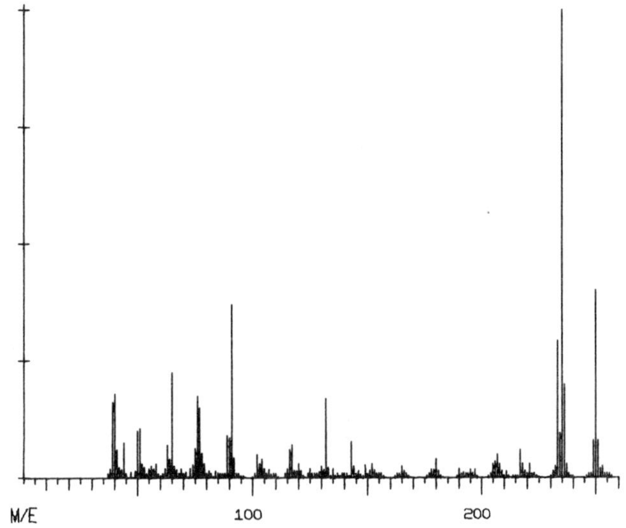
M/E
100
200

METHAQUALONE

$C_{16}H_{14}N_2O$ MW = 250

Base Peak = 235 32 Peaks

Mass	Int.	Mass	Int.	Mass	Int.	Mass	Int.	Mass	Int.	Mass	Int.	Mass	Int.	Mass	Int.	Mass	Int.
39	162	40	180	50	100	51	106	63	70	65	224	76	175	77	150	90	84
91	370	116	60	117	70	120	29	130	25	132	170	143	77	149	28	152	29
165	26	166	15	178	18	180	40	190	20	195	18	206	35	207	50	217	60
218	30	233	292	235	999	249	81	250	400								

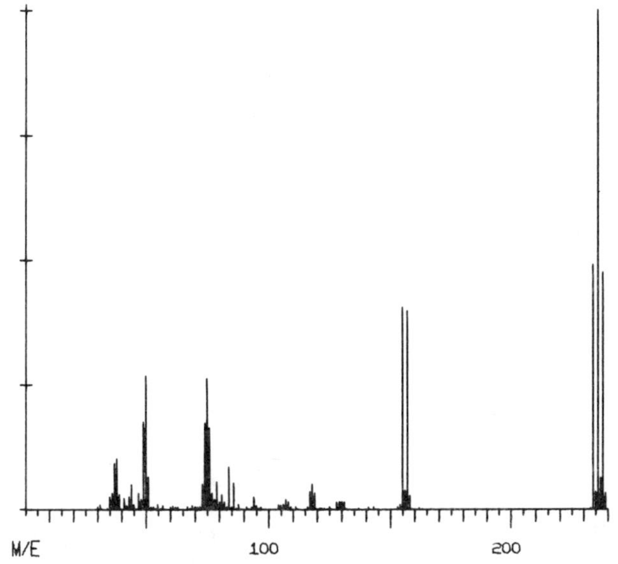
M/E
100
200

p-DIBROMOBENZENE

$C_6H_4Br_2$ MW = 234

Base Peak = 236 23 Peaks

Mass	Int.	Mass	Int.	Mass	Int.	Mass	Int.	Mass	Int.	Mass	Int.	Mass	Int.	Mass	Int.	Mass	Int.
30	4	31	10	37	93	38	103	49	174	50	268	74	173	75	262	76	162
84	84	94	25	95	8	107	19	117	34	118	50	119	33	141	5	143	4
155	406	157	398	162	2	234	489	236	999								

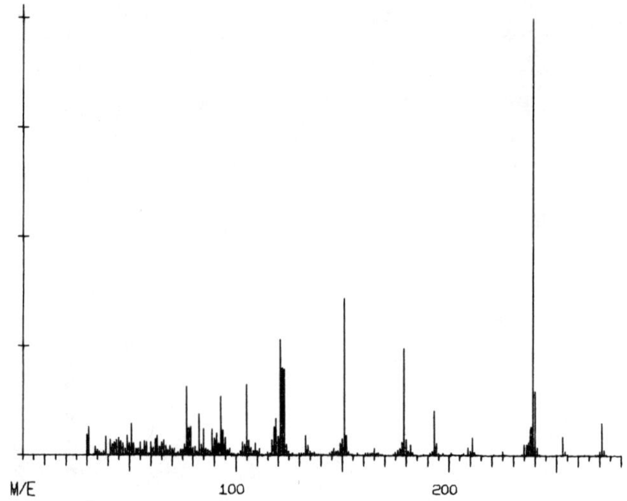
M/E
100
200

N,N-D3-ME-PHENOBARBITAL (D5-ET)

$C_{14}H_5D_{11}N_2O_3$ MW = 271

Base Peak = 239 30 Peaks

Mass	Int.	Mass	Int.	Mass	Int.	Mass	Int.	Mass	Int.	Mass	Int.	Mass	Int.	Mass	Int.	Mass	Int.
30	47	31	64	39	43	45	40	49	44	51	72	62	38	63	46	77	158
83	94	93	135	94	57	105	163	117	34	121	266	122	199	133	45	134	22
151	361	152	45	165	14	166	6	179	244	180	35	193	102	194	28	209	18
211	41	221	2	225	11	239	999	240	148	253	42	254	11	270	9	271	75
272	12	273	3														

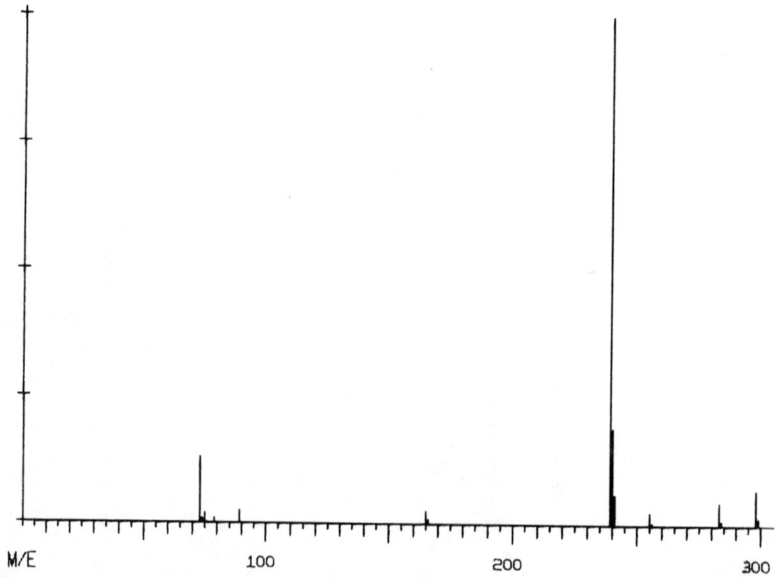
M/E
100
200
300

3,4-DIMETHOXYMANDELIC ACID METHYL ESTER TRIMETHYLSILYL ETHER

$C_{14}H_{22}O_5$ MW = 270

Base Peak = 239 14 Peaks

Mass	Int.	Mass	Int.	Mass	Int.	Mass	Int.	Mass	Int.	Mass	Int.	Mass	Int.	Mass	Int.	Mass	Int.
73	129	75	19	79	9	89	24	165	24	166	9	239	999	240	189	255	24
256	4	283	44	284	9	298	69	299	14								

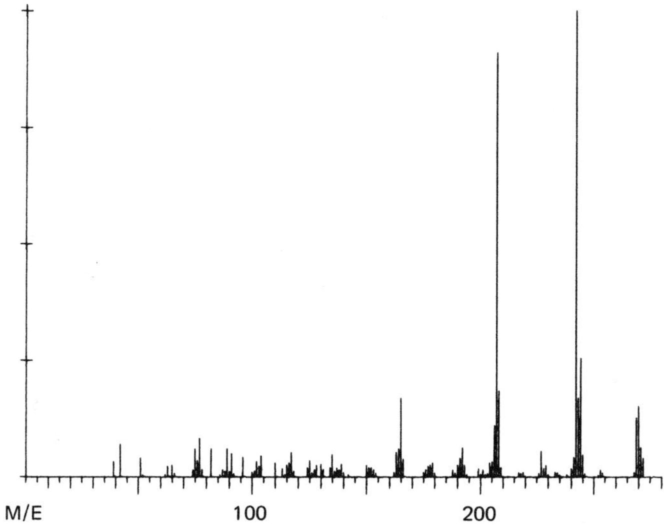
M/E
100
200

MEDAZEPAM

$C_{16}H_{15}CLN_2$ MW = 270

Base Peak = 242 35 Peaks

Mass	Int.	Mass	Int.	Mass	Int.	Mass	Int.	Mass	Int.	Mass	Int.	Mass	Int.	Mass	Int.	Mass	Int.
39	33	42	69	51	39	52	5	65	26	75	59	77	82	82	60	91	51
96	42	104	46	117	52	125	34	130	27	135	48	139	28	150	24	151	21
164	61	165	171	178	24	179	29	191	40	192	62	207	909	208	185	227	54
229	26	242	999	243	169	244	254	245	47	269	128	270	152	272	41		

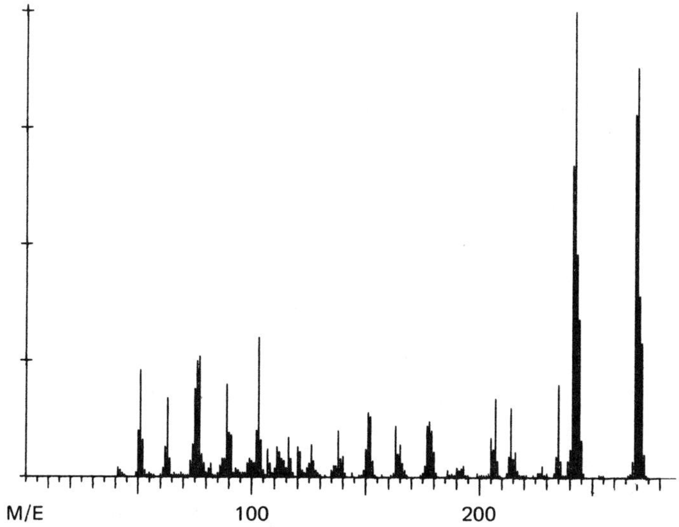
M/E
100
200

N-DESMETHYLDIAZEPAM

$C_{15}H_{11}CLN_2O$ MW = 270

Base Peak = 242 36 Peaks

Mass	Int.	Mass	Int.	Mass	Int.	Mass	Int.	Mass	Int.	Mass	Int.	Mass	Int.	Mass	Int.	Mass	Int.
41	20	42	15	50	100	51	230	63	170	75	190	76	250	77	260	102	100
103	300	104	80	116	85	120	65	126	70	138	100	140	45	151	140	152	130
163	110	165	70	177	110	178	120	190	20	193	25	207	170	214	150	216	55
228	25	241	670	242	999	244	340	245	80	269	780	270	880	272	290	273	50

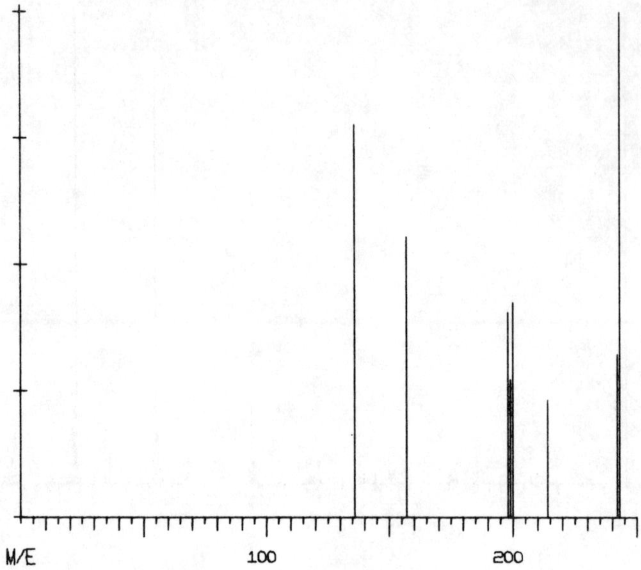
M/E
100
200

NORLEVORPHANOL

$C_{16}H_{21}NO$ MW = 243

Base Peak = 243 7 Peaks

Mass	Int.	Mass	Int.	Mass	Int.	Mass	Int.	Mass	Int.	Mass	Int.	Mass	Int.
136	777	157	555	198	404	200	424	214	232	242	323	243	999

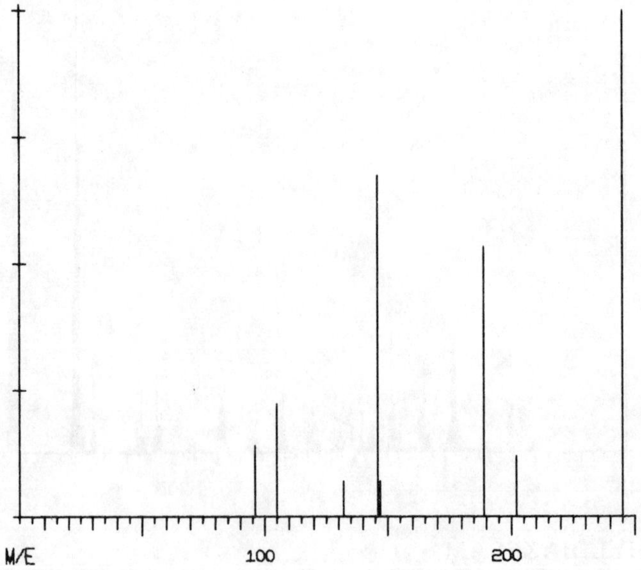
M/E
100
200

FENTANYL

$C_{22}H_{28}N_2O$ MW = 336

Base Peak = 245 8 Peaks

Mass	Int.	Mass	Int.	Mass	Int.	Mass	Int.	Mass	Int.	Mass	Int.	Mass	Int.	Mass	Int.
96	141	105	222	132	70	146	676	147	70	189	535	202	121	245	999

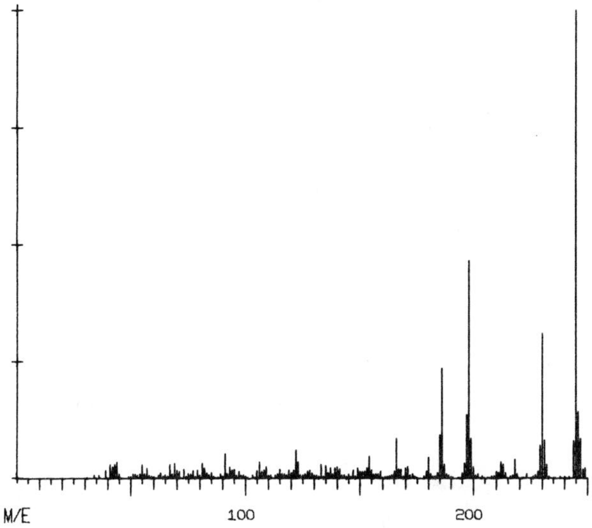
M/E
100
200

THIORIDAZINE MTB. 1

$C_{13}H_{11}NS_2$ MW = 245

Base Peak = 245 33 Peaks

Mass	Int.	Mass	Int.	Mass	Int.	Mass	Int.	Mass	Int.	Mass	Int.	Mass	Int.	Mass	Int.	Mass	Int.
31	3	41	30	44	34	55	30	57	22	67	30	69	33	81	32	82	22
91	52	93	24	106	34	109	25	122	61	123	36	133	29	135	27	153	23
154	48	166	84	171	24	185	92	186	236	197	134	198	465	212	34	213	30
218	39	229	70	230	310	231	82	245	999	246	143						

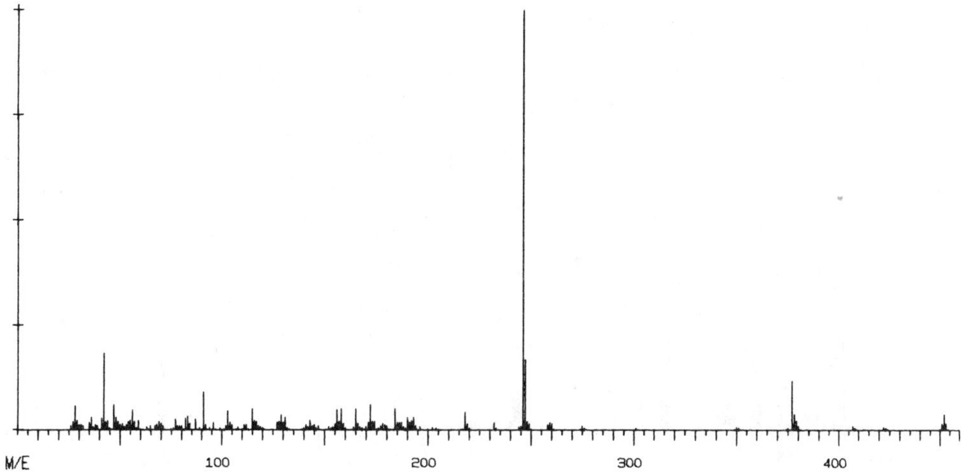
M/E
100
200
300
400

DIPHENOXYLATE

$C_{30}H_{32}N_2O_2$ MW = 452

Base Peak = 246 51 Peaks

Mass	Int.	Mass	Int.	Mass	Int.	Mass	Int.	Mass	Int.	Mass	Int.	Mass	Int.	Mass	Int.	Mass	Int.
28	57	29	23	42	182	47	59	48	30	56	47	69	19	70	15	82	28
83	32	91	91	103	44	115	49	116	23	129	35	131	30	141	12	143	22
156	48	158	49	165	51	172	60	174	19	184	49	190	29	193	29	202	5
204	5	218	43	219	14	232	18	233	5	246	999	247	167	259	18	260	14
275	11	276	5	301	4	350	7	351	4	377	117	378	38	407	9	408	4
422	8	423	4	451	14	452	38	454	3	455	1						

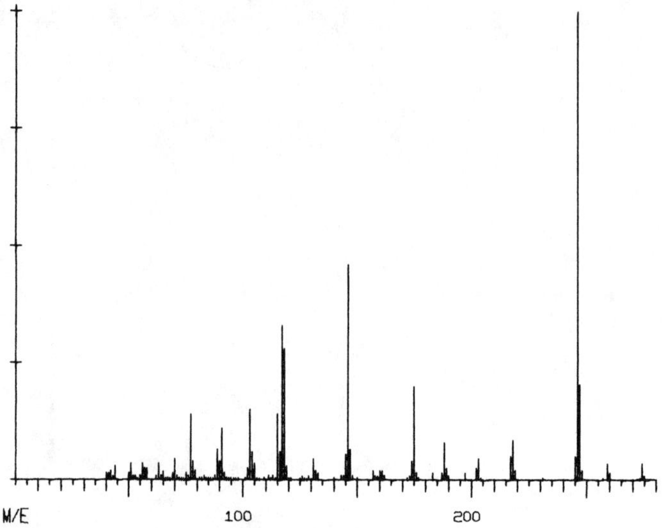
M/E
100
200

MEPHOBARBITAL 3-ETHYL DERIVATIVE

$C_{15}H_{16}N_2O_3$ MW = 274

Base Peak = 246 35 Peaks

Mass	Int.	Mass	Int.	Mass	Int.	Mass	Int.	Mass	Int.	Mass	Int.	Mass	Int.	Mass	Int.	Mass	Int.
42	20	44	30	51	35	56	35	63	35	70	45	77	140	89	65	91	110
103	150	115	140	117	330	118	280	131	45	132	20	145	55	146	460	147	65
160	20	161	20	174	40	175	200	188	80	189	25	202	25	203	45	217	50
218	85	231	5	246	999	247	205	259	35	260	15	274	35	275	10		

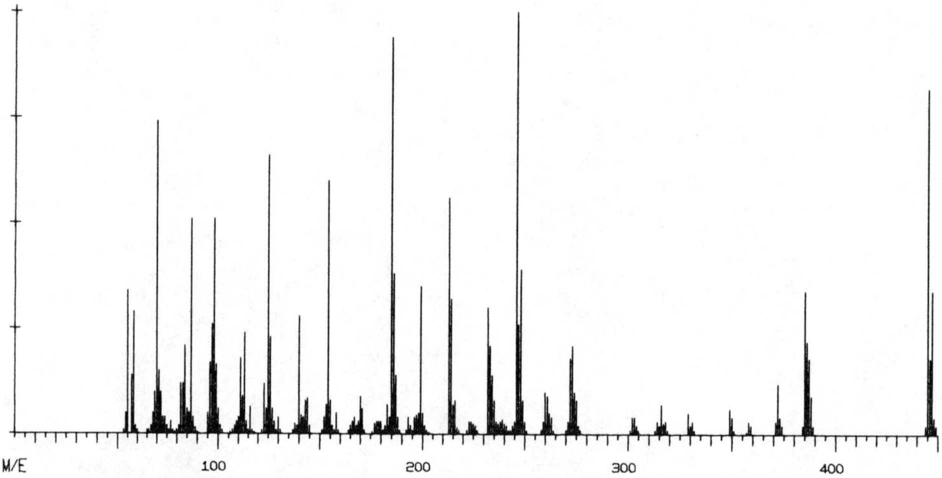
M/E
100
200
300
400

THIOPROPAZATE

$C_{23}H_{28}CLN_3O_2S$ MW = 445

Base Peak = 246 50 Peaks

Mass	Int.	Mass	Int.	Mass	Int.	Mass	Int.	Mass	Int.	Mass	Int.	Mass	Int.	Mass	Int.	Mass	Int.
55	339	58	289	70	739	71	149	84	209	87	509	97	259	98	509	111	179
113	239	125	659	126	229	140	279	144	84	154	599	155	79	170	89	171	59
185	939	186	379	198	49	199	349	213	559	214	319	216	79	223	29	232	299
233	209	246	999	248	389	260	99	261	89	272	179	273	209	302	41	303	39
314	29	316	69	329	49	331	29	349	59	350	39	358	29	359	19	372	119
373	39	385	339	386	219	445	819	447	339								

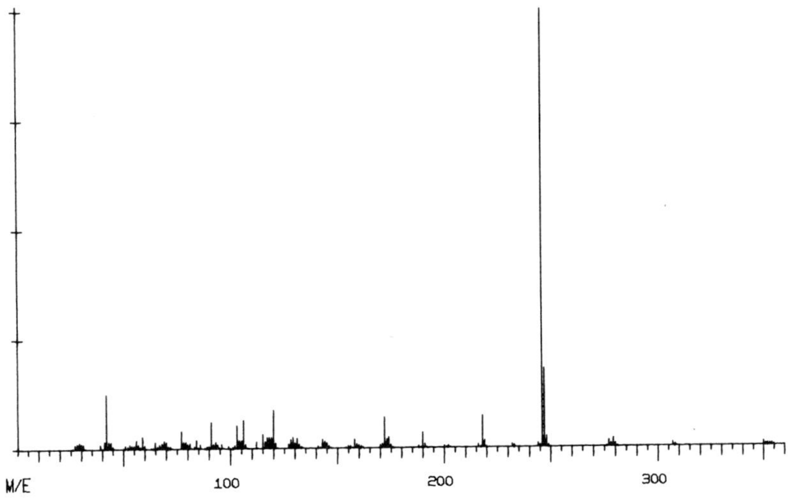
M/E
100
200
300

ANILERIDINE

$C_{22}H_{28}N_2O_2$ MW = 352

Base Peak = 246 39 Peaks

Mass	Int.	Mass	Int.	Mass	Int.	Mass	Int.	Mass	Int.	Mass	Int.	Mass	Int.	Mass	Int.	Mass	Int.
28	12	29	16	41	18	42	124	56	19	59	28	65	16	69	17	77	41
84	20	91	60	103	52	106	65	115	33	120	88	129	26	143	19	144	14
158	21	159	9	172	70	173	21	174	24	175	7	190	36	191	10	202	4
218	72	219	14	232	8	233	4	246	999	247	180	277	14	279	20	307	10
308	4	350	10	351	6												

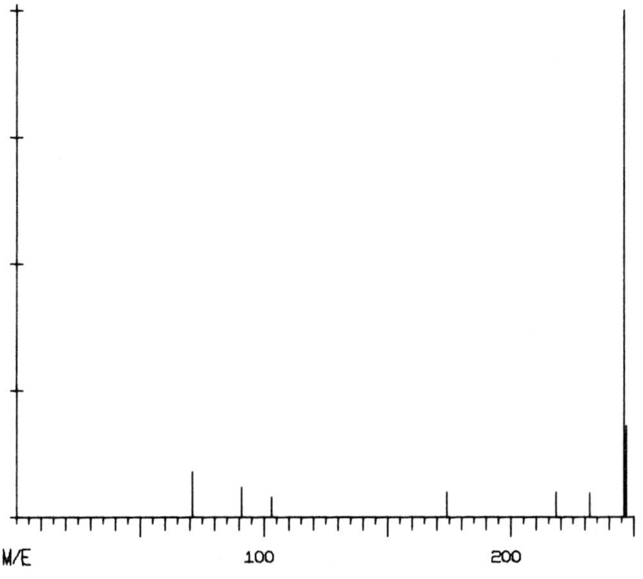
M/E
100
200

FURETHIDINE

$C_{21}H_{31}NO_4$ MW = 361

Base Peak = 246 8 Peaks

Mass	Int.	Mass	Int.	Mass	Int.	Mass	Int.	Mass	Int.	Mass	Int.	Mass	Int.	Mass	Int.
71	90	91	60	103	40	174	50	218	50	232	50	246	999	247	181

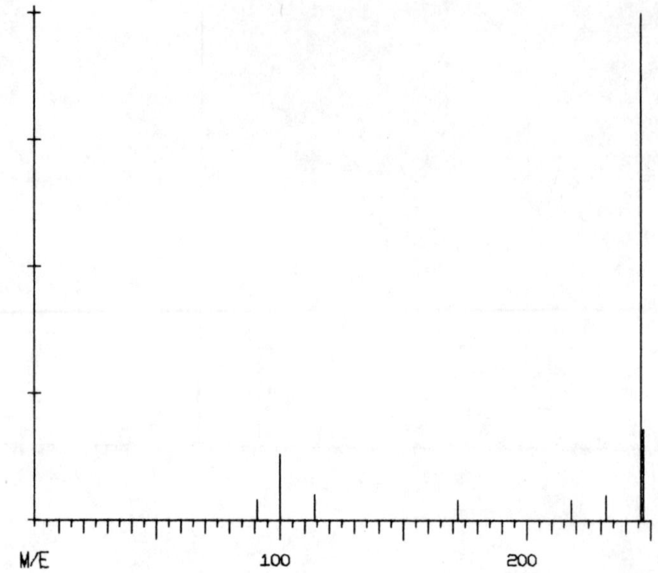
M/E
100
200

MORPHERIDINE

$C_{20}H_{30}N_2O_3$ MW = 346

Base Peak = 246 8 Peaks

Mass	Int.	Mass	Int.	Mass	Int.	Mass	Int.	Mass	Int.	Mass	Int.	Mass	Int.	Mass	Int.
91	40	100	131	114	50	172	40	218	40	232	50	246	999	247	181

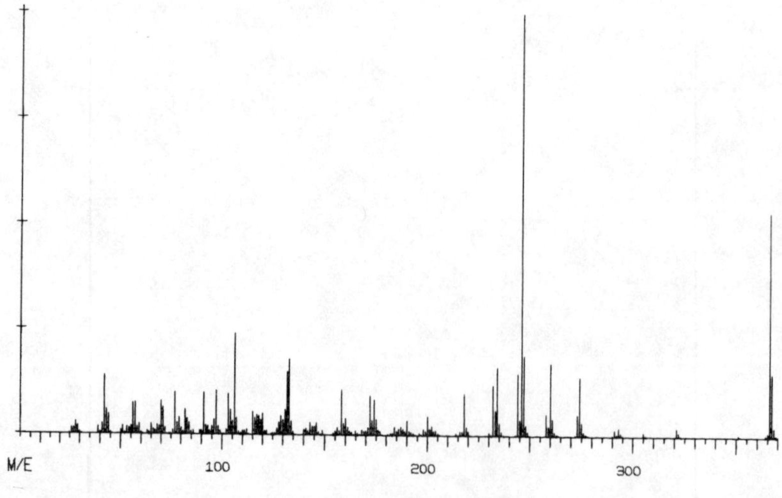
M/E
100
200
300

PIMINODINE

$C_{23}H_{30}N_2O_2$ MW = 366

Base Peak = 246 48 Peaks

Mass	Int.	Mass	Int.	Mass	Int.	Mass	Int.	Mass	Int.	Mass	Int.	Mass	Int.	Mass	Int.	Mass	Int.
28	29	29	19	42	140	43	60	56	74	57	78	70	80	71	65	77	100
82	61	91	100	97	104	104	60	106	240	120	49	131	61	132	149	133	180
146	30	158	108	160	40	172	94	174	84	175	37	190	34	200	46	202	22
204	12	218	99	219	21	232	120	234	162	246	999	247	190	258	52	260	172
273	49	274	140	291	16	293	21	305	9	306	2	321	19	322	11	351	4
366	532	367	150	370	2												

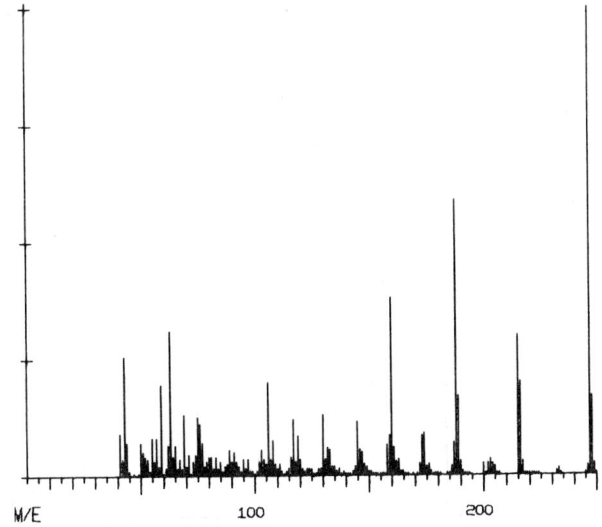

M/E
100
200

1,2-DIMETHYL-5-METHOXYINDOLE-3-ACETIC ACID METHYL ESTER (FROM INDOMETHACIN)

$C_{14}H_{17}NO_3$ MW = 247

Base Peak = 247 32 Peaks

Mass	Int.	Mass	Int.	Mass	Int.	Mass	Int.	Mass	Int.	Mass	Int.	Mass	Int.	Mass	Int.	Mass	Int.
41	90	43	255	55	80	59	195	63	310	69	130	76	110	77	70	91	50
103	55	106	200	117	120	119	85	130	130	132	60	145	115	158	65	159	85
160	380	173	85	174	90	187	70	188	590	189	170	203	35	215	300	216	200
217	30	232	10	233	15	247	999	248	170								

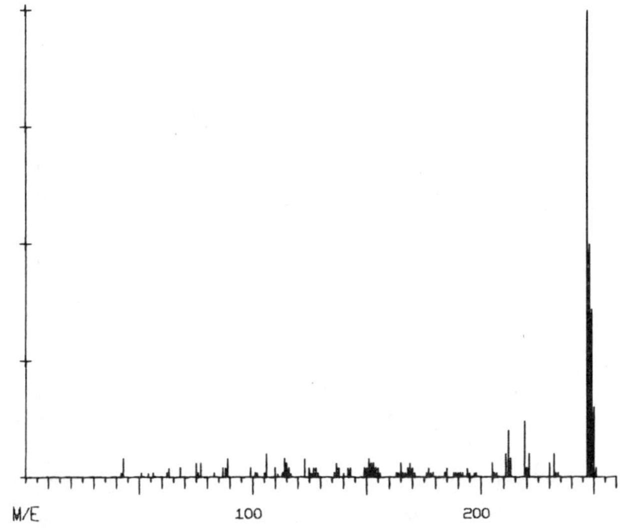

M/E
100
200

PYRIMETHAMINE

$C_{12}H_{13}CLN_4$ MW = 248

Base Peak = 247 32 Peaks

Mass	Int.	Mass	Int.	Mass	Int.	Mass	Int.	Mass	Int.	Mass	Int.	Mass	Int.	Mass	Int.	Mass	Int.
42	10	43	40	51	10	54	10	63	20	75	30	77	30	89	40	99	20
101	10	106	50	114	40	123	40	125	20	137	30	138	20	151	40	152	30
165	30	169	30	177	20	185	20	188	10	194	20	211	50	212	100	219	120
221	50	230	30	232	50	247	999	248	500								

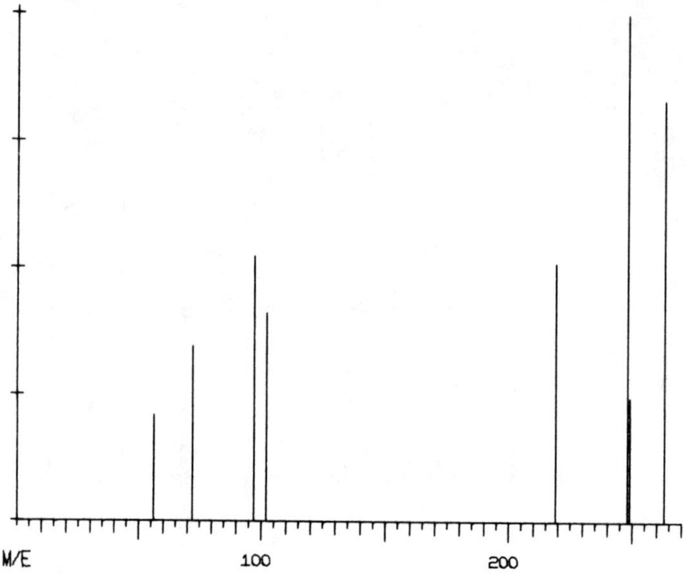
M/E
100
200

DIMETHYLTHIAMBUTENE

$C_{14}H_{17}NS_2$ MW = 263
Base Peak = 248 8 Peaks

Mass	Int.	Mass	Int.	Mass	Int.	Mass	Int.	Mass	Int.	Mass	Int.	Mass	Int.	Mass	Int.
56	211	72	344	97	522	102	411	219	511	248	999	249	244	263	833

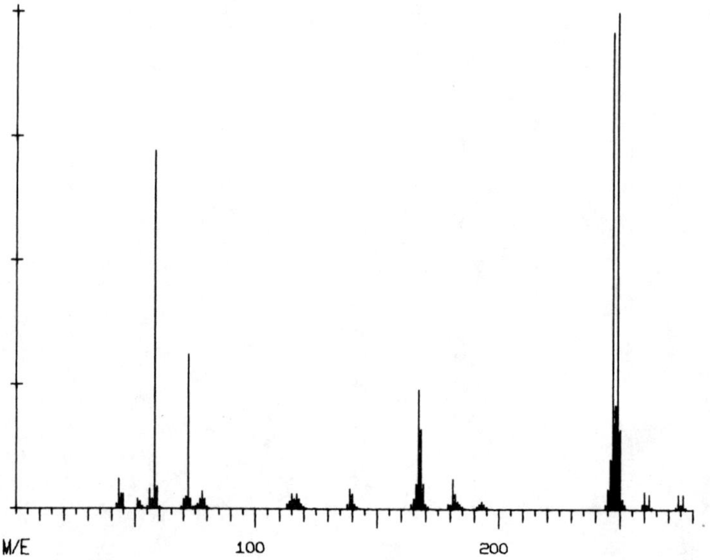
M/E
100
200

PARABROMDYLAMINE

$C_{16}H_{19}Br_1N_2$ MW = 319
Base Peak = 249 26 Peaks

Mass	Int.	Mass	Int.	Mass	Int.	Mass	Int.	Mass	Int.	Mass	Int.	Mass	Int.	Mass	Int.	Mass	Int.
43	59	44	29	58	719	59	44	71	24	72	309	77	19	78	34	115	29
117	29	118	19	119	9	139	39	140	29	167	239	168	159	181	59	182	29
192	9	193	14	247	959	249	999	260	34	262	29	274	29	276	29		

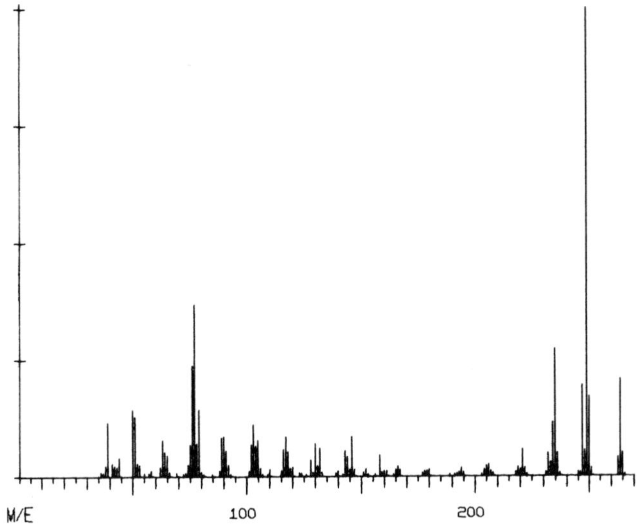
M/E
100
200

ETHINAZONE

$C_{17}H_{16}N_2O$ MW = 264

Base Peak = 249 34 Peaks

Mass	Int.	Mass	Int.	Mass	Int.	Mass	Int.	Mass	Int.	Mass	Int.	Mass	Int.	Mass	Int.	Mass	Int.
39	114	44	40	50	143	51	127	63	78	75	68	76	237	77	367	90	85
103	110	105	77	117	84	118	52	130	69	132	59	143	56	146	86	158	46
166	22	167	15	178	13	180	14	193	10	194	17	205	23	206	24	219	20
221	57	234	115	235	272	247	195	249	999	264	207	265	51				

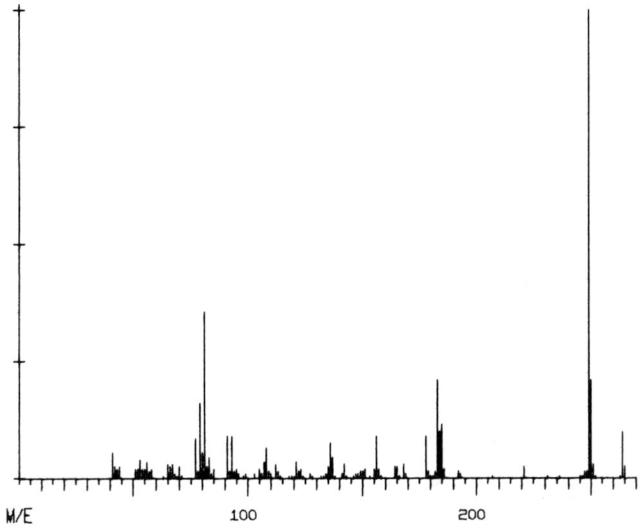
M/E
100
200

HEXOBARBITAL 3-ETHYL DERIVATIVE

$C_{14}H_{20}N_2O_3$ MW = 264

Base Peak = 249 32 Peaks

Mass	Int.	Mass	Int.	Mass	Int.	Mass	Int.	Mass	Int.	Mass	Int.	Mass	Int.	Mass	Int.	Mass	Int.
41	54	42	24	53	39	56	34	65	29	67	29	79	159	81	354	91	89
93	89	107	34	108	64	121	34	123	19	136	74	137	44	151	19	156	89
164	24	168	29	183	209	185	114	192	14	193	9	207	4	221	24	231	4
236	4	249	999	250	209	264	99	265	24								

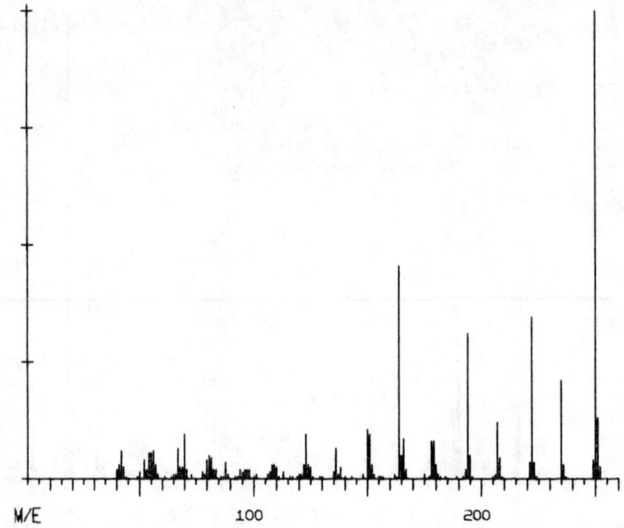
M/E
100
200

8-METHYLXANTHINE 1,3,7-TRI-ETHYL DERIVATIVE

$C_{12}H_{18}N_4O_2$ MW = 250

Base Peak = 250 32 Peaks

Mass	Int.	Mass	Int.	Mass	Int.	Mass	Int.	Mass	Int.	Mass	Int.	Mass	Int.	Mass	Int.	Mass	Int.
41	30	42	60	54	55	56	60	67	65	70	95	81	50	82	45	94	20
96	20	108	30	109	30	122	30	123	95	136	65	138	25	150	105	151	95
164	455	166	85	178	80	179	80	194	310	195	50	207	120	208	45	221	35
222	345	235	210	236	30	250	999	251	130								

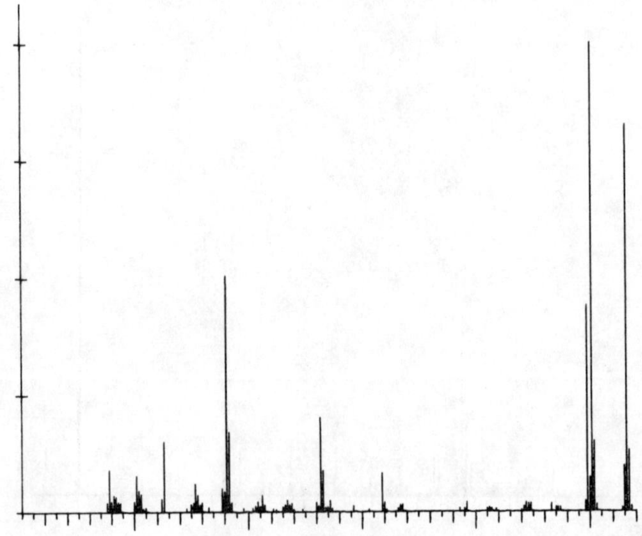
M/E
100
200

METHAQUALONE MTB. 5

$C_{16}H_{14}N_2O_2$ MW = 266

Base Peak = 251 32 Peaks

Mass	Int.	Mass	Int.	Mass	Int.	Mass	Int.	Mass	Int.	Mass	Int.	Mass	Int.	Mass	Int.	Mass	Int.
41	22	44	19	51	50	52	30	63	104	65	155	77	48	89	42	91	304
92	104	106	27	117	26	118	12	130	14	132	151	133	58	146	7	159	61
160	12	167	15	195	6	196	18	205	11	206	13	222	22	223	21	233	27
235	10	249	464	251	999	266	663	267	124								

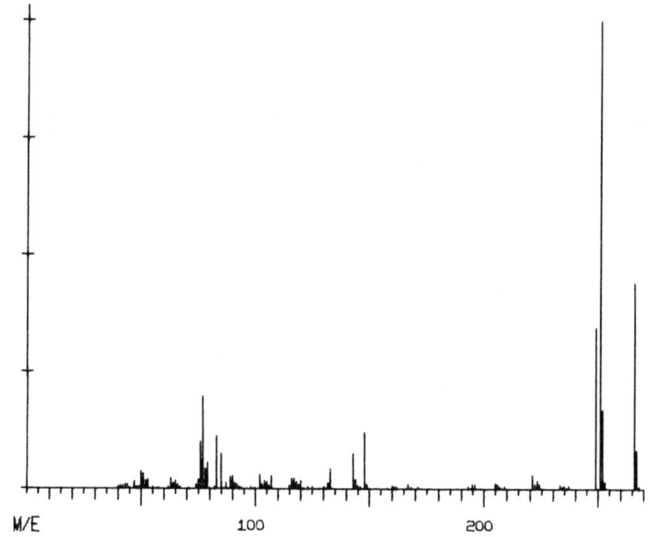
M/E
100
200

METHAQUALONE MTB. 3

$C_{16}H_{14}N_2O_2$ MW = 266

Base Peak = 251 34 Peaks

Mass	Int.	Mass	Int.	Mass	Int.	Mass	Int.	Mass	Int.	Mass	Int.	Mass	Int.	Mass	Int.	Mass	Int.
44	10	47	16	50	38	51	32	63	23	75	20	77	198	83	112	90	28
102	30	107	27	116	23	118	15	120	18	133	42	143	76	148	119	149	9
160	8	167	11	178	1	179	1	195	9	196	9	205	13	206	9	221	30
223	17	233	10	237	8	249	345	251	999	266	441	267	83				

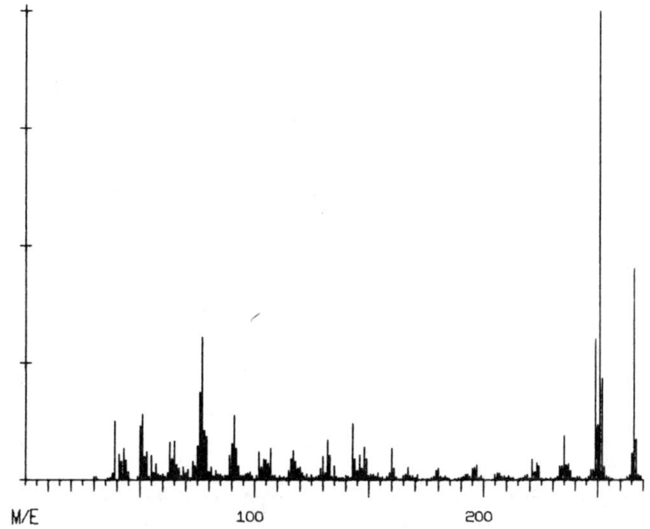
M/E
100
200

METHAQUALONE MTB. 2

$C_{16}H_{14}N_2O_2$ MW = 266

Base Peak = 251 36 Peaks

Mass	Int.	Mass	Int.	Mass	Int.	Mass	Int.	Mass	Int.	Mass	Int.	Mass	Int.	Mass	Int.	Mass	Int.
30	8	31	7	39	124	43	67	50	115	51	141	63	79	65	82	76	187
77	304	90	78	91	138	107	68	117	62	118	41	130	51	132	84	143	121
146	53	148	71	160	67	167	28	179	19	180	24	196	27	197	32	206	14
207	16	221	46	223	37	235	94	237	36	249	303	251	999	266	453	267	88

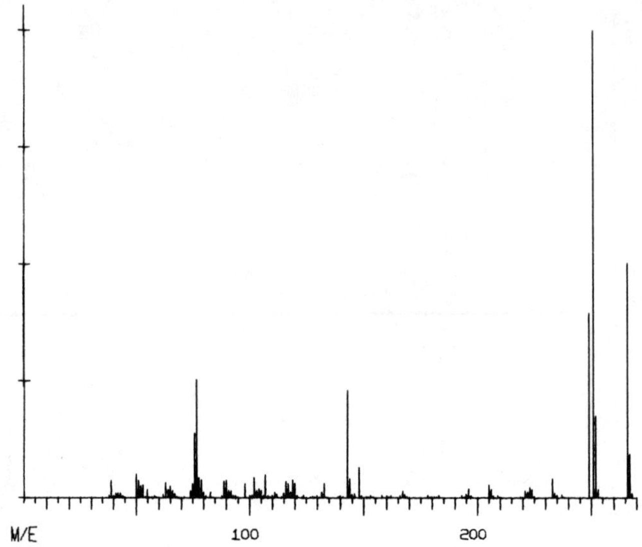
M/E
100
200

METHAQUALONE MTB. 4

$C_{16}H_{14}N_2O_2$ MW = 266

Base Peak = 251 34 Peaks

Mass	Int.	Mass	Int.	Mass	Int.	Mass	Int.	Mass	Int.	Mass	Int.	Mass	Int.	Mass	Int.	Mass	Int.
39	37	41	11	50	51	51	37	63	32	75	31	76	137	77	252	90	37
102	42	107	50	116	35	119	37	120	30	143	231	144	40	146	9	148	64
167	16	168	8	178	4	183	4	195	8	196	21	205	27	206	17	223	23
224	17	233	40	234	11	249	395	251	999	266	502	267	92				

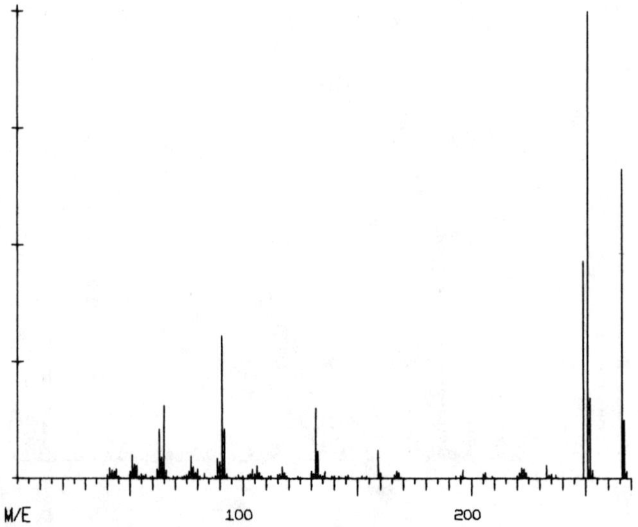
M/E
100
200

METHAQUALONE MTB. 5 (GC)

$C_{16}H_{14}N_2O_2$ MW = 266

Base Peak = 251 32 Peaks

Mass	Int.	Mass	Int.	Mass	Int.	Mass	Int.	Mass	Int.	Mass	Int.	Mass	Int.	Mass	Int.	Mass	Int.
39	89	41	36	51	77	52	44	62	27	63	151	77	60	89	49	91	506
92	169	106	36	117	25	119	18	130	20	132	203	133	88	146	13	159	82
160	20	167	16	195	8	196	19	206	9	207	8	222	19	223	18	233	18
237	11	249	440	251	999	266	824	267	129								

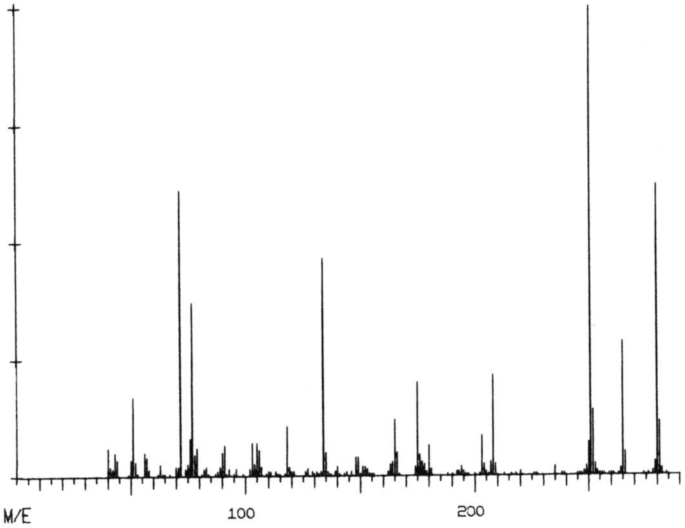
M/E
100
200

DIPHENYLHYDANTOIN 1,3-DIMETHYL DERIVATIVE

$C_{17}H_{16}N_2O_2$ MW = 280

Base Peak = 251 36 Peaks

Mass	Int.	Mass	Int.	Mass	Int.	Mass	Int.	Mass	Int.	Mass	Int.	Mass	Int.	Mass	Int.	Mass	Int.
40	60	43	50	51	170	56	50	63	25	72	610	76	80	77	370	91	65
103	70	105	70	106	55	118	105	119	20	134	465	135	50	148	40	149	40
165	120	166	50	175	200	180	65	192	10	194	20	203	85	208	215	216	5
220	10	235	20	238	5	251	999	252	140	265	285	266	50	280	620	281	115

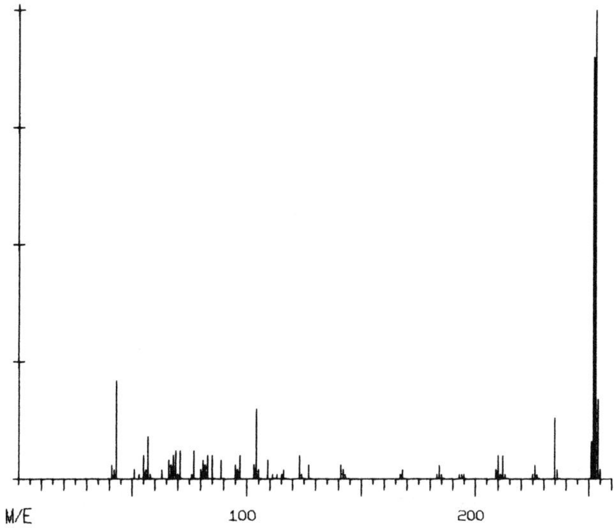
M/E
100
200

TRIAMTERENE

$C_{12}H_{11}N_7$ MW = 253

Base Peak = 253 30 Peaks

Mass	Int.	Mass	Int.	Mass	Int.	Mass	Int.	Mass	Int.	Mass	Int.	Mass	Int.	Mass	Int.	Mass	Int.
41	30	43	210	55	50	57	90	69	60	71	60	77	60	83	50	95	30
97	50	104	150	109	40	123	50	127	30	141	30	142	20	167	10	168	20
183	10	184	30	193	10	194	10	210	50	212	50	225	10	226	30	235	130
236	20	252	900	253	999												

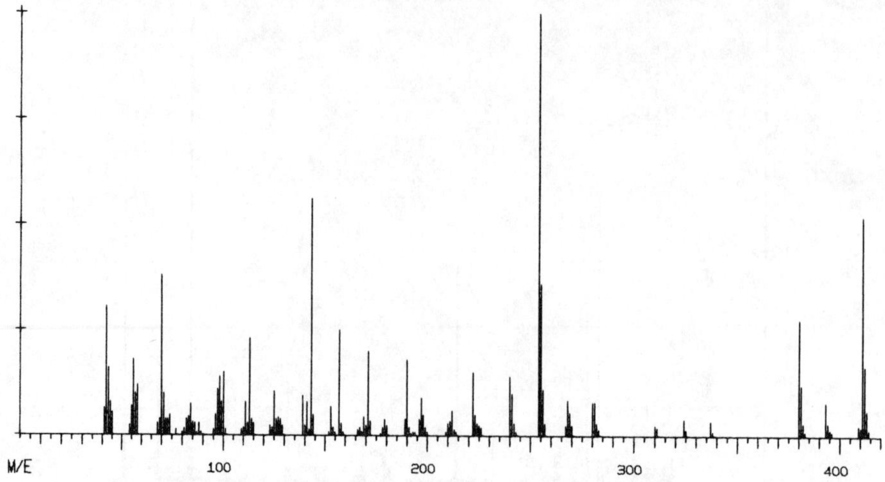
M/E
100
200
300
400

ACETOPHENAZINE

$C_{23}H_{29}N_3O_2S$ MW = 411

Base Peak = 254 50 Peaks

Mass	Int.	Mass	Int.	Mass	Int.	Mass	Int.	Mass	Int.	Mass	Int.	Mass	Int.	Mass	Int.	Mass	Int.
42	304	43	159	56	179	58	119	70	379	71	99	83	44	84	76	98	139
100	149	111	79	113	229	125	104	127	44	139	94	143	559	153	69	157	249
169	44	171	199	179	39	180	26	190	179	197	89	211	34	212	59	222	149
223	49	240	139	241	99	254	999	255	359	268	84	269	54	280	79	281	79
310	24	311	19	324	39	325	14	337	36	338	11	380	274	381	119	393	81
394	30	409	24	411	519	412	164	413	59								

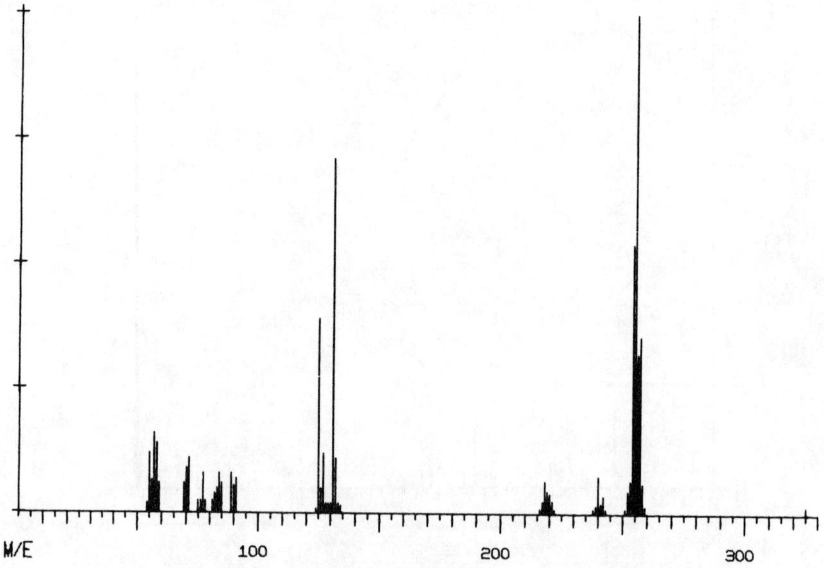
M/E
100
200
300

CLEMIZOLE

$C_{19}H_{20}CLN_3$ MW = 325

Base Peak = 255 21 Peaks

Mass	Int.	Mass	Int.	Mass	Int.	Mass	Int.	Mass	Int.	Mass	Int.	Mass	Int.	Mass	Int.	Mass	Int.
57	159	58	139	70	89	71	109	77	81	84	79	90	54	91	69	125	389
131	709	132	109	133	19	218	64	219	44	240	74	242	34	254	539	255	999
258	61	259	14	325	14												

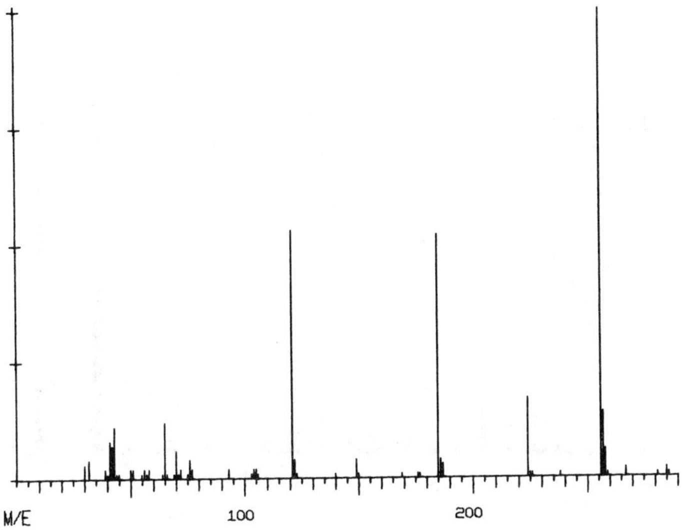
M/E
100
200

PROBENECID

$C_{13}H_{19}NO_4S$ MW = 285

Base Peak = 256 32 Peaks

Mass	Int.	Mass	Int.	Mass	Int.	Mass	Int.	Mass	Int.	Mass	Int.	Mass	Int.	Mass	Int.	Mass	Int.
30	30	32	40	41	80	43	110	50	20	51	20	65	120	70	60	76	40
77	20	93	20	103	10	104	20	105	20	121	530	122	40	140	10	149	40
150	10	169	10	185	520	186	40	224	170	225	10	238	10	256	999	257	140
258	60	267	20	281	10	285	20	286	10								

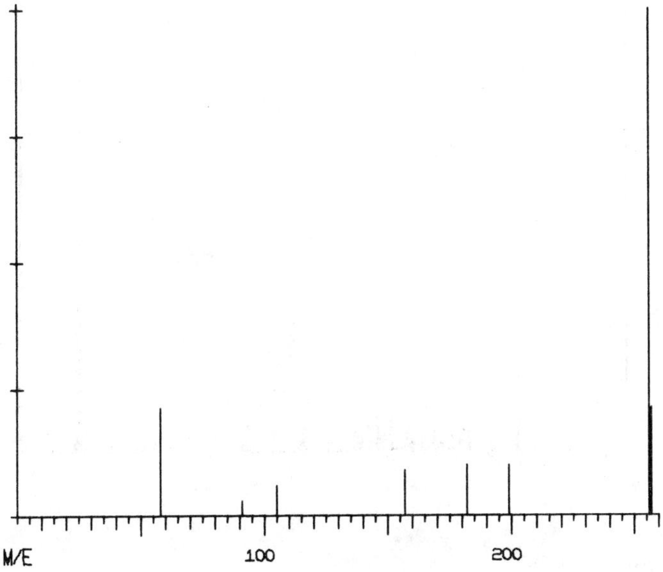
M/E
100
200

PHENOMORPHAN

$C_{24}H_{29}NO$ MW = 347

Base Peak = 256 8 Peaks

Mass	Int.	Mass	Int.	Mass	Int.	Mass	Int.	Mass	Int.	Mass	Int.	Mass	Int.	Mass	Int.
58	212	91	30	105	60	157	90	182	101	199	101	256	999	257	212

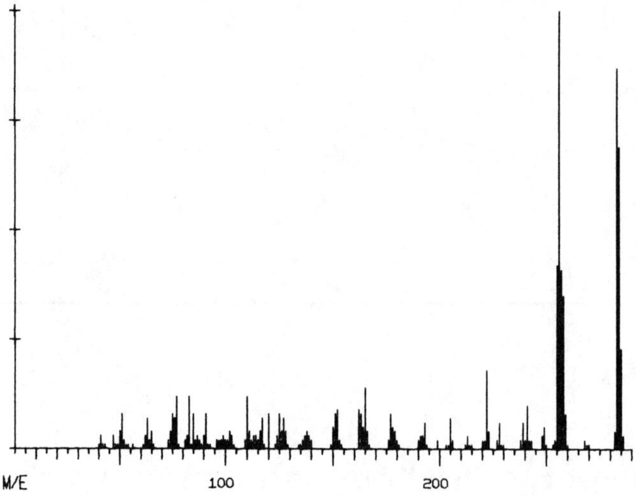
M/E
100
200

DIAZEPAM

$C_{16}H_{13}CLN_2O$ MW = 284

Base Peak = 256 37 Peaks

Mass	Int.	Mass	Int.	Mass	Int.	Mass	Int.	Mass	Int.	Mass	Int.	Mass	Int.	Mass	Int.	Mass	Int.
41	30	47	30	50	40	51	80	63	70	75	80	77	120	83	120	91	80
102	40	110	120	117	70	120	80	125	80	137	30	138	40	151	80	152	90
162	90	165	140	177	80	178	50	191	30	193	60	205	70	213	30	222	180
228	60	239	60	241	100	255	420	256	999	258	350	259	80	283	870	284	690
286	30																

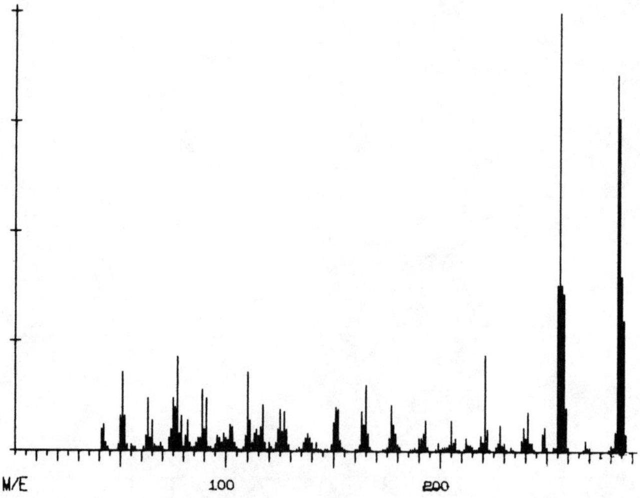
M/E
100
200

DIAZEPAM

$C_{16}H_{13}CLN_2O$ MW = 284

Base Peak = 256 38 Peaks

Mass	Int.	Mass	Int.	Mass	Int.	Mass	Int.	Mass	Int.	Mass	Int.	Mass	Int.	Mass	Int.	Mass	Int.
41	50	42	60	50	80	51	180	63	120	75	120	77	215	89	140	91	120
102	60	110	180	117	105	125	95	127	90	137	30	138	40	151	100	152	95
163	90	165	150	177	105	178	60	192	40	193	70	205	70	207	30	221	220
228	60	239	55	241	90	255	380	256	999	258	360	259	100	283	860	284	760
286	300	287	40														

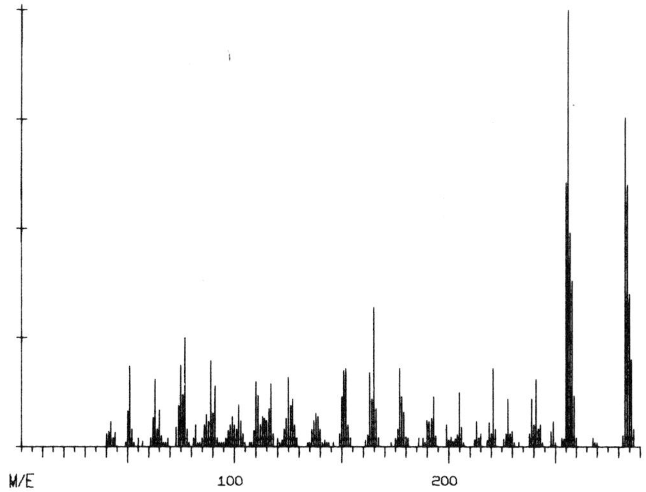
M/E
100
200

DIAZEPAM

$C_{16}H_{13}CLN_2O$ MW = 284

Base Peak = 256 38 Peaks

Mass	Int.	Mass	Int.	Mass	Int.	Mass	Int.	Mass	Int.	Mass	Int.	Mass	Int.	Mass	Int.	Mass	Int.
41	34	42	57	50	82	51	184	63	154	75	188	77	251	89	198	91	139
102	97	110	149	117	144	125	159	127	109	138	77	139	71	151	174	152	179
163	171	165	319	177	181	178	114	192	64	193	114	205	124	213	57	221	179
228	111	239	109	241	154	255	604	256	999	258	381	259	117	283	754	284	599
286	199	287	39														

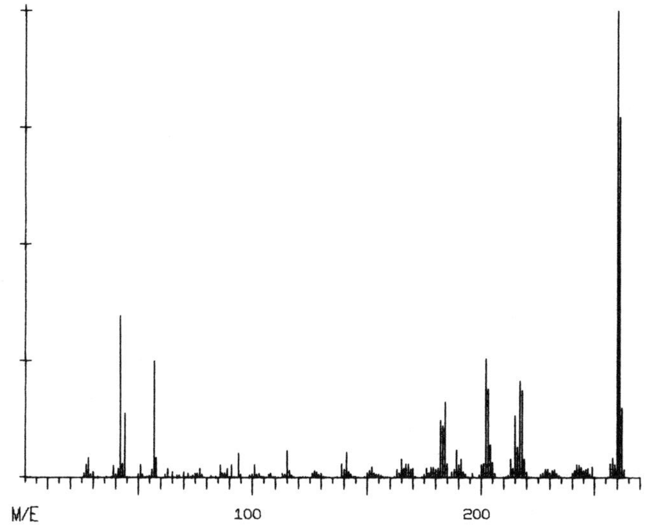
M/E
100
200

PHENINDAMINE

$C_{19}H_{19}N$ MW = 261

Base Peak = 260 36 Peaks

Mass	Int.	Mass	Int.	Mass	Int.	Mass	Int.	Mass	Int.	Mass	Int.	Mass	Int.	Mass	Int.	Mass	Int.
27	28	28	42	42	348	44	137	57	250	58	43	63	21	65	12	77	19
86	28	91	28	94	52	115	58	116	14	127	15	128	13	139	30	141	56
151	15	152	23	165	41	167	31	182	122	184	163	189	59	191	40	202	255
203	190	217	207	218	188	242	27	243	28	244	23	257	31	260	999	261	773

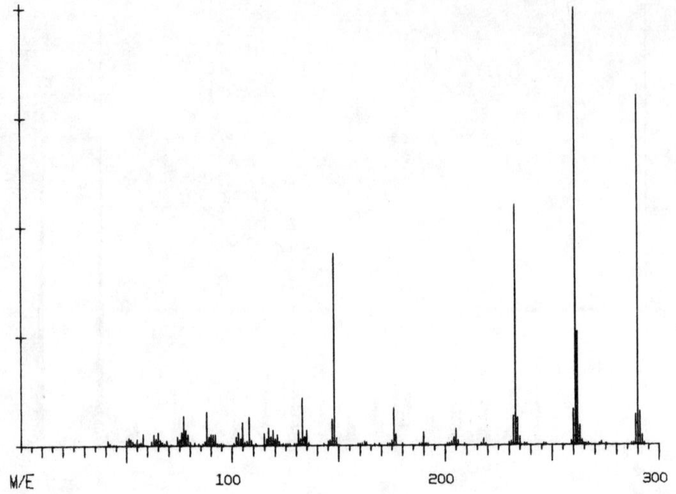
M/E
100
200
300

HYDROXYPHENOBARBITAL 1,3-DIMETHYL DERIVATIVE METHYL ETHER

$C_{15}H_{18}N_2O_4$ MW = 290

Base Peak = 261 38 Peaks

Mass	Int.	Mass	Int.	Mass	Int.	Mass	Int.	Mass	Int.	Mass	Int.	Mass	Int.	Mass	Int.	Mass	Int.
41	12	42	6	51	17	58	28	63	24	65	30	77	68	88	78	90	26
103	30	105	52	108	64	119	34	131	36	133	109	135	36	147	60	148	440
161	8	162	10	176	88	177	24	188	6	190	30	204	18	205	38	218	16
219	6	232	68	233	550	246	5	247	5	261	999	262	260	273	8	275	6
290	800	291	78														

M/E
100
200

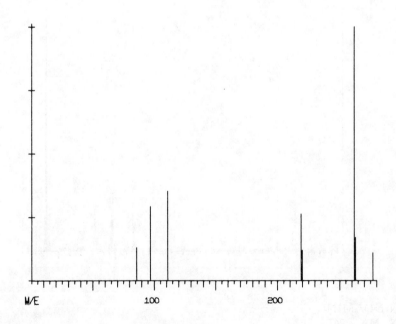

ETHYLMETHYL THIAMBUTENE

$C_{15}H_{19}NS_2$ MW = 277

Base Peak = 262 8 Peaks

Mass	Int.	Mass	Int.	Mass	Int.	Mass	Int.	Mass	Int.	Mass	Int.	Mass	Int.	Mass	Int.
86	131	97	292	111	353	219	262	220	121	262	999	263	171	277	111

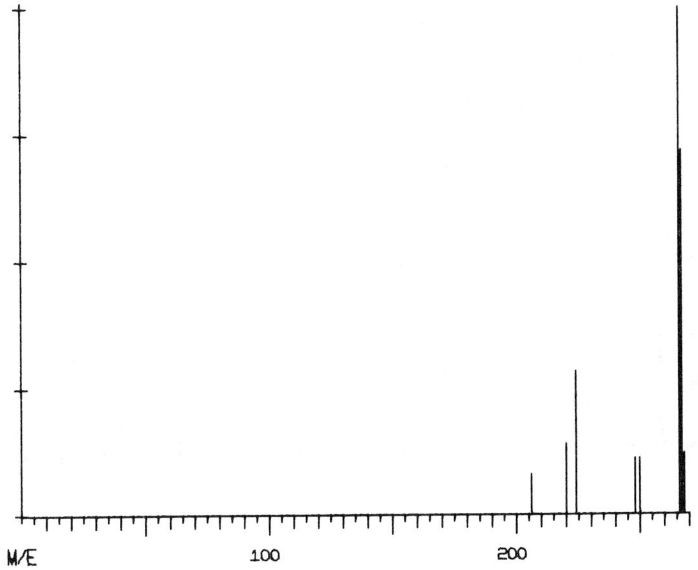
M/E
100
200

APOMORPHINE

$C_{17}H_{17}NO_2$ MW = 267

Base Peak = 266 7 Peaks

Mass	Int.	Mass	Int.	Mass	Int.	Mass	Int.	Mass	Int.	Mass	Int.	Mass	Int.
206	80	220	141	224	282	248	111	250	111	266	999	267	717

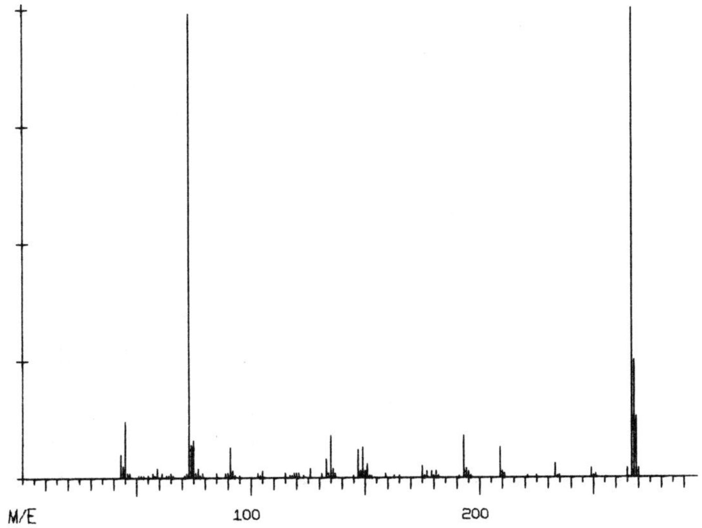
M/E
100
200

SALICYLIC ACID TMS ESTER TMS ETHER

$C_{13}H_{22}O_3Si_2$ MW = 282

Base Peak = 267 34 Peaks

Mass	Int.	Mass	Int.	Mass	Int.	Mass	Int.	Mass	Int.	Mass	Int.	Mass	Int.	Mass	Int.	Mass	Int.
43	50	45	120	57	10	59	20	73	990	75	80	76	10	77	20	91	65
92	15	105	15	115	10	119	10	126	20	133	40	135	90	147	60	149	65
163	5	165	5	175	25	177	15	193	90	194	20	209	65	210	15	221	5
225	5	233	30	234	5	249	20	251	10	267	999	268	250				

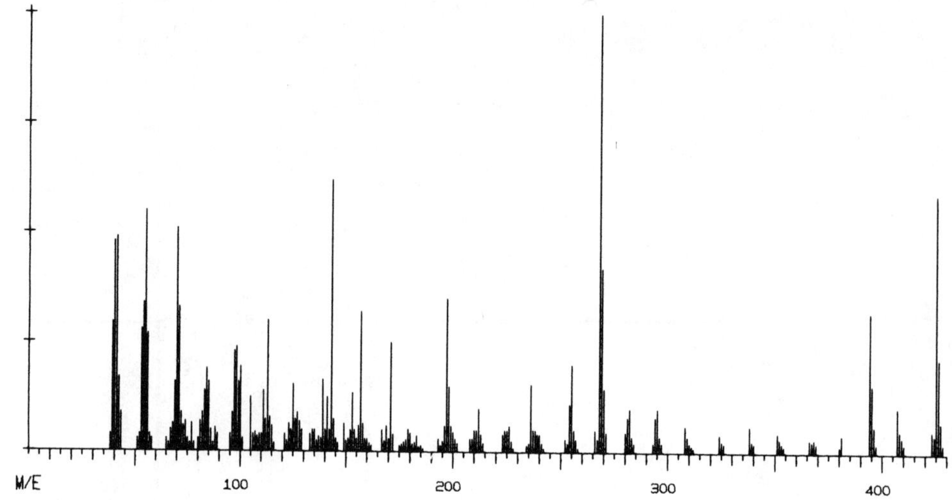
M/E
100
200
300
400

CARPHENAZINE

$C_{24}H_{31}N_3O_2S$ MW = 425
Base Peak = 268 58 Peaks

Mass	Int.	Mass	Int.	Mass	Int.	Mass	Int.	Mass	Int.	Mass	Int.	Mass	Int.	Mass	Int.	Mass	Int.
40	479	41	489	54	339	55	549	70	509	71	329	84	189	85	159	97	229
98	239	111	139	113	299	125	154	127	89	139	164	143	619	153	134	157	319
169	59	171	249	179	53	180	43	197	349	198	149	210	49	212	99	224	49
226	59	236	154	237	49	254	109	255	199	268	999	269	419	281	79	282	99
294	79	295	99	308	59	309	34	324	39	325	24	338	59	339	24	351	44
352	29	366	29	368	29	380	14	381	39	394	319	395	154	407	104	408	49
423	49	425	589	426	214	427	69										

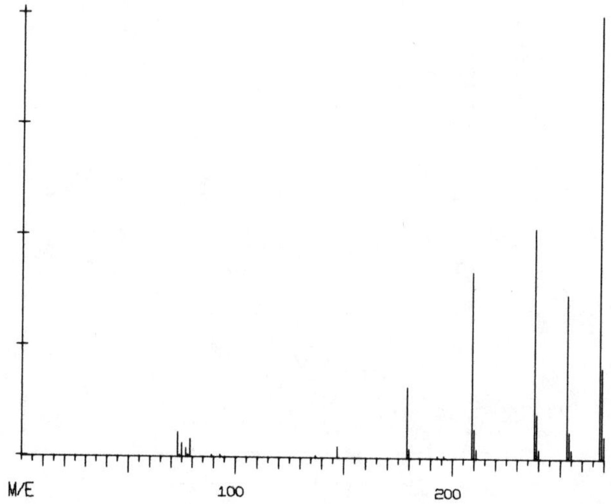
M/E
100
200

3-METHOXY-4-HYDROXYPHENYLACETIC ACID METHYL ESTER TMS ETHER

$C_{13}H_{20}O_4SI$ MW = 268
Base Peak = 268 19 Peaks

Mass	Int.	Mass	Int.	Mass	Int.	Mass	Int.	Mass	Int.	Mass	Int.	Mass	Int.	Mass	Int.	Mass	Int.
73	54	75	29	77	19	79	39	93	4	137	4	147	24	179	159	180	19
193	4	196	4	209	419	210	64	238	519	239	99	253	369	254	59	268	999
269	204																

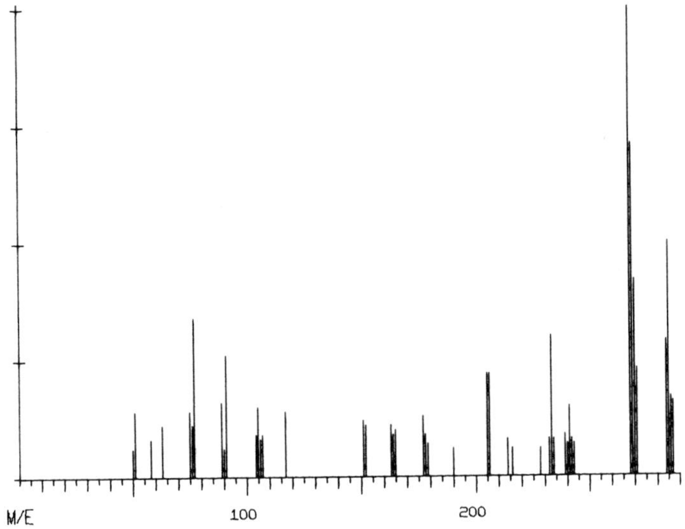
M/E
100
200

CHLORDIAZEPOXIDE MTB. 1

$C_{15}H_{12}ClN_3O$ MW = 285

Base Peak = 268 29 Peaks

Mass	Int.	Mass	Int.	Mass	Int.	Mass	Int.	Mass	Int.	Mass	Int.	Mass	Int.	Mass	Int.	Mass	Int.
51	140	58	80	63	110	75	140	77	340	89	160	90	60	91	260	105	150
117	140	151	120	152	110	163	110	165	100	177	130	178	90	190	60	205	220
206	220	216	60	228	60	233	300	241	150	268	999	269	710	284	290	285	500
286	170	287	160														

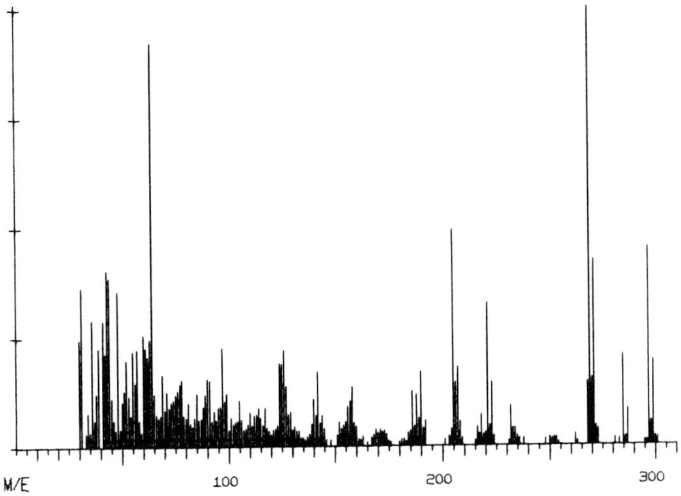
M/E
100
200
300

HYDROCHLOROTHIAZIDE

$C_7H_8CLN_3O_4S$ MW = 265

Base Peak = 269 42 Peaks

Mass	Int.	Mass	Int.	Mass	Int.	Mass	Int.	Mass	Int.	Mass	Int.	Mass	Int.	Mass	Int.	Mass	Int.
30	246	31	362	43	403	44	384	48	354	60	255	63	245	64	922	77	142
78	152	90	156	97	224	105	104	114	88	124	191	126	219	140	107	142	170
157	102	158	134	160	46	171	38	186	126	187	45	188	118	190	170	205	493
207	181	221	324	223	145	232	89	233	41	250	20	252	19	269	999	271	422
272	45	285	207	297	453	299	194	300	19	301	17						

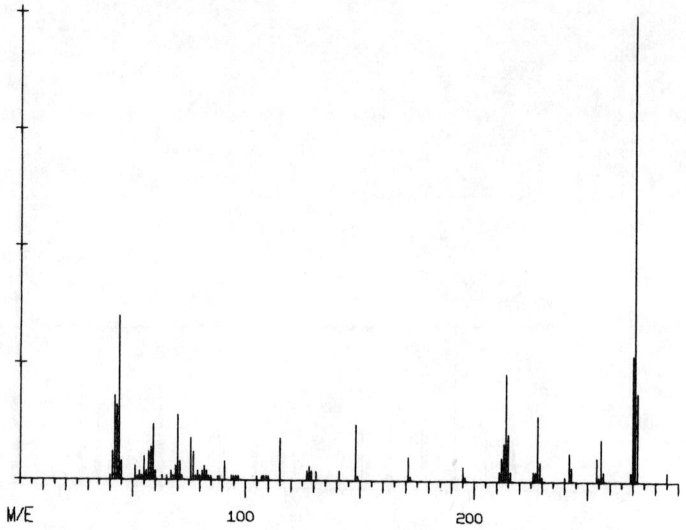
M/E
100
200

NALORPHINE

$C_{19}H_{21}NO_3$ MW = 311

Base Peak = 271 33 Peaks

Mass	Int.	Mass	Int.	Mass	Int.	Mass	Int.	Mass	Int.	Mass	Int.	Mass	Int.	Mass	Int.	Mass	Int.
42	180	44	350	58	70	59	120	70	140	71	40	76	90	77	60	91	40
94	10	105	10	115	90	127	20	128	30	141	20	148	120	149	10	171	50
172	10	195	30	196	10	214	230	215	100	228	140	229	40	242	60	243	30
254	50	256	90	270	270	271	999	272	190	285	20						

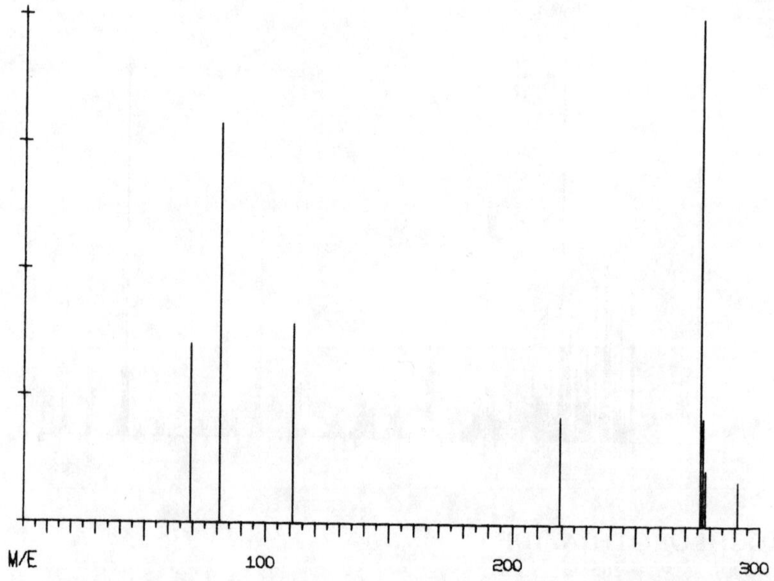
M/E
100
200
300

THIAMBUTENE

$C_{16}H_{21}NS_2$ MW = 291

Base Peak = 276 7 Peaks

Mass	Int.	Mass	Int.	Mass	Int.	Mass	Int.	Mass	Int.	Mass	Int.	Mass	Int.
69	353	81	787	111	393	219	212	276	999	277	212	291	90

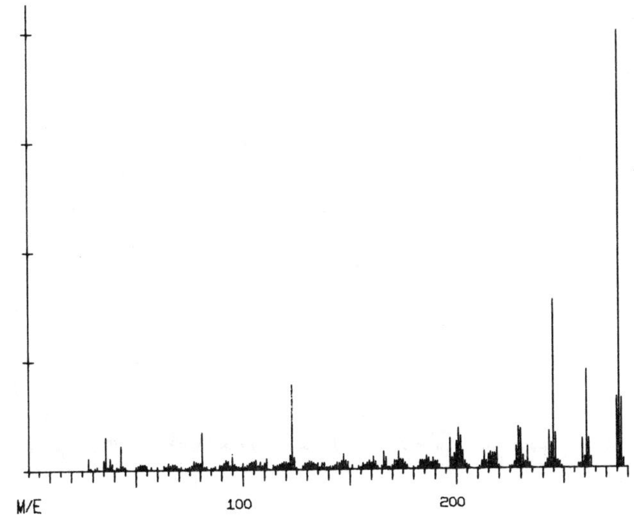

M/E
100
200

TRIMETHOPRIM MTB. 3

$C_{13}H_{16}N_4O_3$ MW = 276

Base Peak = 276 38 Peaks

Mass	Int.	Mass	Int.	Mass	Int.	Mass	Int.	Mass	Int.	Mass	Int.	Mass	Int.	Mass	Int.	Mass	Int.
28	27	32	8	36	75	43	54	52	13	53	13	65	16	68	13	77	21
85	85	92	22	95	35	106	22	111	26	122	33	123	195	132	18	133	16
146	21	147	34	166	39	173	40	186	29	187	26	197	70	201	92	202	74
213	41	228	50	229	95	230	90	243	86	245	384	246	81	259	67	261	225
275	162	276	999														

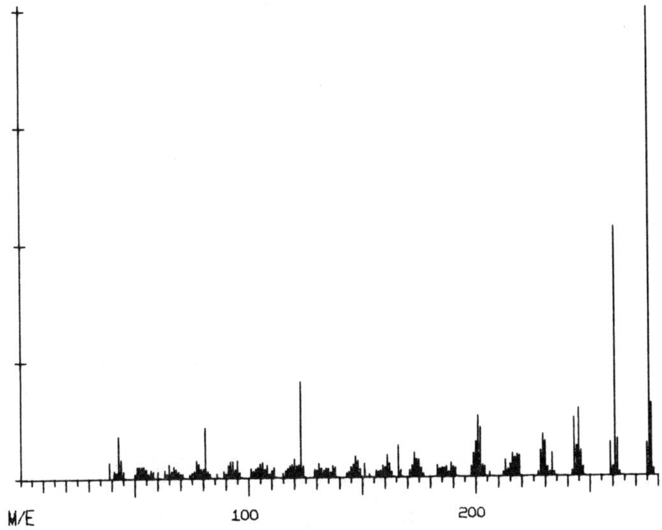

M/E
100
200

TRIMETHOPRIM MTB. 4

$C_{13}H_{16}N_4O_3$ MW = 276

Base Peak = 276 36 Peaks

Mass	Int.	Mass	Int.	Mass	Int.	Mass	Int.	Mass	Int.	Mass	Int.	Mass	Int.	Mass	Int.	Mass	Int.
43	91	44	39	53	26	54	26	65	29	67	26	77	37	81	107	92	34
95	38	105	29	106	32	120	39	123	204	137	26	138	23	147	45	148	34
166	68	173	53	174	37	175	37	200	74	201	130	202	106	213	36	228	56
229	90	230	74	243	125	244	64	245	146	261	532	262	80	276	999	277	155

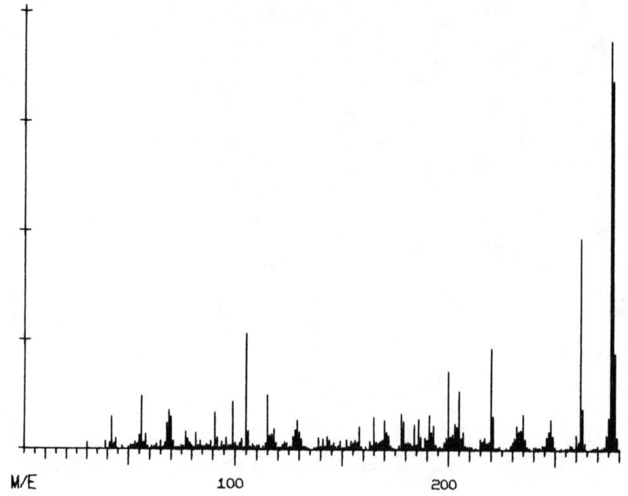
M/E
100
200

METHADONE MTB. 1

$C_{20}H_{23}N$ MW = 277

Base Peak = 276 37 Peaks

Mass	Int.	Mass	Int.	Mass	Int.	Mass	Int.	Mass	Int.	Mass	Int.	Mass	Int.	Mass	Int.	Mass	Int.
30	15	42	76	44	25	56	122	58	36	69	91	70	76	77	41	82	38
91	86	99	109	105	265	115	124	118	47	129	68	139	28	143	30	152	23
158	53	165	76	170	67	178	83	186	71	191	81	200	180	203	59	205	135
220	233	221	77	232	54	235	83	247	42	248	69	262	484	263	96	276	999
277	845																

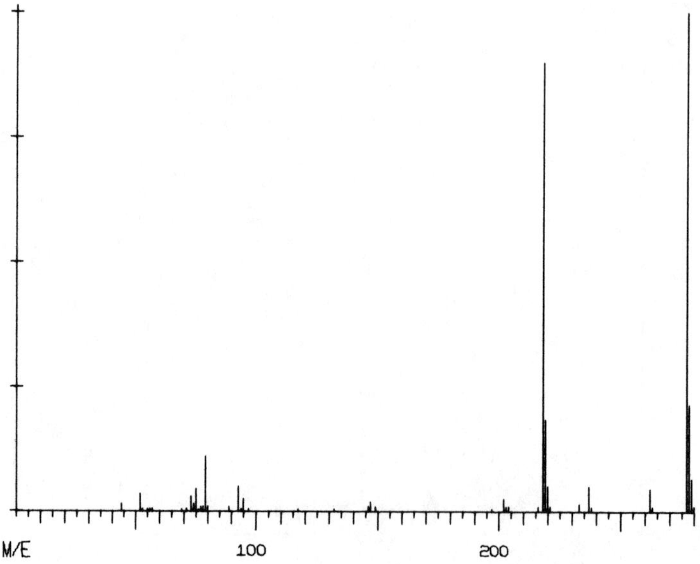
M/E
100
200

5-HYDROXYINDOLEACETIC ACID METHYL ESTER TRIMETHYLSILYL ETHER

$C_{14}H_{19}NO_3$ MW = 249

Base Peak = 277 24 Peaks

Mass	Int.	Mass	Int.	Mass	Int.	Mass	Int.	Mass	Int.	Mass	Int.	Mass	Int.	Mass	Int.	Mass	Int.
44	14	52	34	53	4	73	29	75	44	77	9	79	109	93	49	95	24
117	4	132	4	146	9	147	19	197	4	202	24	203	9	218	899	219	184
233	14	237	49	262	44	263	9	277	999	278	214						

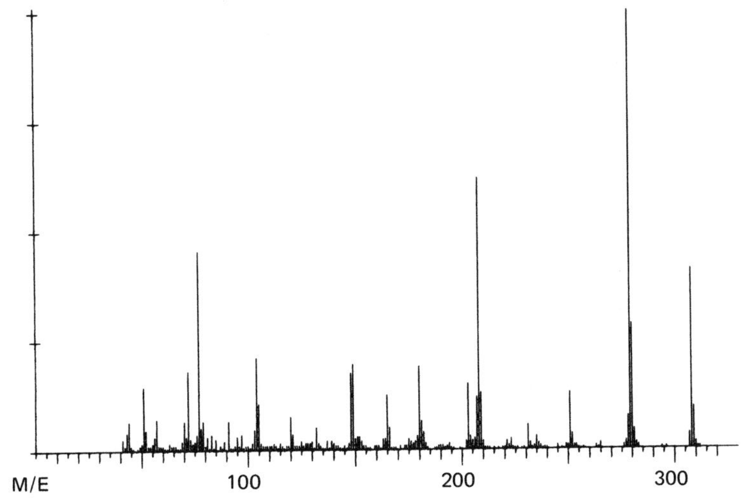
M/E
100
200
300

5,5-DIPHENYLHYDANTOIN 2,3-DIETHYL DERIVATIVE

$C_{19}H_{20}N_2O_2$ MW = 308

Base Peak = 279 40 Peaks

Mass	Int.	Mass	Int.	Mass	Int.	Mass	Int.	Mass	Int.	Mass	Int.	Mass	Int.	Mass	Int.	Mass	Int.
43	40	44	65	51	145	57	70	70	65	72	180	77	455	79	65	91	65
103	45	104	210	105	105	120	75	121	35	132	50	137	20	148	175	149	195
165	125	166	50	180	190	181	65	189	10	194	15	203	150	208	620	221	20
223	25	231	55	235	30	251	130	252	35	263	10	265	15	279	999	280	285
294	5	296	5	308	410	309	95										

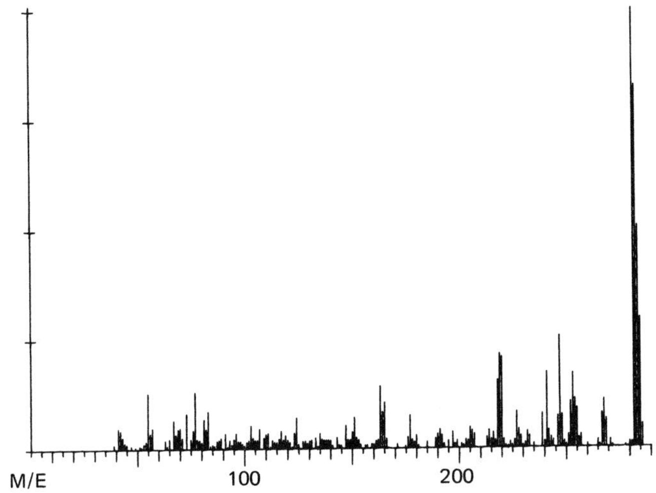
M/E
100
200

CHLORDIAZEPOXIDE (GC)

$C_{16}H_{14}ClN_3O$ MW = 299

Base Peak = 282 37 Peaks

Mass	Int.	Mass	Int.	Mass	Int.	Mass	Int.	Mass	Int.	Mass	Int.	Mass	Int.	Mass	Int.	Mass	Int.
41	47	42	43	55	128	57	47	67	64	73	81	77	129	83	84	91	36
103	53	107	44	117	39	123	36	124	71	133	24	135	34	147	52	151	70
163	142	165	105	177	76	180	30	191	43	197	37	205	48	206	41	219	214
220	207	239	78	241	173	247	254	253	169	267	78	268	111	282	999	283	825
286	52																

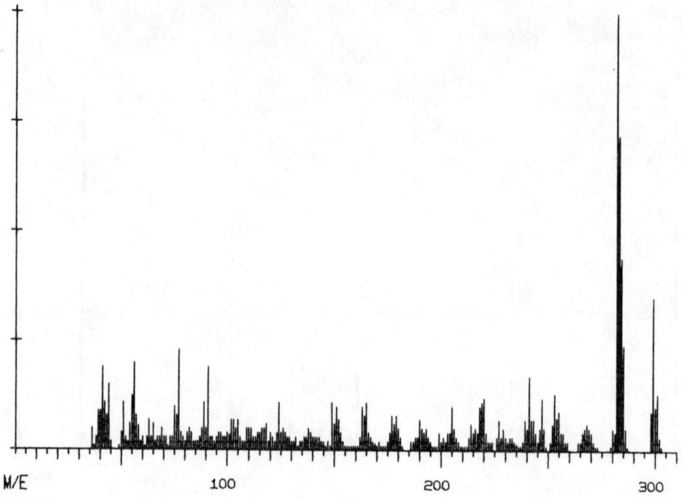
M/E
100
200
300

CHLORDIAZEPOXIDE
$C_{16}H_{14}CLN_3O$ MW = 299
Base Peak = 282 40 Peaks

Mass	Int.	Mass	Int.	Mass	Int.	Mass	Int.	Mass	Int.	Mass	Int.	Mass	Int.	Mass	Int.	Mass	Int.
41	189	44	149	55	124	56	199	63	69	75	99	77	229	89	109	91	189
102	69	104	49	105	69	118	59	124	109	138	49	139	39	149	109	151	99
163	99	165	109	177	79	179	79	190	69	191	49	205	99	214	59	219	109
220	119	239	69	241	169	247	119	253	129	267	49	268	59	282	999	283	719
298	89	299	349	300	99	301	129										

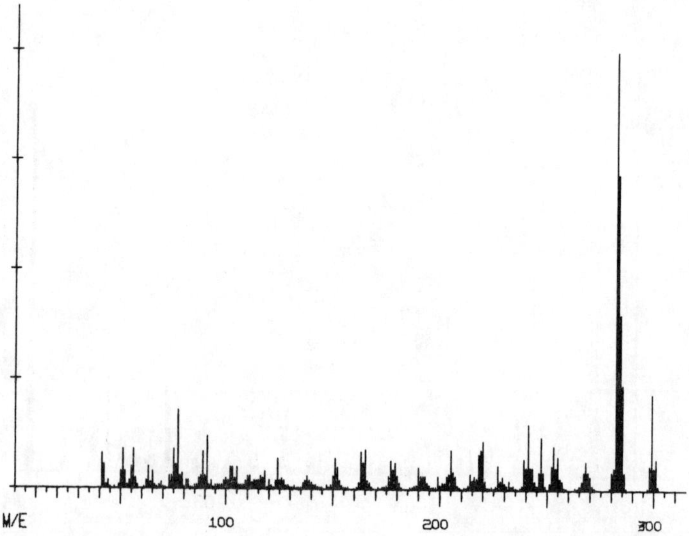
M/E
100
200
300

CHLORDIAZEPOXIDE
$C_{16}H_{14}CLN_3O$ MW = 299
Base Peak = 282 40 Peaks

Mass	Int.	Mass	Int.	Mass	Int.	Mass	Int.	Mass	Int.	Mass	Int.	Mass	Int.	Mass	Int.	Mass	Int.
41	80	42	55	51	90	56	90	63	50	75	90	77	180	89	85	91	120
102	50	105	50	110	30	118	40	124	70	137	20	138	30	151	70	152	50
163	85	165	90	177	65	179	60	190	50	191	30	205	90	206	40	219	90
220	110	239	70	241	150	247	120	253	100	267	40	268	65	282	999	283	720
298	55	299	220	300	50	301	70										

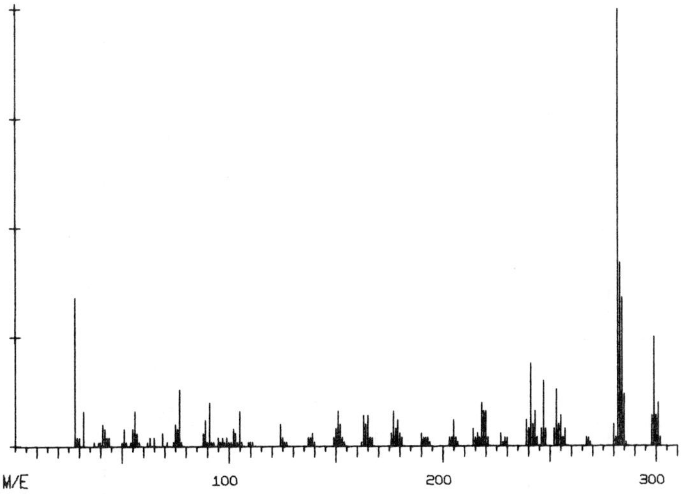
M/E
100
200
300

CHLORDIAZEPOXIDE

$C_{16}H_{14}CLN_3O$ MW = 299

Base Peak = 282 42 Peaks

Mass	Int.	Mass	Int.	Mass	Int.	Mass	Int.	Mass	Int.	Mass	Int.	Mass	Int.	Mass	Int.	Mass	Int.
28	340	32	80	41	50	42	40	51	40	56	80	69	30	75	50	77	130
89	60	91	100	102	40	104	10	105	80	124	50	125	20	137	20	139	30
151	80	152	50	163	70	165	70	177	80	179	60	190	30	191	20	205	60
214	40	218	100	219	80	241	190	243	80	247	150	253	130	267	20	268	20
282	999	283	420	298	70	299	250	300	70	301	100						

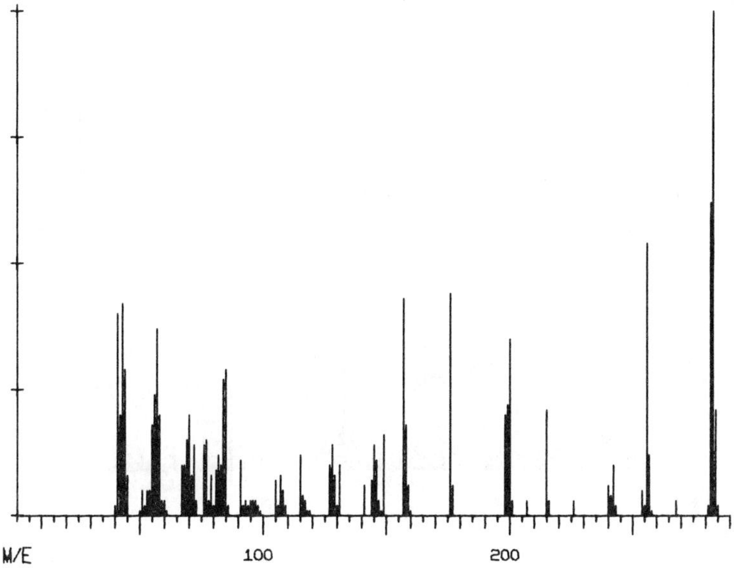
M/E
100
200

LEVALLORPHAN

$C_{19}H_{25}NO$ MW = 283

Base Peak = 283 35 Peaks

Mass	Int.	Mass	Int.	Mass	Int.	Mass	Int.	Mass	Int.	Mass	Int.	Mass	Int.	Mass	Int.	Mass	Int.
41	400	43	420	56	240	57	370	69	150	70	200	84	270	85	290	91	110
93	30	107	80	115	120	127	100	128	140	144	70	145	140	157	430	158	180
160	10	176	440	177	60	199	220	200	350	207	30	215	210	216	30	226	30
240	60	242	100	256	540	257	120	258	10	268	30	282	620	283	999		

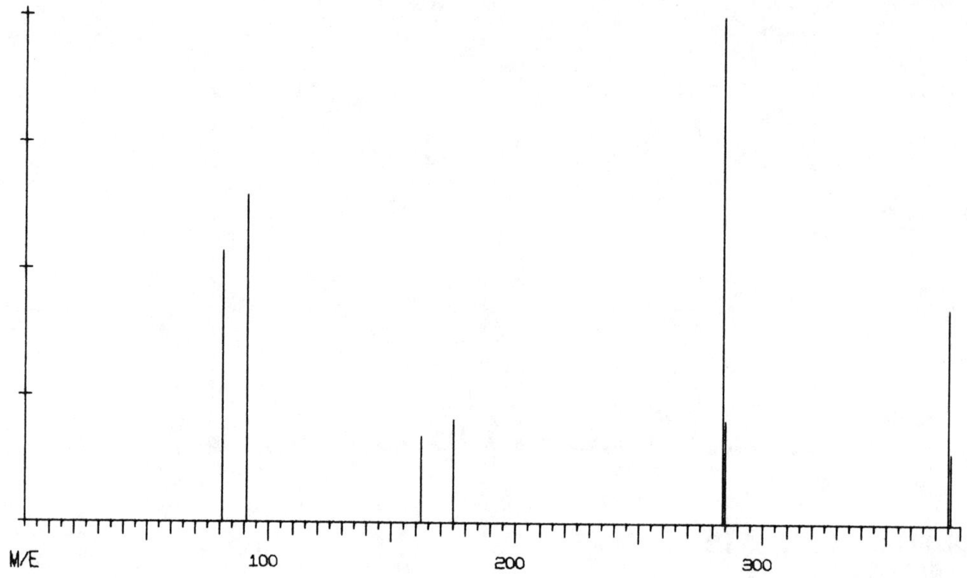
M/E
100
200
300

BENZYLMORPHINE

$C_{24}H_{25}NO_3$ MW = 375

Base Peak = 284 8 Peaks

Mass	Int.	Mass	Int.	Mass	Int.	Mass	Int.	Mass	Int.	Mass	Int.	Mass	Int.	Mass	Int.
81	535	91	646	162	171	175	202	284	999	285	202	375	424	376	141

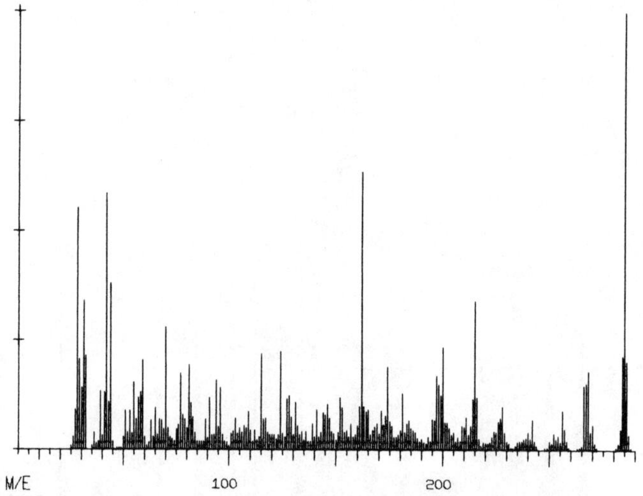
M/E
100
200

MORPHINE

$C_{17}H_{19}NO_3$ MW = 285

Base Peak = 285 40 Peaks

Mass	Int.	Mass	Int.	Mass	Int.	Mass	Int.	Mass	Int.	Mass	Int.	Mass	Int.	Mass	Int.	Mass	Int.
28	552	31	341	42	584	44	381	55	154	59	204	65	98	70	281	77	176
81	194	94	161	96	142	109	87	115	219	124	226	128	124	141	92	144	86
146	104	152	120	162	634	163	100	174	190	181	129	197	170	200	235	214	117
215	341	216	120	228	101	240	40	242	71	256	89	257	47	267	152	268	181
284	214	285	999	286	202	287	36										

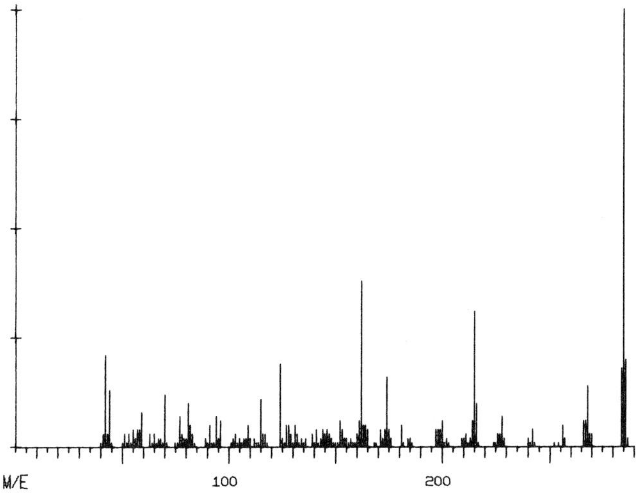
M/E
100
200

MORPHINE

$C_{17}H_{19}NO_3$ MW = 285

Base Peak = 285 38 Peaks

Mass	Int.	Mass	Int.	Mass	Int.	Mass	Int.	Mass	Int.	Mass	Int.	Mass	Int.	Mass	Int.	Mass	Int.
42	210	44	130	55	40	59	80	63	30	70	120	77	70	81	100	94	70
96	60	109	50	115	110	124	190	127	50	141	40	144	40	146	40	152	60
161	60	162	380	174	160	181	50	197	40	200	60	214	60	215	310	216	100
228	70	240	20	242	40	256	50	257	20	266	60	268	140	284	180	285	999
286	200	287	20														

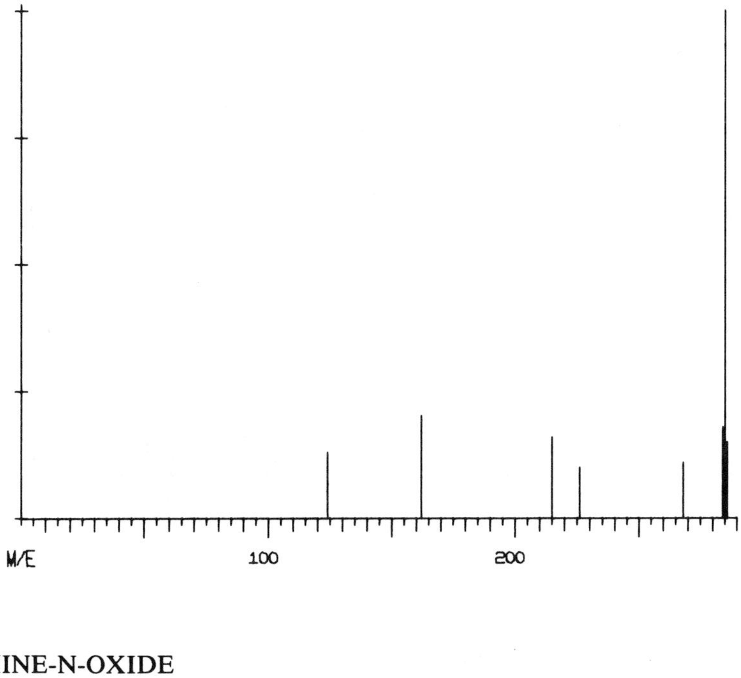
M/E
100
200

MORPHINE-N-OXIDE

$C_{17}H_{19}NO_4$ MW = 301

Base Peak = 285 8 Peaks

Mass	Int.	Mass	Int.	Mass	Int.	Mass	Int.	Mass	Int.	Mass	Int.	Mass	Int.	Mass	Int.
124	131	162	202	215	161	226	101	268	111	284	181	285	999	286	151

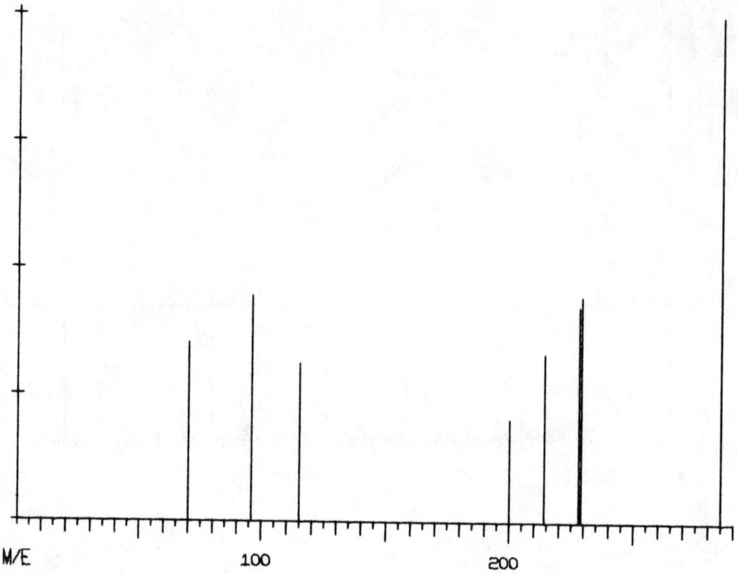
M/E
100
200

HYDROMORPHINE

$C_{17}H_{19}NO_3$ MW = 285

Base Peak = 285 8 Peaks

Mass	Int.	Mass	Int.	Mass	Int.	Mass	Int.	Mass	Int.	Mass	Int.	Mass	Int.	Mass	Int.
70	353	96	444	115	313	200	202	214	333	228	424	229	444	285	999

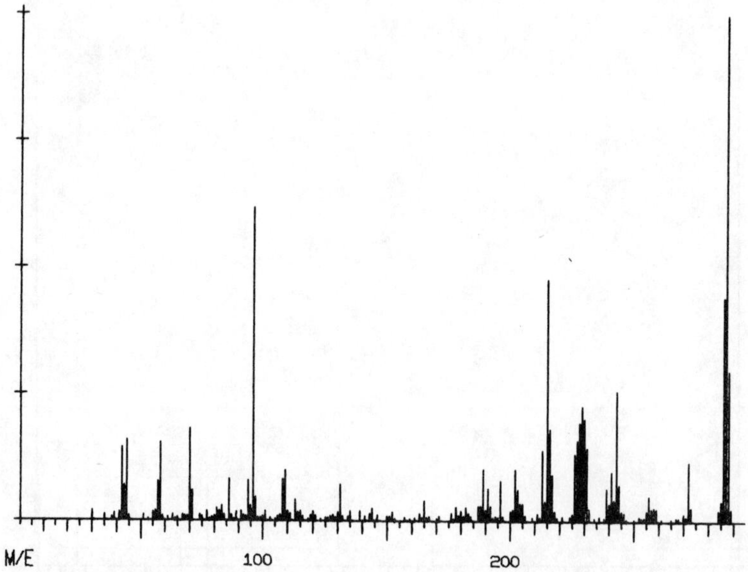
M/E
100
200

CYPROHEPTADINE

$C_{21}H_{21}N$ MW = 287

Base Peak = 287 39 Peaks

Mass	Int.	Mass	Int.	Mass	Int.	Mass	Int.	Mass	Int.	Mass	Int.	Mass	Int.	Mass	Int.	Mass	Int.
30	20	42	145	44	161	57	77	58	155	70	183	71	59	83	30	86	83
94	79	96	617	108	82	109	101	120	19	131	72	139	20	144	26	146	13
152	18	163	16	165	40	178	27	187	30	189	103	196	81	213	141	215	478
228	195	229	227	230	202	243	258	244	73	256	51	258	27	259	24	272	118
285	39	286	442	287	999												

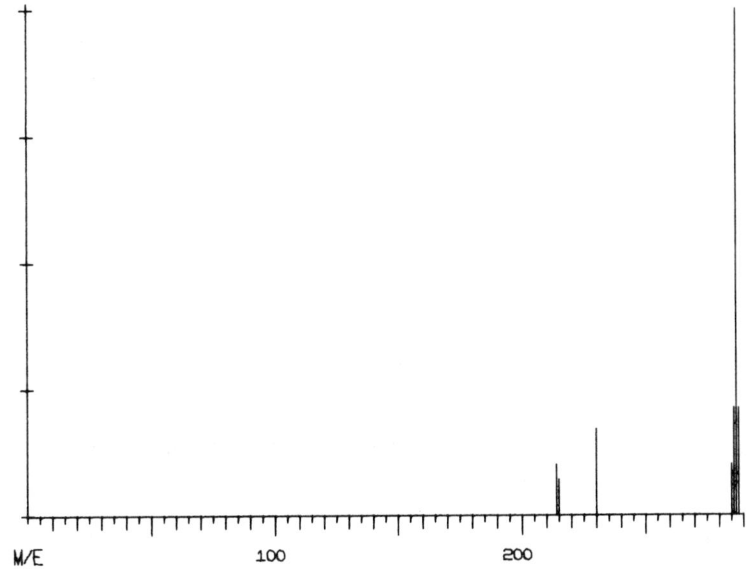
M/E
100
200

DIHYDROMORPHINE

$C_{17}H_{21}NO_3$ MW = 287

Base Peak = 287 6 Peaks

Mass	Int.	Mass	Int.	Mass	Int.	Mass	Int.	Mass	Int.	Mass	Int.
214	101	215	70	230	171	285	101	286	212	287	999

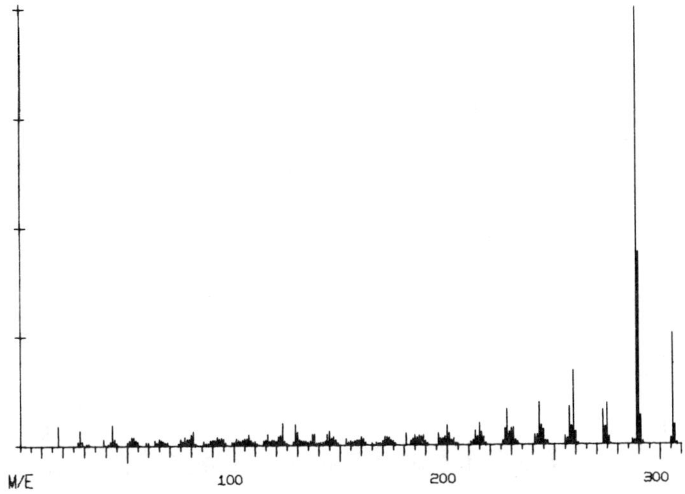
M/E
100
200
300

TRIMETHOPRIM MTB. 1

$C_{14}H_{18}N_4O_4$ MW = 306

Base Peak = 289 42 Peaks

Mass	Int.	Mass	Int.	Mass	Int.	Mass	Int.	Mass	Int.	Mass	Int.	Mass	Int.	Mass	Int.	Mass	Int.
27	10	28	36	39	16	43	48	52	19	53	19	65	15	75	15	80	25
81	33	92	21	94	17	107	24	116	24	123	51	129	48	138	26	145	32
147	20	153	16	160	21	172	20	181	27	185	23	196	27	200	45	213	32
215	50	227	37	228	83	231	40	243	97	244	44	257	87	258	42	259	169
273	79	275	95	289	999	290	441	306	256	307	44						

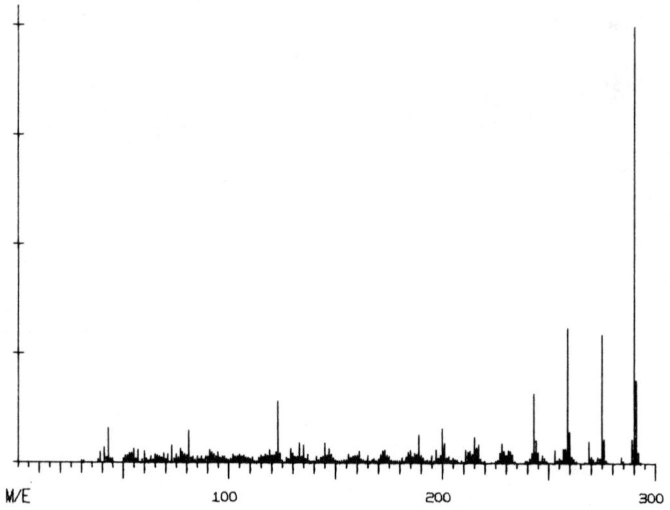
M/E
100
200
300

TRIMETHOPRIM

$C_{14}H_{18}N_4O_3$ MW = 290

Base Peak = 290 40 Peaks

Mass	Int.	Mass	Int.	Mass	Int.	Mass	Int.	Mass	Int.	Mass	Int.	Mass	Int.	Mass	Int.	Mass	Int.
30	4	31	4	41	35	43	80	55	33	57	29	69	22	73	39	77	33
81	75	91	30	95	26	105	21	117	20	123	143	129	32	133	45	145	45
147	33	148	21	161	28	173	29	184	24	185	29	189	64	200	81	211	30
215	60	217	43	228	44	231	29	243	161	244	53	257	33	259	309	260	72
275	295	276	54	290	999	291	191										

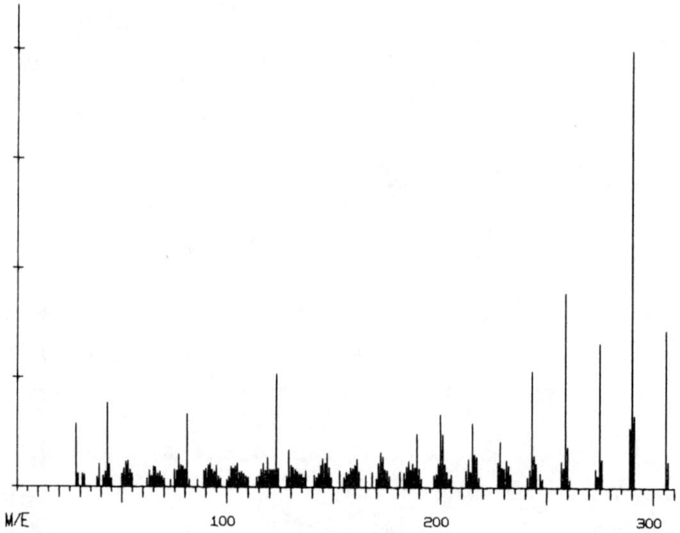
M/E
100
200
300

TRIMETHOPRIM MTB. 2

$C_{14}H_{18}N_4O_4$ MW = 306

Base Peak = 290 42 Peaks

Mass	Int.	Mass	Int.	Mass	Int.	Mass	Int.	Mass	Int.	Mass	Int.	Mass	Int.	Mass	Int.	Mass	Int.
28	146	29	30	43	192	44	53	52	58	53	61	65	48	66	46	77	76
81	167	92	56	95	51	105	56	117	54	123	260	129	84	144	50	145	66
146	56	147	78	172	80	173	70	185	61	187	55	189	122	200	167	213	64
215	148	216	74	228	106	231	62	243	267	244	73	257	60	259	445	260	92
275	330	276	6	290	999	291	165	306	360	307	60						

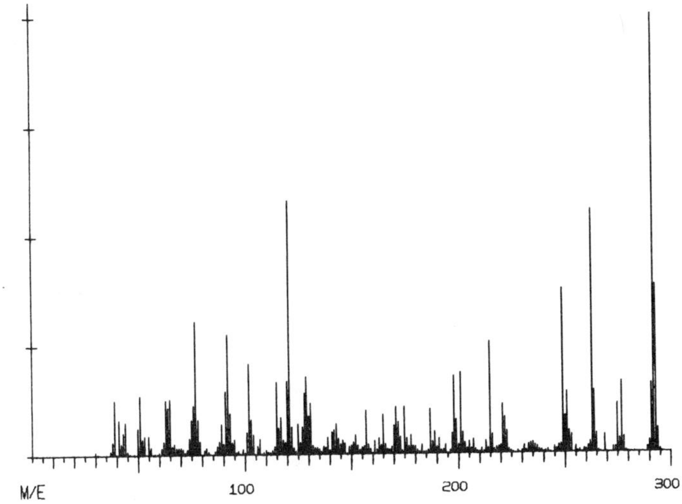
M/E
100
200
300

WARFARIN MTB. 1

$C_{19}H_{16}O_3$ MW = 292

Base Peak = 292 40 Peaks

Mass	Int.	Mass	Int.	Mass	Int.	Mass	Int.	Mass	Int.	Mass	Int.	Mass	Int.	Mass	Int.	Mass	Int.
30	8	31	3	39	125	41	80	50	61	51	134	63	125	65	127	76	112
77	305	92	274	102	207	115	164	117	86	121	579	129	178	142	56	143	69
152	42	157	99	165	89	171	107	175	107	187	102	198	178	201	186	202	48
215	256	221	113	222	83	234	22	235	24	249	374	251	141	263	555	264	143
275	112	277	162	292	999	293	382										

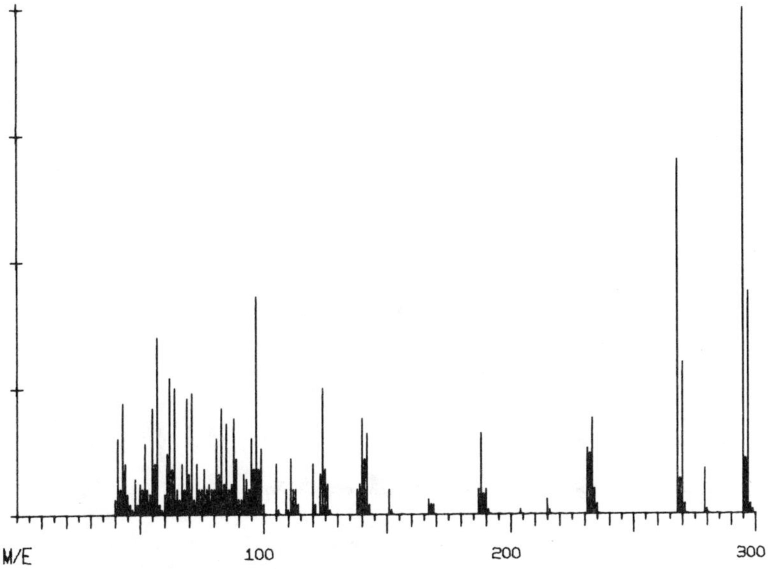
M/E
100
200
300

CHLORTHIAZIDE

$C_7H_6CLN_3O_4S_2$ MW = 295

Base Peak = 295 34 Peaks

Mass	Int.	Mass	Int.	Mass	Int.	Mass	Int.	Mass	Int.	Mass	Int.	Mass	Int.	Mass	Int.	Mass	Int.
41	150	43	220	55	210	57	350	62	270	64	250	83	210	88	190	95	150
97	430	105	100	111	110	120	100	124	250	140	190	142	160	151	50	152	10
167	30	168	20	187	50	188	160	190	50	204	10	215	30	216	10	231	130
233	190	268	700	270	300	279	90	280	10	295	999	297	440				

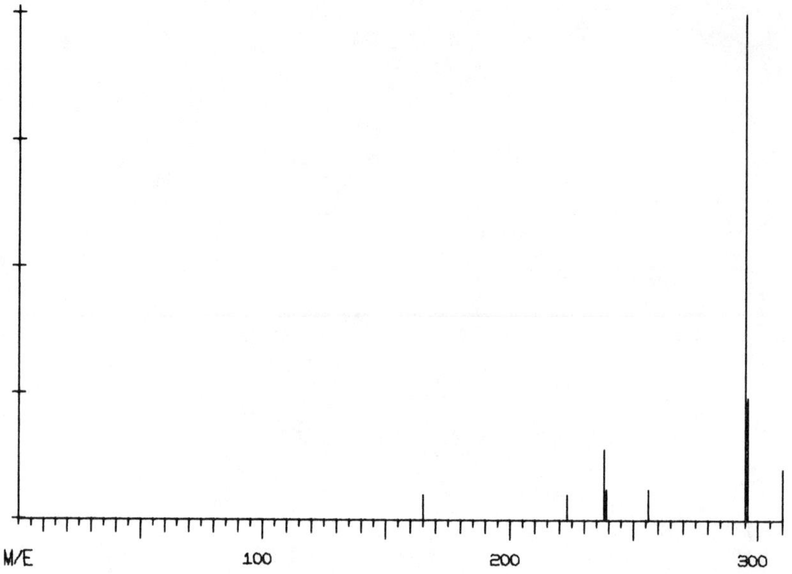
M/E
100
200
300

CANNABINOL

$C_{21}H_{26}O_2$ MW = 310
Base Peak = 295 8 Peaks

Mass	Int.	Mass	Int.	Mass	Int.	Mass	Int.	Mass	Int.	Mass	Int.	Mass	Int.	Mass	Int.
165	50	223	50	238	141	239	60	256	60	295	999	296	242	310	101

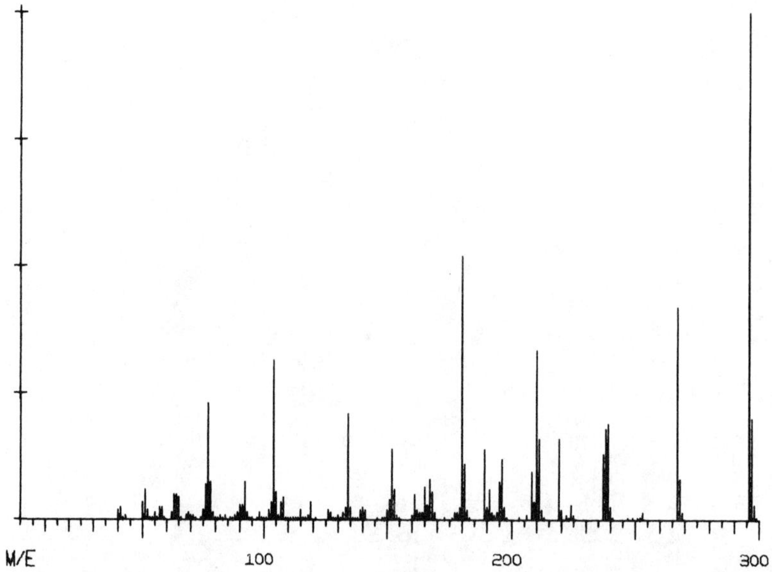
M/E
100
200
300

3-HYDROXYDIPHENYLHYDANTOIN METHYL ETHER 3-METHYL DERIVATIVE

$C_{17}H_{16}N_2O_3$ MW = 296
Base Peak = 296 36 Peaks

Mass	Int.	Mass	Int.	Mass	Int.	Mass	Int.	Mass	Int.	Mass	Int.	Mass	Int.	Mass	Int.	Mass	Int.
40	20	41	25	50	35	51	60	63	50	64	50	77	230	78	75	92	75
103	35	104	315	105	55	119	35	126	20	133	25	134	210	152	140	153	60
165	65	167	80	180	520	181	110	189	140	196	120	210	335	211	160	219	160
224	30	238	180	239	190	247	5	253	15	267	420	268	80	296	999	297	200

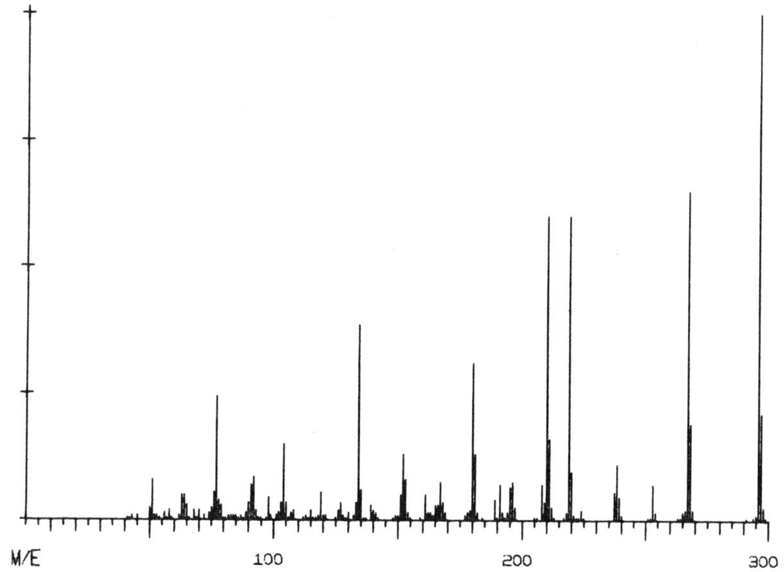
M/E
100
200
300

P-HYDROXYDIPHENYLHYDANTOIN METHYL ETHER 3-METHYL DERIVATIVE

$C_{17}H_{16}N_2O_3$ MW = 296
Base Peak = 296 36 Peaks

Mass	Int.	Mass	Int.	Mass	Int.	Mass	Int.	Mass	Int.	Mass	Int.	Mass	Int.	Mass	Int.	Mass	Int.
43	10	45	10	50	25	51	80	63	50	64	50	76	55	77	245	91	70
92	85	104	150	105	35	119	55	127	35	134	385	135	60	152	130	153	80
161	50	167	75	180	310	181	130	191	70	196	75	210	600	211	160	219	600
220	95	237	55	238	110	253	70	254	15	267	650	268	190	296	999	297	210

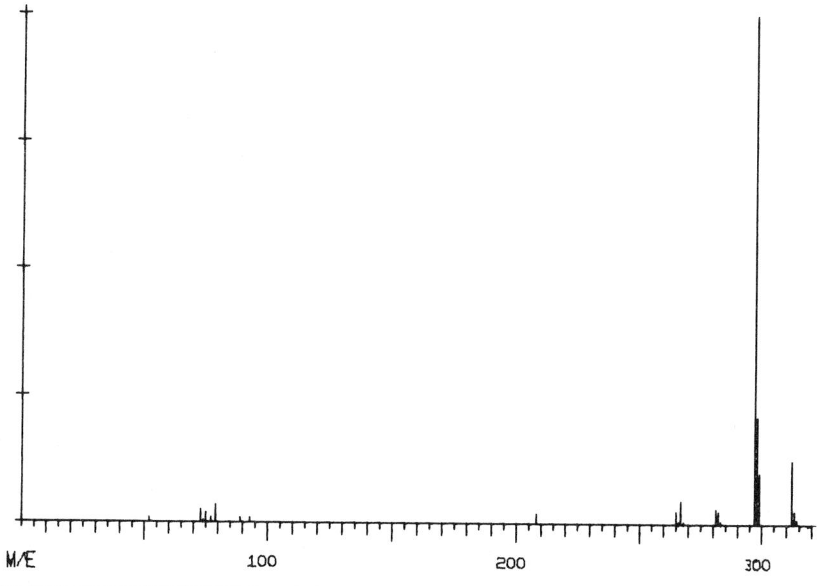
M/E
100
200
300

2,5-DIHYDROXYBENZOIC ACID METHYL ESTER TRIMETHYLSILYL ETHER

$C_{14}H_{24}O_4SI_2$ MW = 312
Base Peak = 297 16 Peaks

Mass	Int.	Mass	Int.	Mass	Int.	Mass	Int.	Mass	Int.	Mass	Int.	Mass	Int.	Mass	Int.	Mass	Int.
52	9	73	24	75	19	77	9	79	34	93	9	208	19	265	24	267	44
281	29	282	24	297	999	298	209	312	124	313	24	314	9				

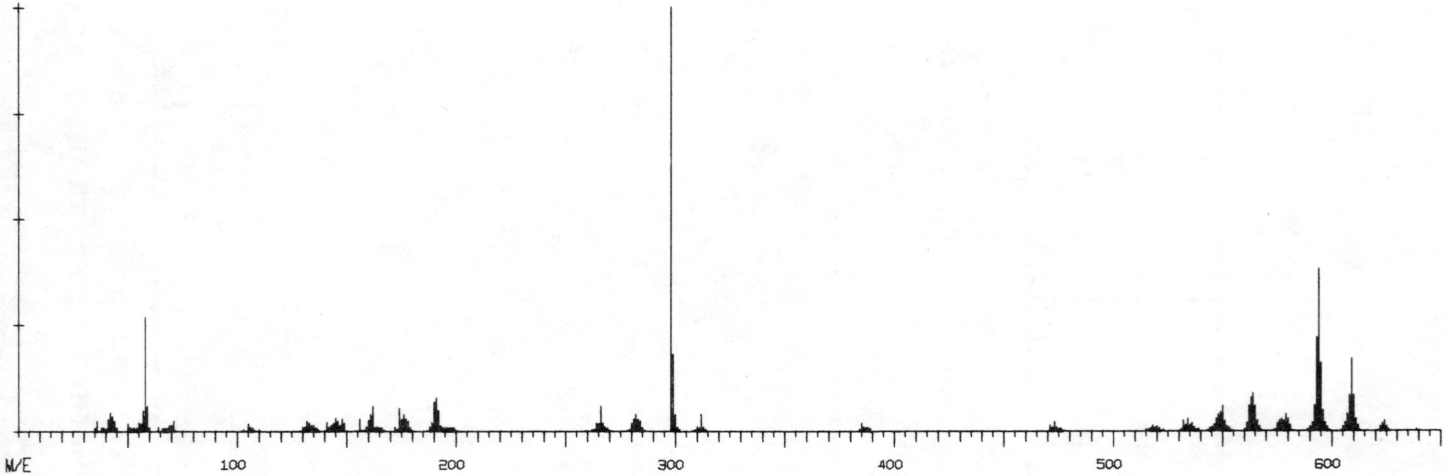
M/E
100
200
300
400
500
600

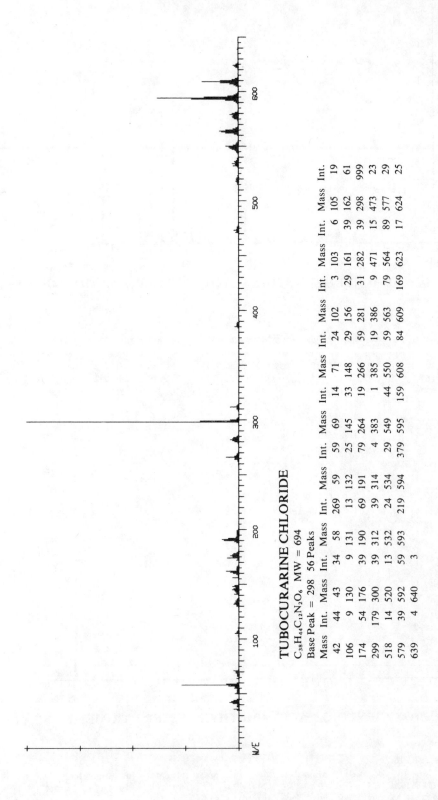

TUBOCURARINE CHLORIDE

$C_{38}H_{44}C_{12}N_2O_6$ MW = 694

Base Peak = 298 56 Peaks

Mass	Int.	Mass	Int.	Mass	Int.	Mass	Int.	Mass	Int.	Mass	Int.	Mass	Int.	Mass	Int.	Mass	Int.
42	44	43	34	58	269	59	59	69	14	71	24	102	3	103	6	105	19
106	9	130	9	131	13	132	25	145	33	148	29	156	29	161	39	162	61
174	54	176	39	190	69	191	79	264	19	266	59	281	31	282	39	298	999
299	179	300	39	312	39	314	4	383	1	385	19	386	9	471	15	473	23
518	14	520	13	532	24	534	29	549	44	550	59	563	79	564	89	577	29
579	39	592	59	593	219	594	379	595	159	608	84	609	169	623	17	624	25
639	4	640	3														

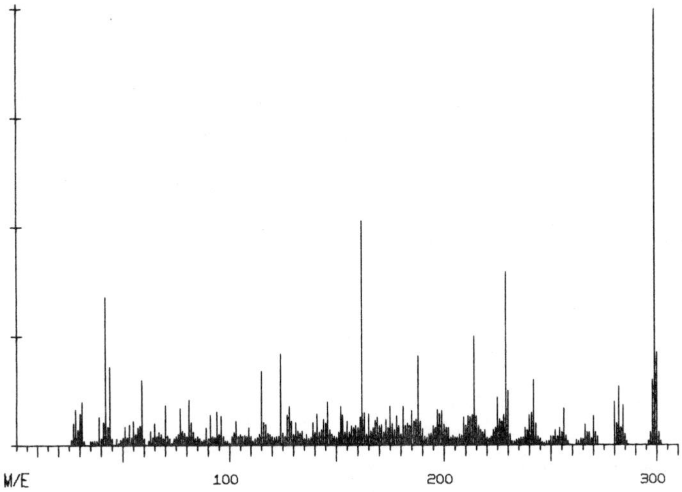
M/E
100
200
300

CODEINE

$C_{18}H_{21}NO_3$ MW = 299

Base Peak = 299 42 Peaks

Mass	Int.	Mass	Int.	Mass	Int.	Mass	Int.	Mass	Int.	Mass	Int.	Mass	Int.	Mass	Int.	Mass	Int.
28	82	31	101	42	341	44	180	55	55	59	150	65	51	70	93	77	84
81	106	91	71	94	77	115	170	116	52	124	210	128	91	141	73	144	59
146	100	152	90	162	514	163	76	175	90	181	91	188	204	197	82	213	71
214	249	225	110	229	398	230	124	242	149	254	40	256	86	266	47	270	68
280	99	282	134	298	151	299	999	300	213	301	30						

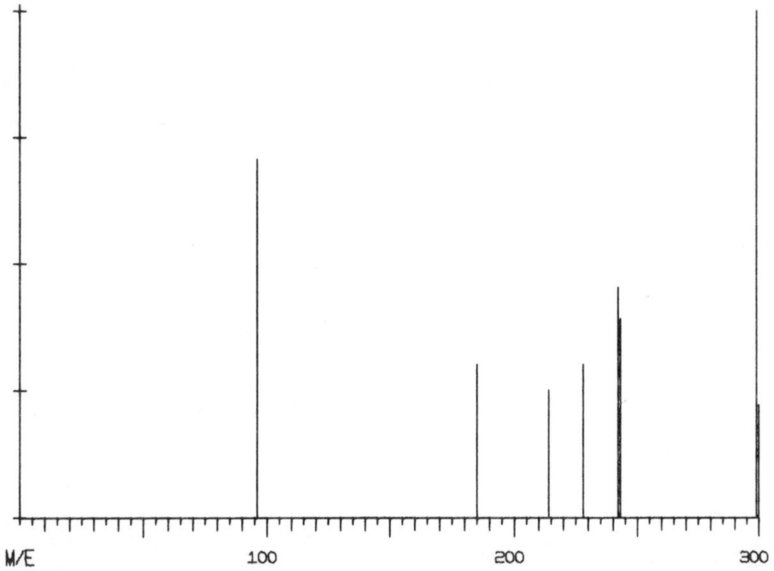
M/E
100
200
300

METOPON

$C_{18}H_{21}NO_3$ MW = 299

Base Peak = 299 8 Peaks

Mass	Int.	Mass	Int.	Mass	Int.	Mass	Int.	Mass	Int.	Mass	Int.	Mass	Int.	Mass	Int.
96	707	185	303	214	252	228	303	242	454	243	393	299	999	300	222

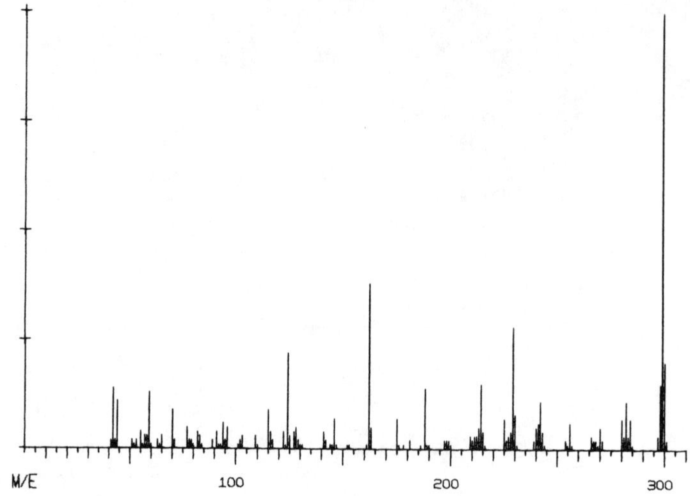
M/E
100
200
300

CODEINE

$C_{18}H_{21}NO_3$ MW = 299

Base Peak = 299 40 Peaks

Mass	Int.	Mass	Int.	Mass	Int.	Mass	Int.	Mass	Int.	Mass	Int.	Mass	Int.	Mass	Int.	Mass	Int.
42	140	44	110	55	40	59	130	65	30	70	90	77	50	82	40	94	60
96	50	115	90	116	40	124	220	128	50	141	40	142	20	146	70	147	10
162	380	163	50	175	70	181	20	188	140	197	20	213	50	214	150	225	70
229	280	230	80	242	110	254	20	256	60	266	30	270	50	280	70	282	110
298	150	299	999	300	200	301	20										

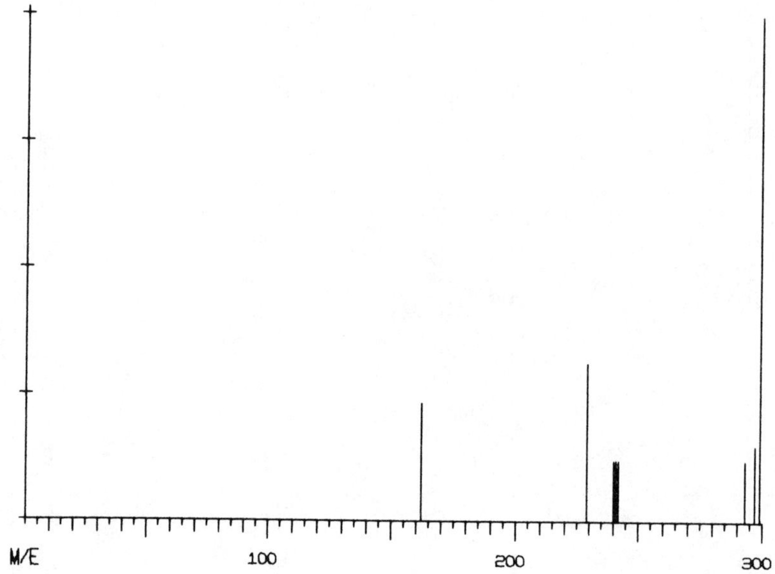
M/E
100
200
300

CODEINE-N-OXIDE

$C_{18}H_{21}NO_4$ MW = 315

Base Peak = 299 6 Peaks

Mass	Int.	Mass	Int.	Mass	Int.	Mass	Int.	Mass	Int.	Mass	Int.
162	232	229	313	240	121	241	121	297	151	299	999

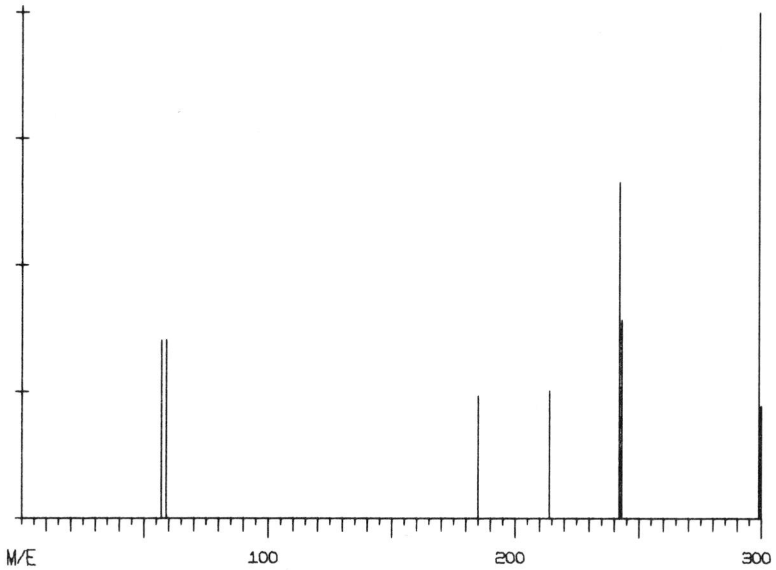
M/E
100
200
300

DIHYDROCODEINONE

$C_{18}H_{21}NO_3$ MW = 299
Base Peak = 299 8 Peaks

Mass	Int.	Mass	Int.	Mass	Int.	Mass	Int.	Mass	Int.	Mass	Int.	Mass	Int.	Mass	Int.
57	353	59	353	185	242	214	252	242	666	243	393	299	999	300	222

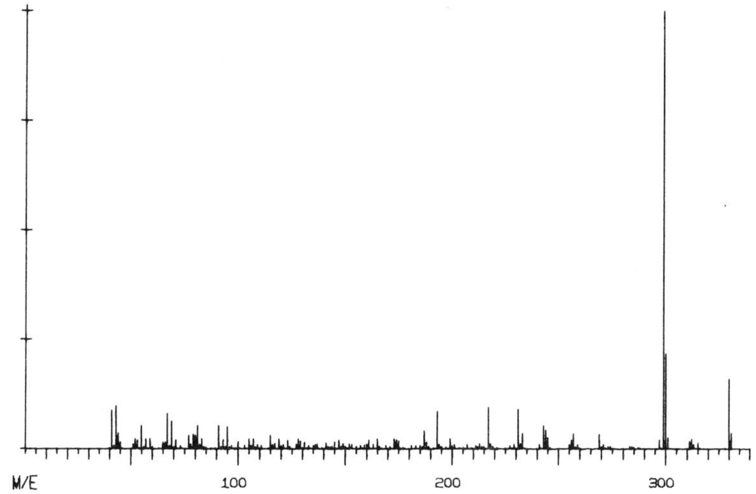
M/E
100
200
300

TETRAHYDROCANNABINOL MTB. 1

$C_{21}H_{30}O_3$ MW = 330
Base Peak = 299 43 Peaks

Mass	Int.	Mass	Int.	Mass	Int.	Mass	Int.	Mass	Int.	Mass	Int.	Mass	Int.	Mass	Int.	Mass	Int.
41	87	43	97	52	22	55	53	67	81	69	63	79	33	81	53	91	52
95	50	105	23	115	31	119	23	128	22	141	13	145	15	147	21	149	13
165	23	173	22	174	21	187	39	193	85	199	22	207	9	213	12	217	94
218	12	231	91	243	53	244	42	257	34	259	11	269	33	274	6	284	6
297	21	299	999	300	218	301	25	315	15	330	161	331	35				

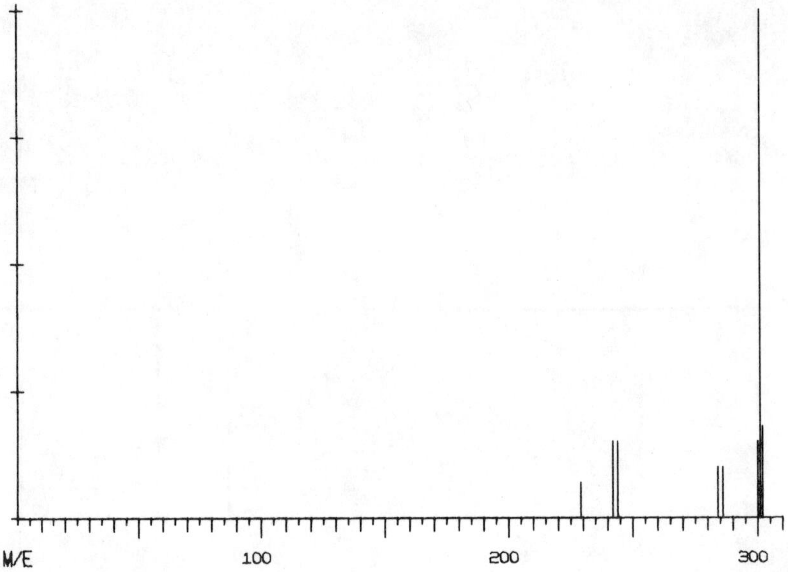

M/E
100
200
300

DIHYDROCODEINE

$C_{18}H_{23}NO_3$ MW = 301

Base Peak = 301 7 Peaks

Mass	Int.	Mass	Int.	Mass	Int.	Mass	Int.	Mass	Int.	Mass	Int.	Mass	Int.
229	70	242	151	244	151	284	101	286	101	301	999	302	181

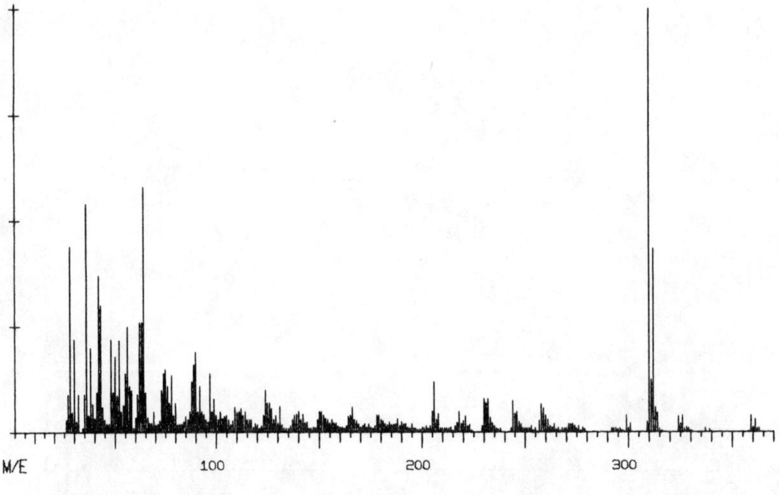

M/E
100
200
300

METHYCHLOTHIAZIDE

$C_9H_{11}CL_2N_3O_4S_2$ MW = 359

Base Peak = 310 48 Peaks

Mass	Int.	Mass	Int.	Mass	Int.	Mass	Int.	Mass	Int.	Mass	Int.	Mass	Int.	Mass	Int.	Mass	Int.
28	440	30	219	36	539	42	369	48	219	56	249	62	259	64	579	78	135
89	159	90	189	97	140	109	59	112	58	124	99	125	70	139	43	140	51
150	49	151	51	165	41	166	60	178	39	179	39	188	33	190	26	205	49
206	119	218	49	221	38	230	80	232	79	244	74	246	49	258	68	259	58
272	20	273	20	294	11	299	41	310	999	312	434	314	47	326	39	328	13
337	11	359	41	361	29												

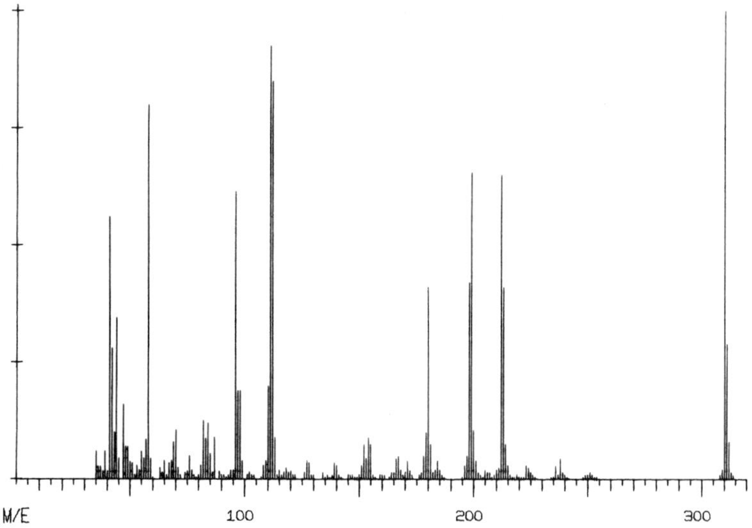
M/E
100
200
300

MEPAZINE

$C_{19}H_{22}N_2S$ MW = 310

Base Peak = 310 35 Peaks

Mass	Int.	Mass	Int.	Mass	Int.	Mass	Int.	Mass	Int.	Mass	Int.	Mass	Int.	Mass	Int.	Mass	Int.
41	559	44	344	57	84	58	799	69	81	70	104	82	124	84	121	96	614
97	189	111	924	112	849	127	39	128	36	139	34	140	29	152	74	154	89
166	44	167	49	179	99	180	409	198	419	199	655	212	649	213	409	223	29
224	24	236	29	238	44	249	9	251	14	310	999	311	289	314	9		

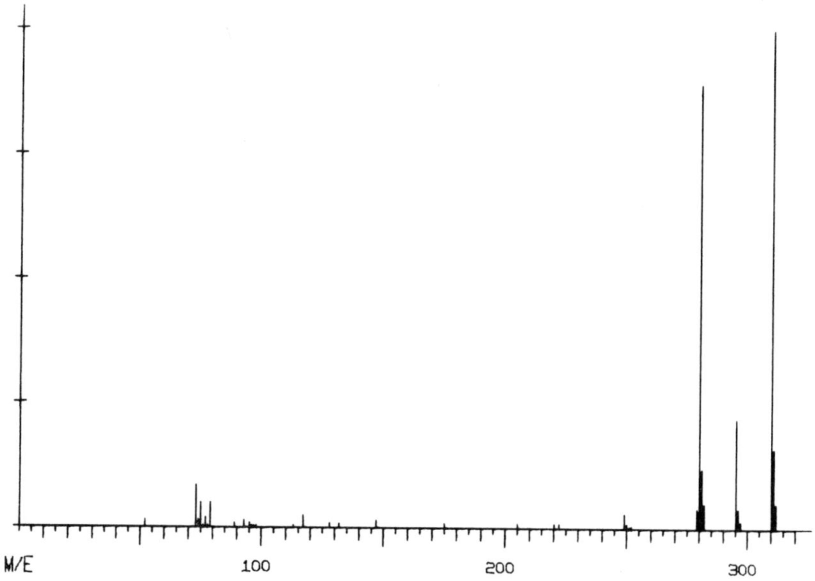
M/E
100
200
300

3,5-DIMETHOXY-4-HYDROXYCINNAMIC ACID METHYL ESTER TRIMETHYLSILYL ETHER

$C_{15}H_{22}O_5SI$ MW = 310

Base Peak = 310 24 Peaks

Mass	Int.	Mass	Int.	Mass	Int.	Mass	Int.	Mass	Int.	Mass	Int.	Mass	Int.	Mass	Int.	Mass	Int.
52	14	73	84	75	49	77	19	79	49	93	14	95	9	113	4	117	24
128	9	132	9	147	14	175	9	205	9	220	9	222	9	249	29	250	9
280	889	281	119	295	219	296	39	310	999	311	159						

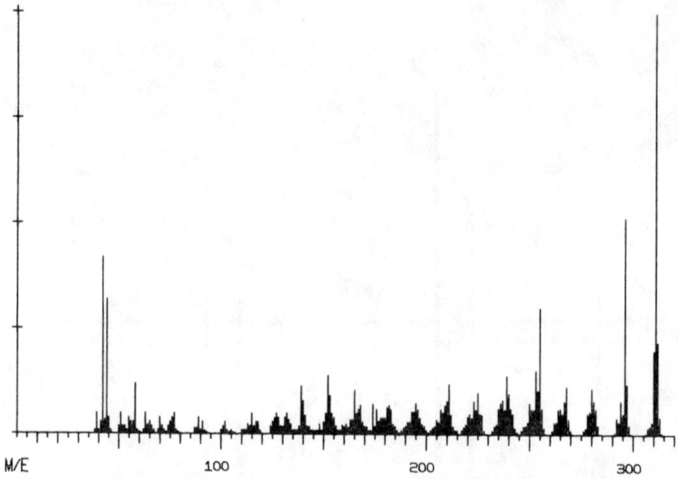
M/E
100
200
300

THEBAINE

$C_{19}H_{21}NO_3$ MW = 311

Base Peak = 311 41 Peaks

Mass	Int.	Mass	Int.	Mass	Int.	Mass	Int.	Mass	Int.	Mass	Int.	Mass	Int.	Mass	Int.	Mass	Int.
42	420	44	320	51	50	58	120	63	49	70	40	76	40	77	50	91	30
102	30	115	50	117	30	126	40	127	49	139	115	140	80	152	140	153	92
165	105	168	70	174	72	182	70	195	75	196	60	210	80	211	121	223	79
225	99	239	140	240	98	253	152	255	300	267	80	268	114	280	110	281	79
296	514	297	120	311	999	312	221	314	4								

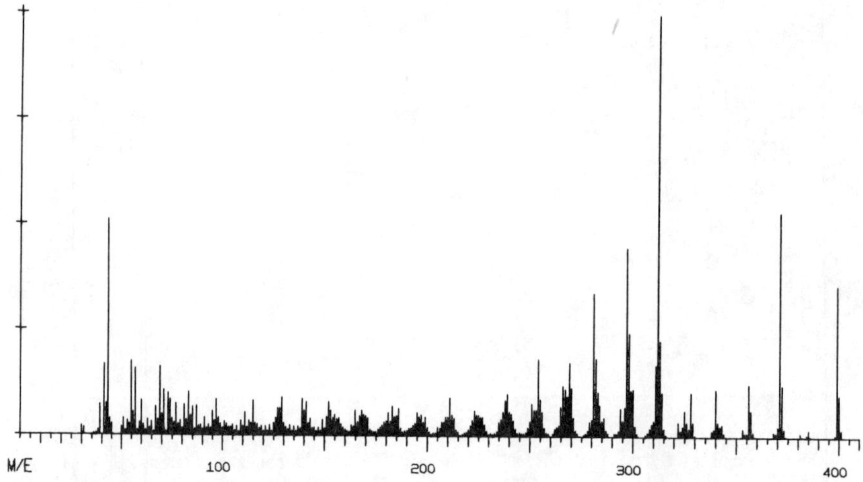
M/E
100
200
300
400

COLCHICINE

$C_{22}H_{25}NO_6$ MW = 399

Base Peak = 312 56 Peaks

Mass	Int.	Mass	Int.	Mass	Int.	Mass	Int.	Mass	Int.	Mass	Int.	Mass	Int.	Mass	Int.	Mass	Int.
30	22	31	18	41	167	43	509	55	175	57	157	69	162	71	107	77	74
83	102	95	58	97	86	111	56	115	83	127	66	129	90	139	88	141	80
152	80	153	59	165	61	168	59	183	69	186	65	195	54	197	51	211	91
212	49	223	59	224	50	238	84	239	101	254	183	255	88	266	119	269	175
281	340	282	186	297	448	298	246	312	999	313	227	314	104	325	63	328	105
340	113	342	25	343	29	356	125	357	62	371	533	372	123	384	6	385	17
399	361	400	101														

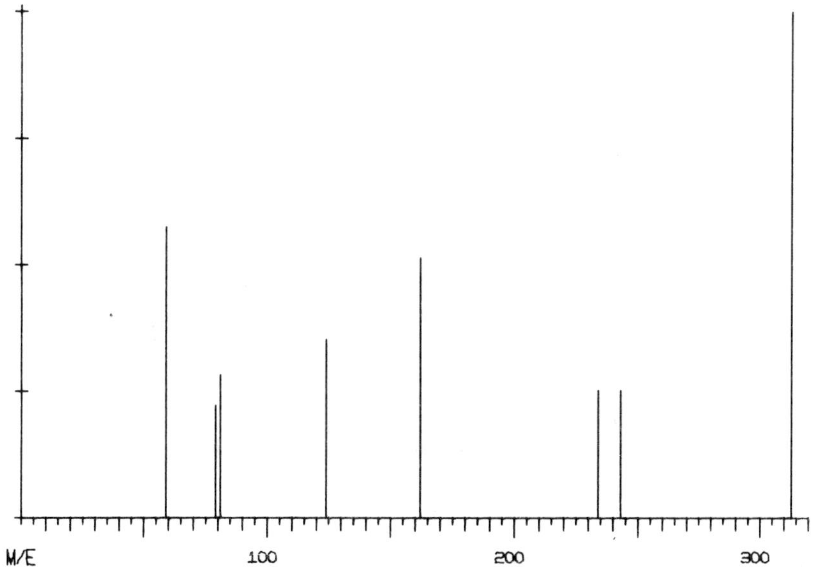
M/E
100
200
300

ETHYLMORPHINE

$C_{19}H_{23}NO_3$ MW = 313

Base Peak = 313 8 Peaks

Mass	Int.	Mass	Int.	Mass	Int.	Mass	Int.	Mass	Int.	Mass	Int.	Mass	Int.	Mass	Int.
59	575	79	222	81	282	124	353	162	515	234	252	243	252	313	999

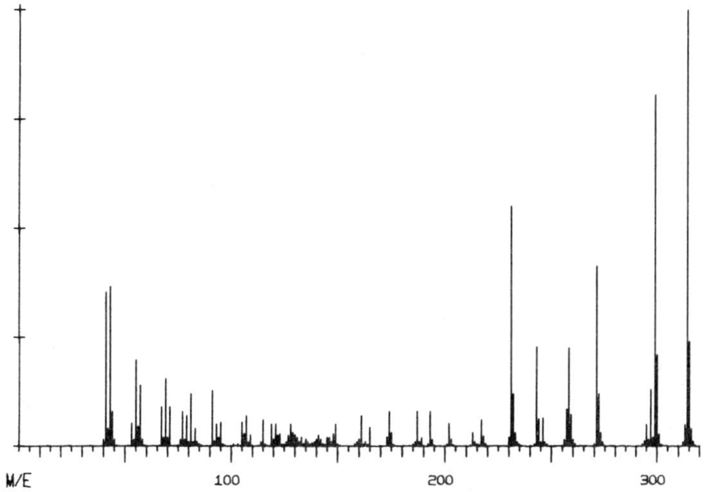
M/E
100
200
300

TETRAHYDROCANNABINOL

$C_{21}H_{30}O_2$ MW = 314

Base Peak = 314 42 Peaks

Mass	Int.	Mass	Int.	Mass	Int.	Mass	Int.	Mass	Int.	Mass	Int.	Mass	Int.	Mass	Int.	Mass	Int.
41	352	43	364	55	198	57	141	69	154	71	91	77	81	81	120	91	127
95	54	107	69	115	59	121	51	128	51	141	24	145	21	148	27	149	49
161	71	165	42	174	81	187	79	189	21	193	81	202	53	213	31	217	61
218	22	231	551	243	228	246	64	257	84	258	226	271	412	272	120	273	29
297	130	299	805	300	210	313	51	314	999	315	241						

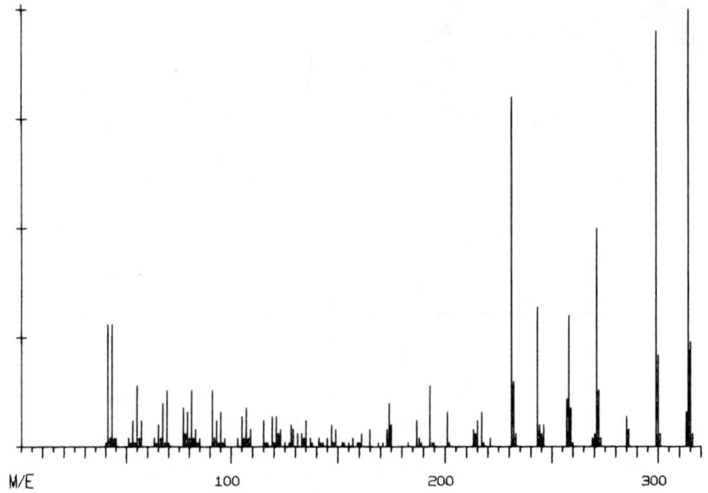
M/E
100
200
300

TETRAHYDROCANNABINOL

$C_{21}H_{30}O_2$ MW = 314

Base Peak = 314 42 Peaks

Mass	Int.	Mass	Int.	Mass	Int.	Mass	Int.	Mass	Int.	Mass	Int.	Mass	Int.	Mass	Int.	Mass	Int.
41	280	43	280	53	60	55	140	67	100	69	130	77	90	81	130	91	130
95	80	105	70	107	90	119	70	121	70	133	30	135	60	147	50	149	40
165	40	173	40	174	100	187	60	193	140	201	80	213	40	215	60	217	80
221	20	231	800	243	320	244	50	257	110	258	300	271	500	272	130	285	70
286	40	299	950	300	210	313	80	314	999	315	240						

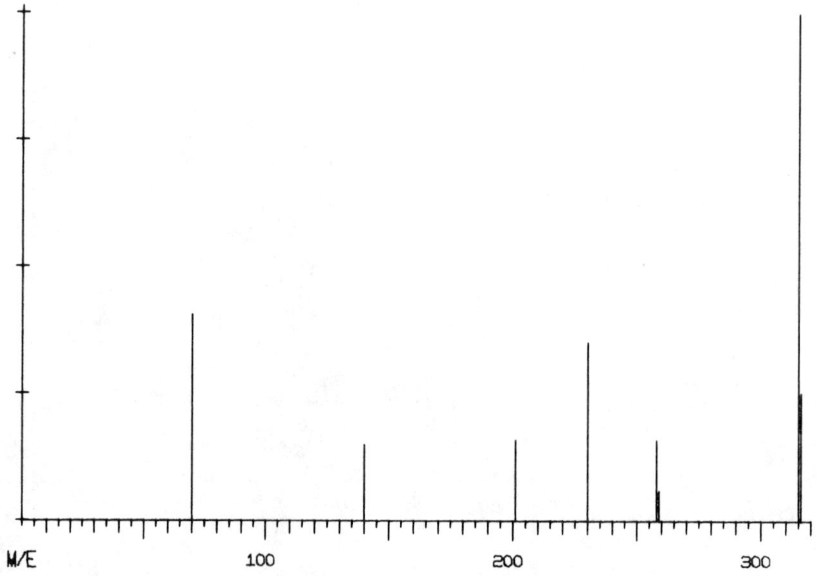
M/E
100
200
300

OXYCODONE

$C_{18}H_{21}NO_4$ MW = 315

Base Peak = 315 8 Peaks

Mass	Int.	Mass	Int.	Mass	Int.	Mass	Int.	Mass	Int.	Mass	Int.	Mass	Int.	Mass	Int.
70	404	140	151	201	161	230	353	258	161	259	60	315	999	316	252

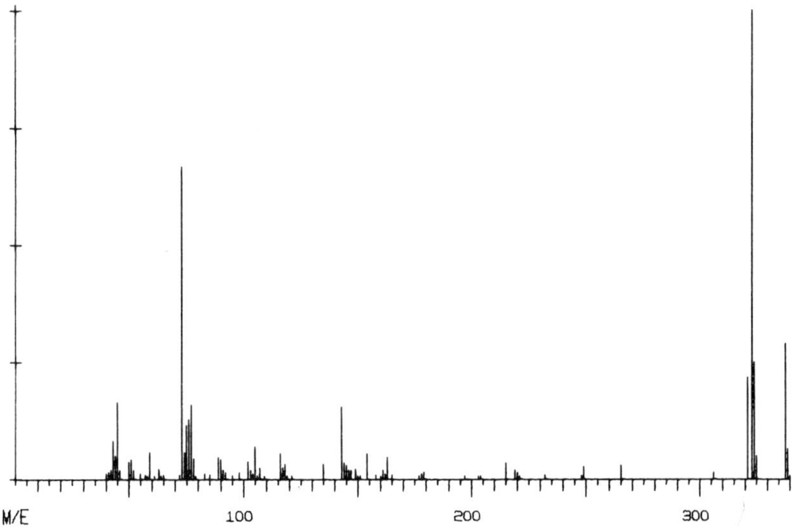

M/E
100
200
300

METHAQUALONE MTB. 3 TMS

$C_{19}H_{22}N_2O_2SI$ MW = 338

Base Peak = 323   36 Peaks

Mass	Int.	Mass	Int.	Mass	Int.	Mass	Int.	Mass	Int.	Mass	Int.	Mass	Int.	Mass	Int.	Mass	Int.
43	82	45	166	51	42	59	58	73	667	75	114	76	128	77	161	90	42
102	37	105	70	116	56	118	33	119	7	143	156	144	35	149	23	154	56
161	19	163	47	178	12	179	14	197	7	203	7	215	35	219	19	220	14
232	9	248	7	249	28	265	30	306	16	323	999	324	249	338	291	339	65

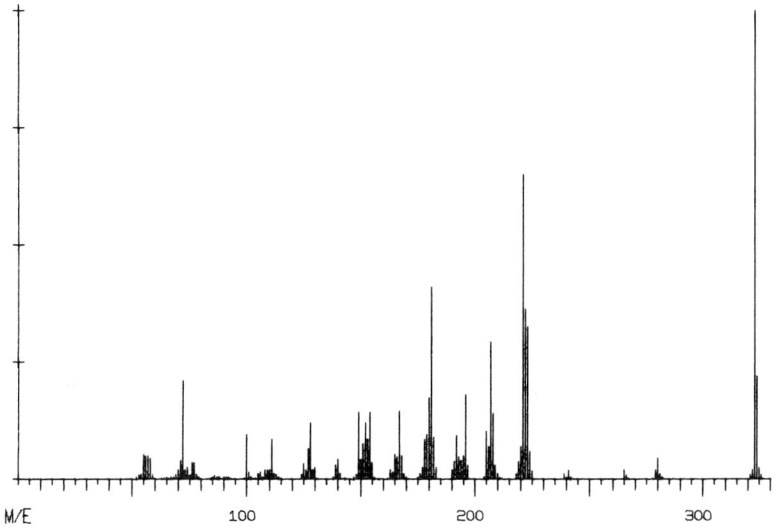

M/E
100
200
300

LYSERGIDE

$C_{20}H_{25}N_3O$ MW = 323

Base Peak = 323   35 Peaks

Mass	Int.	Mass	Int.	Mass	Int.	Mass	Int.	Mass	Int.	Mass	Int.	Mass	Int.	Mass	Int.	Mass	Int.
55	52	57	50	71	41	72	211	76	35	77	34	100	95	101	14	108	21
111	84	127	64	128	120	139	31	140	42	149	142	154	142	165	52	167	145
180	172	181	411	192	92	196	179	207	292	208	140	221	649	222	362	239	12
241	21	244	1	265	19	266	10	279	21	280	45	323	999	324	221		

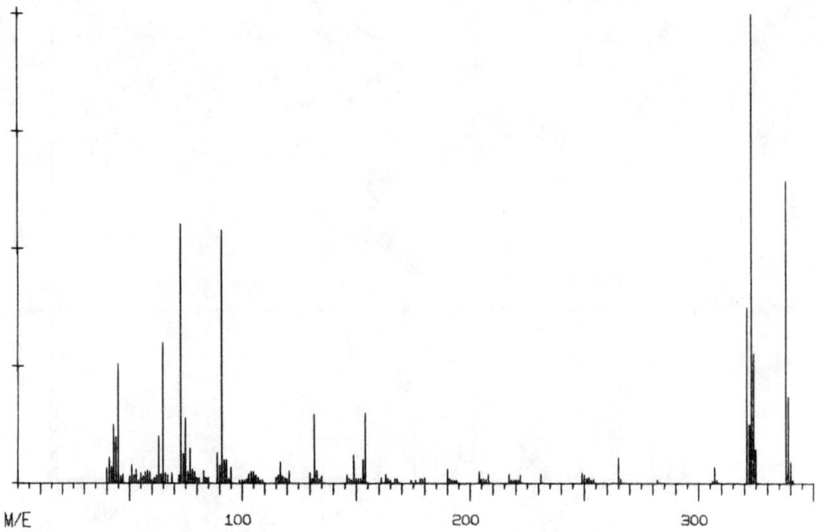
M/E
100
200
300

METHAQUALONE MTB. 5 TMS

$C_{19}H_{22}N_2O_2SI$ MW = 338

Base Peak = 323   40 Peaks

Mass	Int.	Mass	Int.	Mass	Int.	Mass	Int.	Mass	Int.	Mass	Int.	Mass	Int.	Mass	Int.	Mass	Int.
43	126	45	255	51	41	53	31	65	300	73	553	77	74	89	66	91	539
93	50	105	26	117	45	121	27	130	23	132	148	133	27	149	60	154	149
161	13	163	20	178	10	180	13	190	29	191	10	204	26	208	20	217	20
222	17	231	20	249	23	250	17	265	56	266	10	282	8	307	35	308	7
321	376	323	999	338	644	339	185										

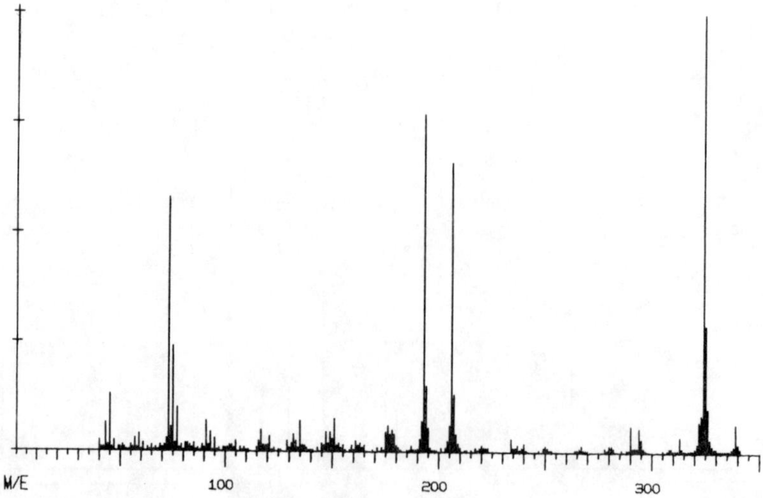
M/E
100
200
300

2-HYDROXYBENZOYLGLYCINE TMS ESTER TMS ETHER

$C_{15}H_{25}NO_4Si_2$ MW = 339

Base Peak = 324   44 Peaks

Mass	Int.	Mass	Int.	Mass	Int.	Mass	Int.	Mass	Int.	Mass	Int.	Mass	Int.	Mass	Int.	Mass	Int.
43	65	45	130	57	30	59	40	73	580	75	240	76	20	77	100	91	70
93	45	105	25	117	50	121	35	129	25	132	40	135	70	147	45	151	75
161	25	163	20	176	60	178	50	193	770	194	150	206	660	207	130	217	10
220	15	234	30	240	20	249	10	250	15	266	10	267	15	280	15	281	15
290	60	294	55	308	10	313	35	324	999	325	290	339	65	340	20		

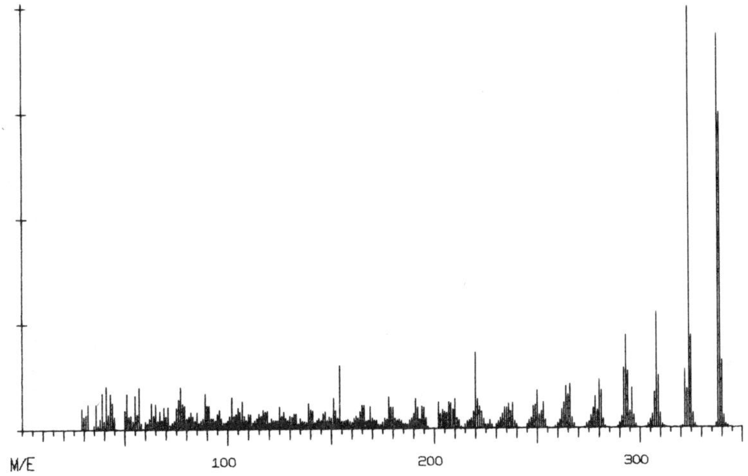
M/E
100
200
300

PAPAVERINE

$C_{20}H_{21}NO_4$ MW = 339

Base Peak = 324   48 Peaks

Mass	Int.	Mass	Int.	Mass	Int.	Mass	Int.	Mass	Int.	Mass	Int.	Mass	Int.	Mass	Int.	Mass	Int.
29	50	32	60	39	88	41	102	51	84	57	99	63	62	65	60	77	100
89	86	90	54	102	76	105	50	107	65	119	42	125	53	139	59	140	44
151	72	154	150	165	56	166	54	178	74	179	50	191	71	194	52	202	62
210	70	220	180	221	70	236	60	238	60	250	90	253	62	264	101	266	105
280	114	281	89	292	142	293	220	308	276	309	126	324	999	325	220	338	935
339	750	342	10	343	4												

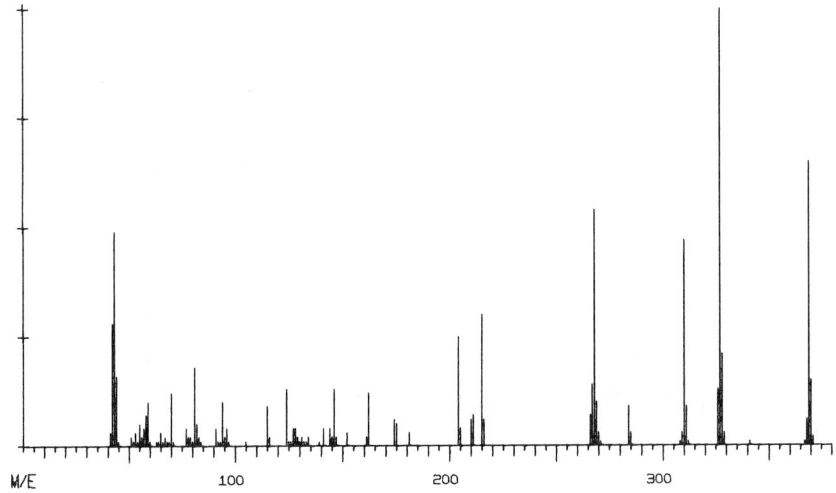
M/E
100
200
300

HEROIN

$C_{21}H_{23}NO_5$ MW = 369

Base Peak = 327   39 Peaks

Mass	Int.	Mass	Int.	Mass	Int.	Mass	Int.	Mass	Int.	Mass	Int.	Mass	Int.	Mass	Int.	Mass	Int.
42	280	43	490	58	70	59	100	65	30	70	120	81	180	82	50	91	40
94	100	115	90	116	20	124	130	127	40	141	40	144	40	146	130	152	30
161	20	162	120	174	60	175	50	204	250	215	300	216	60	267	140	268	540
284	90	285	30	310	470	311	90	326	130	327	999	328	210	329	30	368	60
369	650	370	150	371	20												

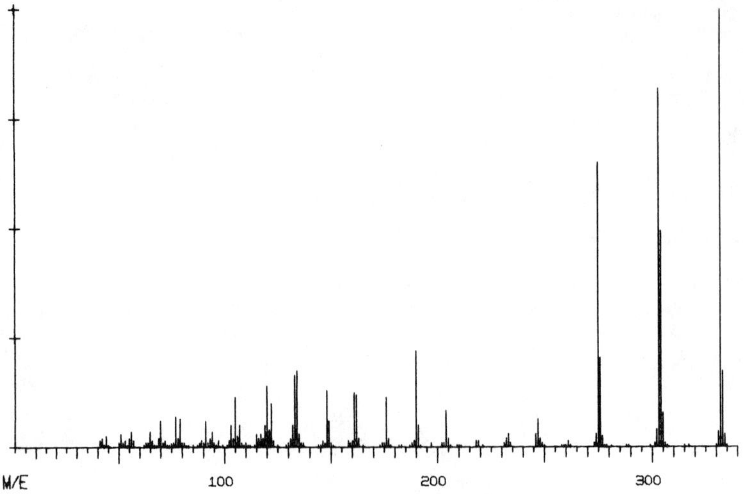
M/E
100
200
300

HYDROXYPHENOBARBITAL 1,3-DI-ETHYL DERIVATIVE ETHYL ETHER

$C_{18}H_{24}N_2O_4$ MW = 332

Base Peak = 332 44 Peaks

Mass	Int.	Mass	Int.	Mass	Int.	Mass	Int.	Mass	Int.	Mass	Int.	Mass	Int.	Mass	Int.	Mass	Int.
42	19	44	24	51	29	56	34	65	34	70	59	77	69	79	64	91	59
103	49	105	114	107	49	120	139	122	99	133	164	134	174	148	129	149	59
161	124	162	119	176	114	177	19	190	219	191	49	204	84	205	19	218	14
219	14	232	19	233	29	246	29	247	64	258	4	261	14	275	649	276	204
288	4	289	4	303	819	304	494	315	4	317	4	332	999	333	174		

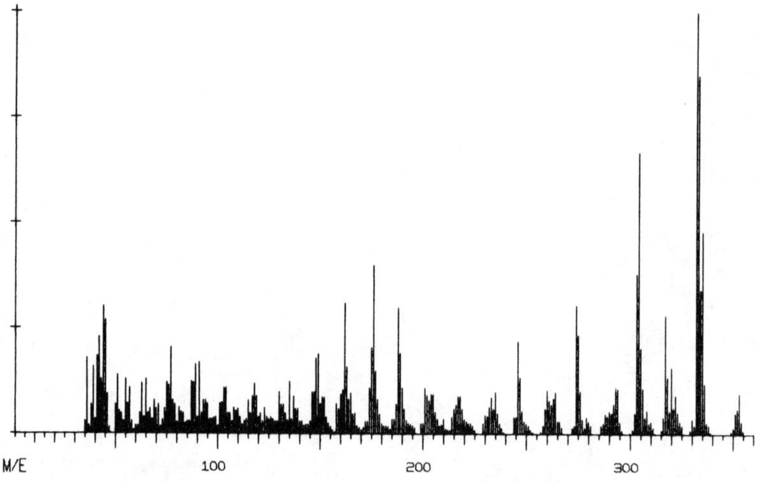
M/E
100
200
300

CHELIDONINE

$C_{20}H_{19}NO_5$ MW = 353

Base Peak = 332 46 Peaks

Mass	Int.	Mass	Int.	Mass	Int.	Mass	Int.	Mass	Int.	Mass	Int.	Mass	Int.	Mass	Int.	Mass	Int.
44	302	45	270	51	141	55	131	65	131	75	122	77	204	89	164	91	171
103	110	104	109	117	90	118	120	130	100	135	124	137	90	148	180	149	190
162	310	163	160	175	204	176	400	188	300	189	192	204	95	205	96	217	89
218	90	233	87	235	100	246	220	247	135	260	104	264	100	274	306	275	236
293	112	294	108	303	380	304	667	317	282	320	159	332	999	333	850	352	60
353	97																

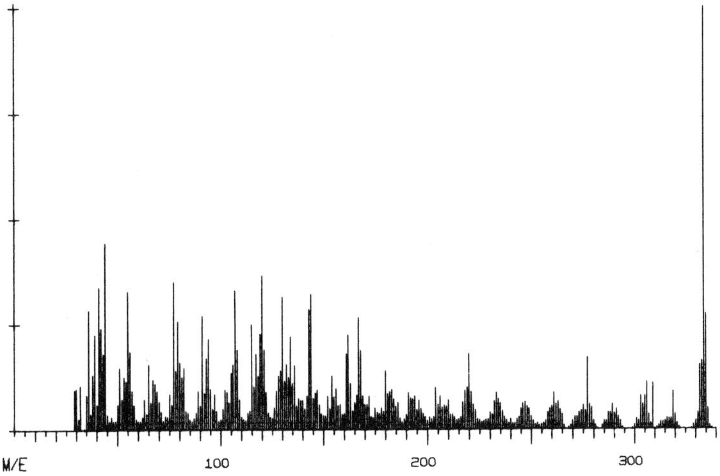
M/E
100
200
300

STRYCHNINE

$C_{21}H_{22}N_2O_2$ MW = 334

Base Peak = 334 46 Peaks

Mass	Int.	Mass	Int.	Mass	Int.	Mass	Int.	Mass	Int.	Mass	Int.	Mass	Int.	Mass	Int.	Mass	Int.
30	97	32	105	41	338	44	442	55	328	56	185	65	154	67	119	77	349
79	257	91	269	94	215	107	329	115	249	120	365	130	314	143	285	144	320
154	127	156	98	162	225	167	264	180	141	183	94	191	87	194	81	204	100
206	81	219	101	220	181	233	87	234	72	246	62	247	66	261	87	263	67
277	170	278	59	289	59	291	50	306	112	309	110	319	91	320	36	334	999
335	272																

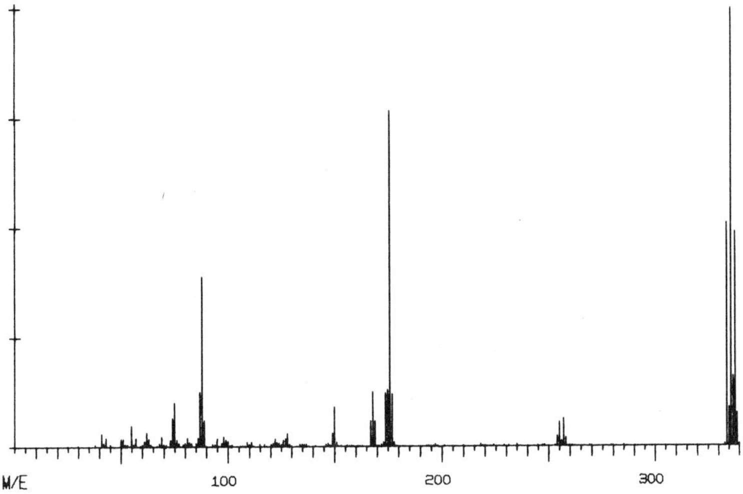
M/E
100
200
300

9,10-DIBROMOANTHRACENE

$C_{14}H_8BR_2$ MW = 334

Base Peak = 336 40 Peaks

Mass	Int.	Mass	Int.	Mass	Int.	Mass	Int.	Mass	Int.	Mass	Int.	Mass	Int.	Mass	Int.	Mass	Int.
30	2	31	1	41	30	43	19	55	47	57	19	74	64	75	101	87	125
88	387	95	17	98	23	109	9	111	9	122	18	128	31	134	5	135	6
149	30	150	89	168	126	169	58	175	129	176	767	195	3	197	5	205	1
210	3	218	5	219	3	231	2	235	4	255	56	257	63	258	20	259	3
279	3	285	3	334	510	336	999										

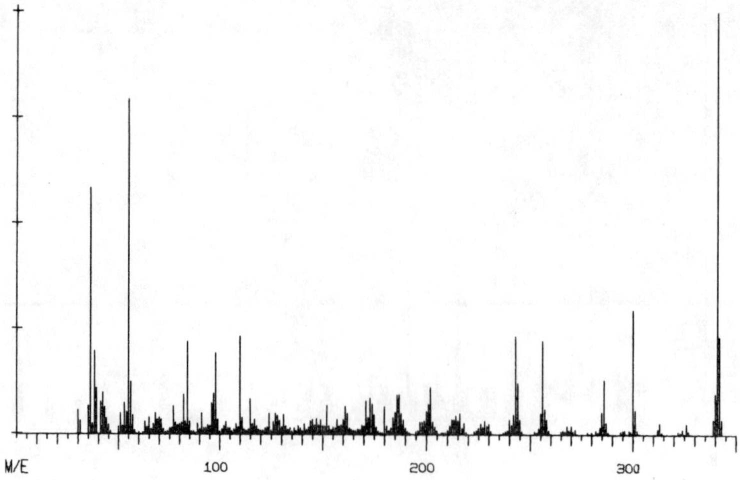
M/E
100
200
300

NALTREXONE

$C_{20}H_{23}NO_4$ MW = 341

Base Peak = 341 48 Peaks

Mass	Int.	Mass	Int.	Mass	Int.	Mass	Int.	Mass	Int.	Mass	Int.	Mass	Int.	Mass	Int.	Mass	Int.
30	57	31	33	36	582	38	198	55	792	56	126	65	41	68	51	82	94
84	219	97	97	98	192	110	232	115	84	124	50	127	50	144	33	145	35
147	38	152	71	171	80	173	87	186	94	187	94	200	56	201	73	202	112
214	45	216	50	228	33	242	48	243	232	244	122	256	222	258	35	270	21
284	18	285	53	286	130	287	28	300	295	301	58	324	13	326	25	340	98
341	999	342	233	343	36												

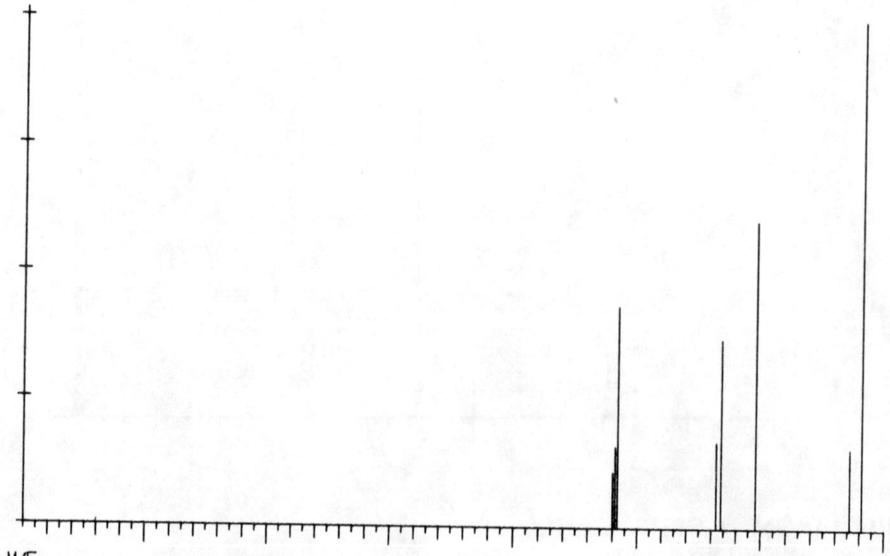
M/E
100
200
300

THEBACON

$C_{20}H_{23}NO_4$ MW = 341

Base Peak = 341 7 Peaks

Mass	Int.	Mass	Int.	Mass	Int.	Mass	Int.	Mass	Int.	Mass	Int.	Mass	Int.
241	161	242	434	282	171	284	373	298	606	336	161	341	999

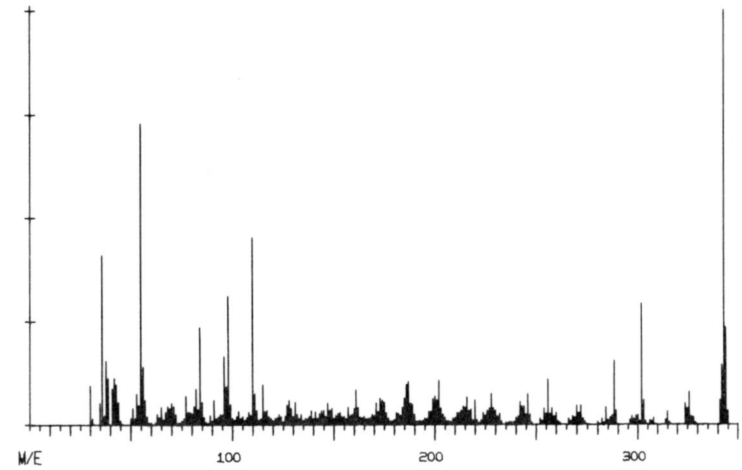
M/E
100
200
300

ALPHA-HYDROXY NALTREXONE

$C_{20}H_{25}NO_4$ MW = 343

Base Peak = 343 48 Peaks

Mass	Int.	Mass	Int.	Mass	Int.	Mass	Int.	Mass	Int.	Mass	Int.	Mass	Int.	Mass	Int.	Mass	Int.
30	94	31	15	36	411	38	155	55	728	56	141	70	53	71	46	82	87
84	236	96	164	98	311	110	452	115	98	128	59	131	55	139	34	145	36
147	52	157	43	161	85	173	66	186	98	187	105	199	63	200	70	202	108
214	46	216	67	228	74	242	55	243	45	246	76	256	111	258	40	270	48
272	48	284	43	288	155	289	34	302	293	303	59	324	52	326	80	328	18
341	59	343	999	344	234												

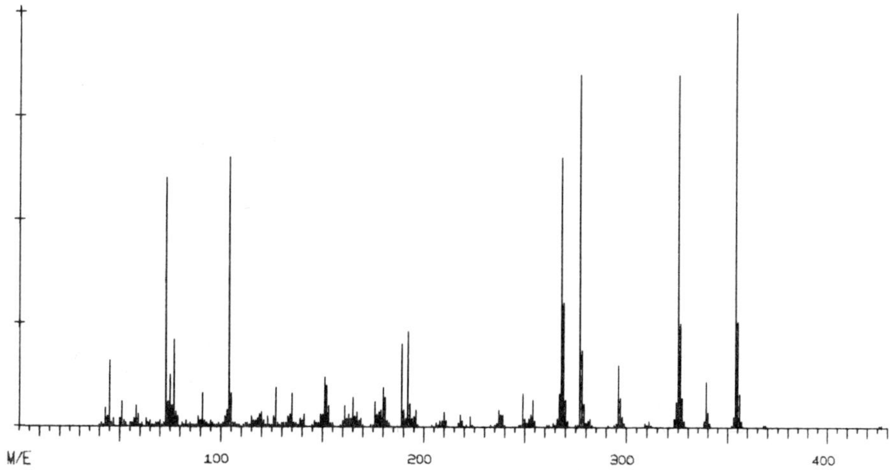
M/E
100
200
300
400

5-(4-HYDROXYPHENYL)-3-METHYL-5-PHENYLHYDANTOIN TMS ETHER

$C_{19}H_{22}N_2O_3Si$ MW = 354

Base Peak = 354 50 Peaks

Mass	Int.	Mass	Int.	Mass	Int.	Mass	Int.	Mass	Int.	Mass	Int.	Mass	Int.	Mass	Int.	Mass	Int.
43	45	45	160	51	60	58	50	73	600	75	125	76	50	77	210	91	80
103	40	104	650	105	80	120	35	127	95	134	30	135	80	151	120	152	100
161	50	165	70	180	95	181	70	189	200	192	230	208	15	210	35	218	30
223	25	237	40	238	30	249	80	254	65	268	650	269	300	277	850	278	185
296	150	297	70	309	10	311	15	325	850	326	250	339	110	340	35	354	999
355	255	356	80	357	15	426	5	427	5								

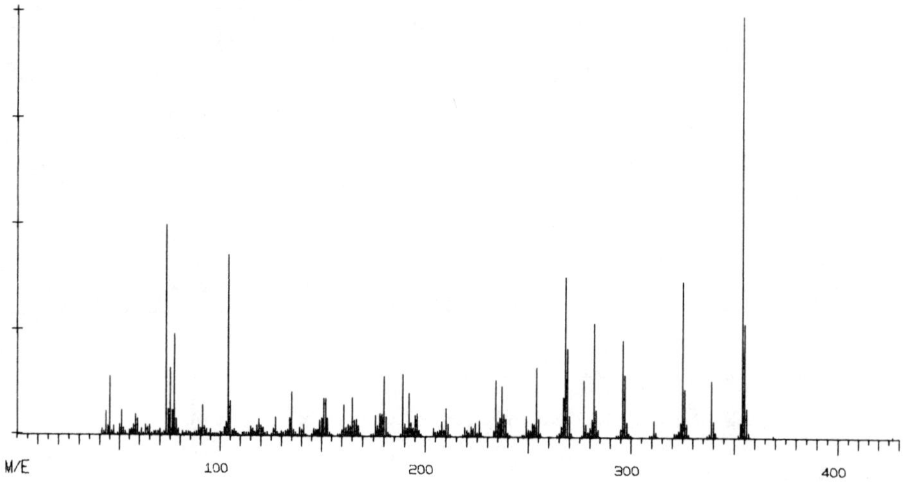
M/E
100
200
300
400

5-(3-HYDROXYPHENYL)-3-METHYL-5-PHENYLHYDANTOIN TMS ETHER

$C_{19}H_{22}N_2O_3Si$ MW = 354

Base Peak = 354 49 Peaks

Mass	Int.	Mass	Int.	Mass	Int.	Mass	Int.	Mass	Int.	Mass	Int.	Mass	Int.	Mass	Int.	Mass	Int.
43	58	45	139	51	59	58	49	73	499	75	159	76	61	77	239	91	73
103	33	104	429	105	83	119	39	127	44	134	42	135	104	151	89	152	91
161	76	165	92	178	56	180	143	189	147	192	103	208	34	210	67	224	32
226	37	234	134	237	119	249	51	254	164	268	379	269	209	277	136	282	269
296	229	297	147	311	39	312	13	325	369	326	114	339	136	340	37	354	999
355	269	356	71	357	13	427	4										

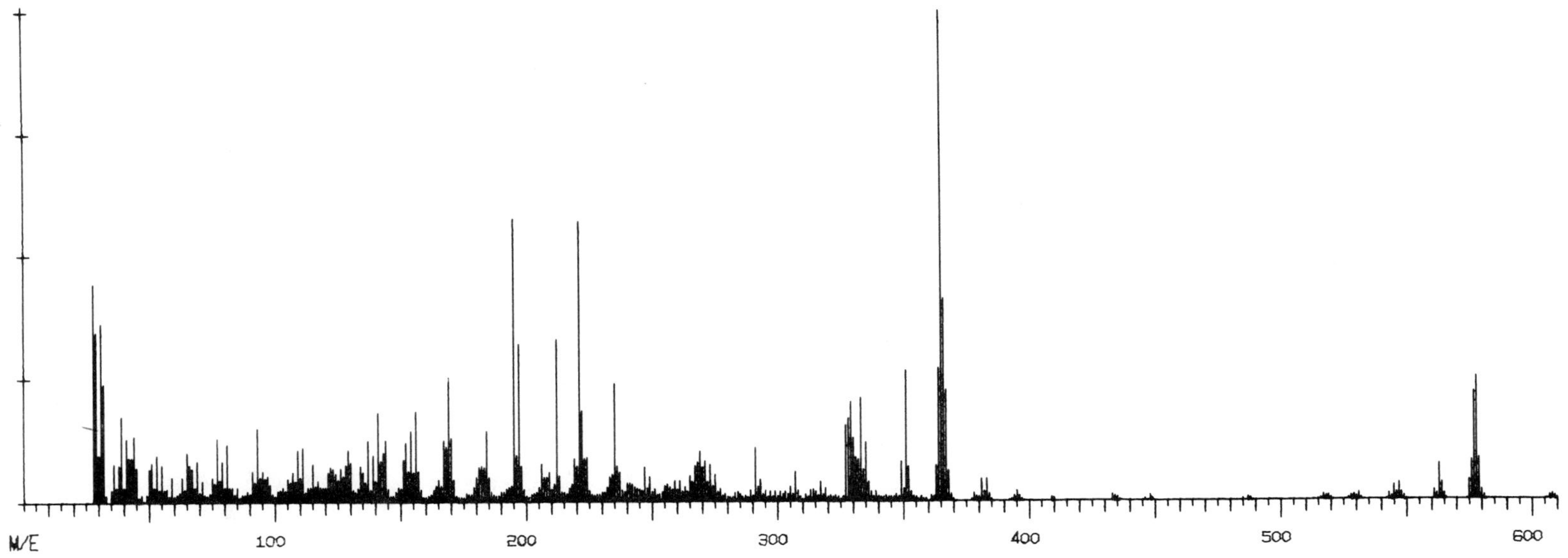
M/E
100
200
300
400
500
600

DESERPIDINE

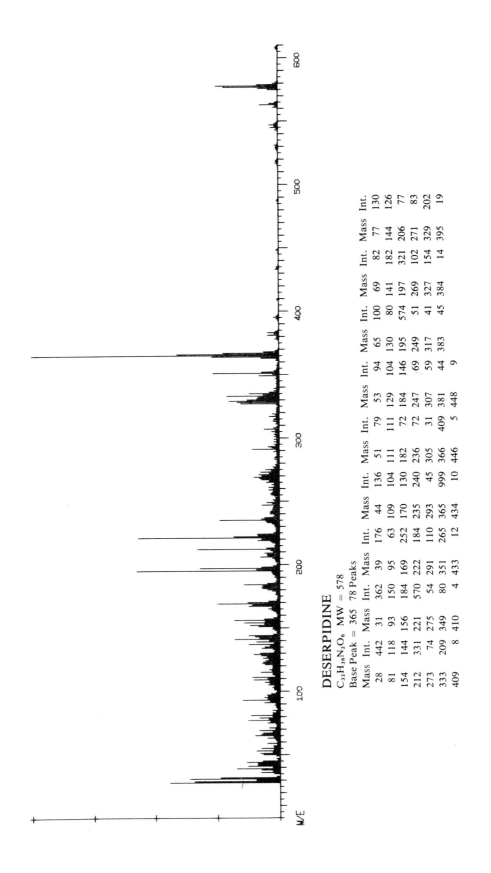

$C_{32}H_{38}N_2O_8$ MW = 578

Base Peak = 365 78 Peaks

Mass	Int.	Mass	Int.	Mass	Int.	Mass	Int.	Mass	Int.	Mass	Int.	Mass	Int.	Mass	Int.	Mass	Int.
28	442	31	362	39	176	44	136	51	79	53	94	65	100	69	82	77	130
81	118	93	150	95	63	109	104	111	111	129	104	130	80	141	182	144	126
154	144	156	184	169	252	170	130	182	72	184	146	195	574	197	321	206	77
212	331	221	570	222	184	235	240	236	72	247	69	249	51	269	102	271	83
273	74	275	54	291	110	293	45	305	31	307	59	317	41	327	154	329	202
333	209	349	80	351	265	365	999	366	409	381	44	383	45	384	14	395	19
409	8	410	4	433	12	434	10	446	5	448	9						

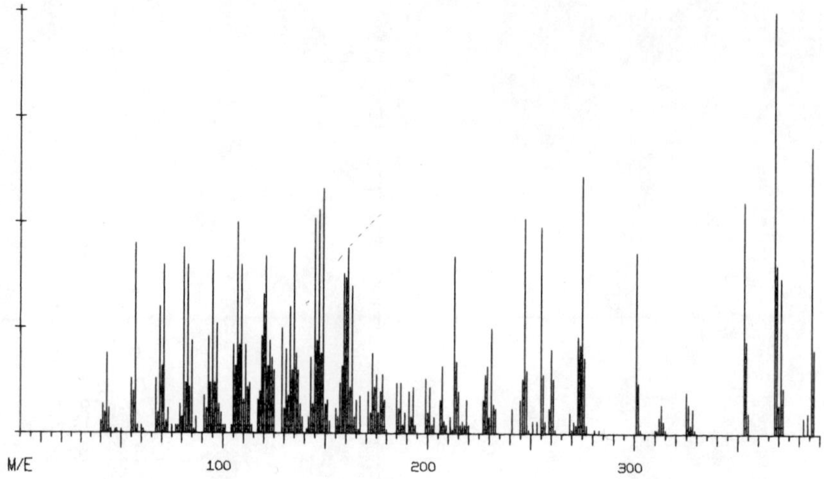
M/E
100
200
300

CHOLESTEROL

$C_{27}H_{46}O$ MW = 386

Base Peak = 368 50 Peaks

Mass	Int.	Mass	Int.	Mass	Int.	Mass	Int.	Mass	Int.	Mass	Int.	Mass	Int.	Mass	Int.	Mass	Int.
41	70	43	190	55	130	57	450	69	300	71	400	81	440	83	400	95	410
97	260	107	500	109	400	120	330	121	420	135	440	145	510	147	530	149	580
160	370	161	440	175	140	178	140	193	110	199	130	213	420	214	170	228	140
229	160	231	250	232	70	247	510	255	490	260	200	261	130	273	230	275	610
301	430	302	120	325	100	326	70	328	60	329	10	353	550	354	220	368	999
369	400	371	370	372	110	386	680	387	200								

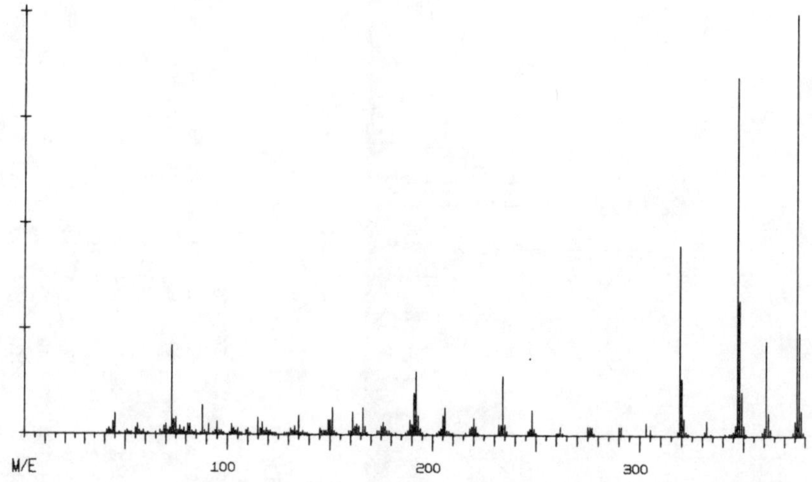
M/E
100
200
300

HYDROXYPHENOBARBITAL 1,3-DI-ETHYL DERIVATIVE TMS ETHER

$C_{19}H_{28}N_2O_4Si$ MW = 376

Base Peak = 376 50 Peaks

Mass	Int.	Mass	Int.	Mass	Int.	Mass	Int.	Mass	Int.	Mass	Int.	Mass	Int.	Mass	Int.	Mass	Int.
44	29	45	49	55	14	56	24	73	209	75	39	81	24	88	69	91	24
95	29	115	39	117	29	119	14	131	14	133	19	135	44	149	34	151	64
161	54	166	64	175	19	176	29	191	99	192	149	205	44	206	64	219	19
220	39	232	24	234	139	247	14	248	59	260	4	262	19	275	19	276	19
290	19	291	19	303	29	305	14	319	449	320	134	331	9	332	34	347	849
348	319	361	224	362	54	376	999	377	244								

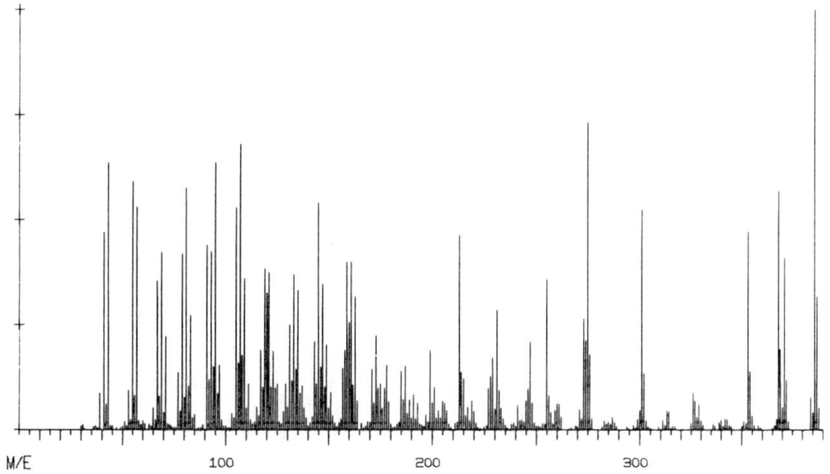
M/E
100
200
300

CHOLESTEROL

$C_{27}H_{46}O$ MW = 386

Base Peak = 386 54 Peaks

Mass	Int.	Mass	Int.	Mass	Int.	Mass	Int.	Mass	Int.	Mass	Int.	Mass	Int.	Mass	Int.	Mass	Int.
30	11	31	13	41	470	43	636	55	589	57	529	67	356	69	422	79	420
81	576	91	439	95	636	105	531	107	681	119	384	121	376	133	370	145	541
147	348	159	400	161	399	163	318	178	154	187	152	199	190	201	103	213	465
214	137	228	128	229	172	231	284	232	95	247	211	255	358	260	66	261	62
273	265	275	732	287	32	299	24	301	525	302	136	326	90	327	71	328	33
329	62	353	472	354	141	368	569	369	192	371	409	372	121	386	999	387	318

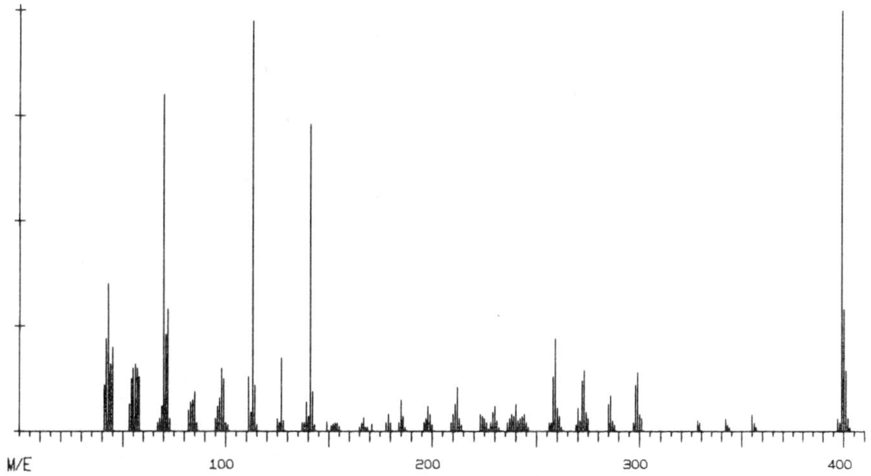
M/E
100
200
300
400

TORECANE

$C_{22}H_{29}N_3S_2$ MW = 399

Base Peak = 399 49 Peaks

Mass	Int.	Mass	Int.	Mass	Int.	Mass	Int.	Mass	Int.	Mass	Int.	Mass	Int.	Mass	Int.	Mass	Int.
42	219	43	349	55	149	56	159	70	799	72	289	84	74	85	94	98	149
99	124	111	129	113	974	125	29	127	174	141	729	142	94	149	23	154	20
166	21	167	33	179	39	185	74	198	59	199	39	211	64	212	104	223	39
229	44	230	59	240	64	244	39	245	19	258	129	259	219	272	119	273	144
298	109	299	139	300	39	301	29	328	24	329	19	342	29	355	39	356	19
357	9	397	29	399	999	400	289										

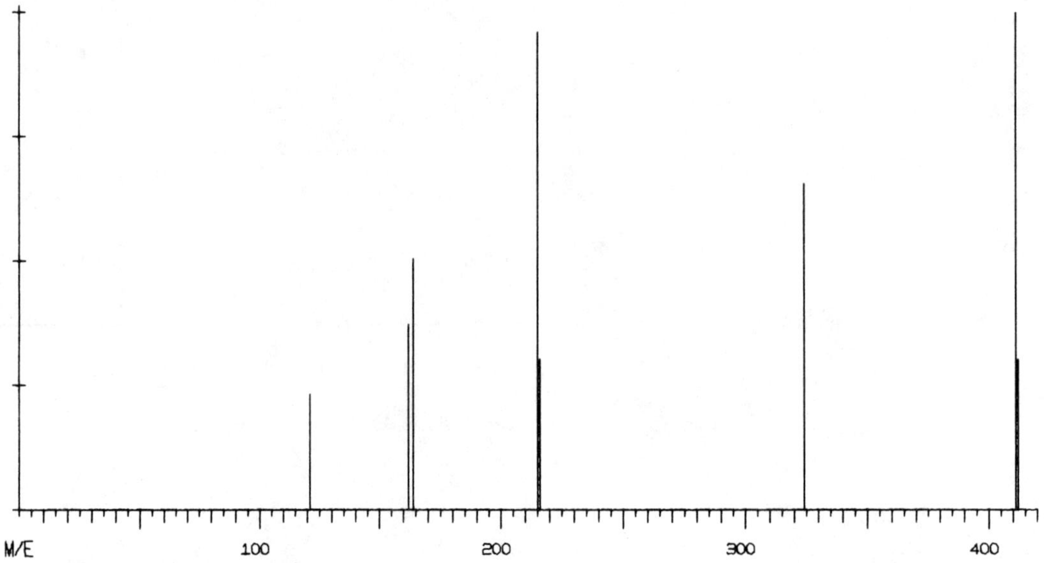
M/E
100
200
300
400

ETORPHINE

$C_{25}H_{33}NO_4$ MW = 411

Base Peak = 411 8 Peaks

Mass	Int.	Mass	Int.	Mass	Int.	Mass	Int.	Mass	Int.	Mass	Int.	Mass	Int.	Mass	Int.
121	232	162	373	164	505	215	959	216	303	324	656	411	999	412	303

INDEX

A

B

C

D

E

F

G

H

I

J

K

L

M

N

O

P

Q

R

S

U

V

W

X

Y

Z